An Introduction to Close Binary Stars

Binary systems of stars are as common as single stars in the universe. They are of fundamental importance to astronomers because they allow stellar masses, radii, and luminosities to be measured directly and can also be used as distance indicators for nearby galaxies. The evolution of binary stars also helps to explain a host of diverse and energetic phenomena such as x-ray binaries, cataclysmic variables, novae, symbiotic stars, and some types of supernovae. This textbook is the first to provide a pedagogical and comprehensive introduction to binary stars. It combines theory and observations at all wavelengths to develop a unified understanding of binaries of all categories – from pre-main-sequence systems through all stages of evolution to systems containing neutron stars and black holes.

Starting with essential orbital theory, the book reviews methods for calculating orbits from radial velocities, pulse-timing observations, speckle and direct interferometry, and polarimetry. It then examines the Roche model and ideas about mass exchange and loss, methods for analysing light and polarisation curves, the masses and dimensions of different binary systems, and recent developments in imaging the surfaces of stars and accretion structures around them.

This textbook provides advanced undergraduate and graduate students with a thorough introduction to binary stars as well as a lucid companion for courses on stellar astrophysics, stellar structure and evolution, and observational astrophysics. Researchers will also find it an invaluable and authoritative reference.

Ronald W. Hilditch is a reader in astrophysics at the School of Physics and Astronomy, University of St. Andrews, Scotland.

An Introduction to Close Binary Stars

R. W. HILDITCH
University of St. Andrews, Scotland

PUBLISHED BY THE PRESS SYNDICATE OF THE UNIVERSITY OF CAMBRIDGE
The Pitt Building, Trumpington Street, Cambridge, United Kingdom

CAMBRIDGE UNIVERSITY PRESS
The Edinburgh Building, Cambridge CB2 2RU, UK
40 West 20th Street, New York, NY 10011-4211, USA
10 Stamford Road, Oakleigh, VIC 3166, Australia
Ruiz de Alarcón 13, 28014 Madrid, Spain
Dock House, The Waterfront, Cape Town 8001, South Africa

http://www.cambridge.org

First published 2001

Printed in the United States of America

Typeface Times Roman 10/13 pt. *System* LaTeX 2_{ε} [TB]

A catalog record for this book is available from the British Library.

Library of Congress Cataloging in Publication Data

Hilditch, R. W.

An introduction to close binary stars / R. W. Hilditch.

p. cm.

Includes bibliographical references and index.

ISBN 0-521-24106-5 – ISBN 0-521-79800-0 (pbk.)

1. Double stars. I. Title.

QB821.H625 2001
523.8′41 – dc21 00-041410

ISBN 0 521 24106 5 hardback
ISBN 0 521 79800 0 paperback

Figures 7.1, 7.2, 7.3, 7.6, 7.7, 7.11, 7.12, 7.17, 7.18, 7.19, and 7.20 are reproduced with permission from Blackwell Scientific Ltd.

To my parents,
who first showed me the stars,
and to Jenny, Lara, and Claire
for their love, their laughter, and their music.

Contents

Preface

The subject of binary stars is always discussed in introductory texts in astronomy and astrophysics. The usual prescription involves the distinctions between visual (or resolved) binaries and the spectroscopic and eclipsing binaries, as well as schematic examples of resolved orbits, radial-velocity curves, and light curves. Examples of interacting binaries are discussed, and there are artists' impressions of Roche-lobe-filling stars sending gas streams across to impact an accretion disc surrounding a black hole, with jets of ejected matter from the inner regions of a thick accretion disc interacting with the local interstellar medium. A brief discussion usually emphasizes the importance of binaries for the determination of stellar masses and other parameters and their central role in explaining the properties and evolutionary states of many unusual stellar objects, such as novae, symbiotic stars, and x-ray binaries.

I have assumed that the reader of this text has already benefited from an introductory course in astronomy, including a careful reading of one of the many excellent introductory texts currently available. The basic ideas of astrophysics, including stellar evolution and the essential ideas about binary stars, should be well understood. I have assumed also that the reader has studied physics and mathematics to a similar level. Beyond these assumptions, I have tried to write a text that will be readily understood by an intermediate-to-advanced-level undergraduate in astrophysics who is interested in the more practical, observational, and data-analysis aspects of studies of close binary stars. Theoretical aspects are discussed as necessary, and some results are simply quoted with a relevant reference for further reading. But the main emphasis is on observational astrophysics. There are many examples of real observational data to illustrate points discussed in the text, rather than reliance on idealized data or schematic illustrations. The most readily accessible data for producing these figures often have been those from my own research programmes or those of research colleagues. In any event, the sources of all observational data have been fully referenced. Figures with captions that include a reference to a publication followed by the words "with permission" were obtained directly from the referenced authors, who happily provided them. All other figures (about 90% of the total) were made by me.

I have never liked any artificial division between undergraduate lecture-course material and research-level topics, and so each subject area is developed into the levels where research students will find many useful methods concerning practical observations and data analysis. Such extensions of essential course work are precisely what university teaching staff like to promote in final-year projects for undergraduate students, sometimes with the excellent outcome of published research papers. Such projects are also very helpful starting points for new graduate students embarking on research, and in that context I hope that this book will prove useful as well. Perhaps it will also help those of my generation who forget things!

I have included a set of problems at the end of the book, with outline answers that should serve as examples, so that the text can be used as a guide for an undergraduate or early postgraduate lecture course with specific problems. I have tried to keep the discussions of all the diverse areas of research involving binary stars up to date, but I doubt that I have truly succeeded in every area. To that end, however, I have referenced all recent research-level monographs, collections of review papers, conferences, and workshops that I have found useful. This text is not a research-level monograph; it is meant to serve as an introduction and a useful reference manual for equations, sources of catalogues, and useful numbers that observational astronomers require.

I would like to thank my research colleagues and friends who have taken time to read parts or all of the original manuscript and have helped me make improvements, namely, Ian Howarth, Graham Hill, Douglas Gies, and Keith Horne; also, thanks to Adam Black for his encouragement and to the production team at Cambridge University Press for their meticulous attention to detail, particularly Camilla Knapp and Ellen Carlin.

My sincere thanks go to all who have helped me along the path of understanding this delightful subject of astronomy in general, and, of course, binary stars. Particular thanks go to the following: Graham Hill, an old friend and research colleague for 30 years, with whom I have published more than 30 papers, despite our being at institutes separated by 10,000 km for most of that time; Dave Kilkenny and Tony Lynas-Gray, friends from postdoctoral days; all my former, and current, teaching and research colleagues at St. Andrews, and others in the UK astronomical community; and, for binary-star research, Alan Batten, John Hutchings, Johannes Andersen, Birgitta Nordström, Slavek Rucinski, Robert Connon Smith, Andrew King, Ian Howarth, David Stickland, Paul Murdin, Andrew Collier Cameron, and Keith Horne. I would also like to thank all my former research students, some of whom are still well employed as astronomers, Brian McLean, Alan McFadzean, Ian Skillen, Majeed Jarad, Tom McFarlane, Steve Bell, Paul Rainger, Al Reynolds, Pierre Maxted, and Paul Bennie, and my former postdoctoral associates, David King, Andy Adamson, Steve Bell, and Tim Harries, all of whom contributed ability, enthusiasm, and hard work to our research efforts so that we always gained super grades from the Research Council assessors. I am also very grateful to the many undergraduate students who have taken my courses over the years and whose supportive comments and enthusiasm have influenced my approaches to teaching and research.

Finally, to my wife, Jenny, and our two daughters, Lara and Claire: Thank you for all your love and unfailing support throughout these times.

St. Andrews, Scotland

1 Close binary stars: A historical review

1.1 Introduction

The points of light in the night-time sky that we call stars can be divided into two categories. There are the truly single stars, like the Sun, which may happen to have a retinue of planets in orbit about them, with planetary masses that are found, at least in our Solar System, to total less than one-thousandth of the mass of the parent star. There are also pairs of stars, with the two components moving in bound orbits about their common centre of mass, which we call *binary systems of stars*, or just *binary stars*. Extensive observational programmes (Abt 1983) have demonstrated that single stars are about as common as binary stars, or, to put it another way, there are about 50% more individual stars in the sky than there are observable points of light. This means that the components of these binary stars are so close together that we cannot visually resolve them spatially into two separate stars. Appropriately, they are referred to as *close* binary stars, as distinguished from the more obvious *visual* binary stars, for which the observer can clearly resolve the two components and measure their apparent motions on the sky around the centre of mass of the binary system. Indeed, we have discovered a substantial number of visual binaries whose components are themselves close binaries, so that some apparently double, or even triple, stars have been found to be quadruple or sextuple systems. For example, the brightest star in the constellation Gemini, α Gem, or Castor, is a visual binary, with the two stars currently at a separation of 2 seconds of arc (arcsec), and with an orbital period of 420 years. Both of those stars are known also to be close binaries, with orbital periods of 2.9 days for α^1 (Castor A) and 9.2 days for α^2 (Castor B). In addition, there is a third component, Castor C, at 73 arcsec separation, that is gravitationally bound to the visual pair and is itself a close, eclipsing binary (YY Gem), with a period of 0.8 day. So Castor is really a system of six stars!

For a given distance d of the binary system from the observer, we can see the two components, at a linear separation a, to be at an angular separation α, measured in arcseconds, in accordance with the small-angle formula

$$a = \alpha d/206{,}265 \qquad (1.1)$$

where the quantity 206,265 is the number of arcseconds in one radian (Figure 1.1).

For visual binaries observed with a long-focus telescope that provides a large scale at the focus, the lower limit to the measurable angular separation is of the order of 0.2 arcsec on the sky, dependent also on the quality of the observing site. For binary stars with $\alpha < 1$ arcsec, we enter the realm of the *interferometric* binary stars, where the components have been

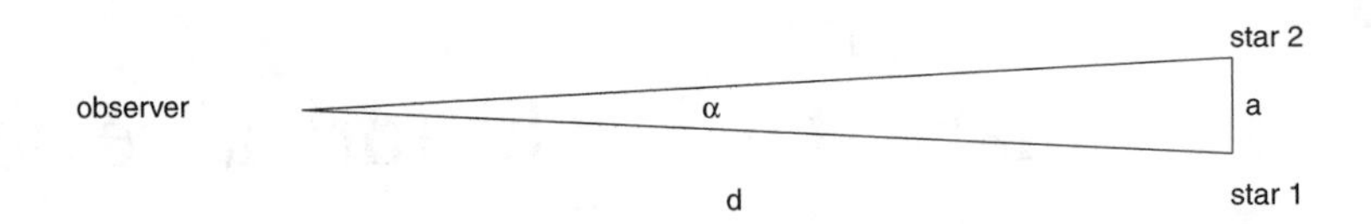

Fig. 1.1. Two stars at a linear separation a are observed to be at an angular separation α when observed from a distance d.

resolved by means of speckle interferometry down to angular separations of ~0.03 arcsec (Hartkopf 1992). There are now links between studies of the interferometric binaries and some of the observational procedures used to study close binaries. But, at the present time, none of the close binary stars has been resolved into two separate stars by direct observation through any telescope, nor by means of interferometry.

For most of the binary systems discussed in this text, the physical separations between the components typically are in the range of 1–20 times the sum of the stellar radii. As an example, to separate the components of a close binary with $a = 10$ solar radii ($R_\odot$) at a distance of only $d = 10$ parsecs (pc), we would need to achieve an angular resolution of 0.005 arcsec. Although such angular resolution has not yet been achieved, it would seem that we are not far from realizing that possibility. The Mark III Interferometer (Shao et al. 1988) and the Navy Prototype Optical Interferometer (Armstrong et al. 1998) have produced resolved orbits during the 1990s for some nearby bright binary systems with longer orbital periods, achieving resolutions of approximately 20 milliarcseconds (mas). Two groups, at the Center for High Angular Resolution Astronomy (CHARA) at Georgia State University (USA) and at the Sydney University Stellar Interferometer (SUSI), are actively pursuing even higher resolutions via interferometric arrays, with expectations of achieving resolutions of 0.0002 arcsec (Davis 1992; McAlister 1992; McAlister et al. 1995; Kelz 1996). At the time of this writing, the first full descriptions of the commissioning and early observations from the SUSI array have been published (Davis et al. 1999a,b). Perhaps by the time this book is published we shall be seeing the first results of optically resolving the close binaries that are nearest to us in the Milky Way galaxy.

At first sight, this range of separations seen in close binaries may appear somewhat restrictive. But note that stellar radii extend over seven orders of magnitude, from the neutron stars at ~10–15 km to the red supergiants at ~1000 $R_\odot$. Large stars require large orbital sizes in order for them to be accommodated in binary systems, and Kepler's third law links orbital size a to orbital period P via $a^3 \propto P^2$. It follows that close binary stars exhibit a large range of orbital periods, from as short as 660 seconds to as long as a few decades. We should not attempt to define the parameter space of close binary stars by imposing an upper limit to the separation, or, equivalently, the orbital period, based upon a particular instrumental capability. Rather, we should equate the term *close binary* with the term *interacting binary*, meaning that the two stars do not pass through all the stages of their evolution independently of each other, but in fact each has its evolutionary path significantly altered by the presence of its companion. The processes of interaction then extend beyond the gravitational effects of simply ensuring bound orbits and non-spherical shapes for the stars, and beyond the mutual irradiation of the components, to extensive exchanges of mass between them such that mass ratios can be completely reversed, as well as to extensive mass losses from the system. The aforementioned range of stellar dimensions then ensures that the realm of close binary stars is a substantial one, and it overlaps with those systems that are now being resolved by speckle and direct interferometry. By contrast, the visual binaries are systems in which the evolutions of the two stars are essentially independent, as illustrated, for example, by the visual binary Sirius A and B, which has an orbital period of 50 years. Here, the currently brighter component (Sirius A) is a normal main-sequence A star, located about half-way through its core hydrogen-fusion phase, whilst Sirius B has already evolved to the white-dwarf stage. Such binaries may appear to be curious at first sight, but their differential paths of evolution are readily explained by standard evolution theory for single stars.

So there is scope for close binaries to be resolved into 'visual pairs' by means of interferometry for suitable combinations of separation and distance from the Earth. Such an achievement will be a remarkable advance both astrophysically and astronomically, for a number of reasons that are noted later. However, none of such expectations should reduce the importance of other techniques used to study these close binary stars. It is these other techniques, some of which now achieve effective resolutions of 10^{-6} arcsec, and all that we have learned about close binary stars by employing them that provide the major reason for writing this text. Before we consider these developments, it will be useful to consider why the study of binary stars is important for all the fields of astronomy and astrophysics and is not just an occasionally exciting side-show. It is also worthwhile to place the subject of binary stars into the historical context of progress in astronomy and astrophysics in general.

1.2 General results from binary stars

Studies of binary stars by all the techniques available to modern astrophysics are able to measure a wide range of parameters pertaining to the component stars, with some of them determined to very high accuracy (e.g., uncertainties of less than 1%). For the *eclipsing* binaries, we determine the masses and radii of the component stars in absolute terms, that is, in Système International (SI) units of kilograms (kg) and metres (m). It is usual, because of their sizes, to express these quantities in solar masses ($M_\odot$) and radii ($R_\odot$). With reliable determinations of the effective temperatures ($T_{\rm eff}$) of the two stars, their luminosities are then determined directly from $L = 4\pi R^2 \sigma T_{\rm eff}^4$, where σ is the Stefan-Boltzmann constant, and L is usually expressed in solar units ($L_\odot$). Because we measure received flux densities and determine luminosities in absolute units, we can then use the inverse-square law to derive distances very accurately, and, most importantly, they are distances that are independent of all other techniques that are used in astronomy, except for the annual parallaxes of the nearby stars that underpin the effective-temperature scale adopted for all stars, as well as all other distance methods. It is here that we expect the new interferometry, mentioned in Section 1.1, to provide the critical data that will 'close the loop'. When the *angular* separations of the components of nearby eclipsing binaries are determined, and those stars already have well-defined *linear* radii and separations, then we have distances from the angular and linear separations and luminosities of stars of known linear radii. Hence, we determine direct effective temperatures that add to those already determined for the nearest single stars and extend the determinations to more of the hot stars that happen to be poorly represented in the solar neighbourhood. Because much of the foregoing, though significantly not all, is now securely defined, it is to be expected that detailed studies of binary stars in other galaxies of the Local Group will provide astronomy with some much-needed support for estimates of the distance scale for the universe (Hilditch 1996; Guinan, Bradstreet, and DeWarf 1996; Kaluzny et al. 1998, 1999; Stanek et al. 1998, 1999).

Such fundamental data for stars can facilitate direct tests of models of the structures of stars of different masses on the main sequence, as well as models for the evolution of such stars throughout all the stages of their ageing process. Virtually all empirical determinations of stellar masses have resulted from studies of close binary stars – all main-sequence stars (Andersen 1991), some giants and supergiants, including classical Cepheids, horizontal-branch stars, central stars of planetary nebulae, subdwarf O and B stars, helium stars, white dwarfs, neutron stars (pulsars), and black holes.

Definitively determined empirical masses for main-sequence stars extend from approximately 30 $M_\odot$ down to about 0.2 $M_\odot$. At the massive O-star end of this range, the stellar initial mass function (IMF) illustrates that O-type stars are extremely rare. The IMF gives the distribution of the numbers (N) of stars of different masses (m) as $N(m) \propto m^{-\gamma}$, with $\gamma \approx 1.5 \pm 0.3$ over about two decades in mass (Miller and Scalo 1979; Silk 1995). So for every O star of 30 $M_\odot$ there are approximately 200 solar-type stars. In addition, the local solar neighbourhood of the Milky Way galaxy is remarkably devoid of massive stars, so nature is not being overly helpful in allowing us to measure the masses of O stars. Despite many years of effort by many astronomers following rigorous observing programmes, the total number of well-determined O-star masses is only about 12. A recent discussion (Schönberner and Harmanec 1995) of precise empirical data on eclipsing binaries composed of main-sequence stars, compared with the latest structural models for stars (Schaller et al. 1992; Claret and Giménez 1992), has demonstrated that there is now excellent agreement between observations of masses and luminosities and the theoretical values up as far as 30 $M_\odot$. The most recent data on O+O-star binaries, reviewed by Harries and Hilditch (1997a), show this excellent agreement extending to somewhat higher masses, and these data are largely resolving the discrepancies in absolute magnitudes (of order 0.5 mag) and effective-temperature scales (of order 2000 K) that have been discussed in recent years. Because these O stars are amongst the most luminous stars in galaxies, it is important for all of astronomy that these issues be fully reconciled. It is for these reasons that we note that not all of our empirical data on unevolved stars are as nearly complete as we would like.

At the low-mass end of the main sequence, the M stars are intrinsically faint, and their small sizes relative to their separations in orbits, with periods of a few days, mean that binary systems will display eclipses only for very favourable orbital inclinations. Nevertheless, there have been a few grand successes, such as the recent determination of the masses of the two components of CM Dra (Metcalfe et al. 1996) to an accuracy of 0.001 $M_\odot$.

Extending such work to other galaxies, where the chemical compositions of the stars are different from those in the Milky Way, will also provide further tests of stellar models and evolution (e.g., Bell et al. 1991, 1993). Note that there are main-sequence stars of high mass in the Magellanic Clouds, with heavy-element abundances that are small fractions of those in the Sun. Such stars have long since disappeared from the Milky Way galaxy, because the massive stars of earlier, metal-deficient generations evolved through to their end states as supernovae many aeons ago. By contrast, at the low-mass end of the main sequence, the binary CM Dra in our own galaxy is a pair of M-type dwarfs with low metallicities and old ages that can be used to test models for the production of helium at the earliest stages of the universe (Metcalfe et al. 1996).

Additionally, there are quite subtle tests for stellar-atmosphere theory, with examinations possible for the details of limb darkening and the proximity effects of gravity darkening and mutual irradiation between the two stars of a binary. Amongst the low-mass stars on the main sequence, where magnetic activity is observed in the form of starspots, plages, and enhanced chromospheric and coronal variability, we can now obtain maps of the photospheres and the higher-atmosphere structures by means of *eclipse mapping*, *Doppler imaging*, and *Zeeman-Doppler imaging*. Stars in close binaries have rotational velocities that typically are about 100 km s^{-1}, or some 50 times greater than that of the Sun, and astrophysical understanding of magnetic activity is based upon the dynamo model – the faster the rotation speed, the greater the magnetic activity. Detailed examinations of such low-mass binaries, repeated over many years, are now revealing evidence of differential rotations of the stellar surfaces, cyclic behaviour in spot activity, and migration of spots and the associated chromospheric and coronal activity. So dynamo theory is now being subjected to findings from more than one laboratory (the Sun).

For the high-mass stars, with their extensive mass losses caused by stellar winds, we can use polarimetry with spectroscopy to investigate the structures within the winds and the collisions between such winds. Certain supergiant stars, with their extensive low-density atmospheres, have also been found to be members of the class of eclipsing binary stars, and the companion can be used as a probe of that atmospheric structure during the eclipses observed via spectroscopy, polarimetry, and photometry. The phenomenon of apsidal motion, the precession of the binary orbit in its own plane, presents an opportunity to establish the density distribution within a star, a substantial test of stellar-structure theory that has only recently been reconciled satisfactorily (Claret and Giménez 1993).

For the *interacting* or *mass-exchanging* binaries, the foregoing quantities are determinable (though perhaps not quite so precisely as in the detached main-sequence star systems, except in favoured cases), together with examinations of the mass-exchange and mass-loss processes that take place and the properties of the *accretion discs* and *streams* that are observed. As we shall see, the latest techniques of imaging the surfaces of the component stars and the accretion components, via *eclipse mapping* and *Doppler tomography*, are now yielding findings of unprecedented accuracy and intepretative power vis-à-vis magnetic activity on stellar surfaces, the mass-transfer streams, discs, and columns, and their interactions with the stellar surfaces. We are now able to examine plasma interactions with the very strong magnetic fields of approximately 10^8 tesla (T) seen in the *intermediate-polar* and *polar* types of *cataclysmic variable* stars.

All of these binaries contribute to tests of binary-star evolution theory. Such models propose that some stars will have their outer layers stripped off during the rapid mass-transfer processes that occur on dynamical time scales. The products of nuclear evolution in the core regions of such stars should then be revealed on the surfaces. These processes are predicted for the originally more massive stars amongst the Algol systems; but after the rapid process of mass-ratio reversal, these stars are very substantially fainter than their companions. The consequence is that separation of the spectrum of the now fainter star from the combined light of the system requires all the sophistication that is possible from *tomography* and *disentangling algorithms*. New successes in this field are needed, because there are conflicting theoretical models for the processes of mass loss and mass accretion in Algol systems involving specific predictions that have not yet been tested (Vanbeveren 1995). Amongst the Wolf-Rayet (WR) binaries, such exposure of nuclearly processed matter is very obvious, but because that is also observed amongst the single WR stars and seemingly is associated with their large-scale wind outflows, a successful interpretation of these findings is proving to be very difficult.

Study of the role of binary stars as sinks of binding energy in the dynamics and evolution of clusters of stars, particularly globular clusters, has developed both theoretically and observationally in the past 20 years into a major research area. By observing in the x-ray and ultraviolet regions of the spectrum, we can now study the properties of binaries in the densely crowded central-core regions of globular clusters (Bailyn 1996), with some remarkable, even bizarre, objects discovered as a result, such as the system 4U1820-30 in NGC 6624 (Stella, Priedhorsky, and White 1987), with an orbital period of 660 seconds! Significant progress in understanding the origins of the *blue stragglers* in old open clusters and in globular clusters has also been made in the past decade, mainly because of extensive optical charge-coupled-device (CCD) photometric surveys, revealing the presence of a rich variety of binary stars amongst the blue stragglers and the turn-off region of the main sequence (Mateo 1996).

Finally, there are all the eruptive and unusual stellar phenomena that have been found to be the direct results of stars being members of binary systems, all the *novae* and *dwarf novae*, now regarded as subsets of the *cataclysmic variables*, the *Type Ia supernovae*, the *symbiotic stars*, and the *binary x-ray sources*, with several subclasses that contain neutron stars or black holes as the gravitational-potential wells into which matter flows from an evolving companion. These binaries are classic examples of the fundamental contribution that stellar astrophysics makes to our general understanding of physical processes in the universe. The processes of accretion in these systems can be investigated extensively because the accretion structures are often the most luminous parts of the systems. Neutron stars and black holes are exotic entities whose physical properties are yet to be completely established. Because of the extremely accurate timing of the arrival of radio pulses from the rapidly rotating neutron stars that we call pulsars, radio astronomy has been able to determine the most precise binary orbits ever and to investigate the changes in those orbital parameters predicted by the general-relativity contribution to apsidal motion, as well as the changes due to gravitational radiation. Yet remember that these exotic binaries, with their curious accretion structures, are still just parts of the whole subject of binary-star evolution. If we can provide a convincing picture of all the evolutionary paths that binary stars can take to arrive at all the diverse types of systems that we observe, then we will have made a major contribution to the fields of astronomy and astrophysics.

I readily admit to being totally biased about the need to study binary stars, simply for the sake of finding out more about these objects that are more common than single stars in space. But their importance to astronomy and astrophysics in general really cannot be over-emphasized.

It is probably because of the need for substantial amounts of telescope time, together with very careful analyses of the observational data, that the use of eclipsing binaries does not often appear amongst the list of fundamental methods for determining distances. But the dedication of that time yields plenty of rewards that are relevant to stellar structure, evolution, and distances, and therefore galaxy evolution as well. There are also all the physical processes mentioned earlier that can be investigated thoroughly with the help of the mapping qualities inherent in binary orbital motion, further justifying the importance of binary-star studies for astrophysics.

1.3 The historical context

In the latter half of the eighteenth century, it was clearly accepted, in the published papers of that time, that the Sun was to be regarded as a normal, though nearby, star, whilst the annual parallax of Sirius, the apparently brightest star in the sky, was known certainly to be less than one second of arc, making that star very distant. Many apparently double stars, with angular separations between their components of only a few arcseconds, had been found. There was no evidence that such double stars were physically associated, and William Herschel attempted to make use of them in establishing distances via the inverse-square law of light propagation. Thus he began a truly long-term observing programme, in collaboration with his sister Caroline, with reports published in 1802 and 1803 that listed their observations over 40 years. Michell (1767) had used a probability argument to conclude that 'it is highly probable in particular, and next to a certainty in general, that such double stars, etc., as appear to consist of two or more stars placed very close together, do really consist of stars placed near together'. But it seems that Herschel was the cautious empiricist there. He was concerned to establish the true cause of the relative angular motions of the components of some double stars that he had observed. Only by pursuing an observational programme for 40 years was he able to demonstrate clearly that the relative motions on the sky of the components of some double stars were due to orbital motion in accordance with Newton's laws. It is clear from his reports in the *Philosophical Transactions of the Royal Society of London* that he wanted to distinguish between that interpretation and other possibilities involving the relative motions, and the distances, of the stars and the observer. The orbital periods of some of his double stars, or visual binary stars, were of the order of 300 years, so that at least a 40-year time span was essential to prove that the position-angle changes were due to orbital motion. For example, Herschel (1803) demonstrated that the binary system involving Castor, the brightest star in the constellation Gemini, displayed incremental changes in position angle amounting to $45^\circ\ 39'$ in 43 years and 142 days, suggesting an orbital period of 342 years. The modern value for the orbital period is 420 years (Rabe 1958), which illustrates the difficulties of determining complete orbits from relatively short sets of data. So another two centuries will pass before the brightest components of Castor return to the relative position first recorded by Herschel.

Herschel (1802) seems to have been the first to introduce the term *binary star*, with the following:

> 'If a certain star should be situated at any, perhaps immense, distance behind another, and but very little deviating from the line in which we see the first, we should then have the appearance

of a double star. But these stars, being totally unconnected, would not form a binary system. If, on the contrary, two stars should really be situated very near each other, and at the same time, so far insulated as not to be materially affected by the attractions of neighbouring stars, they will then compose a separate system, and remain united by the bond of their own mutual gravitation towards each other. This should be called a real double star; and any two stars that are thus mutually connected, form the binary sidereal system which we are now to consider.... It is easy to prove, from the doctrine of gravitation, that two stars may be connected together as to perform circles, or similar ellipses, round their common centre of gravity. In this case, they will always move in direction opposite and parallel to each other; and their system, if not destroyed by some foreign cause, will remain permanent.'

Extensive searches for other visual pairs were conducted throughout the nineteenth century, notably by J. South, J. Herschel, F. Bessel, F. G. W. Struve, O. Struve, and S. W. Burnham, the last publishing his *Catalogue of Double Stars within 121° of the North Pole* in 1906, containing information on 13,665 double stars. R. G. Aitken, amongst others, continued these surveys and continual monitoring of established visual binaries into the twentieth century, and he published a successor to Burnham's catalogue entitled the *New General Catalogue of Double Stars within 120° of the North Pole* in 1932. Aitken's book *The Binary Stars* (1935) provides a substantial survey of the historical development of the subject of visual binaries up to that time. Subsequent catalogues have been the *Index Catalogue of Visual Double Stars* (IDS), from Jeffers, van den Bos, and Greeby (1963), and the *Fourth Catalogue of Orbits of Visual Binary Stars*, by Worley and Heintz (1983), who maintain a regularly updated data base held at the U.S. Naval Observatory. Dommanget and Nys are working on a new catalogue containing more than 62,000 systems that will extend their investigations for the *Hipparcos* satellite project. Many useful reviews of the current status of visual-binary and multiple-system research can be found in *Complementary Approaches to Double and Multiple Star Research* (McAlister and Hartkopf 1992).

Once the orbiting nature of some of these visual pairs of stars had been established by means of careful observations of their relative positions, it was recognized that they could be exploited for tests of the universal nature of Newton's law of gravitation and his laws of motion. The mathematical method required to determine the geometry of the relative orbit of one component with respect to its companion from a set of observations of position angle and separation was first developed by Savary in 1827. Many other methods and improvements followed, perhaps the most influential being that developed independently by Thiele (1883) and by Innes (1926) and now commemorated in the use of the Thiele-Innes constants.

The visual binaries provided the first true determinations of the masses of stars other than the Sun, once their distances from the Earth had been established with some degree of certainty. Kepler's third law relates the orbital period P and the size of the relative orbit, strictly the semimajor axis a, to the sum of the masses of the two stars $(m_1 + m_2)$ by

$$G(m_1 + m_2) = 4\pi^2 a^3 / P^2 \tag{1.2}$$

where G is the universal gravitational constant. Analyses of the astrometric data of angular position of one component of a binary relative to its companion provided the angular size of the relative orbit, together with the orbital period, and the three angles that specify the three-dimensional orientation of the orbit in space. Then measured parallaxes for the nearest visual binaries to the Sun permitted conversion of the angular size of the orbit to a linear measurement, and hence the sum of the masses from the foregoing equation. But the individual masses of the two stars could not be determined from these purely relative orbits. Developments in astrometry were required first in order to allow the determination of the motion of *each* component projected onto the sky, and hence providing directly a value for the mass ratio m_2/m_1. Several such determinations of individual masses were published by Boss in 1900, whilst Aitken (1935) listed only 16 visual binaries whose individual masses had been established with some precision. Despite the large numbers of recognized visual binaries, their orbital periods of tens to hundreds of years meant that reliable determination of masses would be a slow and painstaking process. Real progress in that area required a faster procedure, provided by the discovery of close binary stars with orbital periods measured in hours and days rather than decades and centuries.

The existence of *close, unresolved* binary stars was first recognized on the basis of the particular form of variations in brightness of a star caused by the mutual eclipses of the two components. The first identification of this aspect of eclipsing binaries is generally credited to John Goodricke in 1783 for his interpretation of the known variability of β Persei, or Algol, the 'demon'. It is interesting to note that Hoskin (1982) has examined the historical records more closely and has demonstrated that Goodricke's associate, Edward Pigott, deserves much of the credit, though it seems that Pigott himself went to some lengths to give unqualified public credit to Goodricke. Thus, Goodricke (1783) demonstrated that the brightness of Algol decreased to a minimum at about two magnitudes below its normal level over an interval of 3.5 hours and then increased again over the same time interval, and that variability was periodic, with $P = 2$ days 20.8 hours. He went on to propose that 'if it were not perhaps too early to hazard even a conjecture on the cause of this variation, I should imagine it could hardly be accounted for otherwise than either by the interposition of a large body revolving around Algol, or some kind of motion of its own, by which part of its body, covered with spots or such like matter, periodically turned towards the earth'. So we see that he was not entirely convinced that the light variations had to be due to Algol being an eclipsing system. In fact, modern studies have demonstrated that Goodricke was correct on both counts, but with the starspots located on the cool secondary component of the Algol binary system (see Chapter 5).

During the nineteenth century, the first steps were being taken along the path toward understanding the importance of spectroscopy in astronomy. Beginning with the work of Fraunhofer (1815) on the spectrum of the Sun, and progressing through the detailed laboratory studies by Kirchhoff and Bunsen in the 1840s, the technique of spectroscopy blossomed in astronomy, particularly because parallel developments in photography provided the means to make a permanent record of a stellar spectrum. The first photograph of the spectrum of a star (Vega, which is bright to the naked eye) was taken by Huggins in 1872. This technological development opened the way to exploitation of the Doppler-Fizeau effect, first discussed by C. Doppler in 1842, and especially with reference to light by H. Fizeau in 1848, for determining the line-of-sight velocity V of a source of light from the displacement at wavelength λ of spectral lines from their laboratory (rest) positions at λ_0 via the simple relationship $(\lambda - \lambda_0)/\lambda_0 = V/c$, where c is the speed of light. (This equation is the non-relativistic approximation to a more rigorous equation required when V is a substantial fraction of c.) In the visible part of the electromagnetic spectrum ($\lambda \approx 500$ nm), a velocity of $V = 70$ km s^{-1} is required to make $(\lambda - \lambda_0)$ as large as 0.1 nm! Obviously, such subtle changes in wavelength can be measured only by means of painstaking techniques applied to a permanent record of a spectrum, and that spectrum must be secured via a spectrometer, or spectrograph, that is extremely stable. Overcoming these technological difficulties took years of effort by many astronomers, notably Sir William Huggins, E. C. Pickering, H. Vogel, and W. W. Campbell. But the inspiration was clear: the possibility of determining the three-dimensional motions of the stars in space from combinations of measurements of the line-of-sight or *radial* velocity and the proper or transverse motions across the sky, and the distances from parallaxes and brightnesses. It seems appropriate here to quote Sir William Huggins, from an article published just over 100 years ago (Huggins 1897), who had the following comment on the value of measuring radial velocities of astronomical sources: 'It would be scarcely possible, without the appearance of great exaggeration, to attempt to sketch out even in broad outline the many glorious achievements which doubtless lie before this method of research in the immediate future'. A great deal of the information that appears in this book is due solely to our ability to measure line-of-sight velocities, never mind the recent revelations from galaxy redshifts concerning the large-scale structure of the universe, so it seems that Huggins' fears of exaggeration have been obviated by the realities of the past 100 years.

The breakthrough for study of close binary stars came in 1889, when Pickering noted, on objective-prism spectra, that the star Mizar sometimes showed double lines, and other times just single lines. But it was Vogel (1890) who first proved that the star Algol was indeed a binary star, by demonstrating that the primary star was receding from the observer before primary eclipse, and approaching the observer after primary eclipse. The modern value for the orbital velocity of the primary star, projected into our line of sight, is just 44 km s^{-1}. In the blue region (4200 Å) of the electromagnetic spectrum, this velocity produces a Doppler shift of just 0.62 Å, or 0.62×10^{-10} m. It is worthwhile remembering these factors when we are considering velocities determined to uncertainties of about 1 km s^{-1} for solar-type stars observed with ordinary equipment, or to uncertainties of about 1 m s^{-1} in the recently successful searches for planets around stars other than the Sun (Mayor and Queloz 1995)! The term *spectroscopic binary* has been used ever since Vogel's work to describe a binary system discovered as a result of radial-velocity studies of an apparently single star in space. For a binary where the spectrum of only one component is observed, we use the term *single-lined spectroscopic binary* (SB1), and where both spectra are detected, the term *double-lined spectroscopic binary* (SB2). We shall see later (Chapter 3) that the SB2 systems provide the maximum information about a binary system with the minimum of assumptions. As technology and methods of analysis of spectra have improved, particularly during the past 15 years, the number of established SB2 systems has increased significantly, with previously intractable SB1 systems revealing their secrets because of careful use of new detectors and procedures of analysis. The *Eighth Catalogue of the Orbital Elements of Spectroscopic Binary Systems*, compiled by Batten, Fletcher, and McCarthy (1988), contains details and discussions of 1469 close binary stars.

The development of the photographic process also permitted the establishment of systematic searches for variable stars, from which many eclipsing binaries were discovered. Prior to 1900, discoveries of stars that varied in apparent brightness had been due, primarily, to visual observations in which the observers happened to note changes within a night, or between nights, and Argelander (1844) provided the first summary discussion of 22 variable stars. It was his method of making visual estimates of relative brightnesses of stars that was followed subsequently, and by 1900 about 700 variable stars had been discovered. But from then on, the major technique for discovering variable stars of all types has been the use of photographic sky-patrol cameras, pioneered for the Magellanic Clouds by Leavitt, and then systematically used in the Milky Way galaxy by the Harvard group beginning in 1907 (Shapley, Gaposchkin) and by the German observatories at Berlin-Babelsberg, Sonneberg, and Bamberg beginning in 1929 under the leadership of Prager, Hoffmeister, and Zinner. In the USSR, Kukarkin and Parenago established a sky-patrol system in 1942, which led to their extensive involvement in successive editions of the *General Catalogue of Variable Stars* (GCVS). By the late 1950s, Sonneberg and Leiden observers were patrolling selected Milky Way fields (for galactic-structure purposes), and Strohmeier at Bamberg extended the long-term commitment of that observatory from searches for variable stars in the Northern Hemisphere sky to patrols of the Southern Hemisphere sky as well. Of course, chance has played its part as well as in finding variables, as have detailed studies of stars that were considered, from other observations, to be likely variable sources. The amateur observer community, exemplified by the AAVSO (American Association of Variable Star Observers), the BAV (Bundesdeutsche Arbeitgemeinschaft für Veranderliche Sterne), VSNET (Japan), and the Variable Star Sections of the British Astronomical Association and the Royal Astronomical Society of New Zealand, has also continued to provide invaluable data on variable stars in general. In addition, the technology developed from the photoelectric effect, firstly in the form of photomultiplier tubes with linear responses and fast recording capabilities, and lately in the form of charge-coupled devices (CCDs) with the same linear responses but in the format of area detectors, has resulted in major advances in detecting variables of low amplitude, or at faint magnitudes. The result has been continuing expansion in the number of known variable stars, and by the time of the fourth edition of the GCVS (Kholopov et al., 1985) there were 28,450 confirmed variables in the Milky Way galaxy, of which roughly 70% were classed as pulsating variables, 10% as eruptive objects, and 20% as eclipsing variables.

As soon as a catalogue of objects is considered to be necessary, it follows that some system of classification must be introduced in order to place those objects into appropriate subgroups. For the eclipsing binaries discovered in variable-star surveys, a simple method was adopted based upon the overall shape of the light variation. Note that to produce a light curve, we require a determination of the orbital period of the binary, which in itself is no simple procedure (see Chapter 5). The classification scheme that was adopted and continues to be used now for the great majority of eclipsing variables is the following:

EA An eclipsing variable of the Algol type, meaning that its light curve has similarities to that of Algol itself, namely, clearly defined eclipses, with obvious start and end times, and light variations outside the eclipses that are almost negligible, or minimal.

There is no suggestion that an eclipsing binary with an EA light curve must have the same physical characteristics as Algol, and probably, in most cases, they will not. The classification is based purely upon the shape of the light curve, without any further interpretation.

EB An eclipsing variable of the β Lyrae type, meaning well-defined eclipses, together with considerable variations outside eclipses, due to the non-spherical shapes of the component stars. An EB system does not need to have the same physical characteristics as β Lyrae.

EW An eclipsing variable of the W Ursae Majoris type, meaning a continuous variation of brightness, with no distinction between the eclipse phases and the out-of-eclipse phases. Both component stars are strongly distorted from spherical shapes, and their physical characteristics may well be similar to those of the prototype, W Ursae Majoris (W UMa).

Figure 1.2 shows examples of these types of light curves, which can be recognized even from incomplete sets of data.

Beginning in the 1990s, new surveys have been undertaken with automated telescopes and CCD cameras to search for evidence of dark matter in the halo and bulge regions of the Milky Way galaxy. Massive compact halo objects (MACHOs) that have no luminosity would not be detectable save for their effects upon the passage of photons from more distant sources towards

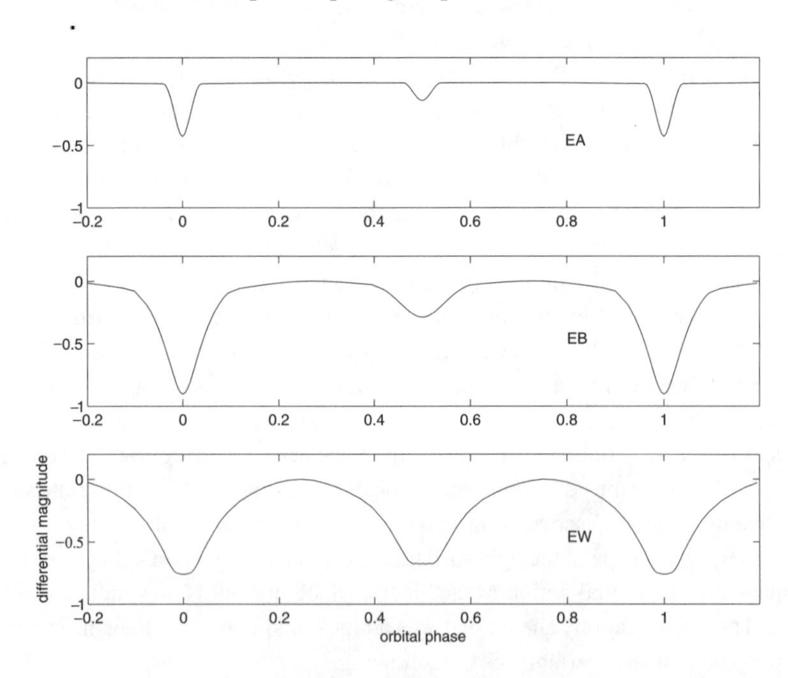

Fig. 1.2. Illustrative examples of the three types (EA, EB, EW) of light curves of eclipsing binaries used for classification in the *General Catalogue of Variable Stars*. The range of amplitudes for light curves in each class is considerable, dependent principally upon the orbital inclination, the relative sizes of the two stars, and their ratio of surface brightnesses.

an observer. The gravitational 'lensing' of light by a massive body is predicted precisely by the general theory of relativity and is expected to yield clear evidence for MACHOs in the halo. Whilst gravitational microlensing events have certainly been identified in the MACHO, EROS, and OGLE surveys, it would seem that the numbers recorded thus far have been less than anticipated. But a major bonus from these monitoring surveys has been the huge number of variable stars detected, with many of them being eclipsing binaries or cataclysmic variables. The advantage of these automated searches over earlier surveys for variable stars lies in their continual monitoring at high precision throughout a night, extending over entire observing seasons (months) for the selected sky fields, and then being repeated for several years. The consequence for binary-star studies is that for the first time we have access to a least-biased survey of the distribution of the orbital periods of eclipsing binaries and of the relative numbers of binaries with different types of light curves. The EROS project discovered 79 eclipsing binaries (Grison et al. 1995) located in the bar region of the Large Magellanic Cloud (LMC), from one season of observations. The MACHO collaboration has published its first set of light-curve classifications for 611 eclipsing binary stars in the LMC (Alcock et al. 1997), based upon the first 400 days of observation by the MACHO project over 22 LMC fields. These latter binaries are all brighter than $V \approx 18.0$ mag, which, with the LMC distance modulus of $(V - M_v) \approx 18.5$, shows that they consist of stars of spectral type earlier than about A0 on the main sequence. The OGLE survey of the galactic-bulge region of the Milky Way revealed many eclipsing binaries with EW light curves, and Rucinski (1997a,b) has discussed the statistics of this remarkable data set with reference to the evolution of contact binaries. The astrometric and photometric survey of the entire sky by the *Hipparcos* satellite (Perryman et al. 1997) has also identified a number of new eclipsing binaries amongst stars brighter than $V \sim 9$ mag.

In addition to these 'classically recognized' binary stars, many other types of stars that exhibit unusual phenomena have been found, on the basis of careful observations and analyses, to be members of binary systems. Athough a great deal of observational and interpretative effort was expended on the nova phenomenon in the first half of the twentieth century, it was not until the 1960s that Kraft (1963) felt confident enough to be able to propose that all dwarf nova, and nova, events were the results of particular types of stars undergoing mass-transfer interactions in very close binary systems. This particular subject area, the study of a subset of the close binary stars now called the *cataclysmic variables*, or CVs, has developed into an enormous industry involving many large research groups throughout the astronomical community, for several interconnecting reasons, both observational and theoretical. Because these binaries have orbital periods of just a few hours, they are tailor-made for guaranteed research-publication success with allocation of just a few nights of heavily over-subscribed large-telescope time. In addition, they are *eruptive* systems, so that observations secured during outbursts that last a few days, and between outbursts, reveal different physical conditions that may explain the causes of the outbursts themselves. But perhaps most importantly, it is the process of *accretion* of matter by one star from its companion that drives the enthusiasm to understand these systems. It is the *accretion disc* that provides the most luminous part of the whole interacting binary system, so that the accretion process can be studied *in extenso* via theoretical models in a well-defined dynamical situation and via observations secured simultaneously over enormous wavelength ranges at multiple observing sites. It is both amusing and interesting to note that this subject area of cataclysmic variables has experienced substantial advances because of observations that have been secured, for example, by the Hubble Space Telescope, the world's most expensive astronomical instrument, simultaneously with observations from some of the world's more humble ground-based telescopes at college observatories. As one example of the explosion of interest in this subject, it is worth noting that a recent book by B. Warner (1995), *Cataclysmic Variable Stars*, is 700 pages long and is a reduced version of the original manuscript. Just to make sure that observers would have no difficulties in distinguishing these interesting points of light from all the others in the sky, Downes and Shara (1993) provided an excellent catalogue, together with identification charts, for 751 CVs. It has recently been updated (Downes et al. 1997) and now contains 1020 such systems.

Ever since the classification of stars according to the appearance of their spectra was initiated, astronomers have been identifying unusual objects. One of those classes seemed perhaps to be more heterogeneous than most, namely, the *symbiotic stars*, so called because they exhibited certain spectral characteristics attributable to hot objects, as well as those of very cool objects. The obvious suggestion was that such composite spectra were due to the binary nature of the source; yet it took much careful interpretation of the observational data to establish that these objects are indeed binaries, albeit in a decidedly interactive state, such that the extended envelopes around these systems are able to mask the evidence for the orbital motions of the components. The very long orbital periods of some of these symbiotic systems also make it difficult to establish their binary nature. Kenyon (1986) has stimulated much renewed interest in these systems by providing a helpful survey of their properties.

From the beginning of the first observations during the 1960s at x-ray wavelengths from outside the Earth's atmosphere, it was found that some x-ray sources were unresolved and were variable in x-ray flux density. The source Sco X-1 (Giacconi et al. 1962) was established as a binary system by Gursky et al. (1966) and Sandage et al. (1966), but it took until the early 1970s, after the launch of the *Uhuru* satellite, to establish that some of the bright x-ray 'stars' were pulsing regularly at periods of a few seconds and that they underwent periodic eclipses, with inferred orbital periods of days, such as Cen X-3 (Schreier et al. 1972) and Her X-1 (Tananbaum et al. 1972). It was quickly recognized that an important discovery had been made of a new class of binary systems. The resultant coordinated campaigns of observations at x-ray, ultraviolet, optical, infrared, and radio wavelengths demonstrated that the sources were binary systems composed of a quite ordinary star transferring mass to the deep gravitational-potential well of a compact companion, usually, as was later shown, a neutron star, a star typically of 1.4 $M_\odot$ and a radius of only 10–15 km. Whether or not some of these x-ray binaries contained black holes, rather than neutron stars, remained an open question for the subsequent 20 years or so (e.g., Cyg X-1) (Webster and Murdin 1972; Bolton 1972), but recent investigations of systems like V404 Cyg (Casares, Charles, and Naylor 1992) have resolved the issue in the affirmative. Cowley (1992), in her review of the evidence for black holes in stellar binary systems, still felt obliged to remark that 'the fact that even the most likely examples are still referred to in the literature as "candidates" shows our hesitancy to accept their existence'. Thus, in parallel with the explosion of interest in cataclysmic variables came the related studies of *x-ray binaries*, now divided broadly into the *high-mass* and *low-mass* subgroups, as well as the recently identified *supersoft* x-ray sources. Major recent reviews of the properties of x-ray binaries can be found in an article by Verbunt (1993) and in the compilation of excellent review articles edited by Lewin, van Paradijs, and van den Heuvel (1995). Also contained in those two sources are discussions of the binary radio pulsars, the first of which was discovered by Hulse and Taylor (1975), which earned them the Nobel Prize in physics in 1993. The evolutionary paths that link the unevolved main-sequence binaries, through the x-ray binary stage, to the binary pulsar stage are also investigated in those reviews.

One of the direct consequences of major advances in detector technologies at x-ray, optical, infrared, and radio frequencies has been the ability of astronomers to investigate the crowded-field central regions of the globular clusters. Among the many results, the discoveries of compact binaries with extremely short orbital periods, as low as $P \approx 11$ minutes, are nothing short of amazing! There is only enough room in such binary systems to accommodate neutron stars. The fact that these stars were also discovered to be emitting pulsed radiation at periods of milliseconds, that is, rotating at about 36,000 rpm, heightened the enthusiasm to find sensible explanations for their existence in these old stellar clusters. Searches for such millisecond pulsars demanded substantial improvements in the technology of radio detectors, and particularly in the speed of computers to handle the incoming data stream. Whilst about 25 such sources were discovered in globular clusters between 1987 and 1990, improvements in sampling rates up to about 1 million per second (Lyne 1996) were required before successful all-sky searches could be implemented. By 1995, the number of known millisecond pulsars in binary systems that were not members of globular clusters had risen to 29, whilst those in globular clusters were up to many dozens. Phinney (1996) noted that extrapolation from the available data would suggest that the approximately 200 globular clusters probably contain more than 10^7 binary stars and more than 10^3 pulsars. These single pulsars, and the pulsar binaries, are to be found in the dense cluster-centre environments, where we can expect them to have been strongly influenced by gravitational encounters with other stars. The millisecond-pulsar binaries that have been found in the general field of the galaxy (at all galactic latitudes, and not concentrated towards the galactic plane) and those in globular clusters are linked in an evolutionary sense with the low-mass x-ray binaries. They are old neutron stars that have been recycled into activity again as a consequence of mass transfer from their binary companions.

Optical and ultraviolet (UV) investigations of globular clusters were made possible by the development of CCD detectors and by the spatial resolving power of the Hubble Space Telescope (HST). The field moved from an embarrassing shortage of known binaries in globular clusters in 1980 (one eclipsing binary in ω Cen, and two old novae) to a veritable plethora by 1996: A list of 38 was given by Mateo (1996), and he commented that his list would soon be 'terribly incomplete' because so many surveys of globular clusters were under way. Such searches involve time-series observations with CCD cameras over many nights, followed by intensive analyses to discover the photometric variables amongst the crowded images of the stars, even in the outer regions of these clusters. Mateo remarked that those surveys were not for 'the faint-hearted', since they usually resulted in finding one eclipsing binary with a photometric amplitude greater than 0.1 mag per 1300 main-sequence stars. These binaries in the outer regions of the clusters probably are primordial systems, that is, formed at the same time as the whole cluster, and investigations are under way to see if they can plausibly explain the presence in globular clusters of the so-called *blue stragglers*, stars that are found loosely scattered around the main sequence in the Hertzsprung-Russell (HR) diagram of a globular cluster, but above the cluster turn-off region. Improvements in the resolution of the HST are now allowing investigations of binaries in the central regions of some globulars, which may contribute to tests of the theories of gravitational interactions between stars in such dense environments (10^6 stars pc^{-3}), as do the low-mass x-ray binaries and the pulsar binaries.

If we have identified binary stars at every stage of their evolution from the main sequence to their various end states of white dwarfs, neutron stars, and black holes, and even to *recycled* end states, what evidence can we present for the existence of binary stars at the proto-star and pre-main-sequence (PMS) stages? Once again, it has been the development of new detectors, specifically at the infrared and millimetre wavelengths, that has come to the rescue. Whilst several visual binaries were recognized more than 50 years ago during the pioneering investigations of T Tauri stars (Joy and van Biesbroeck 1944), most of the discoveries of PMS binaries have been made during the past decade. In the star-formation region of Taurus-Auriga, several investigations have shown that in the projected separation range of 0.13–13 arcsec there are 44 multiple (mostly binary) systems, providing a frequency of multiple systems compared with single stars of $42 \pm 6\%$ (Mathieu 1994). Such a value is remarkably similar to the values determined for the frequency of main-sequence binaries, at least amongst those of approximately solar masses (Duquennoy and Mayor 1991). Mathieu (1994) concluded that the ratio of multiple systems to single PMS stars was at least 50%, with binary formation being the *primary route* in the star-formation process. The distributions of orbital eccentricities are remarkably similar to those found amongst field G-star binaries, with circular orbits at periods of a few days, and a wide range of eccentricities at longer periods. We have to find the mechanisms that allow some of these binary stars, which have separations to be measured in astronomical units (AU) rather than the few solar radii ($R_\odot$) of our *close* binaries, to lose angular momentum and develop into the short-period systems that we observe on the main sequence. Tidal interactions between accretion discs that are found around PMS stars, and the consequent loss of *orbital* angular momentum, offer one possible explanation. Another involves gravitational encounters between these wide binaries and single stars in a young-cluster environment, so that the binary becomes *harder* (has a shorter orbital period), and the interloper gains kinetic energy from the exchange to escape from the cluster. It is also possible that binaries can exchange partners in such interactions.

This brief discussion of the historical development of the subject of binary stars illustrates how the research area has blossomed and diversified, particularly in the past 25 years, to a level where it is difficult for an individual to keep track of all the developments in the field. Nevertheless, it seems to me that the particular type of binary that is being investigated and the technology that is being used to observe it do not specifically dictate how the data should be analysed. Rather, there are common procedures of analysis and interpretation, and it is these methods that I wish to emphasize throughout this text. There will, of course, be many examples given, drawn deliberately from different types of binaries, but I shall try to avoid becoming embroiled in the details of specific evolutionary models for a particular subset of the binary stars.

1.4 Methods of analysis for close binary stars

The variations in the projected radial velocities of the component stars in a binary system, due to their orbital motions about their common centre of mass, can be plotted against the orbital phase once the orbital period has been determined. Such velocity curves need to be analysed to determine the quantities that describe the geometry of the orbits and the quantities that are related to the masses of the two stars. A number of methods, principally graphical, were devised soon after Vogel's discovery of Algol's orbit in 1890. That developed by Lehmann-Filhés (1894) seems to have been the most popular and was well described by Aitken (1935); with the advent of powerful computers, it is now largely of only historical interest. Lehmann-Filhés (1908) also developed a method of differential corrections for the preliminary orbital elements, determined graphically. It is based on the method of a least-squares solution for n simultaneous equations with m unknown quantities ($m < n$) (see Chapter 3). That procedure is followed in the modern-day codes and is now completed in a matter of seconds, whereas a single iteration of a least-squares procedure conducted by hand must have taken many hours. For binary systems with small values of the orbital eccentricity ($e \leq 0.1$), alternative methods developed by Sterne (1941) must be used. In Chapter 3 we discuss the techniques of observation and analysis that provide determinations of the orbital parameters for spectroscopic binaries, and for systems containing pulsars, we present analyses of the pulse arrival times. We also discuss the determination of orbits for resolved binaries.

The first eclipsing-binary *light curves*, graphs of brightness variations versus orbital phase, that were analysed successfully were those presented by spherical stars revolving in circular orbits. The principal technique of the time was the analytical method developed by Russell and Shapley (Russell 1912a,b; Russell and Shapley 1912a,b), which was later developed into the Russell-Merrill method that was much used up to about 1970 (Irwin 1947; Merrill 1950; Russell and Merrill 1952). Whilst the technique was analytically sound for such spherical stars, including the effects of limb darkening, it was a time-consuming exercise involving graphical constructions and extensive use of nomograms and tables. The orbital inclinations and the stellar dimensions, expressed in terms of the orbital separation, were determined with this method for many eclipsing binaries. Other procedures were also developed, notably by Kopal (1950), who used the n photometric observations directly, together with extensive tabulated functions, to establish sets of n simultaneous equations that could be solved by least squares for a small number m of corrections to preliminary values (determined graphically) of orbital inclinations, stellar radii, and relative surface brightnesses.

It was Kopal's extensive use of his method, together with results from other researchers, that allowed him to define his celebrated classification scheme for binary stars that is based on the Roche model for the surfaces of constant gravitational potential around two mass points (see Chapter 4). Kopal (1955, 1959) was able to demonstrate, from the results of light-curve and velocity-curve analyses, that binary stars could be placed into three broad classes: *detached*, *semidetached*, and *contact* binaries. These terms continue in use to the present day because of their importance with respect to the stages of evolution of particular binary systems, as well as their importance for aspects of the interacting or mass-exchanging binaries. We discuss the terminology in the next section, and the Roche model and other factors concerning non-spherical stars in Chapters 4 and 5.

During the 1960s and 1970s there developed a divergence of procedures. On the one hand, Kopal (1979) and Kitamura (1965, 1967) independently explored Fourier techniques to analyse light curves, particularly for systems that were composed of stars that were clearly non-spherical and suffered from other mutual interactions such as the (misnamed) *reflection effect*. Their approach was similar to that which had gone before in the Russell-Merrill method, in the sense that they attempted to separate the eclipse effects from the mutual-interaction effects of non-spherical shape and reflection. On the other hand, the growing power of computers allowed researchers to consider the global problem of eclipsing non-spherical stars with non-uniform distributions of surface brightness producing light curves as they revolved around their common centres of mass. This latter approach of *synthesizing* close-binary light curves from numerical models that took account of all the known astrophysics of the problem won the argument, with the work of Lucy (1968a,b), Mochnacki and Doughty (1972a,b), and Rucinski (1973) on contact binaries and that of Hill and Hutchings (1970), Hutchings and Hill (1971a,b), and Wilson and Devinney (1971) on all types of binaries leading the enormous change in methods of analysis. These procedures have dramatically altered the precision with which we can determine the physical characteristics of binary stars of many types, and they have led to much more exacting tests of the models for the evolution of binary stars in general. We discuss the light variations of binary stars and the modern methods for their analysis in Chapter 5. Chapter 6 provides a summary of our knowledge of the masses, radii, temperatures, and luminosities of stars in binary systems, determined from the techniques described in Chapters 3–5.

The past decade has seen the emergence of new procedures that extend the synthesis approach of comparing a model directly with observational data into the realm of securing images of the surfaces of the stars in binaries, and the accretion streams and discs as well. These techniques are variously called *eclipse mapping*, *Doppler tomography* or *imaging*, and *Zeeman-Doppler imaging* because they rely on different procedures with different types of data. These developments have been made possible through the work reviewed, for example, by Piskunov and Rice (1993) on surface imaging of stars in general. Horne (1985), Marsh and Horne (1988), and Rutten et al. (1994) established and refined the techniques of Doppler tomography and maximum-entropy eclipse mapping applied to the accretion discs of cataclysmic variables, whilst Bagnuolo and Gies (1991) showed how tomography could be used to separate combined spectra into two or more components. Eclipse mapping has been developed by Vincent et al. (1993), Strassmeier (1994), and Collier Cameron (1997) to determine surface images of magnetically active stars. The effective resolution of the images that are attained is at the level of 10^{-6} arcsec for a nearby binary star, a long way from the aspirations of 'merely' resolving the two stellar points of light in interferometric binaries! These new techniques are only now beginning to have the major impact on the subject of binary stars that is bound to follow. If we can image a point source of light into its component parts with the resolution noted earlier, then we can begin to address completely new areas of the astrophysics of close binary stars. We discuss these new procedures and their results in Chapter 7.

1.5 Types of binaries: Present-day terminology

The classification, introduced by Kopal, of detached, semidetached, and contact binaries was based on analyses of photometric light curves of eclipsing systems, aided by results from spectroscopic observations to determine mass ratios, masses, separations, and temperatures. Because of the impact that the scheme has had on our subsequent understanding of the evolution of binary stars, the terms have remained in extensive use, and they are used also for systems that do not show eclipses, but whose properties are inferred from other procedures to be in those geometrical states. The Roche model for binary stars identifies two volumes or lobes, each centred on the centre of mass of one component, that define the upper limiting volume that each star may occupy. That is, any particle lying inside one lobe belongs gravitationally to that lobe alone; particles lying outside the two lobes belong gravitationally to the complete binary. Binaries whose component stars lie well inside these Roche lobes are described as *detached* systems; those in which one component completely fills its Roche lobe, whilst its companion lies inside its Roche-lobe volume, are *semidetached* systems; and those for which both components fill, or even overfill, their Roche lobes are *contact* systems. Illustrative examples of such systems are shown in Figures 1.3–1.5, and Figure 1.6 shows an idealized model of a cataclysmic variable, or an x-ray binary system.

In the hope of providing a useful guide and summary of the names and overall properties of all the different types of binary stars that are studied today, I have compiled Table 1.1:

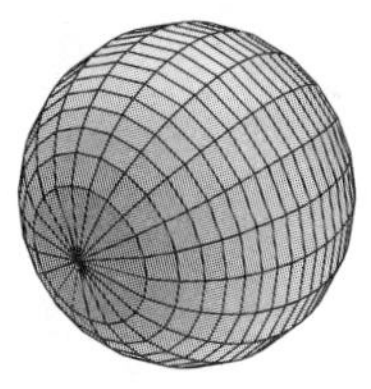

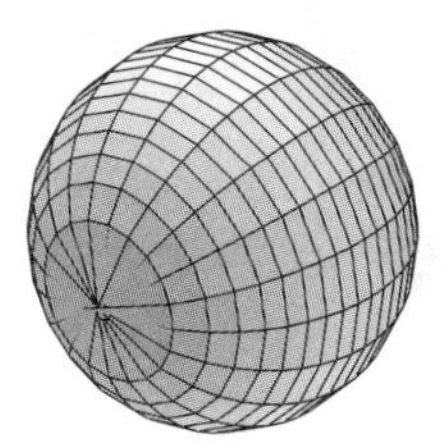

Fig. 1.3. A *detached* binary system, where the radii of the components, expressed in terms of the separation between the centres of mass of the two stars, are 0.116 and 0.100 for the primary (more massive) star and the secondary (less massive) star, respectively. The mass ratio (secondary/primary) is $q = 0.85$. Orbit sizes for detached binaries range from about 1 $R_{\odot}$ up to those of resolved binaries (astronomical units, AUs).

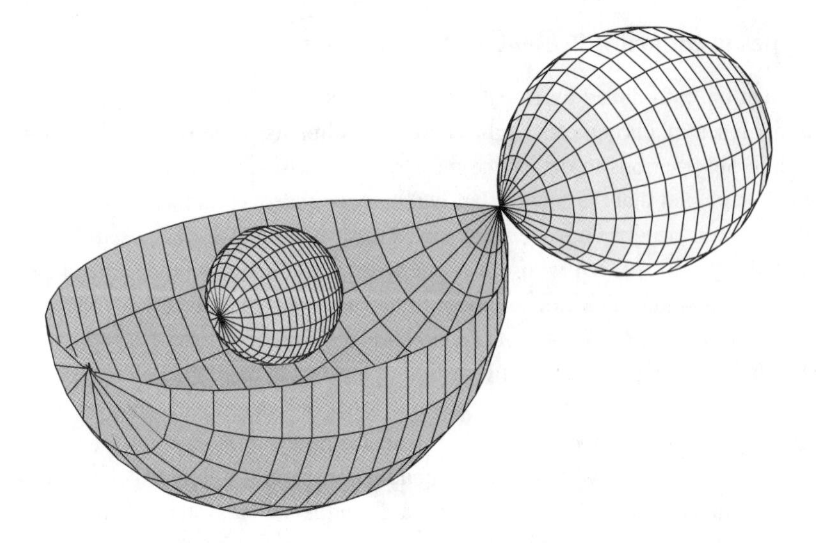

Fig. 1.4. A *semidetached* binary system, with a mass ratio of $q = 0.30$. The primary component has a radius of $0.15a$, where a is the separation between the centres of mass of the two stars, and is well detached from its Roche lobe, as illustrated by the lower outline of that lobe. The secondary component fills its Roche lobe and has a *mean* or *volume* radius of $0.28a$. Orbit sizes for semidetached binaries range from about 1 $R_{\odot}$ up to AUs.

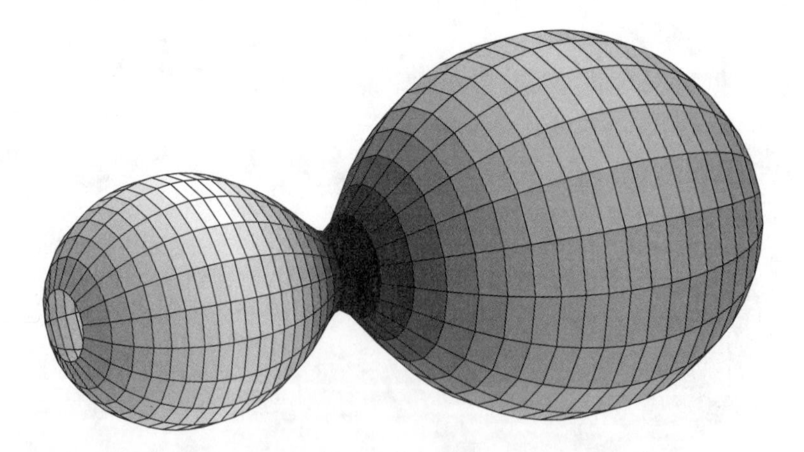

Fig. 1.5. A *contact* binary system, with a mass ratio of $q = 0.40$. In this illustration, the two stars overfill their respective Roche lobes, so that a common envelope surrounds both stellar cores, and there is an adjoining neck around the inner Lagrangian point (L_1). The fill-out factor $F = 1.16$ (see Chapter 5). Orbit sizes for typical W-UMa-type contact binaries are about 2 $R_{\odot}$.

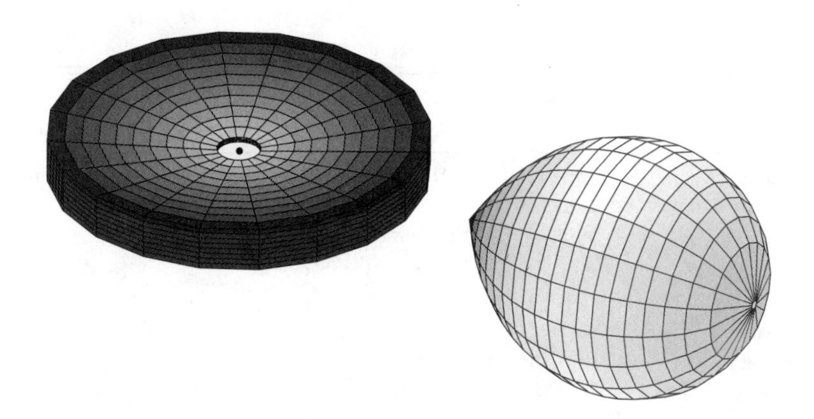

Fig. 1.6. The canonical view of a *cataclysmic variable* (CV), or an *x-ray binary*, with one star filling its Roche lobe, and its compact companion (the black dot) surrounded by an accretion disc. For a CV, the compact star is a white dwarf, usually the primary, and the secondary is a low-mass main-sequence star. For an x-ray binary, the compact object is a neutron star or a black hole, whilst the companion is a mid- or low-mass main-sequence star (an LMXB) or a high-mass supergiant (an HMXB). Amongst HMXBs, the x-ray source is powered by means of an accretion disc from Roche-lobe overflow (RLOF), together with stellar-wind outflow in some cases. The accretion disc is fed by a narrow accretion stream (not shown) from the inner Lagrangian point on the Roche-lobe-filling component. It is deflected by some 10–20° from the line of centres by the Coriolis effect but is still within the orbital plane. At the site of impact of the stream and the disc, a region of enhanced brightness is formed. Orbit sizes for typical CVs and small LMXBs are about 1–2 $R_{\odot}$, whereas typical HMXBs have orbits about 20 times larger.

Categories of Binary Stars. Its purpose is to serve as a quick reference guide when readers are involved in details later in the book or when they need rapid access to information such as orbital period ranges, typical masses, evolutionary states, and evidence of interactions or mass exchange.

As illustrated earlier by the description of the development of the subject of binary stars, this text is, for a change, biased towards the observational data, the methods of analysis, and the empirical results, and then lastly come the comparisons of those results with the relevant theoretical models for the structure and evolution of binary stars. Of course, the theoretical models for stars and their atmospheres have been used in developing the methods to analyse the data, but there are always critical observations that can test the correctness of these models, and I hope that these tests will be seen clearly in the following chapters.

1.6 Evolutionary processes for binary stars

Because stellar-evolution time scales are generally to be measured in millions or billions of years, we are always searching for binary systems that can be found in those crucial stages of rapid evolution between one equilibrium state and the next, hoping to be able confidently to place a particular observed system into a particular stage of evolution. Those crucial stages

Table 1.1. *Categories of binary systems*

Type of system; mass ratio ($q = m_2/m_1$)	Spectral range; orbital period	Photometric features	Spectral features
Two main-sequence stars, not necessarily equal (SB1, SB2) Detached $q \approx 0.1$–1.0	O–M days to 100s of days	Normal eclipses of spherical stars, except some O–B stars with some distortion	Normal absorption-line spectra
Algol systems: Classical (e.g., Algol) O–B-type (e.g., u Her) Semidetached Classical: $q \approx 0.1$–0.3 O–B-type: $q \approx 0.3$–0.7	pr/sec: B–AV/F–KIV O–B/B–A Days	Very different depths of eclipses; reflection effect + distortion; often asymmetric light curves due to gas streaming	Emission lines of H, Ca II in optical; C IV etc. in UV; accretion streams; occasional discs; mass loss; radio flares
W UMa systems (e.g., W UMa) Contact $q \approx 0.08$–0.8	F–K ≤ 0.7 day	Continuous light variations due to ellipticity; sometimes asymmetric due to starspots; soft x-ray sources	rotationally broadened and blended absorption lines; emission lines in UV (chromospherically active)
High-mass x-ray binaries (HMXB) (e.g., Cyg X-1) Semidetached/detached $q \leq 0.1$	pr: O–BI–II, BeIII–IV sec: neutron star or black hole Days–years	ellipsoidal light variations due to distorted O–B star; x-ray pulses from rapidly spinning secondaries (neutron stars); transients for eccentric orbit Be/ns systems	Hα and He II emissions offset in phase to O–B absorption lines; O–B star RLOF/wind powers x-ray source
Low-mass x-ray binaries (LMXB) (e.g., Sco X-1) Semidetached $q \approx 0.5$–1.5	pr: A–MV sec: neutron star or black hole Hours	ellipsoidal light variations and reflection effect; x-ray heating of primary from accretion disc + neutron star; x-ray bursts, dips + QPO behaviour from disc	H, He II, C III–N III emission; x-ray sources, all with variable x-ray spectra; RLOF powers x-ray source
Polars/intermediate polars eg AM Her Semidetached $q \approx 0.5$	pr: white dwarf sec: K-MV Hours	High linear and circular polarisation; high and low luminosity states, like CVs; accretion columns on to magnetic poles of white dwarf	emission line spectra; x-ray sources; cyclotron features; magnetic fields of 10^8 T for white dwarfs

(*Contd.*)

Table 1.1. *(Continued)*

Type of system; mass ratio ($q = m_2/m_1$)	Spectral range; orbital period	Photometric features	Spectral features
Cataclysmic variables (e.g., U Gem); many subgroups Semidetached $q \approx 0.5$–1	pr: white dwarf sec: K–MV Hours	Very variable and complex optically; flickering; high and low luminosity states; accretion disc, stream, hot-spots dominate	Blue continuum; emission lines in optical/UV; some soft x-ray sources from accretion disc
Supersoft x-ray (e.g., Cal 83) Semidetached q(wd/comp) < 1	pr: white dwarf (wd) Hours–days (?)	Some eclipses; reflection effect	Emission-line spectra of H and He II; evidence of bipolar outflows
RS CVn systems (e.g., RS CVn) Detached $q \approx 0.7$–1.2	pr: GV sec: KIV Days	Two well-defined eclipses, but distortions due to extensive starspots	SB2 with optical/UV emission; soft x-ray + radio emission; chromospherically active
β Lyrae systems, W Serpentis systems Semidetached	pr: O–B Days	Continuous light variations due to ellipticity; often asymmetric due to gas-streaming effects	Complex, very broad line profiles; optical/UV emission; extensive mass loss
ζ Aurigae systems (e.g., ζ Aur, VV Cep) Detached/Semidetached?	pr: O–BV sec: MI–II 2–20 years	Very long eclipses; variable light curves	Complex; optical/UV chromospheric emission; mass loss
Pulsar binaries (e.g., PSR1913+16) Detached Wide range of q	pr: neutron star sec: mostly white dwarfs or neutron stars Hours to 100s of days	Radio pulsars with pulse periods of 15 ms to 4 seconds	
Ba and CH stars Detached/Semidetached?	pr: red giant sec: white dwarf 100s of days–years		Unusual overabundances of Ba and C
Symbiotic stars	pr: red giant sec: m-s, wd, or sd 10s-100s of days	Variable outbursts and declines	Composite spectra nebular emission lines

occur when stars undergo changes in size due to their own evolutionary (single-star) processes and/or when their binary nature imposes limits on those changes. In Chapter 4 we discuss the evidence for mass exchange and the methods of analysis used in determining the rates of mass exchange between the components of interacting binary stars and the mass losses from the systems. In their book *Structure and Evolution of Single and Binary Stars*, de Loore and Doom (1992) have provided a very helpful discussion of the techniques used to model the evolutionary behaviour of mass-exchanging binary stars. The following sketch, however, should serve as a useful guide to the main aspects of the models and the terminology in use today.

Kippenhahn and Weigert (1967) introduced a simple classification scheme for binary stars undergoing exchanges of mass between their components, resulting from the evolutionary expansion of each star. They noted that single-star evolution theory identifies three main stages of evolutionary expansion of a star: during the main-sequence stage, during the transition to the red-giant stage, and during the transition to the asymptotic-giant-branch (AGB) or red-supergiant stage, dependent upon the initial mass of the star. Figure 1.7 illustrates these cases by plotting the changes in radius of a star of 7 $M_\odot$ as a function of time.

Note that whilst the main-sequence stage exhibits, typically, a change in radius by a factor of about 2 on long time scales, during the transition to the red-giant stage that factor is approximately 50, and that change occurs very quickly relative to stellar-evolution time scales; then, during the transition to the AGB or red-supergiant stage there is an additional factor of

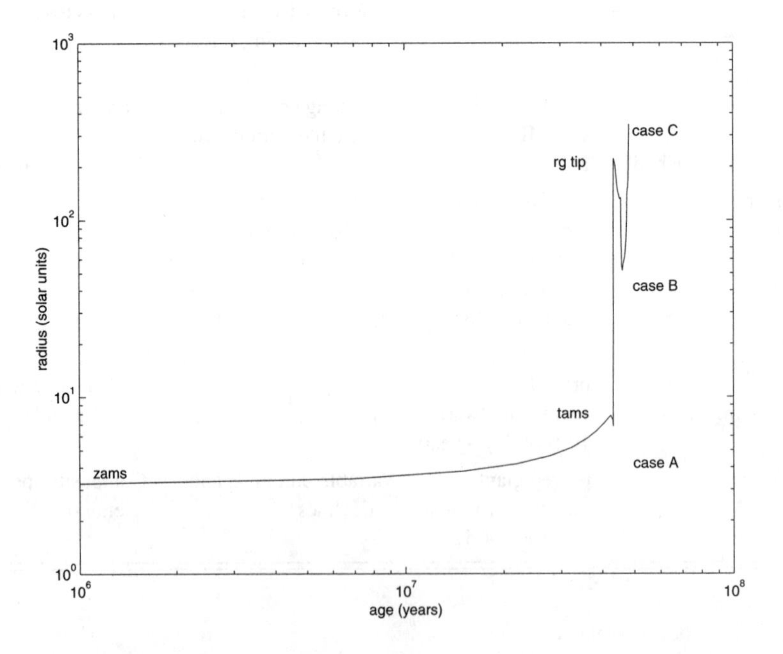

Fig. 1.7. The change in radius of a star of 7 $M_\odot$ plotted as a function of time, from the zero-age main sequence (zams), through the terminal-age main sequence (tams), the tip of the red-giant branch (rg tip), and into the main core-He-fusion phase; data from the models of Schaller et al. (1992).

about 10, with the change again occurring quickly. For a single star, such changes pose no difficulties, but for a star in a binary system, its expansion may be limited by the size of its available Roche lobe, which in turn depends primarily upon the separation between the two components (or the orbital period, through Kepler's third law), and secondly upon the mass ratio. The three cases applicable to mass transfer in binary systems are as follows:

Case A The initial orbital period is short enough (approximately a few days) for a star to reach its Roche lobe at some time during its expansion across the main-sequence band, that is, as a normal core hydrogen-burning star.

Case B The initial orbital period lies in the range of several days to about 100 days, so that a star will reach its Roche lobe during the rapid transition phase between the main-sequence stage and the red-giant stage. When the mass loss is interrupted by core-helium ignition in the red giant, then the mass transfer is referred to as case BB.

Case C The initial orbital period exceeds 100 days, so that there is sufficient volume for a star to evolve through to the supergiant stage without hindrance.

There are also hybrid situations, where the mass transfer may continue through the transition stages, resulting in cases AB and BB. Next we proceed to a summary of the various evolutionary stages that are considered appropriate for each of the types of binary systems listed in Table 1.1.

The *detached* binaries that are composed of unevolved main-sequence components provide the basis for empirical calibration of mass–luminosity and mass–radius relationships. These systems serve, therefore, as tests of single-star structure and evolution theory. Just how far these tests can be extended beyond the main-sequence stage depends upon the room available within a given binary system. That is, is there sufficient volume for a star to evolve to red-giant dimensions, and beyond, before its surface encounters the limiting Roche equipotential and forces a mass-transfer or mass-loss event? Observers have found sufficient numbers of binaries that are eclipsing systems with reasonably short orbital periods (days) to be able to satisfy the requirements of these tests for main-sequence stars. In addition, the RS CVn systems, which typically contain one subgiant component, are still in detached states. But for stars of red-giant dimensions, whose evolutionary paths have been unaffected by their presence in a binary, we have to study systems of long orbital periods, typically hundreds of days. These factors impose enormous practical difficulties for following eclipses, though some amazing achievements have been reported, such as those by Andersen et al. (1991) on the F7III+G8III system TZ For ($P = 75.7$ days) and by Griffin and Duquennoy (1993) on the F2Ib+G8IIb system OW Gem ($P = 1260$ days). The probability of a system having observable eclipses depends upon the sizes of the stars relative to their separation. That probability will be low for a long-period system containing a red giant and a main-sequence star. Consequently, to obtain complete astrophysical descriptions for evolved, but unperturbed, stars in binaries, we need to combine spectroscopic observations with the interferometric techniques that can render such systems resolvable binaries and hence permit determinations of distances, orbital inclinations, masses, linear radii, and so forth. Such studies are still in their early stages of achieving substantial numbers of results. There are a few eclipsing systems that do contain supergiants, the ζ Aurigae systems, with orbital periods of years, that may still be in detached states, not yet having undergone mass-transfer events, but some of them may well be in the case-C stage of mass transfer.

The only other types of binary system listed in Table 1.1 that can be described as *detached* systems are those containing two compact objects, white dwarfs or neutron stars, that are the core remnants of highly evolved stars. These systems undoubtedly have undergone mass-transfer and/or mass-loss events that may have been spectacular. They are at the latest stages of the evolutionary paths for binary stars.

Nearly all the other systems listed in Table 1.1 can be described as *semidetached* binaries. Beyond that geometrical label, the evolutionary paths followed to reach the various observed states may have been quite different and may have involved more than one mass-transfer event. The most simple example is the Algol-type system, where the originally more massive star (the loser) evolves more quickly, in accordance with standard single-star evolution theory, overfills its Roche lobe, and initiates a *rapid* phase of mass transfer leading to complete mass-ratio reversal. The currently observed state of these systems is that the mass loser, which is now the less massive component in the binary, still fills its Roche lobe and transfers mass more slowly to its mass-gainer companion, which we observe as a normal-looking main-sequence star. Most observed Algol and β Lyrae systems have properties that can reasonably be explained by case-B or case-AB evolution models that include loss of mass from the binary system – so-called *non-conservative* evolution. There are specific predictions, from competing models, about the degree of chemical inhomogeneity that may be displayed on the stellar surfaces in these systems, but these predictions have not yet been properly tested.

The *high-mass* and *low-mass x-ray* binaries (HMXB, LMXB) each contain a neutron star or a black hole, the result of a Type II supernova event (SNe). The evolutionary path for HMXBs requires a case-B mass-transfer event to make the original primary (more massive) star the loser and to reverse the mass ratio. But the evolution time scale for the core-helium-burning loser is much shorter than the main-sequence lifetime of the gainer, and so the loser reaches the stage of a Type II SNe first. Because the SNe occurs on the less massive star in the system, the chances of the binary being disrupted are much reduced, though not excluded, and the consequence is expected to be a binary composed of a radio pulsar and an unevolved main-sequence-star companion, in a highly eccentric orbit. Two such systems have been discovered in the past decade, namely, PSRB1259-63 (Johnston et al. 1992) and PSRJ0045-7319 (Kaspi et al. 1994). These provide examples of the expected link to the HMXBs, where the non-degenerate star has now evolved and transfers mass to the neutron star companion, thus powering an x-ray source from the accretion of matter, via an accretion disc around the neutron star, into a deep gravitational-potential well. The mass transfer may occur by means of Roche-lobe overflow (RLOF) or by means of a radiatively driven stellar wind from an O-type supergiant or by means of episodic transfers in the transient HMXBs that contain emission-line Be stars.

In the cases of LMXBs and the cataclysmic variables (CVs), RLOF occurs in case C, ensuring mass-ratio reversal and a *common-envelope phase* of evolution where orbital angular momentum is lost from the binary because of the two stellar cores orbiting within the expanding envelope of the mass-losing supergiant. The orbital period shrinks to values of hours, as a result of the foregoing, in concert with the loss of orbital angular momentum by means of *magnetic braking* from the stellar wind of a magnetically active low-mass main-sequence star. The result can be a neutron star, after a Type II SNe, orbiting a low-mass star in an LMXB; or it can be a white dwarf, the core remnant of a mass-losing supergiant if its initial mass was less than about 8 $M_{\odot}$ or the core remnant of an older AGB star orbiting a low-mass star in a CV. In both cases, theory has to explain why we see the low-mass star filling its Roche lobe in order to transfer mass to the compact companion via an accretion disc, because the main-sequence lifetimes of low-mass stars are about 10^{10} years. The explanation seems to be that magnetic braking removes orbital angular momentum at the necessary rate to shrink the binary orbit, and hence the Roche-lobe sizes, onto the surface of the main-sequence star, thereby forcing mass transfer to occur. This theory also can explain why the orbital periods should decrease, even when the mass-transferring star is the component of lower mass (see Chapter 4). The *intermediate polars* and *polars* are subsets of the CVs, where the white-dwarf component has a substantial or very strong ($B \approx 10^8$ T) magnetic field. For polars, the axial rotation period of the white dwarf is locked to the orbital period, whereas for intermediate polars the rotation period of the white dwarf can be rapid, leading to beat phenomena between these frequencies. For intermediate polars, the field is strong enough to disrupt the inner parts of the accretion disc and channel the accretion flow onto one or two magnetic poles on the white dwarf. For polars, the field is strong enough for its magnetosphere to reach the companion star, so that the accretion flow is completely controlled by the field lines, resulting in *accretion columns/curtains* on the magnetic poles. The *supersoft x-ray sources* have now also been determined to be LMXBs, with white-dwarf accretors that are undergoing hydrogen nuclear-fusion reactions on their surfaces, either steadily or in some repetitive manner. The differences between the usual LMXBs and these supersoft sources lie in the Roche-lobe-filling mass donor being the more massive star (e.g., in the range 1.3–2.5 $M_{\odot}$) and in the mass transfer occurring on a thermal time scale, so that the emitted x-ray spectrum peaks at only 15–80 eV, rather than the 1000-eV range of typical LMXBs (Kahabka and van den Heuvel 1997).

For the remaining systems in Table 1.1, the W-UMa-type *contact* binaries and other contact systems throughout the main sequence seem to result from case-A evolution directly into contact states, as a consequence of their achieving very short orbital periods at early stages in their evolution. Magnetic braking is favoured to produce the W UMa stars with ages of about 8×10^9 years (Maceroni and van't Veer 1996), whilst one may need to appeal to binary-star+single-star gravitational encounters in young star clusters to explain the presence of contact systems with O-type components and orbital periods of less than a day (e.g., V701 Sco, Bell and Malcolm 1987). The W Serpentis systems seem to be semidetached binaries, where the primary component fills its Roche lobe and is transferring mass to its companion, as well as losing mass from the system. These seem to be case-B systems undergoing a rapid phase of evolution that is difficult to study precisely because the material ejected by the binary succeeds in hiding the details of the underlying system. The same comment applies to the hydrogen-poor and helium-rich binaries, perhaps case-BB objects, and the *symbiotic* stars, perhaps case-C objects with an already evolved subdwarf or white-dwarf companion. The Ba- and CH-star binaries show abundance anomalies on the currently primary components that seem to be the consequences of their gaining nuclearly processed material from their post-AGB companions, now white dwarfs.

1.7 Summary

It should be clear from the foregoing that the study of close binary stars has not reached any sort of peak from which interest will decline in the future. There are continuing developments in the methods of observation, analysis, and simulation that are now permitting us to ask new questions and to test hypotheses and detailed models that predict specific interactions between the components of binary stars. As a result, our observational data are much more precise than those available a generation ago, at least for relatively simple binaries, and we have begun to open the door to detailed studies of some of the most intractable types of interacting binaries. The evolutionary models, based upon the latest revisions and improvements for representing fundamental physical processes of opacity, energy transport, and so forth, within single stars, will now be required to become more sophisticated to explain the evolutionary links between the various subgroups of binary stars. The main purpose of this text is to explain how observations of binary stars are made, analysed, and interpreted, and then to see if our overall understanding of all the diverse phenomena displayed by binary stars allows us to extend the success in understanding single-star evolution into these interactive evolutionary situations, and thence to achieve a more secure description of the astronomical universe.

2 Two-body orbital motion

2.1 Introduction

Newton first showed that a spherical body of mass m has a gravitational-potential field equivalent to a point mass m located at the centre of the sphere; see Binney and Tremaine (1987) for a lucid discussion of this topic. So we should be able to consider a system of two spherical stars as though it were composed of two point masses orbiting each other in their mutual gravitational field. This approximation will be entirely valid provided that the stars are not too close, which is implicit in the description of the shapes of the stars as spherical. In Chapter 4 we shall discuss what happens when the point-mass approximation breaks down. But, for now, it is reasonable to derive the fundamental relationships among the stellar masses, their orbital speeds, and the shapes and sizes of the orbits.

It should be pointed out that this point-mass approximation is much more generally valid than might be thought at first sight. The reason is simply that stars are 'centrally condensed' objects, with some 85–90% of a star's total mass contained within the inner 50% of its radius, at least on the zero-age main sequence (Kippenhahn and Weigert 1991). So the point-mass approximation can be a sufficient description for many binary stars. Furthermore, as stars evolve, they become *more* centrally condensed, and thus the validity of the approximation is extended further.

2.2 Newtonian gravitation

2.2.1 Central conservative forces

A force F is described as a *central* force if it is always directed towards a fixed point, the *centre* of force, and is parallel to the radial direction $\vec{r}$. It is a *conservative* force if the motion of a body in that force field obeys the law of conservation of energy, namely, that kinetic energy + potential energy = total energy = a constant. It follows that for such a central conservative force, the law of conservation of angular momentum can be written as follows: angular momentum $\vec{J} = m\vec{r} \times \dot{\vec{r}} =$ constant, for a body of mass m. If we take the *moment* of the force $\vec{F}$, namely, $G = \vec{r} \times \vec{F}$, we see immediately that $\vec{G} = 0$, and because $\dot{\vec{J}} = G$, it follows that the angular momentum $\vec{J}$ is constant. The constancy of the angular momentum means that the motion of a body under the action of the force $\vec{F}$ is confined to the plane containing the vectors $\vec{r}$ and $\dot{\vec{r}}$, and the direction of the angular-momentum vector $\vec{J}$ is perpendicular to that plane.

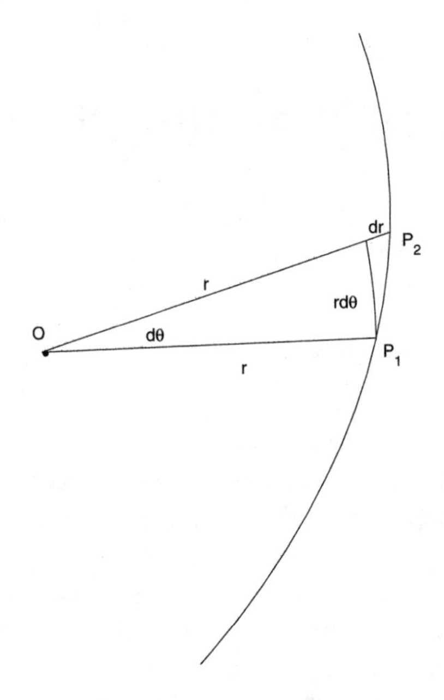

Fig. 2.1. The motion of a body under the action of a central force directed towards the centre of the reference frame O. The polar coordinates of the body at P_1 are (r, θ) at time t, and those at P_2 are $(r + dr, \theta + d\theta)$ at time $t + dt$.

In Figure 2.1 we consider the motion of a body of mass m, acted upon by a central force directed towards the origin O of a reference frame. In a time dt, the body moves from P_1 to P_2, its radial position changes from r to $r + dr$, and its angular position changes from θ to $\theta + d\theta$. The *velocity* $\dot{\vec{r}}$ can be separated into two components, the *radial* component along the direction of $\vec{r}$, and the *transverse* component at a right angle to $\vec{r}$.

The radial component of velocity is $dr/dt = \dot{r}$, and the transverse component is $r d\theta/dt = r\dot{\theta}$. Thus the velocity is $\dot{\vec{r}} = \dot{r}\hat{r} + r\dot{\theta}\hat{\theta}$, where $\hat{r}$ and $\hat{\theta}$ are the unit vectors in the radial and transverse directions, respectively. The *speed* of the body is then given by the expression $V^2 = \dot{r}^2 + r^2\dot{\theta}^2$.

From studying Figure 2.1, we see that $d\hat{r}/dt = (d\hat{r}/d\theta)(d\theta/dt) = \hat{\theta}\dot{\theta}$ and $d\hat{\theta}/dt = (d\hat{\theta}/d\theta)(d\theta/dt) = -\hat{r}\dot{\theta}$.

Then the *acceleration* experienced by the body, due to the action of the force $\vec{F}$, is

$$\ddot{\vec{r}} = \ddot{r}\hat{r} + \dot{r}\dot{\theta}\hat{\theta} + \dot{r}\dot{\theta}\hat{\theta} + r\ddot{\theta}\hat{\theta} + r\dot{\theta}\dot{\theta}(-\hat{r}) = (\ddot{r} - r\dot{\theta}^2)\hat{r} + (2\dot{r}\dot{\theta} + r\ddot{\theta})\hat{\theta} \tag{2.1}$$

The first term in equation (2.1) is the *radial* component of the acceleration, and the second term is the *transverse* component. This second term is linked directly to Kepler's second law of planetary motion, namely, that the radius vector $\vec{r}$ sweeps out equal areas in equal intervals of time. The magnitude of the angular-momentum vector $\vec{J}$ is given by $J = mr^2\dot{\theta}$. In a time interval dt, the area swept out by the radius vector is $dA = (1/2)r^2 d\theta$, so that $dA/dt = (1/2)r^2\dot{\theta} = J/2m =$ constant. Thus we obtain the famous equation $r^2\dot{\theta} =$ constant, a succinct description of Kepler's second law. But note also, now, that

$$\frac{1}{r}\frac{d(r^2\dot{\theta})}{dt} = 2\dot{r}\dot{\theta} + r\ddot{\theta} = 0 \tag{2.2}$$

because $r^2\dot{\theta}$ = constant. So the transverse component of the acceleration is zero, in accordance with the purely radial dependence of the force field.

The laws of conservation of energy and of angular momentum can now be written as

$$\frac{1}{2}m\dot{\vec{r}}^2 + Q(r) = C = \text{constant} \tag{2.3}$$

$$m\vec{r} \times \dot{\vec{r}} = \vec{J} = \text{constant} \tag{2.4}$$

where $Q(r)$ is the potential energy of the body. The motion is confined to a plane defined by the direction of $\vec{J}$. We can write these equations in scalar form as

$$\frac{1}{2}m(\dot{r}^2 + r^2\dot{\theta}^2) + Q(r) = C \tag{2.5}$$

$$mr^2\dot{\theta} = J \tag{2.6}$$

To establish what *form* the motion of the body will take under the influence of the force $\vec{F}$, first we eliminate the $\dot{\theta}$ dependence in these conservation equations to provide the *radial energy equation*

$$\frac{1}{2}m\dot{r}^2 + \frac{J^2}{2mr^2} + Q(r) = C \tag{2.7}$$

so called because all the terms are functions of the radial distance r. For an attractive gravitational force acting between a central mass m_c and the body of mass m under consideration, the force is $\vec{F} = -Gm_c m\hat{r}/r^2$, and the corresponding potential energy of the mass m is given by $Q(r) = -Gm_c m/r$. We need to eliminate the time derivatives from this radial energy equation in order to describe the motion that takes place in terms of the position coordinates (r, θ).

The time derivatives are removed by using the variable $u = 1/r$. Then, $du/d\theta = (-1/r^2)dr/d\theta$, and $\dot{r} = (dr/d\theta)\dot{\theta} = -r^2\dot{\theta}(du/d\theta)$. That is, $\dot{r} = (-J/m)du/d\theta$, so that we have

$$\frac{J^2}{2m}\left(\frac{du}{d\theta}\right)^2 + \frac{J^2u^2}{2m} - Gm_c mu = C \tag{2.8}$$

If we make the substitution $l = J^2/Gm_c m^2$ and add unity to each side of equation (2.8), then we obtain

$$l^2(du/d\theta)^2 + l^2u^2 - 2lu + 1 = 2(Cl/Gm_c m) + 1 = e^2 \tag{2.9}$$

Furthermore, putting $x = lu - 1$, so that $dx/d\theta = l du/d\theta$, we obtain

$$(dx/d\theta)^2 + x^2 = e^2 \tag{2.10}$$

and because the left-hand side is positive, we have $dx/d\theta = (e^2 - x^2)^{1/2}$. This equation is readily integrated to give

$$\int_{x_0}^{x} \frac{dx}{(e^2 - x^2)^{1/2}} = \int_{\theta_0}^{\theta} d\theta = \arcsin(x/e) - \arcsin(x_0/e) = \theta - \theta_0 \tag{2.11}$$

when $|x| \leq |e|$. The maximum value of x is e when $\theta - \theta_0 = \pi/2$, and $x = x_0 = 0$ when $\theta - \theta_0 = 0$. If we allow the arbitrary constant of integration θ_0 to equal 0 and make x have a maximum when $\theta = 0$, so that r has a minimum at that point, then $x = e\cos\theta$.

We can now return to our original variables, r and θ, and note that the *form* of the orbit is described by the equation

$$r = l/(1 + e\cos\theta) \tag{2.12}$$

which is the standard polar equation for a conic section of eccentricity e and parameter l, and r is a minimum at $\theta = 0$. Thus the motion followed by a body in the attractive inverse-square force field of Newtonian gravitation is in general a conic section defined by the values of e and l. For the two-body motion of any persistent interest for the topics in this book, such conic sections will be *ellipses* of eccentricity $e < 1$.

2.2.2 The two-body problem

Two stars of masses m_1 and m_2 ($m_1 \geq m_2$) orbit their common centre of mass in their mutual gravitational field according to Newton's law of gravitation and the laws of motion. In Figure 2.2, the stars are at positions P_1 (mass m_1) and P_2 (mass m_2) at time t, with position vectors, relative to an arbitrary inertial reference point O, of $\vec{r_1}$ and $\vec{r_2}$, respectively. Then the separation between the two stars is the vector $\overrightarrow{P_2P_1} = \vec{r} = \vec{r_1} - \vec{r_2}$, and the unit vector in that direction is $\hat{r} = \vec{r}/r$.

For an isolated two-body system, Newtonian gravitation shows that the forces acting on the two point masses are equal but of opposite sign, $\vec{F_1} = -\vec{F_2}$. The magnitude of each force is $F = Gm_1m_2/r^2$, where G is the gravitational constant; also, the force F equals the mass of the body multiplied by the acceleration that it experiences at time t. Accordingly, we can write

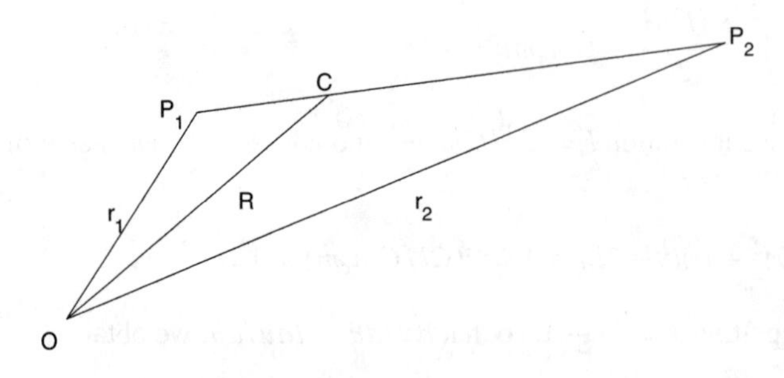

Fig. 2.2. Vector diagram for the positions of the two masses, m_1 at P_1, m_2 at P_2, and the centre of mass, C, relative to a reference point O.

the *equations of motion* for the two stars as

$$m_1\ddot{\vec{r}}_1 = -\frac{Gm_1m_2}{r^2}\,\hat{r}; \qquad m_2\ddot{\vec{r}}_2 = -\frac{Gm_1m_2}{r^2}\,(-\hat{r}) \tag{2.13}$$

Adding the two equations gives $m_1\ddot{\vec{r}}_1 + m_2\ddot{\vec{r}}_2 = 0$, and integrating twice with respect to time provides $m_1\dot{\vec{r}}_1 + m_2\dot{\vec{r}}_2 = \vec{A}$ and $m_1\vec{r}_1 + m_2\vec{r}_2 = \vec{A}t + \vec{B}$, where the constants of integration are the vectors $\vec{A}$ and $\vec{B}$. If, in Figure 2.2, C is the *centre of mass* of the system, and $\overrightarrow{OC} = \vec{R}$, then, by definition of the centre of mass, we have

$$m_1\vec{r}_1 + m_2\vec{r}_2 = (m_1 + m_2)\vec{R} \tag{2.14}$$

and hence $(m_1 + m_2)\vec{R} = \vec{A}t + \vec{B}$ and $(m_1 + m_2)\dot{\vec{R}} = \vec{A}$. Thus the centre of mass moves through space at constant velocity, unless acted upon by other external forces.

2.2.3 The relative motion

It becomes important, in various applications, to consider the motion of one component of a binary system relative to its companion rather than with respect to the centre of mass of the system. That relative motion is described simply by subtracting the two equations of motion, equations (2.13), to obtain

$$\ddot{\vec{r}}_1 - \ddot{\vec{r}}_2 = -\frac{G(m_1 + m_2)}{r^2}\,\hat{r} = \ddot{\vec{r}} \tag{2.15}$$

If we then multiply that equation by a mass term μ, so that

$$\mu\ddot{\vec{r}} = -\frac{G(m_1 + m_2)}{r^2}\,\mu\hat{r} = -\frac{Gm_1m_2}{r^2}\,\hat{r} \tag{2.16}$$

we find that the term $\mu = m_1m_2/(m_1 + m_2)$, which is called the *reduced mass*. The motion of either body in its *relative orbit* is as though the orbiter has a mass μ, and the central mass is the total mass of the system.

2.2.4 Barycentric orbits

The other type of motion that we must describe is the motion of each body relative to the centre of mass, the *barycentre*, of the system. Referring back to Figure 2.2, we can write the vectors $\overrightarrow{CP_1} = \vec{R}_1$ and $\overrightarrow{CP_2} = \vec{R}_2$, so that

$$\vec{r}_1 = \vec{R} + \vec{R}_1; \qquad \vec{r}_2 = \vec{R} + \vec{R}_2; \qquad \vec{r} = \vec{R}_1 - \vec{R}_2$$

and thus $m_1\vec{R}_1 + m_2\vec{R}_2 = 0$. Hence, we can write two alternative expressions for the vector $\vec{r}$ connecting the two stars as

$$\vec{r} = +\frac{(m_1 + m_2)}{m_2}\,\vec{R}_1 \quad \text{or} \quad \vec{r} = -\frac{(m_1 + m_2)}{m_1}\,\vec{R}_2 \tag{2.17}$$

Because the accelerations $\ddot{\vec{r}}_1 = 0 + \ddot{\vec{R}}_1$ and $\ddot{\vec{r}}_2 = 0 + \ddot{\vec{R}}_2$, we can then write the two barycentric equations of motion as

$$\ddot{\vec{R}}_1 = -\frac{Gm_2}{r^3}(\vec{R}_1 - \vec{R}_2); \qquad \ddot{\vec{R}}_2 = -\frac{Gm_1}{r^3}(\vec{R}_2 - \vec{R}_1) \tag{2.18}$$

These two equations are not satisfactory, because they have different vectors on each side of each equation. But we can use the two expressions for the vector $\vec{r}$, equations (2.17), to remedy the problem and provide the final versions of the barycentric equations of motion:

$$\ddot{\vec{R}}_1 = -\frac{Gm_2^3}{(m_1 + m_2)^2}\frac{\vec{R}_1}{\vec{R}_1^3}; \qquad \ddot{\vec{R}}_2 = -\frac{Gm_1^3}{(m_1 + m_2)^2}\frac{\vec{R}_2}{\vec{R}_2^3} \tag{2.19}$$

A comparison of the equations of motion, equations (2.16) and (2.19), shows that the multiplier of the vector on the right-hand side of each equation is a combination of the gravitational constant and a variable expression involving the masses of the two stars. We shall see these particular combinations in a number of relationships that are found to be useful in binary-star analysis.

2.3 Relationships between barycentric and relative orbits

The following relationships among the three orbits are found to be essential in many applications, in addition to providing a convenient summary of what the orbits actually look like in space, that is, their relative sizes and orientations, and providing a useful mental picture of the motions of the two stars in their orbits. We consider the three parameters orbital period P, semimajor axis a, and orbital eccentricity e at some time t. We label the quantities for the barycentric orbit of mass m_1 with the subscript 1, and those of mass m_2 with the subscript 2, leaving those of the relative orbit without subscripts.

It follows directly from Figure 2.3 that if C is the centre of mass of the binary system, then it always lies on the straight line joining the two masses, and consequently the orbital

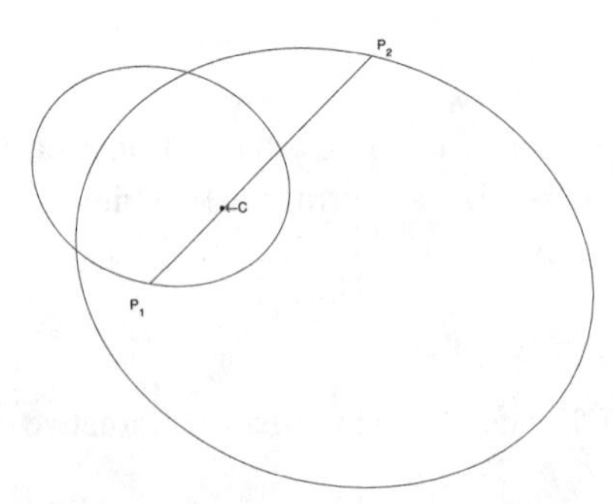

Fig. 2.3. The barycentric orbits of two masses about their common centre of mass, C, when the mass ratio $m_2/m_1 = 0.5$ and the orbital eccentricity is $e = 0.6$.

periods of the three orbits are the same $P_1 = P_2 = P$. Kepler's third law can be written as $M = 4\pi^2 a^3/P^2$, and when it is applied to the three orbits, we must use the following:

for the *relative* orbit: $a = a$; $M = G(m_1 + m_2)$
for the *barycentric* orbit of mass m_1: $a = a_1$; $M = Gm_2^3/(m_1 + m_2)^2$
for the *barycentric* orbit of mass m_2: $a = a_2$; $M = Gm_1^3/(m_1 + m_2)^2$

Then it follows that the sizes of the three orbits are in the proportions $a_1 : a_2 : a = m_2 : m_1 : (m_1 + m_2)$, and hence the semi-major axis a of the relative orbit is simply the sum of the semimajor axes of the two barycentric orbits, $a = a_1 + a_2$. The orbital eccentricities are all equal: $e = e_1 = e_2$. The three orbits are co-planar, and the orientations of the two barycentric orbits differ by 180° within that plane.

These relationships among the barycentric orbits and the relative orbit are important for linking together the data from different observing techniques. The radial-velocity variations due to orbital motion in binary systems give information on the barycentric orbits; the light curves of eclipsing binary stars have to be studied in terms of the relative orbit; polarimetric studies also provide information in terms of the relative orbit.

2.4 Orbital speed as a function of position

For the particular case of circular orbits, the speed of a body in its motion around the centre of mass is constant. But for the general case of eccentric orbits, speed is a continuously varying function of position and time. We consider the motion in polar coordinates (r, θ), where r is the radius vector from the focus of the ellipse, and θ is the position angle between the radius vector and a reference direction. Kepler's second law states that the radius vector sweeps out equal areas in equal intervals of time. In an infinitesimal time interval dt, the radius vector r sweeps out an infinitesimal area $dA = r^2 d\theta/2$; for an entire orbit, $dA = \pi ab$, the area of an ellipse, and $dt = P$, the orbital period. So Kepler's second law can be written very succinctly as $r^2\dot{\theta} = 2\pi ab/P = L$, and L is a constant for that ellipse. The formal definition of *angular momentum* for a mass m is $m\vec{r} \times \dot{\vec{r}}$, whose magnitude is $mr^2\dot{\theta}$ in polar coordinates. Thus $r^2\dot{\theta} = L$ is the angular momentum per unit mass, or the *specific angular momentum*. That simple equation describes the essentials of elliptical orbital motion in their entirety! For large radial distances, the value of $\dot{\theta}$, the rate of change of the position angle, must be small; whilst for small values of r, $\dot{\theta}$ must be large in order to satisfy the foregoing equation. So bodies speed up in their orbits as they fall down the gravitational-potential well towards the centre of mass, reaching a maximum in both angular speed and linear speed at the minimum distance from the focus of the ellipse; they slow down as they climb out of the potential well and reach their slowest angular and linear speeds at the point farthest from the focus.

To establish an equation that describes the linear speed of a body in its eccentric orbit as a function of orbital position (r, θ), we can use the standard polar equation for a conic section: $r = l/(1 + e\cos\theta)$, where l is variously called the *parameter* or the *semi–latus rectum* of the ellipse. The older Latin name is very descriptive, namely, the half-length of the line through the focus that is perpendicular to the major axis of the ellipse. So $r = l$ when $\theta = \pi/2$. Figure 2.4

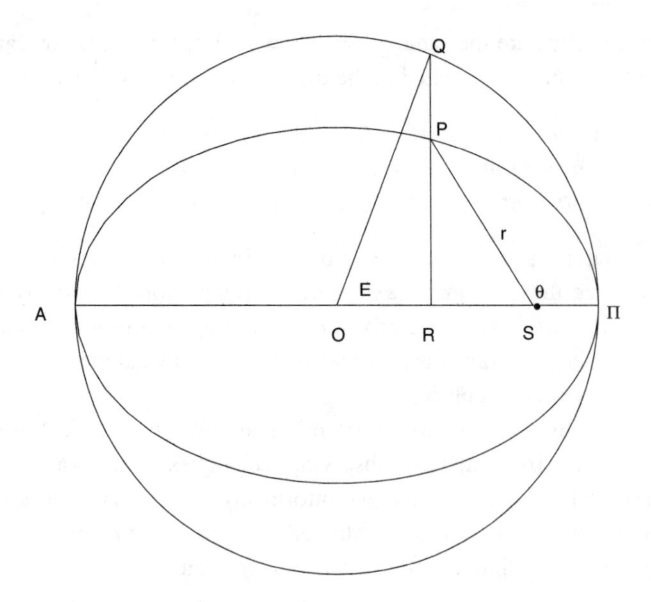

Fig. 2.4. The properties of an ellipse: O is the centre of the ellipse, and S is the focus that contains the centre of gravitational attraction for that elliptical-orbit motion. Π is the pericentre, the point of closest approach of the orbiting body to S, and A is the apocentre, the point farthest away in the orbit from S. Π and A are respectively called *periastron* and *apastron* in binary-star systems. The *auxiliary circle* touches the ellipse at Π and at A.

illustrates the geometry of an ellipse and the various quantities that we need to introduce in these subsections.

Differentiating the preceding polar equation with respect to time t, we obtain

$$-e \sin\theta\, \dot{\theta} = -\frac{l}{r^2}\dot{r} \tag{2.20}$$

We use $r^2\dot{\theta} = L$ to determine the following equations for the radial ($\dot{r}$) and tangential ($r\dot{\theta}$) components of the velocity:

$$\dot{r} = \frac{L}{l} e \sin\theta; \qquad r\dot{\theta} = \frac{L}{l}(1 + e\cos\theta) \tag{2.21}$$

Then the square of the linear speed V is found to be, after a little algebra,

$$V^2 = \dot{r}^2 + r^2\dot{\theta}^2 = \frac{L^2}{l^2}\left[\frac{2}{r} - \frac{(1-e^2)}{l}\right] \tag{2.22}$$

For an ellipse, the semi–latus rectum $l = a(1 - e^2)$, and the specific angular momentum is given by

$$L^2 = \frac{4\pi^2 a^2 b^2}{P^2} = \frac{4\pi^2 a^3}{P^2} a(1 - e^2) = Ml \tag{2.23}$$

where the quantity M is that given earlier for the individual orbits. The final, convenient expression for the square of the linear speed as a function of position in an ellipse of semimajor axis a is

$$\text{ellipse:} \quad e < 1; \qquad l = a(1 - e^2); \qquad V^2 = M\left[\frac{2}{r} - \frac{1}{a}\right] \tag{2.24}$$

$$\text{circle:} \quad e = 0; \qquad l = a; \qquad V^2 = M/a \tag{2.25}$$

It follows that for the three elliptical orbits of a two-body bound system, the various linear speeds are in the proportions $V_1 : V_2 : V = m_2 : m_1 : (m_1 + m_2)$, where we have used the same subscript terminology as in Section 2.3. Note that these various relationships among the two barycentric orbits and the relative orbit in a bound system are making sensible statements. The more massive star (m_1) is closer to the barycentre and has a smaller orbit size than the less massive star (m_2), and consequently it does not require as high a speed to complete its orbit in the same time interval (P) as the less massive star. It is worthwhile remembering these simple facts when we study the data from real binary stars.

2.5 Position as a function of time: Kepler's equation

As we have seen in Section 2.4, the speed in an elliptical orbit is a strong function of position, particularly for orbits of considerable eccentricity. The radius vector r ranges from the minimum value at *periastron* of $r = a(1 - e)$ to the maximum value at *apastron* of $r = a(1 + e)$, and the values of eccentricity found for the orbits of binary stars extend from $e = 0$ to $e > 0.9$ (e.g., the pulsar binary PSRB1259-63 has $e = 0.87$). It follows that the position of a star in an elliptical orbit is not a simple function of time, as for a circle, and the fundamental equation that describes this dependence is called *Kepler's equation*. It is derived in the following manner, by means of the geometry of ellipses that is demonstrated in Figure 2.4.

In Figure 2.4, O is the centre of the ellipse with a focus at S, periastron at Π, and apastron at A. The *auxiliary circle* touches the ellipse at Π and A. The semimajor axis of the ellipse is $O\Pi = a$, and periastron is at a distance $S\Pi = a(1 - e)$ from the focus, and hence $OS = ae$. We consider a star at P at a time t, with position vector $SP = r$ and position angle $\Pi SP = \theta$, which is given the particular name of the *true anomaly*. It is a property of the ellipse and the auxiliary circle that the ratio $PR/QR = b/a$, where b is the *semiminor axis* of the ellipse, and $b = a(1 - e^2)^{1/2}$. The angle $QOR = E$ is also given a particular name: the *eccentric anomaly*. As we have illustrated already, the angles θ and E clearly will not be simple functions of time, unless the orbit is circular.

To derive Kepler's equation, we use the specific-angular-momentum equation $r^2\, d\theta = L dt$ and find alternative expressions for r and $d\theta$ that involve the eccentric anomaly E. From Figure 2.4 we see that

$$RS = -r\cos\theta = OS - OR = ae - a\cos E, \quad \text{so } r\cos\theta = a(\cos E - e) \tag{2.26}$$

$$PR = r\sin\theta; \qquad QR = a\sin E; \qquad PR/QR = b/a \tag{2.27}$$

$$r = (r^2\cos^2\theta + r^2\sin^2\theta)^{1/2} = a(1 - e\cos E) \tag{2.28}$$

Also,

$$\sin\theta = b\sin E/a(1 - e\cos E) \tag{2.29}$$

so that

$$\cos\theta d\theta = b(\cos E - e)\, dE/a(1 - e\cos E)^2 \tag{2.30}$$

But

$$\cos\theta = a(\cos E - e)/a(1 - e\cos E) \tag{2.31}$$

and so

$$d\theta = b\, dE/a(1 - e\cos E) \tag{2.32}$$

We can insert these new expressions for r and $d\theta$ into the specific-angular-momentum equation to obtain the result

$$(1 - e\cos E)\, dE = \frac{L}{ab}\, dt \tag{2.33}$$

which can be integrated immediately to

$$E - e\sin E = \frac{L}{ab}\, t + k \tag{2.34}$$

where k is the constant of integration. To evaluate k, note that at the periastron position in the orbit, $\theta = 0$ and $E = 0$, and we call the time $t \equiv T$ the *time of periastron passage*. Then $k = -(L/ab)T$. Recalling that $b = a(1 - e^2)^{1/2}$ and that $L = [Ma(1 - e^2)]^{1/2}$, we find that $L/ab = [M/a^3]^{1/2} = 2\pi/P$, from Kepler's third law. Thus the final equation that specifies the angular position of the star as a function of time in its elliptical orbit is *Kepler's equation*:

$$E - e\sin E = \frac{2\pi}{P}\,(t - T) \tag{2.35}$$

The right-hand side of this equation is determined directly from observations, once the orbital period and the time of periastron passage have been determined, at least in a preliminary sense; the right-hand side is simply the fractional orbital phase expressed in radian measure.

It is clear that no analytic solution of Kepler's equation to determine E is possible, since we need to know E to evaluate $\sin E$. Such *transcendental equations* must be solved numerically, which is now a simple computer task, but which used to require graphical aids such as *nomograms* for their solution. The simplest numerical technique, which is effective for most practical cases involving binary stars, is to use a Newton-Raphson iterative solution of equation (2.35). That method provides an iterative formula for an improved estimate, x_n, from a previous solution, x_{n-1}, of $x_n = x_{n-1} - f(x_{n-1})/f'(x_{n-1})$, where f' stands for the first derivative of a function $f(x)$ with respect to the quantity x. To apply the method to solve Kepler's equation, we put $f(E) = E - e\sin E - 2\pi(t - T)/P$ and note that $f'(E) = 1 - e\cos E$. Then, for a given observation at a time t, we evaluate the orbital phase in radians and use that value as a preliminary estimate of the eccentric anomaly E. Several iterations are required to

Table 2.1. *Basic relationships for elliptical motion*

Kepler's second law: $r^2\dot{\theta} = L$
Kepler's third law: $M = 4\pi^2 a^3/P^2$, where M is the mass term relevant to the particular orbit under consideration
Relative orbit: $M = G(m_1 + m_2)$
Barycentric orbit of m_1: $M = Gm_2^3/(m_1 + m_2)^2$
Barycentric orbit of m_2: $M = Gm_1^3/(m_1 + m_2)^2$
Specific angular momentum: $L^2 = Ml$, with $l = a(1 - e^2)$
Periastron and apastron distances: $r_p = a(1 - e)$, $r_a = a(1 + e)$
Orbital speeds: $V^2 = M[(2/r) - (1/a)]$, so that $V_1 : V_2 : V = m_2 : m_1 : (m_1 + m_2)$
Semimajor axes: $a_1 : a_2 : a = m_2 : m_1 : (m_1 + m_2)$, and $a = a_1 + a_2$
Radius vector: $r = a(1 - e^2)/(1 + e\cos\theta) = a(1 - e\cos E)$
Relationship between the true and eccentric anomalies: $\tan(\theta/2) = [(1 + e)/(1 - e)]^{1/2}\tan(E/2)$
Kepler's equation: $E - e\sin E = 2\pi(t - T)/P$

make the solution converge, their number depending on the orbital eccentricity e. This is a very simple and brief task for present-day computers.

It is helpful to provide a summary list of the equations that are most used in investigating motions in elliptical orbits (Table 2.1).

2.6 Total angular momentum and energy; orientation of the orbit

The orientation of the relative orbit of a binary system in three-dimensional space is given by the direction of the *total-angular-momentum vector* $\vec{J}$. We shall see that the size and the shape of the relative orbit are determined fundamentally by the magnitude of J and by the *total energy* C of the binary system.

The total energy is given by the sum of the *kinetic energy* K and the *potential energy* Q of the system. For the kinetic energy, we can write

$$K = \frac{1}{2}m_1 V_1^2 + \frac{1}{2}m_2 V_2^2 = \frac{m_1 m_2}{2(m_1 + m_2)}V^2 \tag{2.36}$$

and the speed V in the relative orbit is given by $V^2 = G(m_1 + m_2)[(2/r) - (1/a)]$. For the potential energy, we have

$$Q = -\int_r^\infty \frac{Gm_1 m_2}{r^2}\,dr = -\frac{Gm_1 m_2}{r} \tag{2.37}$$

so that the total energy is

$$C = K + Q = -\frac{Gm_1 m_2}{2a} \tag{2.38}$$

The total energy of the binary system is negative, as is appropriate for a bound orbit, and that is why the term *binding energy* is used for the quantity $-C$, which is then positive. Note that for a given pair of masses, the semimajor axis a of the relative orbit depends solely on the total energy. The larger the orbit, the less negative, and hence larger, the total energy becomes.

The magnitude of the total angular momentum of the system is given by $J = m_1 L_1 + m_2 L_2$, where the specific angular momentum for each barycentric orbit is given by $L_1^2 = [Gm_2^3/(m_1+m_2)^2]a(1-e^2)$, and similarly for L_2. Also, we have, for the orbit of m_1, that $a_1/a = m_2/(m_1+m_2)$, and again similarly for m_2. Hence

$$L_1 = \frac{m_2^2}{(m_1+m_2)^2} L; \qquad L_2 = \frac{m_1^2}{(m_1+m_2)^2} L; \qquad L = G(m_1+m_2)a(1-e^2) \tag{2.39}$$

so that the final expression for the magnitude of the total angular momentum for a binary system is

$$J = m_1 m_2 \left[\frac{Ga(1-e^2)}{(m_1+m_2)}\right]^{1/2} \quad \text{or} \quad J = \frac{2\pi\, a^2(1-e^2)^{1/2} m_1 m_2}{P(m_1+m_2)} \tag{2.40}$$

For a given pair of masses and total energy, the total angular momentum determines the eccentricity of the orbit, or perhaps more instructively, it determines the value of $l = a(1-e^2)$, the semilatus rectum, which is perpendicular to a, and hence defines the *shape* of the elliptical orbit. For a given value of C, circular orbits have the most angular momentum, whilst as $e \to 1$, $J \to 0$, and the orbit degenerates towards a rectilinear ellipse. These quantities, C and J, will be of great importance much later when we investigate the evolution of binary stars and their associated processes of mass exchange between the components and mass loss from the system (Chapter 4).

Because the total-angular-momentum vector $\vec{J}$ is constant for an isolated two-body system, the two stars will continue to orbit the barycentre of the system in the same plane. The orientation of that plane is defined by the direction of $\vec{J}$, which can be defined in terms of a practical coordinate system used by observational astronomy. We describe the orbit in space by means of Figure 2.5.

In Figure 2.5, the mass m_1 is at the origin O of the *xyz* coordinate system; O is not labelled in Figure 2.5 in order to allow the three angles (Ω, ω, θ) to be illustrated clearly. The mass m_2 is at P_2 at time t in the relative orbit about O. The observer's line of sight is along Oz, from *below* the figure, and Oz is perpendicular to the tangent plane of the sky, denoted by *xNy*, and containing O. The arc through N that is parallel to the near side of the orbit is the projection of that part of the orbit onto the celestial sphere, illustrated in the figure by the quadrisphere *xNyzO*. For visual binaries, Ox is taken to point towards the North Celestial Pole. The angular-momentum vector $\vec{J}$ is perpendicular to the orbit plane, denoted by $O\Pi P_2$. Then N, on the intersection of the projection of the orbit plane with the tangent plane of the sky, is defined as the *ascending node* and is the point in the orbit where the star is *receding* from the observer *most rapidly*. Equally, the line *NO* extends through to the other side of the hypothetical sphere to N', the *descending node*, which is the point in the orbit where the star is *approaching* the observer *most rapidly*. The line *NON*$'$ is sensibly called the *line of nodes*, and its orientation is defined by the angle $xON = \Omega$, which is called the *longitude of the ascending node*. To retain

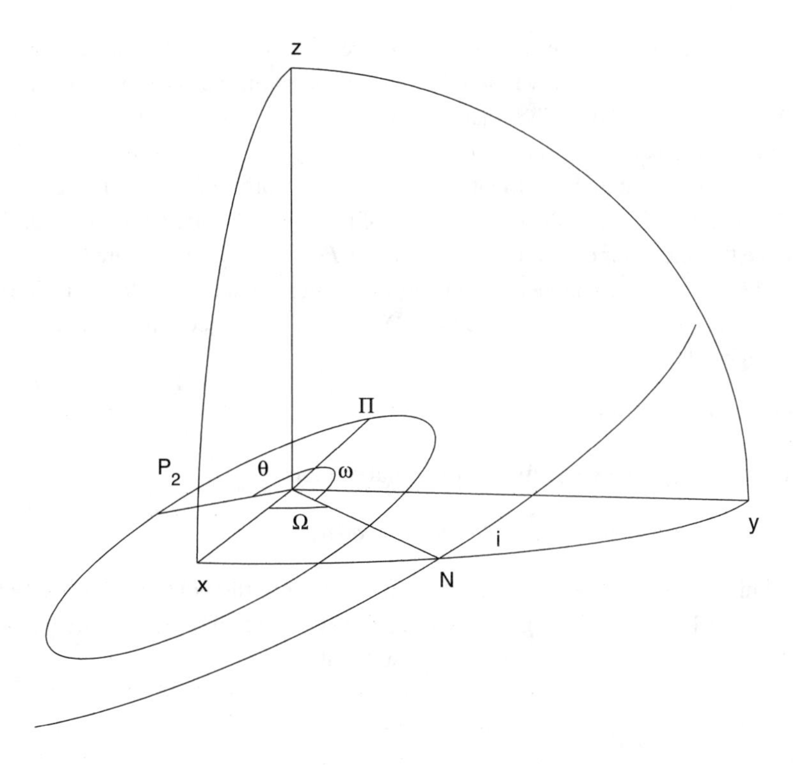

Fig. 2.5. The relative orbit of a binary located in three dimensions and defined by the angles Ω, i, and ω. The tangent plane of the sky is denoted by the plane containing xNy, and the observer views the binary along the direction Oz from below the figure.

clarity in Figure 2.5, the extension of NO to N' is not shown. The *inclination of the orbit* to the tangent plane of the sky is given by the angle i, so that $i = 90°$ means that the observer's line of sight is in the orbit plane. The orientation of the orbit *within its own plane* is given by the angle $NO\Pi = \omega$ and is called the *longitude of periastron*, measured from the ascending node N in the direction of orbital motion and within the orbital plane. In the figure, the mass m_2 at P_2 has a *true anomaly* $\Pi OP_2 = \theta$ at time t that is also measured within the orbital plane.

The three components of the angular momentum are

$$J_x = J \sin i \sin \Omega; \qquad J_y = -J \sin i \cos \Omega; \qquad J_z = J \cos i \tag{2.41}$$

Thus there are six quantities that can be used to define the orbit in three-dimensional space and are called the *elements of the orbit*, namely, $(a, e, i, \omega, \Omega, T)$, providing its size, shape, and orientation.

2.7 Applications to spectroscopic binaries

The first application of the foregoing equations lies in interpreting the measurements of *radial (line-of-sight) velocities* of binary stars to provide the parameters, or *elements*, of the orbits.

These data yield quantities related to the absolute sizes of the orbits and to the masses of the stars in binary systems. We shall derive the relationships between measured radial velocities and the orbital parameters and describe what we would expect to see in a sequence of observations of a spectroscopic binary.

In Figure 2.5 we presented the orientation of the binary orbit in space relative to the line of sight of the observer, Oz, and the tangent plane of the sky at the star being studied. From that figure we see that the polar coordinates of the star at P_2 with respect to O are $(r, \theta + \omega)$, which can be resolved into two components in the orbital plane: $r\cos(\theta + \omega)$ along the line of nodes NON', and $r\sin(\theta + \omega)$ at a right angle to NON'. When we project this second quantity into the line of sight Oz, we obtain

$$z = r\sin(\theta + \omega)\sin i \tag{2.42}$$

Hence, the observed radial velocity due to orbital motion is

$$V_{\rm rad} = \dot{z} = \sin i[\sin(\theta + \omega)\dot{r} + r\cos(\theta + \omega)\dot{\theta}] \tag{2.43}$$

To determine appropriate expressions for $\dot{r}$ and $\dot{\theta}$, we use the polar equation for an ellipse, $r = a(1 - e^2)/(1 + e\cos\theta)$, to give $\dot{r} = e\sin\theta r\dot{\theta}/(1 + e\cos\theta)$, and Kepler's second law, $r^2\dot{\theta} = 2\pi a^2(1 - e^2)^{1/2}/P$, to give the final result that

$$V_{\rm rad} = \frac{2\pi a\sin i}{P(1 - e^2)^{1/2}}[\cos(\theta + \omega) + e\cos\omega] \tag{2.44}$$

The final expression for radial velocity is usually written in the form

$$V_{\rm rad} = K[\cos(\theta + \omega) + e\cos\omega] + \gamma \tag{2.45}$$

where $K = (2\pi a\sin i)/[P(1 - e^2)^{1/2}]$ is called the *semiamplitude* of the velocity curve, and γ is the *systemic velocity*, or the radial velocity of the centre of mass of the binary system.

For constant values of the quantities γ, K, e, and ω, the radial velocity $V_{\rm rad}$ of one component of a binary is a maximum, A, or a minimum, B, when $(\theta + \omega) = 0$ or π, respectively, that is, when the star is at the *ascending* node N or the *descending* node N', respectively. So $A = \gamma + Ke\cos\omega + K$, $B = \gamma + Ke\cos\omega - K$, and hence $K = (A - B)/2$. Because the quantity B is negative (Figure 2.6), K is sensibly called the *semiamplitude* of the velocity curve for that star. If $e = 0$, equation (2.45) gives a cosine curve, as shown in Figure 2.6; but as e increases, the velocity curve becomes increasingly skew-symmetric, as illustrated in Figures 2.7–2.10. It is valuable for the reader to study the different forms of velocity curves that result from the different shapes and orientations of the orbits given in these figures.

The shape and the semiamplitude of the velocity curve should be accurately determined from a set of observations that are well distributed in orbital phase, together with extra emphasis on the two regions of maximum and minimum velocities. These regions are known as the *quadrature* phases because the stars are then near the nodal points N and N', and the radius vector of each star from the barycentre is perpendicular to the observer's line of sight.

For a *double-lined* spectroscopic binary, where the velocity curves of both components can be determined from observations, we can derive the following quantities. For the star of

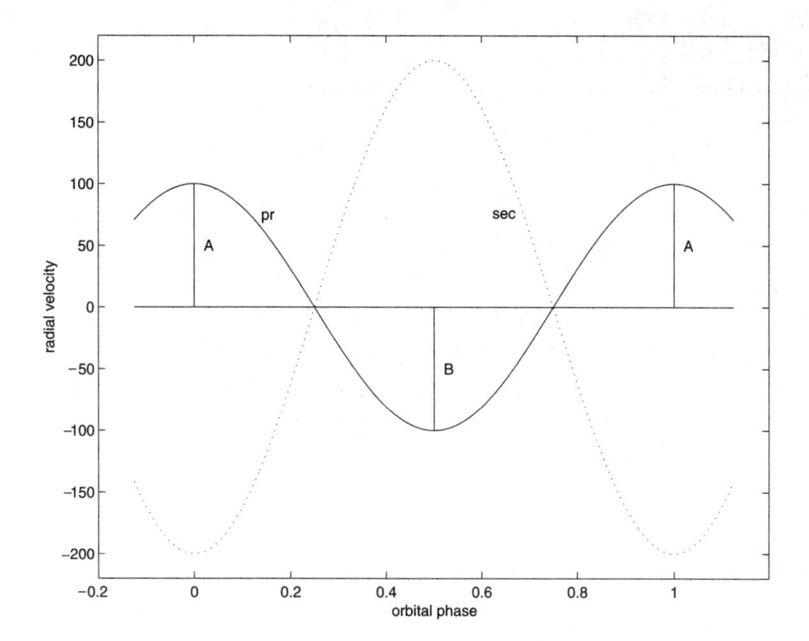

Fig. 2.6. The radial-velocity curves for the two stars in a binary system with a circular orbit. The semiamplitudes are $K_1 = 100$ km s^{-1} and $K_2 = 200$ km s^{-1}, giving a mass ratio $q = m_2/m_1 = 0.5$. The systemic velocity is $\gamma = 0$ km s^{-1}.

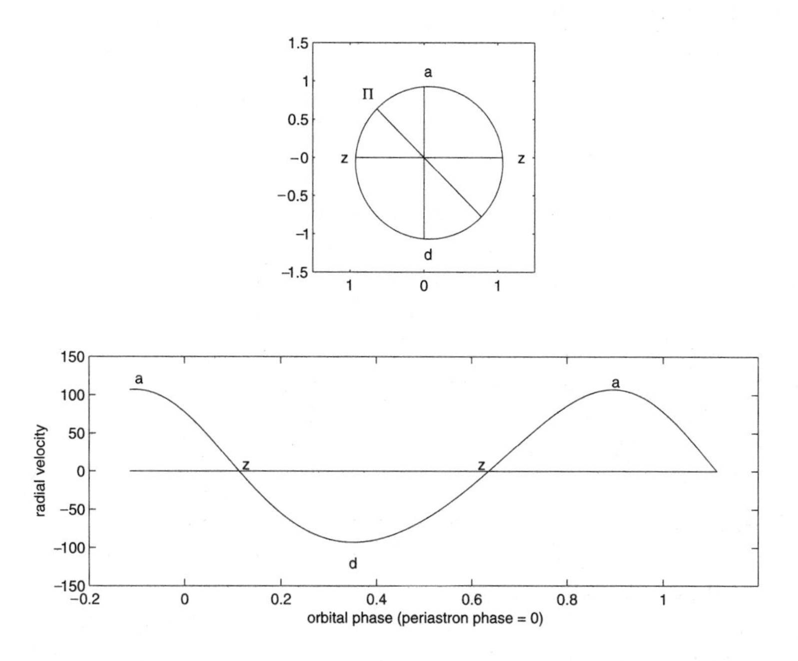

Fig. 2.7. An elliptical orbit of eccentricity $e = 0.1$ and $\omega = 45°$ and the corresponding radial-velocity curve with a semiamplitude $K = 100$ km s^{-1}. The positions of the ascending (a) and descending (d) nodes are shown, together with periastron (Π) and the points of zero (z) orbital radial velocity. In the upper panel, the observer views the binary orbit from the right along zz.

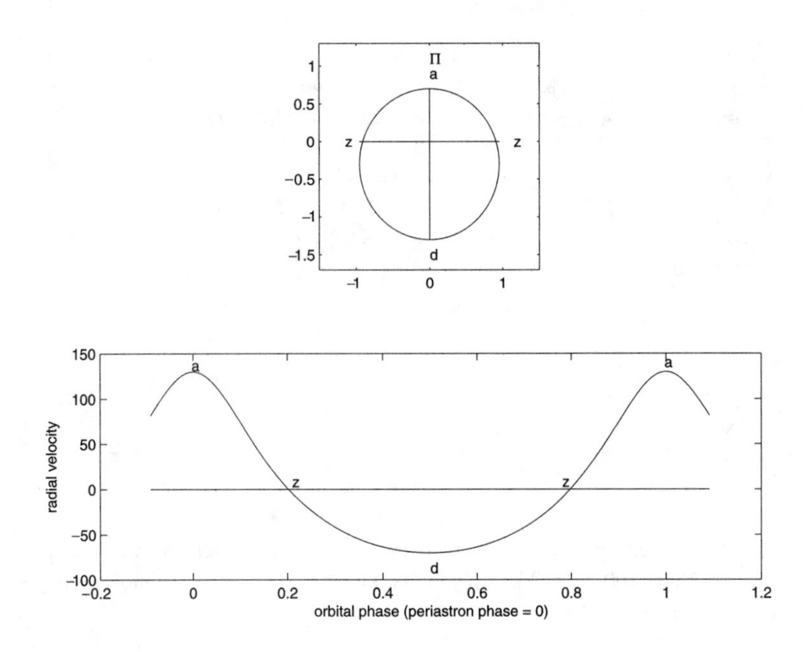

Fig. 2.8. As for Figure 2.7, but with $e = 0.3$ and $\omega = 0°$.

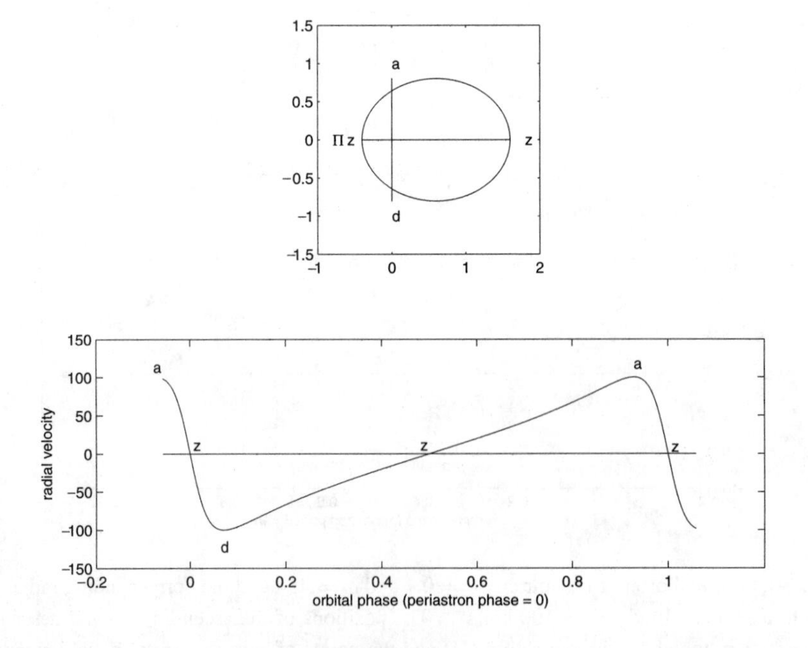

Fig. 2.9. As for Figure 2.7, but with $e = 0.6$ and $\omega = 90°$.

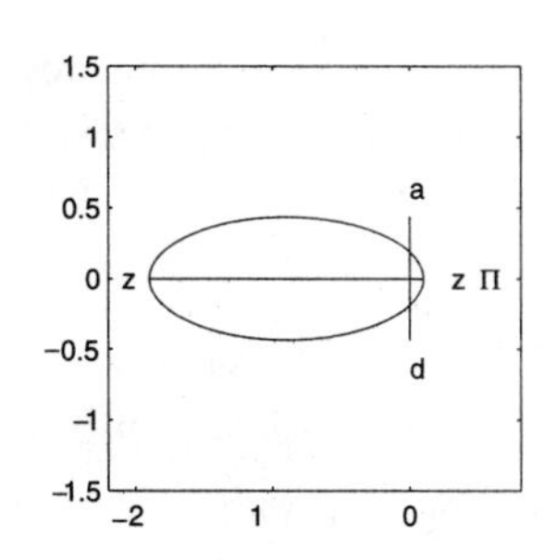

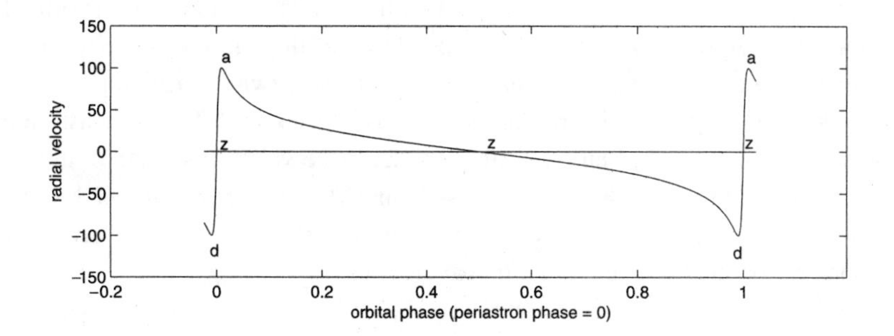

Fig. 2.10. As for Figure 2.7, but with $e = 0.9$ and $\omega = 270°$.

mass m_1 or m_2, the *projected semimajor axis* $a_1 \sin i$ or $a_2 \sin i$ is derived directly from the semiamplitude K_1 or K_2 from

$$a_{1,2} \sin i = \frac{(1 - e^2)^{1/2}}{2\pi} K_{1,2}\, P \tag{2.46}$$

To derive the expressions for the important *minimum masses*, $m_{1,2} \sin^3 i$, note that $m_1 a_1 = m_2 a_2$ and $G(m_1 + m_2) = 4\pi^2 a^3 / P^2$. Then, substituting for m_2, we write

$$(m_1 + m_1 a_1 / a_2) = 4\pi^2 a^3 / G P^2 \tag{2.47}$$

But $a \sin i = a_1 \sin i + a_2 \sin i$, which are both determinate from before, and hence

$$m_1 \sin^3 i = \frac{4\pi^2}{GP^2} \frac{a^3 \sin^3 i}{1 + (a_1 \sin i)/(a_2 \sin i)} \tag{2.48}$$

We then use the foregoing expressions for $a_{1,2} \sin i$ to obtain, for the minimum masses:

$$m_{1,2} \sin^3 i = \frac{1}{2\pi G} (1 - e^2)^{3/2} (K_1 + K_2)^2 K_{2,1} P \tag{2.49}$$

Note the order of the subscripts for the masses and for the semiamplitudes in equation (2.49). The minimum masses are so called because these expressions provide the true masses directly only for orbits whose planes are in our line of sight, when $i = 90°$. For inclined orbits, these expressions provide only the lower limit for the mass of each component.

Normally, we determine velocities in units of kilometres per second for most binary stars, and most orbital periods are measured in days. So the working expressions for the projected semimajor axes in kilometres and *solar radii*, $R_\odot$, and the minimum masses in *solar masses*, $M_\odot$, are

$$a_{1,2} \sin i = (1.3751 \times 10^4)(1 - e^2)^{1/2} K_{1,2} P \text{ km} \tag{2.50}$$

$$= (1.9758 \times 10^{-2})(1 - e^2)^{1/2} K_{1,2} P \, R_\odot \tag{2.51}$$

$$m_{1,2} \sin^3 i = (1.0361 \times 10^{-7})(1 - e^2)^{3/2} (K_1 + K_2)^2 K_{2,1} P \, M_\odot \tag{2.52}$$

The linear dependence of the semimajor axes on the semiamplitudes and the orbital period is to be expected straightforwardly, but the dependence of the minimum masses on the cube of one semiamplitude and the square of the other may be somewhat surprising. Certainly, this very strong dependence is a major reason why it has proved so difficult for astronomers to determine accurate masses for stars. As an example, if the velocity semiamplitude of a star of mass m_1 is uncertain by 10%, then its mass is uncertain by 30%! We need to determine semiamplitudes with uncertainties of less than 1% in order to claim that we have determined empirical masses for stars with uncertainties of less than 3%.

For *single-lined* binaries, we can measure only K_1, and hence we can determine only $a_1 \sin i$ and the quantity known as the *mass function*, $f(m)$, which is defined as

$$f(m) = \frac{m_2^3 \sin^3 i}{(m_1 + m_2)^2} = (1.0361 \times 10^{-7})(1 - e^2)^{3/2} K_1^3 P \, M_\odot \tag{2.53}$$

To estimate the mass m_2 for an SB1 system, we need to know the orbital inclination and the mass m_1 from other information, such as observational data determined via different techniques (see Sections 2.8 and 2.9), or empirical calibrations of masses with other observable quantities, or data from theoretical models for stars of the appropriate type.

2.8 Applications to pulsar binaries

The second application of these equations is to a binary system that contains a star that emits radiation in a *pulsed* manner, a sequence of mostly regular bursts of radiation whose *times of arrival* at the observer can be determined quite accurately. These are the radio pulsars and the x-ray pulsars. The continuous emission of radiation by most astronomical sources does not allow us to determine when a particular photon was emitted. But for *pulsars*, their pulsed radiation provides a means to determine the time interval between pulses, which will vary as we follow the motion of the pulsar in its orbit around its companion. The situation is illustrated schematically in Figure 2.11, where we consider the motion of a pulsar in its relative orbit about its companion. If we had a solitary pulsar with a constant pulse period P_p, such that there were equal intervals of time between arrivals of successive pulses of radiation, the graph of *pulse arrival time* t versus *pulse number* n would show a straight line, as in Figure 2.11. If the pulse period were changing continuously, then the graph would become convex or concave, depending on whether the period were decreasing or increasing, respectively. But if the pulsar is a member of a binary system, then we shall see two additional phenomena

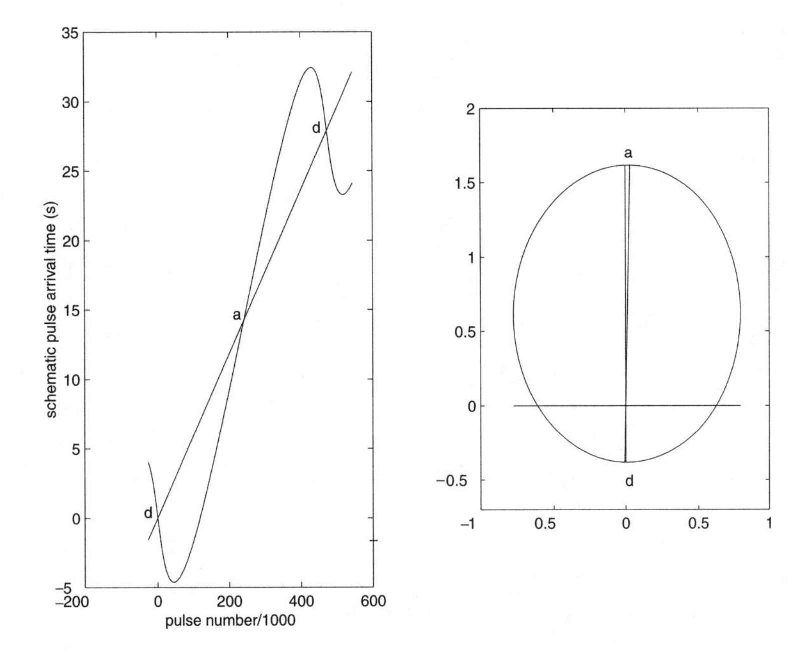

Fig. 2.11. A schematic plot of pulse arrival time versus pulse number and the corresponding orbit for the first discovered pulsar binary PSR1913+16 (Hulse and Taylor 1975). Here $e = 0.617$, $\omega = 178.9°$, the orbital period of the binary is 27,907 second, and the pulse period is 0.05902 second. In order to show the effects clearly, the vertical scale of *schematic pulse arrival time* is given in seconds, with the orbital period reduced by a factor of 1000, and the orbital modulation increased by a factor of 10.

displayed in such a graph. The more obvious factor involves the periodic advances and delays of the times of arrival of the pulses, relative to the straight line for a pulsar of constant period, reflecting the changes in location of the pulsar in its binary orbit. This is the *light-travel time*, simply the time it takes for the pulsed light to travel across the projection of the orbit into the observer's line of sight from the instantaneous position of the pulsar, that is, the quantity

$$z/c = (r/c)\sin(\theta + \omega)\sin i \tag{2.54}$$

where c is the speed of light. The second of the two phenomena due to orbital motion is more subtle and much more difficult to measure, namely, the change, Δf, in frequency, f, of the emitted radiation due to the Doppler-Fizeau effect, namely,

$$\Delta f/f = V/c = \dot{z}/c \tag{2.55}$$

Because typical orbital speeds in binary systems are $V \approx 300$ km s^{-1}, the change in frequency $\Delta f \approx 0.001 f$, so that this second effect of orbital motion on pulse arrival times can be determined only in favourable situations, such as those displayed by the radio pulsars, where the pulses have sharp, narrow profiles.

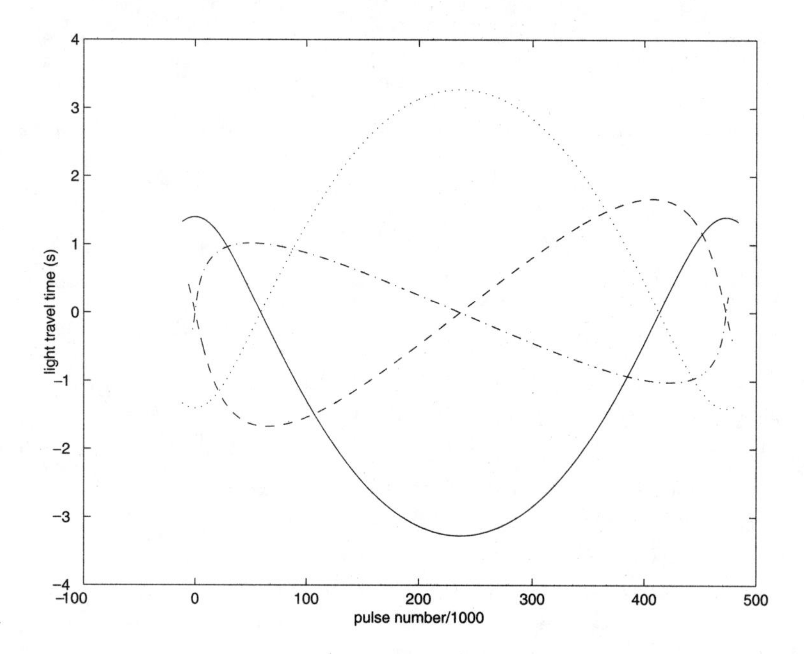

Fig. 2.12. Curves of light-travel time due to orbital motion only plotted against pulse number, using the same values of orbital period, pulse period, and projected semimajor axis as in Figure 2.11, but with different values of the eccentricity and longitude of periastron: solid line, $e = 0.4$, $\omega = 270°$; dotted line, $e = 0.4$, $\omega = 90°$; dash line, $e = 0.7$, $\omega = 180°$; dot-and-dash line, $e = 0.9$, $\omega = 0°$.

For the pulsar orbit, we have

$$r_p = a_p(1 - e^2)/(1 + e\cos\theta) \tag{2.56}$$

so that the light-travel-time equation, equation (2.54), becomes

$$z/c = \frac{a_p \sin i}{c} \frac{(1 - e^2)\sin(\theta + \omega)}{(1 + e\cos\theta)} \tag{2.57}$$

Some representative examples of light-travel-time curves are given in Figure 2.12 for different values of orbital eccentricity and longitude of periastron.

For circular orbits, $e = 0$ and $\omega = 0$, so that

$$z/c = \frac{a_p \sin i}{c} \sin\left[\frac{2\pi(t - T_0)}{P}\right] \tag{2.58}$$

This is the form of the light-travel-time equation that has been used so successfully in interpreting the pulse-timing measurements from the x-ray pulsars, such as SMC X-1, which has a circular orbit. The pulse-timing data provide an accurate determination of the amplitude of the light-travel-time curve, $a_p \sin i/c$. The radial-velocity curve for the optically bright companion to an x-ray pulsar, which is an O–B star in the case of HMXBs, provides a determination of the mass function $f(m)$:

$$f(m) = \frac{m_p^3 \sin^3 i}{(m_c + m_p)^2} \tag{2.59}$$

where m_c is the mass of the companion star. Then the mass ratio of the binary system is given by

$$q = \frac{m_p}{m_c} = \frac{a_c \sin i}{a_p \sin i} \tag{2.60}$$

and hence the minimum masses of the two stars can be calculated. If the orbital inclination i can be found – for example, from the duration of the x-ray eclipses that are displayed by some systems (Chapter 5) – then the masses follow directly.

For the radio pulsars, and generally for measurements of pulse timing at very high precision, we must take account also of the pulse frequency changes noted earlier. Accordingly, the treatment of the problem is somewhat more difficult and is bound up with the methods of reducing the observational data and finding preliminary values for the pulse periods. It will be more appropriate to consider all of these issues together in Section 3.3.

2.9 Applications to spatially resolved binaries

The binary systems that are sufficiently near to the Sun and that have sufficiently large semi-major axes for their relative orbits can be resolved into two separate sources of radiation. Those with the largest angular separations were discovered first, by visual observers using long-focus refractors, and are appropriately called *visual binaries*. As the techniques of *speckle interferometry* and optical interferometry via arrays of telescopes have been developed, the lower limit at which binaries can be regarded as *resolvable* has reached levels of the order of milliarcseconds. These resolved binaries all require the same prescription in order for us to analyse the observational data and determine the parameters of their orbital motion.

We refer again to Figure 2.5 to note that the observer viewing the binary along Oz from below the figure will see the orbital motion of the binary system projected onto the tangent plane of the sky at the position of the binary on the celestial sphere, namely, the plane defined by $OxNy$. The standard procedure is to consider the *relative motion* of star 2 about star 1 at the origin O. The axis Ox is directed towards the North Celestial Pole. A careful study of Figure 2.5 reveals that the radius vector r, which is seen at time t to be at an angular position θ relative to the position of periastron, is projected onto the three coordinate axes as follows:

$$\begin{aligned} x &= r[\cos\Omega\cos(\theta+\omega) - \sin\Omega\sin(\theta+\omega)\cos i] \\ y &= r[\sin\Omega\cos(\theta+\omega) + \cos\Omega\sin(\theta+\omega)\cos i] \\ z &= r[\sin(\theta+\omega)\sin i] \end{aligned} \tag{2.61}$$

Then, on the tangent plane of the sky, we see part of that binary orbital motion shown in Figure 2.5 as illustrated in Figure 2.13. The observer is able to make a sequence of observations, at many epochs, of the projected separation ρ between the two components, P_1 and P_2, and the

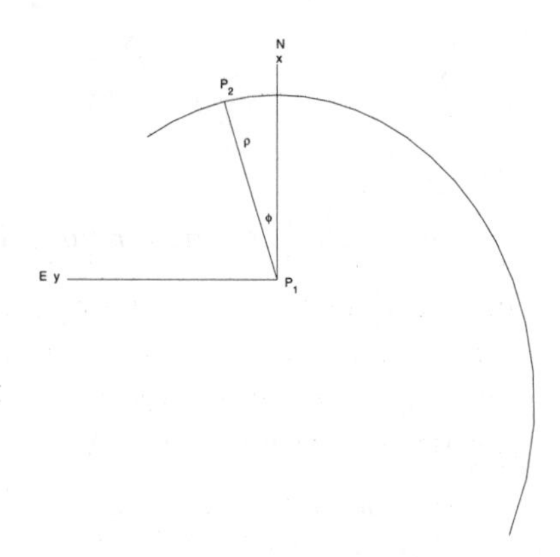

Fig. 2.13. The two components of a visual binary, P_1 and P_2, are observed projected onto the tangent plane of the sky, denoted by xP_1y, at an angular separation ρ and position angle ϕ.

position angle ϕ of the line joining the two components, measured from north through east. These angular separations are measured in arcseconds and can have uncertainties as little as 0.001 arcsec for speckle observations of bright systems with equal components, but more typically are accurate to about ± 0.01 arcsec. The position angles are usually quoted to a precision of one-tenth of a degree for speckle data, and to about $1°$ for visual/photographic observations.

The orbital periods for resolved binaries currently range from about 100 days to years, decades, and centuries, and accordingly it follows that observations of a given binary are likely to be obtained with many different types of instrumentation. These observations carry with them the accuracies inherent in the methods used, and care must be taken in combining these data to determine sensible orbits.

These measured separations and position angles (ρ, ϕ) at time t are related to the projected quantities (x, y) by the simple relationships $\Delta\delta = x = \rho\cos\phi$ and $\Delta\alpha\cos\delta = y = \rho\sin\phi$. Equations (2.61) for (x, y) can be re-arranged to yield

$$\begin{aligned} x &= r\cos\theta[\cos\Omega\cos\omega - \sin\Omega\sin\omega\cos i] \\ &\quad + r\sin\theta[-\cos\Omega\sin\omega - \sin\Omega\cos\omega\cos i] \\ y &= r\cos\theta[\sin\Omega\cos\omega + \cos\Omega\sin\omega\cos i] \\ &\quad + r\sin\theta[-\sin\Omega\sin\omega + \cos\Omega\cos\omega\cos i] \end{aligned} \tag{2.62}$$

We now introduce the *Thiele-Innes constants*, A, B, F, G, which are defined as

$$\begin{aligned} A &= a[\cos\Omega\cos\omega - \sin\Omega\sin\omega\cos i] \\ B &= a[\sin\Omega\cos\omega + \cos\Omega\sin\omega\cos i] \\ F &= a[-\cos\Omega\sin\omega - \sin\Omega\cos\omega\cos i] \\ G &= a[-\sin\Omega\sin\omega + \cos\Omega\cos\omega\cos i] \end{aligned} \tag{2.63}$$

We also define the quantities called the *elliptical rectangular coordinates*, $X = \cos E - e$ and $Y = (1 - e^2)^{1/2}\sin E$, so that $r^2 = a^2(1 - e\cos E)^2 = a^2(X^2 + Y^2)$, and $r\cos\theta = aX$,

$r \sin\theta = aY$. Then

$$\begin{aligned} \Delta\delta &= \rho\cos\phi = AX + FY \\ \Delta\alpha\cos\delta &= \rho\sin\phi = BX + GY \end{aligned} \tag{2.64}$$

Thus, observations of (ρ, ϕ) over a complete orbit, or at least a substantial fraction of an orbit, provide the orbital period P and the time of periastron passage T. To calculate the quantities X and Y for each observation, we also need to know the orbital eccentricity e which is determinable from the *time–displacement curves* that are described in Chapter 3 or from the following relationships. At periastron, where $E = 0$, $X = 1 - e$ and $Y = 0$, so that $x_p = A(1 - e)$ and $y_p = B(1 - e)$, whilst at apastron, $X = -1 - e$ and $Y = 0$, so that $x_a = A(-1 - e)$ and $y_a = B(-1 - e)$. Thus, the coordinates of periastron, referred to the *centre* of the apparent orbit, are (A, B), and the coordinates of the centre of the apparent orbit are $x_c = -Ae$ and $y_c = -Be$. Hence, $x_p/x_c = (1-e)/(-e) = y_p/y_c$, giving another estimate of the eccentricity.

The quantities P, T, and e are traditionally referred to in visual-binary studies as the *dynamical elements*, whilst the Thiele-Innes constants provide the *geometrical elements* Ω, ω, i, and a. Equations (2.64) provide the links between the observational data (ρ, ϕ) and these elements, and we need to see how these elements are derived from the Thiele-Innes constants.

The *true elliptical orbit* of the star P_2 about its companion P_1 is projected onto the plane of the sky such that the quantity $\rho^2\dot{\phi}$ = constant. But the star P_1 is no longer at the focus of the *apparent orbit*, nor is it at the centre of the apparent orbit. The *auxiliary circle* of the *true orbit* also projects onto an ellipse on the plane of the sky, but because the auxiliary circle has a diameter of $2a$, its projection onto xP_1y ensures that the major axis of this *auxiliary ellipse* is also $2a$. Its minor axis is then of length $2a\cos i$.

Now the coordinates of periastron, referred to the centre of the apparent orbit, are (A, B), which is also one point where the apparent orbit and the auxiliary ellipse touch. In addition, at the point on the apparent orbit where the eccentric anomaly $E = \pi/2$, $X = -e$ and $Y = (1 - e^2)^{1/2}$. Then $x = -Ae + F(1 - e^2)^{1/2}$ and $y = -Be + G(1 - e^2)^{1/2}$. So the coordinates of the point $E = \pi/2$ on the apparent orbit, referred to the centre of that ellipse, are $[F(1 - e^2)^{1/2}, G(1 - e^2)^{1/2}]$, and the coordinates of the corresponding point on the auxiliary ellipse are (F, G).

There are two theorems of geometry, due to Apollonius, that state relationships between the major and minor axes of an ellipse and pairs of *conjugate diameters*. Diameters of an ellipse are called *conjugate* if all chords parallel to one diameter are bisected by the other, and vice versa. These two theorems are as follows:

1. *The sum of the squares of any two conjugate diameters is constant and equals the sum of the squares of the two axes.*
2. *The area between the parallelograms on any two conjugate diameters is also constant and equals that of the rectangle of the axes.*

The diameter of the auxiliary ellipse that passes through the centre and the periastron point is conjugate to that passing through the centre and the point with $E = \pi/2$. The sum of these squares is $A^2 + B^2 + F^2 + G^2$, whilst the sum of the squares of the major and minor axes is

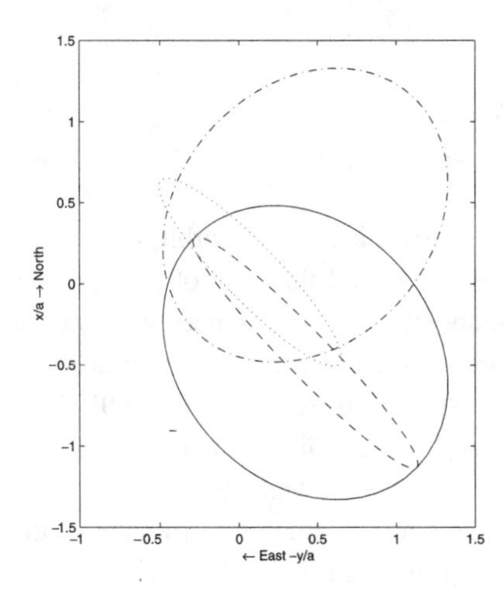

Fig. 2.14. Four projected orbits, all with $e = 0.6$ and the following geometrical elements: solid line, $i = 0°,\ \omega = 0°,\ \Omega = 45°$; dash line, $i = 80°,\ \omega = 0°,\ \Omega = 45°$; dot-and-dash line, $i = 0°,\ \omega = 90°,\ \Omega = 45°$; dotted line, $i = 80°,\ \omega = 90°,\ \Omega = 45°$.

$a^2 + a^2 \cos^2 i$, so $a^2 + a^2 \cos^2 i = A^2 + B^2 + F^2 + G^2 = 2p$, and, from the second theorem, $a^2 \cos i = AG - BF = q$. Then we find that $a^2 = p + (p^2 - q^2)^{1/2}$ and $\cos i = q/a^2$.

Reference back to equations (2.63) shows that

$$\begin{aligned} A + G &= a(1 + \cos i)\cos(\omega + \Omega) \\ A - G &= a(1 - \cos i)\cos(\omega - \Omega) \\ B - F &= a(1 + \cos i)\sin(\omega + \Omega) \\ -B - F &= a(1 - \cos i)\sin(\omega - \Omega) \end{aligned} \tag{2.65}$$

Then $\tan(\omega + \Omega) = (B - F)/(A + G)$, where $\sin(\omega + \Omega)$ has the same sign as $(B - F)$, because $\cos i$ is always positive $(0 \le i \le \pi/2)$. Also, $\tan(\omega - \Omega) = (-B - F)/(A - G)$, where $\sin(\omega - \Omega)$ again has the same sign as $-B - F$ for the same reason. These relationships yield the values of (ω, Ω), as well as possibly $\omega \pm \pi$ and $\Omega \pm \pi$. Without additional independent observations, such as radial velocities, we cannot distinguish between these two possibilities, and convention demands that we adopt the solution for which $\Omega < \pi$.

In Figures 2.14 and 2.15, some representative orbits are projected onto the tangent plane of the sky by means of equations (2.61) to illustrate the variety of possible apparent orbits.

2.10 Applications to heliocentric corrections

There remains one other set of applications of these considerations of basic two-body motion, namely, the motion of the Earth about the Sun and that of a satellite about the Earth. Because all of our astronomical observations are made from such continuously moving platforms, we need to know what those motions are, and we must correct all our Earth-based or satellite-based observations for the effects of those motions. The Sun can be adopted as a convenient

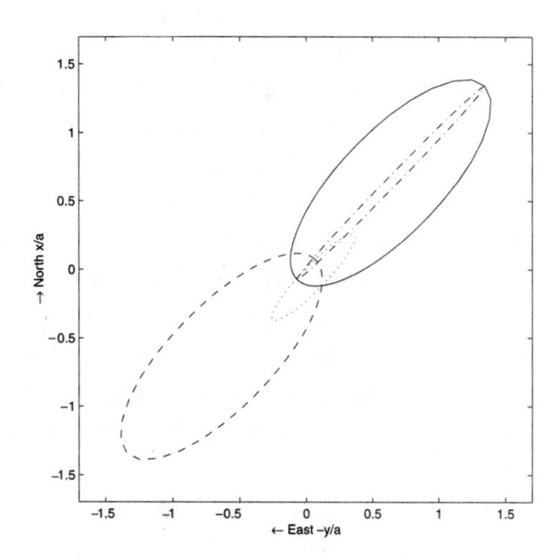

Fig. 2.15. Four projected orbits, all with $e = 0.9$ and the following geometrical elements: solid line, $i = 30°,\ \omega = 0°,\ \Omega = 135°$; dash line, $i = 30°,\ \omega = 180°,\ \Omega = 135°$; dot-and-dash line, $i = 85°,\ \omega = 0°,\ \Omega = 135°$; dotted line, $i = 85°,\ \omega = 270°,\ \Omega = 135°$.

reference point or origin for a reference frame, and all astronomical observations that involve measurements of time or measurements of velocity require *heliocentric corrections* to be calculated for each particular observation. For observations obtained from an Earth-orbiting satellite, it is necessary to correct for the satellite's orbital motion as well as that of the Earth. For observations demanding extremely high accuracy of the timing of observations, the barycentre of the Solar System should be used.

If the observing platform is a satellite orbiting the Earth, then we must first correct all measured velocities and times to those that would be observed from the centre of the Earth. Hence we again use Figure 2.5, with P_2 as the satellite, and the Earth at the origin O of the coordinate system. The equations for the projected x, y, and z coordinates of the satellite are those given in equations (2.61). The corresponding velocity vectors for the satellite's motion are derived by differentiating the equations for x, y, and z and utilizing the standard expressions for elliptical-orbit motion, specifically, the *mean daily motion* $n = 2\pi/P$ (where P is the orbital period), Kepler's equation, and the relationships $r\sin\theta = a(1-e^2)^{1/2}\sin E$ and $r\cos\theta = a(\cos E - e)$.

The resultant expressions for the three velocity components are

$$\begin{aligned}
V_x &= \frac{n}{(1-e\cos E)}[a\sin E(\cos i\sin\Omega\sin\omega - \cos\Omega\cos\omega) \\
&\quad - a(1-e^2)^{1/2}\cos E(\cos\Omega\sin\omega + \cos i\sin\Omega\cos\omega)] \\
V_y &= \frac{n}{(1-e\cos E)}[a(1-e^2)^{1/2}\cos E(\cos i\cos\Omega\cos\omega - \sin\Omega\sin\omega) \\
&\quad - a\sin E(\cos i\cos\Omega\sin\omega + \sin\Omega\cos\omega)] \\
V_z &= \frac{n}{(1-e\cos E)}[a(1-e^2)^{1/2}\cos\omega\cos E - a\sin\omega\sin E]
\end{aligned} \tag{2.66}$$

These equations give the velocity components for the satellite relative to the centre of the Earth, with the orbital elements referred to the equatorial system of coordinates. (The equatorial system is based upon the orientation of the Earth's rotation axis and equatorial plane and is the one used in astronomy for locating objects in the sky by means of right ascension and declination.)

The elements of the satellite's orbit $(a, e, i, \omega, \Omega, T, P)$ must be well known in order to calculate these velocity components accurately, and the motions of all satellites around the Earth are subject to *perturbations* because they are not moving in the idealized two-body force field that we have been considering thus far. These perturbations are caused by, for example, the non-spherical shape of the Earth, frictional drag forces exerted on the satellite by the upper atmosphere of the Earth, and the gravitational fields of the Moon and the Sun. Fortunately, most of these factors are small-scale perturbations on an essentially two-body closed elliptical motion, and accordingly they can be represented by allowing the elements of the orbit of the satellite to vary slowly with time, so-called *secular variations*. In such circumstances, the values of the orbital elements, at a given epoch, are referred to as the *osculating elements*.

For the Earth's orbital motion about the Sun, these same equations can be simplified considerably by making use of the ecliptic system of coordinates instead, because in the ecliptic system the Earth's orbit has $i = 0$, and Ω is arbitrary, so that we can put $\Omega = 0$. Then

$$\begin{aligned}
V_{xe} &= \frac{-n}{(1 - e\cos E)}[a\cos\omega\sin E + a(1 - e^2)^{1/2}\sin\omega\cos E] \\
V_{ye} &= \frac{n}{(1 - e\cos E)}[a(1 - e^2)^{1/2}\cos\omega\cos E - a\sin\omega\sin E] \\
V_{ze} &= 0.0
\end{aligned} \tag{2.67}$$

To transform these velocity components to the equatorial system, we can simply write

$$\begin{aligned}
V_x &= V_{xe} \\
V_y &= V_{ye}\cos\epsilon \\
V_z &= V_{ye}\sin\epsilon
\end{aligned} \tag{2.68}$$

where ϵ is the *obliquity of the ecliptic*.

Hence to determine the correction (due to the orbital motion of the observing platform) to a measured radial velocity of an astronomical source at a position of right ascension α and declination δ, we use the direction cosines of the source, namely,

$$l = \cos\alpha\cos\delta, \qquad m = \sin\alpha\cos\delta, \qquad n = \sin\delta \tag{2.69}$$

and write that the velocity correction to be added to the measured velocity is

$$V_{\rm orb} = lV_x + mV_y + nV_z \tag{2.70}$$

This correction term is simply the projection of the satellite's orbital motion and/or the Earth's orbital motion into the line of sight to the astronomical source at the time of the observation. The range of values of this correction for the Earth's orbital motion reaches a maximum range for sources that lie on the ecliptic, namely, ± 30 km s^{-1}, and so care must be taken in evaluating these heliocentric corrections for velocities.

In addition to the Earth's annual, or orbital, motion around the Sun, we also have to correct for the axial rotation of the Earth, its daily motion. For an astronomical source located at a position (α, δ) in the equatorial system and observed at a *local sidereal time* β from an observatory at a latitude ϕ on the surface of the Earth at a radial distance r from its centre, the correction for rotational velocity is

$$V_{\mathrm{rot}} = \frac{-2\pi R_E}{P_E} \frac{r}{R_E} \cos\phi \cos\delta \sin(\beta - \alpha) \tag{2.71}$$

where $R_E = 6378.166$ km is the equatorial radius of the Earth, and $P_E = 86{,}164$ mean solar seconds is the Earth's rotational period. The maximum/minimum value for V_{rot} is ± 0.46 km s^{-1} for an observer on the equator recording the velocity of a source at $\delta = 0$ and *hour angle* $= (\beta - \alpha) = \pi/2 = 6$ hours.

The foregoing considerations applied to the orbital motion of the Earth about the Sun are based on that motion being considered as purely two-body orbital motion. But, just as for the motion of an Earth-orbiting satellite, we have to take account of the perturbations to the Earth's motion caused by the gravitational influence of the Moon and all the planets of the Solar System (including Pluto!), as well as any additional effects, such as corrections for Newtonian gravitation being only an approximation to the more accurate description of gravitation provided by the general theory of relativity. These summed perturbations can be represented by means of formulae describing the secular variations of the two-body orbital elements (a, e, i, etc.) for the Earth, and instantaneous values of these elements can be calculated for the times of observation. Then the two-body representation, with instantaneous values for those orbital elements, provides an adequate description for the calculation of the velocity corrections. Alternatively, all the components of these velocity corrections for the Earth's orbital motion are tabulated for each day of the calendar year in the *Astronomical Almanac*, together with the equatorial rectangular coordinates of the Earth, X_E, Y_E, Z_E. These latter quantities can also be used to correct the times of the observations made on the Earth for the light-travel time across the Earth's orbit and projected into the line of sight to the astronomical source. The *heliocentric correction* for a time of observation is given by $t_{\mathrm{corr}} = (lX_E + mY_E + nZ_E)/c$, where c is the speed of light.

It is now quite standard practice for various packages of astronomical reduction software to include a code for calculating accurate heliocentric corrections to times of observation and to velocity measurements. These are entirely adequate for most astronomical observations, but for observations of radio pulsars, for example, where accurate timing of the arrivals of pulses is crucial for interpretation of the data, care must be taken to ensure that these heliocentric corrections are calculated with sufficient precision. It is also necessary, in this context, to include the *relativistic clock correction* for the motion of the Earth (Clemence and Szebehely 1967; Blandford and Teukolsky 1976), namely, in units of seconds of time,

$$\Delta t_r = 0.001661\left[\left(1 - \frac{e^2}{8}\right)\sin\eta + \frac{e\sin 2\eta}{2} + \frac{3e^2\sin 3\eta}{8}\right] \tag{2.72}$$

where $\eta = 2\pi(t - T)/P$ is the *mean anomaly* for the Earth at time t, or the orbital phase expressed in radian measure. The correction is zero on January 1, when $t = T$ and $\eta = 0$. The correction is at its greatest positive value around $\eta = 89°$ on April 1, and its greatest negative value around $\eta = 271°$ on October 30.

In the next chapter we shall consider the practical aspects of observations, reductions, and analyses of data in the form of velocities, times of arrival of pulsed radiation, astrometry on resolved binaries, and polarimetry. We shall illustrate how we can use the foregoing equations to determine the *elements of the orbits* from observational data, together with the *mass function* and the *minimum masses* of the component stars.

3 The determination of orbits

3.1 Introduction

The fundamental equations, relationships, and definitions that we considered in Chapter 2 can be applied directly to interpretation of observational data on binary stars from four types of experiments. The first is spectroscopy, which yields measurements of line-of-sight (or radial) velocities, followed by derivation of the elements of the orbits from those observed velocities. These data furnish the quantities related to the absolute sizes of the orbits and to the masses of the stars in binary systems. The second is pulse timing: measurements of the times of arrival of short pulses of radiation from x-ray and radio pulsars that are found to be members of binary systems, followed by derivation of the relevant orbital elements. Nature has been kind enough to provide us with the kinds of binary stars that will permit both of these independent types of observations to be carried out, with the result that quite complete descriptions of the systems are possible, and observational astronomy can yield directly determined masses for the intriguing end states of stellar evolution: neutron stars and black holes. The third experiment is astrometry: accurate determination of the positions of the components of resolved binaries, both relative to each other and relative to a fundamental astrometric reference frame over the whole sky. Once again, we find some binary systems for which both astrometric data and radial-velocity data are available, so that complete descriptions can be established. Lastly, because of some recent developments, we are able to provide descriptions of the orbital motions of binaries that emit some fraction of their radiation in a polarized state, descriptions that are similar to those for resolved binaries.

The following sections will discuss, in turn, aspects of the planning for and the procedures of observational programmes, measurements of observable quantities from these experiments, and the procedures of data analysis that are appropriate to each experiment. It is perhaps inevitable that my detailed knowledge of observing procedures and techniques of data analysis is restricted to a subset of the topics in this chapter. But each section contains at least a summary of the procedures and gives references to expert discussions of these matters. Accordingly, the reader will see that Section 3.2 is substantially larger than any of the subsequent sections.

3.2 Orbits from radial velocities: Spectroscopic binaries

The spectrum of a star provides us with a wide range of information about the properties of the stellar surface, such as temperature, surface gravity, chemical composition, and rotational velocity, as well as evidence of unusual aspects, such as extended atmospheres or interactions with an orbiting companion. These factors will be discussed later, in their appropriate contexts, in Chapters 5 and 7, which involve photometry and the imaging of stellar surfaces, respectively. In this chapter, we restrict ourselves to measurements of the radial velocities of stars, and measurements of binary orbital motion in particular.

In order to produce a *radial-velocity curve* for a component of a binary-star system, we must develop sensible methods both to measure radial velocities from spectra and to determine orbital periods from such a set of spectra secured over some time interval.

3.2.1 Spectroscopic observations

The radiation from an astronomical source can be dispersed, by means of a reflective diffraction grating, into a *spectrum* of the received flux density versus the wavelength of the radiation. The spectral resolution required for determination of radial velocities of stars is substantial: $\lambda/\Delta\lambda > 5000$. In many cases, this level of resolution is not adequate to separate the spectral lines of the two components in a binary, and we must use values in excess of 10,000. As a general rule, the higher the resolution employed, the more accurate are the radial velocities obtained, and the less likely are those velocities to be affected by serious systematic errors. But most scientific investigations are carried out from positions of compromise, and studies of binary stars are not exempt. Invariably, the resolution employed is a compromise dependent upon the available telescope aperture, the available spectrograph and detector, the apparent magnitude of the source, the orbital period of the binary star, and the spectral types of the component stars.

Essentially, these factors are all accounted for in one parameter, the *signal-to-noise ratio* (S/N) that can be achieved in a given time of integration of the spectrum onto the detector. Before investigating the details of what is meant by the term *signal-to-noise ratio*, we should examine the issue of *integration time*. The integration time may have an upper limit imposed by the orbital period of a binary star. For example, if the orbital period of a binary is only 6 hours, there would not be much value in obtaining a spectrum over an integration time of 1 hour, for during that time interval the components of the binary would have changed their radial velocities (because of orbital motion) by very large amounts. The recorded spectrum would have spectral lines that would have become blurred because of orbital motion, in just the same way as slow shutter speeds on ordinary cameras produce blurred images of fast-moving objects. Integration times should not exceed 1–2% of the orbital period if we are to ensure that *motion blur* is not significant. So, for orbital periods in the region of days, integration times can be substantial, and in practice they have to be limited by other considerations, such as the susceptibility of the detector to contamination by cosmic rays. In these cases, it is better observational practice to limit the exposures to typically 1000–2000 seconds and then add consecutive spectra together in order to build up a larger value of S/N. For short orbital periods (minutes to hours), or in binary systems where additional physical processes, like stellar pulsations, are occurring on short time scales, it is necessary to adopt very short integration times (e.g., even as short as a few seconds!) and then add the spectra together in *phase bins* that are defined in accordance with some precisely known period of the physical process, or the orbital period.

The *signal* is clearly some measure of the flux density received by each of the flux-sensitive components, or pixels, of the detector. That received signal comes from the astronomical source, but it is accompanied by additional flux density from the sky background, as well as additional unwanted contributions from the detector itself, scattered infrared flux from warm surfaces on the telescope, and so on. The *noise* has three components: the *shot noise* in the signal from the source and in the background contributions, and the noise from the detector itself, which includes both *readout* noise and *dark counts* from a charge-coupled device (CCD) and *dark counts* only from a photon-counting device. The shot noise in a signal follows a Poisson distribution, so that the variance in the signal is equal to the mean signal itself; so, for a signal S, the associated noise is simply $S^{1/2}$. Hence, for an observed total signal from the source S_{source} per pixel, and from the background S_{back} per pixel, with a CCD readout noise of σ per pixel, we can write, for the S/N ratio,

$$\text{S/N ratio} = \frac{S_{\text{source}}}{[S_{\text{source}} + S_{\text{back}} + \sigma^2]^{1/2}} \tag{3.1}$$

Note that the contributions to the total noise value are added in quadrature, in accordance with standard practice in calculating the uncertainties of measured quantities. If the observed signal from the source is much larger than the sum of the background and the readout noise, then equation (3.1) is reasonably approximated by S/N $\approx (S_{\text{source}})^{1/2}$.

To calculate S_{source}, we need to know some of the optical properties of the telescope and the spectrograph, the sensitivity of the detector employed, and an estimate of the apparent brightness of the source in absolute flux-density units. For a typical Cassegrain reflector, the geometrical collecting area of the telescope is easily calculated: the area of the primary mirror minus the amount occulted by the secondary mirror. But a small amount of light is lost at each surface–air interface in the telescope and in the spectrograph, and so the geometrical collecting area has to be multiplied by an *efficiency* factor giving the fractional amount of the incoming signal that actually reaches the detector. For example, if each interface returns 95% of the incident signal, then a telescope/spectrograph system with 10 interfaces will return only $0.95^{10} \approx 0.60$ of the incident signal. In addition, for spectroscopic observations, there will be losses of incident light around the edges of the spectrograph slit at the focal plane of the telescope. These *slit losses* are not easy to estimate, because they depend on the size of the 'seeing disc' of the image and the degree of image motion that may occur during an integration; but losses of 20% are typical. Lastly, the efficiency of the reflective grating employed must be included in the calculations, and that efficiency can be a significant function of wavelength, typically in the range 50–75%. In overall terms, therefore, the efficiency of an optical system for spectroscopy is unlikely to be much greater than 35%.

The proportion of the signal incident on the primary mirror that reaches the detector is then converted to an electronic signal, according to the *quantum efficiency* of the detector. Typical spectral-response curves for modern CCDs have a broad range, with efficiencies on the order of 60–80% over the spectral range of 400–800 nm. So our incident 100% signal is finally

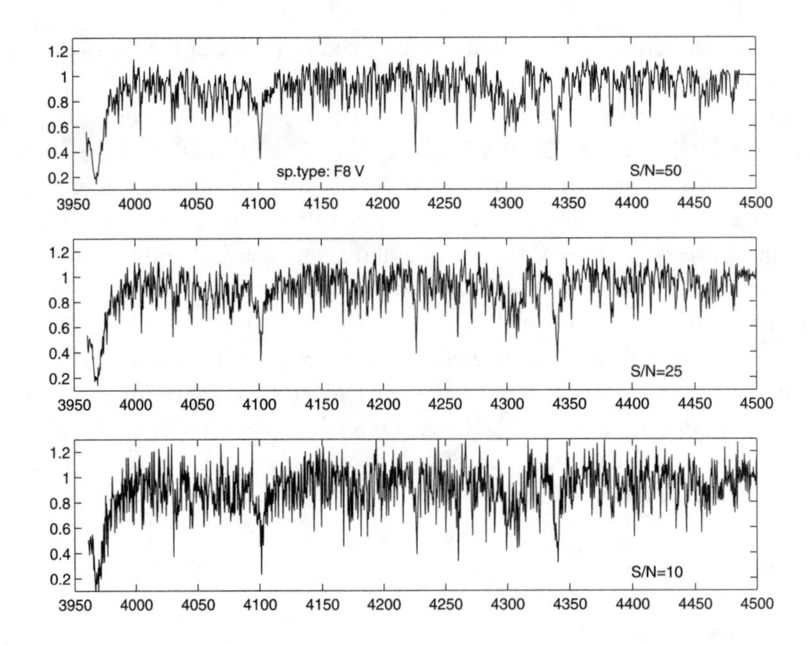

Fig. 3.1. Spectra of an F8V star with S/N values of 10, 25, and 50.

converted into an electrical charge in a CCD at a level of about 25%. It is worth noting that CCDs are about 30 times more efficient than photographic emulsions, so the net response of a modern system is an enormous improvement over that which prevailed in the pre-CCD era.

In Chapter 5 we shall discuss absolute calibration of photometry for astronomical sources. Suffice it to state here that the flux density f received at the telescope from a star of apparent magnitude $m = 0$, at a wavelength $\lambda \approx 5500$ Å, is $f \approx 1000$ photons s^{-1} Å^{-1} cm^{-2}. Then the signal S_{source} received from a source of apparent magnitude m over a total integration time t is given by

$$S_{\text{source}} = \text{aperture area } \times \text{efficiency} \times \text{quantum efficiency} \times f \times 10^{-0.4m} \times t \quad (3.2)$$

Figures 3.1 and 3.2 show the spectra of an F star and a B star at different S/N ratios. It is clear that at $\text{S/N} \approx 10$, the spectra of B stars are unusable for any form of analysis, but those of F stars do show a few major spectral features that would be discernible and measurable for crude determinations of radial velocities. But $\text{S/N} \approx 25$ is needed for any serious work on radial velocities of binary stars, and we would need $\text{S/N} \approx 50$–100 before we could describe the spectra as 'nice data' and likely to give unequivocally good results for radial velocities. Furthermore, in order to use sets of spectra in very precise surface-imaging studies of stars, S/N values substantially in excess of 100 are required.

Further details about signal-to-noise calculations can be found, for example, in *Electronic Imaging in Astronomy*, by McLean (1997), and in *The Observation and Analysis of Stellar Photospheres*, by Gray (1992).

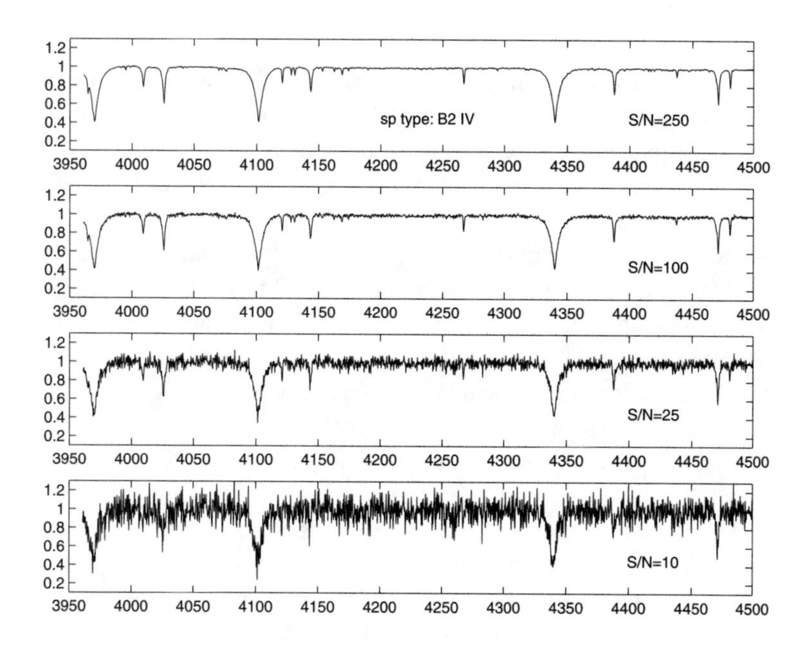

Fig. 3.2. Spectra of a B2IV star with S/N values of 10, 25, 100, and 250.

3.2.2 Selection of spectral range

It was noted earlier that the quantum efficiency of a detector is a significant function of wavelength, though the most recently produced CCDs do have broad responses over more than 400 nm at levels exceeding 60%. Because the spectra of stars display a wide variety of features that can be used to determine radial velocities, it is important to include in the planning of an observing programme a careful consideration of which spectral features will be most suitable. Those spectral features will then dictate which part of the spectral-response function of the detector is to be preferred for that particular observing programme, and the expected integration times and S/N values will have to be calculated according to the quantum efficiency of the detector at that wavelength range.

Radial velocities are determined from the Doppler-Fizeau effect, more commonly called simply the Doppler effect, as given by the standard equation:

$$\frac{\lambda - \lambda_0}{\lambda_0} = \frac{\Delta\lambda}{\lambda_0} = \left[\frac{1 + (V/c)}{1 - (V/c)}\right]^{1/2} - 1 \approx V/c \quad \text{for } V \ll c \tag{3.3}$$

For a spectral line at $\lambda = 4500$ Å, the Doppler shift corresponding to a velocity $V = 50$ km s^{-1} is $\Delta\lambda = 0.75$ Å. It follows that if we wish to determine the velocities of stars accurate to about ± 1 km s^{-1}, then, in addition to appropriate dispersions, we shall need to select a substantial number of spectral lines that are very well defined. This latter qualitative statement requires some elaboration.

The shape and strength of a spectral line depend on a number of factors, principally the temperature and local gravity in the region of the stellar atmosphere where that line was formed, the number of atoms contributing to that line, and the rotational velocity of the star. Most stars have similar chemical compositions, with ratios, by mass, of hydrogen : helium : other elements ('metals') $\equiv X : Y : Z \sim 0.73 : 0.25 : 0.02$. The abundance differences between the various stellar populations in the Milky Way galaxy (MWG) and in other Local Group galaxies are significant only for Z, with values ranging from $Z \leq 0.001$ for globular clusters in the MWG, through $Z = 0.004$ and 0.008 for the Large and Small Magellanic Clouds, respectively, to $Z = 0.04$ for a few open clusters in the MWG. These abundance differences are fairly subtle, and the major factors that control the strengths of spectral lines across the range of spectral types of stars are temperature and local surface gravity. There are additional factors – local turbulence within the stellar atmosphere, magnetically active regions (starspots) – but, again, these are quite subtle effects. We consider, firstly, the spectra of stars on the *main sequence*.

From the middle of the stellar spectral sequence towards cooler temperatures, that is, stars of spectral types F–K, the spectral lines that are most suitable for determination of radial velocities are the substantial numbers of Fe I, Fe II, and other 'metal' lines that are present over wide ranges of wavelength. These lines are generally weak, with central depths below the continuum level that typically are 10% or less, because the number of Fe atoms, for example, is about 10^{-5} of the number of H atoms, and there are many possible transitions for these multi-electron atoms and ions. There is a plethora of such lines in the blue region, 4000–4500 Å, but there are also many in the yellow region and in the red (6000–6800 Å), around Hα at λ6563. Although these 'metal' lines are weak in late-A and early-F stars, their strengths increase as the stellar effective temperature decreases. Because there are so many of them, their combined effects in determining radial velocities by means of *cross-correlation*, even from spectra of poor quality, are quite remarkable, as we shall see. The intrinsic shape of these lines is symmetric and narrow, and the only difficulties that have been experienced in determining radial velocities from them have been due to the *blending* of individual lines into broader features at some wavelengths (e.g., the atomic-line blend at λ6497 shown in Figure 3.3).

The H lines are quite strong at F0, and they decrease steadily through the subtypes of the F–G range. But because there are so many atoms contributing to these lines, and because the atomic transitions of H atoms are susceptible to *pressure broadening* by the *Stark effect*, these lines are notably deeper and broader than the Fe lines. Their strengths decrease substantially through the K subtypes and become lost amongst the increasing numbers of *molecular bands* through the later K subtypes and into the M spectral class. These H lines were used extensively in the earlier investigations of radial velocities of stars, because they are strong and easily identified, particularly on under-exposed photographic spectra. But their intrinsic line shape is so different from that of the weaker but plentiful 'metal' lines that it is not good practice to make use of them in radial-velocity studies conducted by means of cross-correlation. This remark applies equally to the Ca II H and Ca II K lines at λ3934 and λ3970, the CN band at λ4216, the G band of CH at λ4300, and the Na D lines at λ5890 and λ5896. The M-star spectra are characterized by molecular bands, particularly of TiO, strong Na D lines, and CaOH bands. But there are many much weaker lines of atomic 'metals' that realize accurate velocities

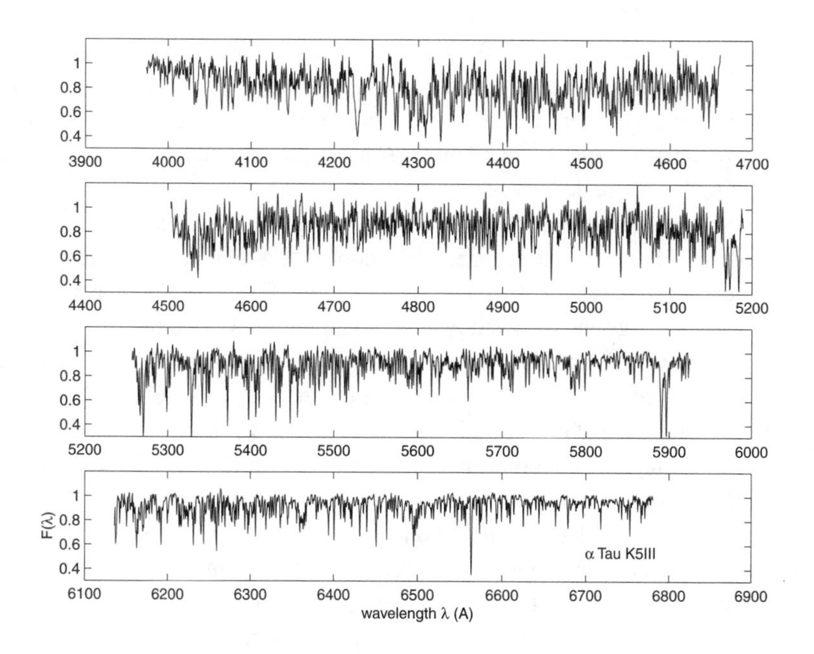

Fig. 3.3. Spectra of the K5III star α Tau obtained with the 0.5-m Leslie Rose Telescope and a spectrograph at the University Observatory, St. Andrews, with a diode-array detector. The original resolution was 0.38 Å px^{-1}. The spectra illustrate the richness of the entire visible spectrum, with thousands of lines due to 'metals', as well as some of the major molecular bands and very strong lines, such as the G band of CH at λ4300 and the Na D lines near λ5890.

from cross-correlation. Figure 3.3 provides an illustration of the richness of late-type spectra throughout the visible spectrum.

These very red K–M stars can be observed more readily in the far-red part of the spectrum, 6000 Å to 1 μm, particularly so since the advent of CCDs with their wide spectral range. However, the Earth's atmosphere is not transparent at all wavelengths, and there are substantial numbers of molecular-absorption bands and atomic lines superposed on stellar spectra from the interactions of the starlight with molecules and atoms in the Earth's atmosphere. There are some *telluric lines* in the region around Hα, and the *telluric absorption bands*, principally due to H_2O vapour, are substantial beyond 6800 Å, with the result that studies of far-red spectra require careful removal of these telluric bands from the recorded spectra. The adopted procedure is to observe a star with a nearly featureless continuous spectrum, such as some *subdwarf* and *white-dwarf* stars that have been adopted as spectrophotometric standards (Massey et al. 1988; Massey and Gronwall 1990), at the same air mass as the programme star. Then the telluric features can be divided out of the programme-star spectra. This procedure is effective at high-altitude observatories where the relative humidity is low. But for sea-level sites, these telluric bands are so strong that it is not realistic to attempt to use the far-red part of the spectrum.

The richness of the absorption-line spectra seen in the cooler stars diminishes as one moves up the spectral sequence from F0 towards the hotter stars. Nearly all of the 'metal' lines have disappeared by spectral subtypes A2–A0, whilst the H lines have increased in strength, reaching their maximum at this region. The only narrow lines that are suitable for velocity measurements in the blue region of the spectrum are the weak lines of Ca II K, C II λ4267, and Mg II λ4481. So determinations of radial velocities around this narrow range of spectral subtypes are often difficult. In addition to this major decrease in the numbers of usable lines, we find that the intrinsic rotational velocities of these hotter stars are generally quite high ($V_{rot} > 50$ km s^{-1}), because they do not have the outer convection zones of the cooler stars and their concomitant magnetic fields that ensure magnetic braking by means of stellar winds. So even in binary systems with orbital periods that are large enough to allow the stellar rotational velocities to be unaffected by tidal synchronization (see Chapter 4) we find A stars and hotter stars with rotational velocities that are substantially greater than those amongst the cooler and generally more slowly rotating F–M stars. The consequence of more rapid rotational velocities is that the spectral lines become broader and weaker and hence more difficult to measure accurately. Indeed, there are many potentially useful absorption lines that are intrinsically weak amongst the B and O spectral types, and they become completely 'washed out' by rapid rotation and therefore are unusable.

The B-type spectral class is defined by the presence of He I lines that increase in strength from B9, reaching a maximum at B2. The numbers of He atoms that contribute to these lines are about 10% of the numbers of H atoms, so the line strengths are accordingly weaker than those of the H lines. Amongst the 'metal' lines, Mg II λ4481 is particularly useful for the B9–B5 subtypes, together with C II λ4267, and sets of Si II and Si III doublets and triplets become increasingly useful from B6 to B2 before decreasing in strength again by B0. The shapes of these absorption lines of He and 'metals' are largely dominated by rotation, and they can all be used in cross-correlation procedures, with good accuracy being attained. The spectra of stars with subtypes in the range B0–B3 are really quite rich in absorption lines from He, Si, O, and N atoms and ions, but stellar rotation can broaden and weaken these lines such that only the much stronger He lines can be used. For the O-type stars, we see the introduction of He II lines, as both the H lines and the He I and Si II and IV lines fade.

The absorption lines of He I and He II are distributed widely through the visible spectrum, just as are the lines of H. In order to use these lines for radial velocities, it is necessary to record a wide range of the spectrum, much more than is necessary for work on F–K stars. The old photographic emulsions, particularly those like Kodak IIIaJ, were capable of recording good spectra at resolutions of $\lambda/\Delta\lambda \sim 10{,}000$ over a wavelength range of 1500 Å, thereby ensuring that many He I and He II lines were measurable from a single spectrum when the blue part of the spectrum was utilized (λ3800–5300) (Figure 3.4). The first CCD detectors, whilst dramatically improving the S/N ratio, could record only about 200 Å of spectrum because of their small linear size, which led to very serious limitations on radial-velocity work. The newer CCDs, with silicon acreage increased by factors of 16–64, have restored that wide wavelength range, and for the most recent échelle spectrographs have far exceeded it, to about 3000 Å.

That difficulty of finding sufficient spectral lines with which to determine radial velocities of O and B stars was obviated by moving to the ultraviolet (UV) part of the spectrum, where there are many lines of more highly ionized species (He II, C III and IV, N III and IV, Fe IV and V, etc.) that are available over only 500 Å. Very large numbers of UV spectra of O- and B-type binary stars were recorded by the International Ultraviolet Explorer (IUE) satellite over

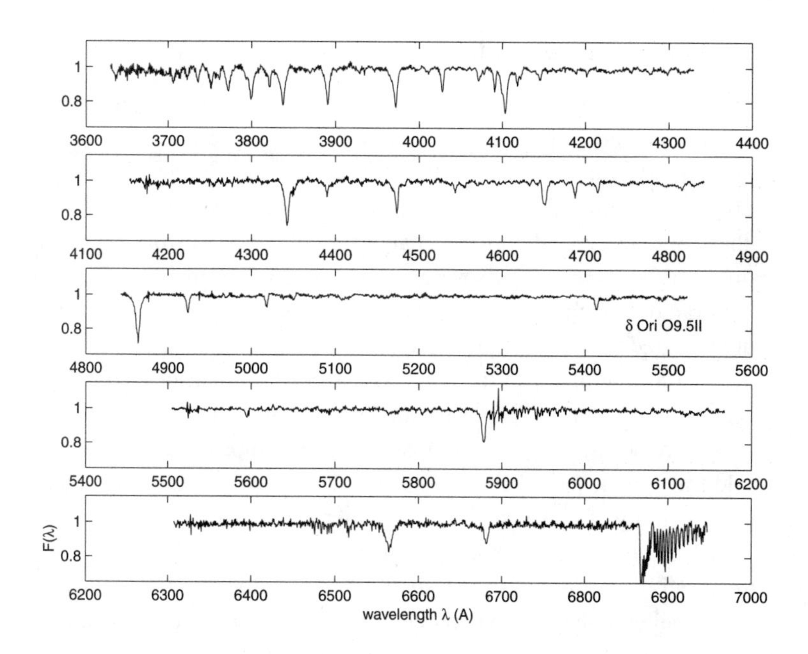

Fig. 3.4. Spectra of the O9.5II star δ Ori obtained with the 0.5-m Leslie Rose Telescope and a spectrograph at the University Observatory, St. Andrews, with a diode-array detector. The original resolution was 0.38 Å px^{-1}. The spectra illustrate the obvious Balmer lines of H and the major He I and He II lines. Note the paucity of lines usable for radial velocities in the yellow region, λ5000–6500, the presence of Na D-line emission close to the He I λ5876 line from local street lamps, and the onset of H_2O bands around λ6900 from the Earth's atmosphere at this sea-level site.

its 18 years of operation, resulting in a major archive of good data. But, in general, we do not enjoy the ready access to the UV spectrum that we take for granted in the visible regions. Recent improvements in the size (real estate!) of CCDs have restored our ability to make high-quality radial-velocity studies of hot stars.

The foregoing descriptions and examples of the spectra of main-sequence stars apply also to *giant* and *supergiant* stars, with some additional factors. The principal change is the reduction in surface gravity from the main-sequence stars through the giants to the supergiants, with the result that the pressure broadening of spectral lines is reduced, dramatically so for the Stark-effect-broadened H lines. This effect on the H lines is well known as a good indicator of stellar luminosity. In addition, the rotational velocities of such large stars are generally low, so that many of the previously mentioned weak lines of various elements become more narrow and better defined, resulting in the potential for improved accuracy for radial velocities. By contrast, the *subdwarf* stars, which have higher surface gravities than main-sequence stars, have more troublesome spectra, with any H lines being broadened dramatically, and many weaker lines being lost into small-amplitude variations of the stellar continuum. The reader is referred to the text *The Classification of Stars*, by Jaschek and Jaschek (1987), for a detailed discussion of the empirical spectra of all subtypes of stars.

3.2.3 Observations and data extraction

Standard observing procedures with CCDs require the recording of *bias frames* and *flat-field frames* at the beginning and end of each night of observing. The bias frames are exposures of zero seconds, with all shutters in the instrument closed, so that the CCD is not recording any flux density. These frames define the bias, or positive-charge offset from zero, on each pixel, so that the charge integrated during an exposure will not start from a negative value. A mean bias frame is usually defined by several zero-second exposures and is subtracted from each observation frame. The flat-field frames are exposures of the CCD to light from, usually, a tungsten source located before the entrance slit to the spectrograph. These frames define the relative responses of all the pixels in the array to the same amount of incoming light. Most pixels in a good CCD chip with 10^6 pixels respond to a given light level to within 1–2% of the mean, though there are occasional defective pixels in many chips. From several exposures, an average flat-field frame is defined that has a mean value of unity and records the small-scale variations in the sensitivity of each pixel in the array about that mean value. The mean flat-field frame is then *divided out* of each observation frame.

Each spectroscopic observation of a star should be immediately preceded by an exposure of the CCD to a wavelength-calibration source or comparison source. That source is usually a discharge tube, filled with argon gas or with a mixture of gases like helium, neon, and thorium and located in the acquisition head of the telescope. The light from the discharge tube enters the spectrograph at the slit in the focal plane of the telescope, thereby passing through the same optical path in the spectrograph as the light from an astronomical source. The resulting narrow *emission-line* comparison spectrum that is recorded by the CCD can be used to provide a calibration of the pixel numbers along the spectrum with the wavelength of the light that is dispersed to each pixel by the spectrograph. To guard against any possibility of flexure of the spectrograph or other unwanted difficulties, it is sensible practice to secure a wavelength-calibration exposure after the telescope has been slewed and set to the position of the selected star and immediately before the exposure on that star. It was not so long ago that the comparison source was, literally, an iron arc mounted on the telescope, and the bright sparks and associated fumes were at least quite exciting in the middle of a long, cold night in a dome. For that reason, these exposures for calibration of wavelengths often are still referred to as *arc spectra*. Sophistication now rules, however, and so we sit in warm rooms, keying commands into a computer that controls all the functions of the modern spectrograph – from the variety of available discharge tubes, to slit width, length, and orientation, selection of the collimator and the grating angle, and the focussing controls with their Hartmann masks.

Figures 3.5 and 3.6 show examples of CCD frames from a comparison spectrum and a stellar spectrum, respectively. The comparison spectrum extends across the *spatial* dimension of the image, because the light from the comparison source fills the spectrograph slit entirely. But the stellar spectrum fills only a fraction of that dimension, being limited to the size of the *seeing disc* of the stellar image on the slit. For observing conditions with 'good seeing', where the image size is about 1 arcsec and the image motion is very small, the stellar image is well defined, and the corresponding spectrum fills only a few rows/columns of the CCD frame, with the edges of the spectrum being quite sharp. Under conditions of 'poor seeing', the

Fig. 3.5. An unprocessed CCD frame of a comparison, or arc, spectrum showing the emission lines filling the spatial direction of the frame (x axis) and dispersed in the wavelength direction (y axis).

Fig. 3.6. An unprocessed CCD frame of a star spectrum showing the size of the seeing disc in the spatial direction (x axis) and dispersed in the wavelength direction (y axis). Note also the substantial number of cosmic-ray events (small dark streaks) recorded in this 1200-second exposure.

spectrum is spread over many pixels in the spatial direction and has poorly defined edges. The impact of poor seeing on the achieved S/N ratio for a given integration time is obvious: Each pixel receives less flux from the source, and less of the image is transmitted by the slit, and yet the readout noise from each pixel remains the same. Poor seeing can prevent an observing programme from taking place.

It is appropriate to note optimum methods for extracting the spectrum of a source from the recorded CCD frame. Once the spatial and wavelength limits of the spectrum have been determined, both the stellar spectrum and the comparison spectrum need to be extracted from the same rectangular box of pixel numbers in the rows and columns of the CCD. A section of the same size and offset, but parallel to the stellar spectrum, can be used to determine the contribution of the background sky to the total signal (star + sky). A simple extraction procedure is to sum the contributions of successive pixels in the spatial direction at each given pixel number in the wavelength direction, and subtract the appropriate sky background level. If a sufficiently wide spatial extent is selected, then spectrophotometric accuracy is preserved, but likely at the expense of including pixels that carry very little contribution from the source but that still add more readout noise to the total signal at each wavelength. Whereas this may not be important for bright sources, it may well be crucial to the success of the observing programme in extracting a spectrum with the minimum of associated readout noise. Accordingly, a method for *optimal extraction* of stellar spectra was proposed by Horne (1986), and it has become the standard technique used for CCD spectroscopy. The complete algorithm is described in detail in Horne's paper (1986), and it is sufficient here to note that the procedure adopts weights for each contributing pixel in the spatial direction that reflect the statistical contribution of each pixel to the total signal [(star + sky) − sky] such that the S/N ratio will be maximized at each wavelength.

The extracted comparison spectra need to be measured to provide the necessary wavelength calibration. The wavelengths of emission lines from argon gas, for example, are known to a precision of 10^{-3} to 10^{-4} Å from detailed laboratory spectroscopy. Accordingly, a detailed calibration of pixel numbers in the wavelength direction versus wavelength is achievable. The positions of the lines in the comparison spectrum are usually measured, interactively, via a spectroscopic reduction code that fits Gaussian profiles to each selected comparison line by a least-squares procedure. The resultant set of line positions in terms of pixel numbers and accurate wavelengths can then be matched via least squares to a polynomial expression, typically of order 5. Because spectrographs are, or should be, very stable instruments at all orientations of the telescope, it follows that all comparison spectra should provide very similar values for the coefficients of the polynomial expression. Accordingly, it is possible to establish, quickly, a standard set of coefficients for a particular specification of the spectrograph, and all recorded comparison spectra should be very similar to that standard. Traditionally such a standard has been called the *standard plate*, as described fully for the visual measurement, via travelling microscopes, of photographic spectra by J. H. Moore in Aitken's text *The Binary Stars* (1935). Modern interactive codes for measurement of comparison spectra adopt the same basic principles of using a standard plate and fitting least-squares polynomials to the *deviations* of an individual spectrum from that standard – codes such as REDUCE (Hill 1982a) and FIGARO (Shortridge et al. 1997). Then any difficulties with the spectrograph or the detector are immediately recognizable as substantial deviations away from the standard. The accuracy of fit for the comparison line positions about best-fitting polynomials typically are about

±0.1 pixel (px), or ±0.05 Å, for spectra with a resolution of 7500. Care should be exercised in these procedures to ensure that the data continue to demonstrate stability of the spectrograph. During continued monitoring of a binary system over many hours, the sequence of comparison spectra may reveal a slow drift of the 'zero point' of the standard plate. If that drift is deemed significant within an integration time (i.e., the spectrograph is not ideally stable), it may well be prudent to interpolate the values of the wavelength scale to the mid-time of each of the exposures on the binary star, as well as to be concerned about the stability of the spectrograph!

The pixel-number values for the stellar spectra in the wavelength direction can be converted to the appropriate wavelength scale by means of the comparison spectrum associated with each particular stellar spectrum. Note that for the digital spectra from CCDs or other diode arrays, we have sequences of flux density as a function of pixel number, and the wavelength calibration provides the conversion of pixel number to wavelength, which is generally somewhat non-linear. To establish a digital spectrum that provides flux density as a function of wavelength that advances in linear increments, it is necessary to interpolate in the observed value of the flux density to each linear increment in wavelength. This procedure is variously referred to as *scrunching* or, more prosaically, as *wavelength linearization*.

3.2.4 Standard stars for radial velocities

The development of new techniques for astronomical spectroscopy through the earlier part of the twentieth century was intimately linked with major research efforts in laboratory spectroscopy and atomic physics, and with parallel developments of more sensitive photographic emulsions. The result of those collaborative efforts in astronomy and physics was the establishment of lists of accurate empirical wavelengths for the atomic transitions of many elements. It was found that the rest wavelengths of spectral lines from many different atoms measured in the laboratory were in agreement with the results from astronomical sources when the line-of-sight motions of those sources were known independently of the Doppler effect, and measured wavelengths could be corrected to the rest frame. Those developments progressed very much in parallel and led to a rapid increase in our knowledge of which atoms, ions, or molecules produced which spectral features at which wavelengths in the spectrum of an astronomical source.

With known rest wavelengths for spectral features, the Doppler effect could be exploited to determine the radial velocities of astronomical sources. A necessary part of the development of the subject was to test the accuracy of those measured radial velocities against the motions of sources that were known from other astronomical observations. The planets and minor planets of the Solar System were the ideal candidates, because their light is mostly reflected sunlight, and their orbital motions had been well established from very long series of astrometric observations. There were some complications due to the resolved angular sizes of the planets and the consequent possible inclusion of a component of rotational motion in the measured radial motion, unless care was taken to ensure the correct orientation of the spectrograph slit relative to a planet's rotation axis. The minor planets were better, because most of them were unresolved, and their spectra were purely reflected sunlight, but they were also fainter than the planets, so that good-quality spectra were more difficult to obtain. Nevertheless, in that way, detailed knowledge of the solar-type spectrum was developed, including empirical wavelengths for blends of atomic transitions seen in astronomical spectra that could not be reproduced in the laboratory. The similarities in many spectral features between G-type stars and F- and K-type stars then allowed the system to be extended. From carefully executed observations conducted at several observatories – principally at Lick Observatory, Dominion Astrophysical Observatory, Victoria, and others – there was developed a set of bright F–K stars whose radial velocities were found to be constant to within $\pm$1–2 km s^{-1}. That set of *radial-velocity standard stars* has been maintained and revised ever since, and it now includes standards at apparent magnitudes of 9–10, as well as many of the original stars of naked-eye brightness.

All spectroscopic observations that are expected to be used for determination of the radial velocities of programme stars should be accompanied by observations with the same instrumentation of these radial-velocity standard stars, as defined by Commission 30 of the International Astronomical Union (IAU). These standard stars typically are of spectral types F–K, and their radial velocities have been monitored and found constant to within $\pm$1 km s^{-1} over decades. The precision of the radial-velocity system is monitored continually by Commission 30, whose triennial reviews are published in the *Reports on Astronomy* by the IAU. A regularly updated list of these stars is published annually in the *Astronomical Almanac*. It is usual practice to secure a spectrum from each of three or four standard stars during any one night of observation. Most spectrographs on modern telescopes are very stable, but it is incumbent upon the observer to ensure that the radial velocities determined from an observing programme are genuinely in agreement with the IAU standard system of radial velocities.

Extension of the radial-velocity system to stars of earlier spectral types had to be carried out through the late-A stars (where there are still many spectral lines due to 'metals'), then to early-A stars, and then into the B and O stars, by means of combinations of known laboratory wavelengths, and by empirical methods. One of these latter methods was pursued by R. M. Petrie at the Dominion Astrophysical Observatory, Victoria, Canada, where he observed A- and B-type stars that were known members of visual binaries and had companions of spectral types F–K. The orbital motions (projected into the observer's line of sight) of the two stars in a typical visual binary with an orbital period of, say, 100 years differ by less than 2 km s^{-1}. The radial velocity of the late-type component, measured on the standard system, can be used to define the radial velocity of the B star and therefore determine the rest wavelengths of the spectral features in the B-type spectrum. Whereas some spectral lines from B stars did already have known wavelengths, other well-defined features seemed to be blends of different transitions whose rest wavelengths depended upon the spectral resolution employed for the observation. Such was the case, for example, with the triplet series of lines due to He I, where 'forbidden' components were blended with the 'permitted' transitions. Additionally, it is generally the case that the early-type stars have higher rotational velocities than their late-type counterparts, which means that radial velocities for O and B stars are not determinable with the same precision as for F–K stars. Consequently, there exist only a few 'secondary standards' for radial velocities amongst the O and B spectral classes. Holmgren et al. (1990a) have provided a small selection of five stars between O9V and B5V. Their investigation of the constancy of radial velocities amongst O–B stars was based on detailed investigations by Andersen and Nordström (1983) of the effects of stellar rotation on the standardization of radial velocities. Their discussion presents lists of recommended wavelengths of spectral lines for different spectral classes from K to mid-B type. Further details of the development of the radial-velocity system can be found, for example, in the work of Petrie (1962), Batten (1978), and Andersen and Nordström (1983).

3.2.5 Radial velocities from cross-correlation

The traditional techniques for determining radial velocities were based on measurements of photographically recorded spectra by means of some travelling microscope or projection device with a cross-wire to set on individual spectral lines. The comparison spectrum from an iron arc, recorded on both sides of the stellar spectrum, had sharp, well-defined emission lines whose positions on a linear scale could be measured to a precision of 1–2 μm. The wavelength calibration was therefore well established. Measurements of lines in the stellar spectrum could be nearly that accurate for narrow, well-defined lines that were not blended with neighbouring lines, but decidedly poorer for rotationally broadened lines. The shift of each line from its known rest wavelength then provided a measure of the radial velocity of the star, from the Doppler effect, equation (3.3), and the average of all those measures on the stellar spectrum gave a final average velocity for the star. The principle involved in this procedure is that each spectral line is treated *independently*, apart from the wavelength-calibration procedure based on the iron-arc source in the rest frame, which is accurately known.

An alternative consideration is to view the stellar spectrum, once it has been placed onto a well-defined wavelength scale, as a *pattern* of spectral features that remains unchanged regardless of the radial velocity of the source and is simply shifted in wavelength in accordance with equation (3.3). But a perusal of that equation shows that the shift $\Delta\lambda$ is a function of both the velocity V of the source and the rest wavelength λ_0 of a particular spectral line. When a substantial section of a spectrum is viewed at different radial velocities, it is seen that the pattern of the spectrum is altered, relative to the rest pattern, in a wavelength-dependent manner. But if we make the simple change of converting from a linear wavelength scale to a linear scale in the natural logarithm of the wavelength, $\ln\lambda$, then this problem of altering the pattern disappears. Equation (3.3) can be rewritten in logarithmic form as

$$\ln\left(\frac{\lambda}{\lambda_0}\right) = \ln\lambda - \ln\lambda_0 = \frac{1}{2}\ln\left[\frac{1+(V/c)}{1-(V/c)}\right] \approx \ln[1+(V/c)] \qquad \text{for } V \ll c \tag{3.4}$$

So a radial velocity V causes a shift in $\ln\lambda$ space that is independent of wavelength. Accordingly, the intrinsic rest-frame pattern of the spectrum of the source remains unchanged in $\ln\lambda$ space, and it is simply shifted along the $\ln\lambda$ axis by an amount that is directly proportional to the radial velocity between the source and the observer. With this view of the spectrum as an unchanging but displaced pattern, we can use the powerful mathematical technique known as *cross-correlation* to determine the radial velocities of astronomical sources relative to radial-velocity standard stars.

Quite generally, the *cross-correlation function* $c(x)$ of the independent variable x between two functions $f(k)$ and $g(k)$ of the independent variable k is defined by the following

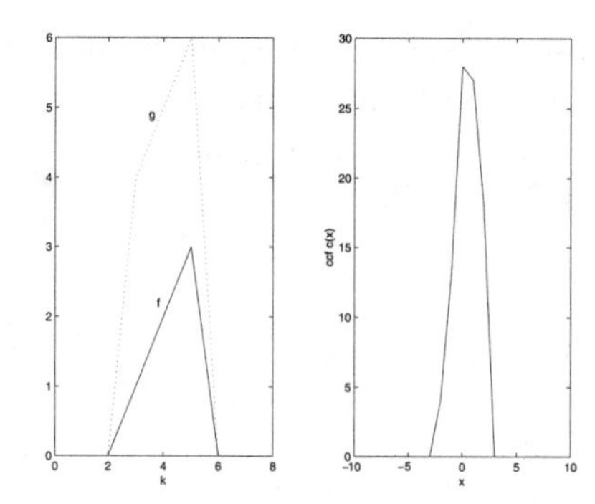

Fig. 3.7. Two simple functions, $f(k)$ and $g(k)$, are convolved or cross-correlated to produce the cross-correlation function $c(x)$ for a range of values of x. The best match between the two functions $f(k)$ and $g(k)$ is obtained at $x = 0$, as expected in this simple demonstration, but note how strongly dependent on x is the value of $c(x)$.

deceptively simple integral:

$$c(x) = \int_{-\infty}^{\infty} f(k)g(k - x)\,dk \tag{3.5}$$

Equation (3.5) represents the action of *convolution* of the two functions $f(k)$ and $g(k)$. For each value of the independent variable x, the cross-correlation function $c(x)$ takes a value that is equal to the integral of the product of the two functions $f(k)$ and $g(k - x)$. When the integral is evaluated for a range of values of x, then we obtain a corresponding range of values of $c(x)$, which provides a numerical description of how well matched, or correlated, the two functions $f(k)$ and $g(k)$ are over that range of values of x. It turns out that the value of the cross-correlation function can be particularly sensitive to the value of x, depending on the form of the two cross-correlated functions. Figure 3.7 illustrates the power of the cross-correlation function (CCF) in determining when two functions, or patterns, are best correlated.

The application of cross-correlation to astronomical spectroscopy was first proposed by Fellgett (1953), and it was first implemented by Griffin (1967) with his *radial-velocity spectrometer*. That instrument performed an analogue form of cross-correlation by stepping a photographic negative of the spectrum of the K2III star Arcturus across the focal plane of the spectrometer. When all the absorption lines of the observed source and Arcturus were coincident, the total amount of light recorded by a photomultiplier behind the focal plane of the spectrometer would be a minimum. Any displacement of the Arcturus mask from that position would allow more light through from the source, and a larger signal would be recorded. That *analogue* technique proved to be extremely effective for stars of late spectral types F5–K, and because it used all the incoming light from the source that passed through the Arcturus mask to produce a single measured flux density at each stepped position of the mask, it was also very efficient. A moderate-sized telescope could be used to determine radial velocities of F–K stars in short time intervals and with accuracies of order ± 0.5 km s^{-1}. Griffin's instrument was used on the 36-inch telescope at the Cambridge Observatories for about 25 years, and others have followed his lead and developed more highly automated spectrometers, notably the CORAVEL instruments made by Baranne, Mayor, and Poncet (1979) for L'Observatoire de Haute Provence and the European Southern Observatory. Similar instruments have been introduced elsewhere, such as that attached to the extremely stable coudé instrument on the 1.2-m telescope at the Dominion Astrophysical Observatory (Fletcher et al. 1982).

Although these radial-velocity spectrometers are very efficient instruments for F5–K stars, they are limited by the need to match the Fe absorption lines in the mask and in the stellar source and by the essential definition of those lines, which depend on the rotational velocity of the star being observed. Because most stars of spectral types earlier than about F5 are more rapidly rotating, the efficacy of the CORAVEL kind of instrument declines for types earlier than that.

The more general problem of performing cross-correlations with *digitized* spectra was considered in detail by Simkin (1974) and by Tonry and Davis (1979), amongst others. Here the spectra are recorded as usual by means of a CCD or other diode array, and the analysis of those spectra to reveal radial velocities is performed later. The procedure can also be used, of course, on photographic spectra that have been scanned by means of a digitized microdensitometer to yield digital spectra. The advantages of post-processing the data are that cross-correlation studies can be performed on spectra of any type (provided that there are sufficient spectral features that are suitable) and that the spectra can be used for other analyses too.

Applications of these procedures to close, interacting binary stars followed quickly, with McLean (1981), McLean and Hilditch (1983), and Hill (1982b) presenting results from CCF studies of contact binaries. G. Hill's (1982b) VCROSS code was developed to determine radial velocities for stars of any type, and he has pioneered many improvements in the best procedures for analysis, depending on the type of binary system studied. His wide-ranging review of those developments (Hill 1993), covering applications to detached, semidetached, and contact binaries, is particularly valuable, and we shall consider some of these factors later. The cross-correlation technique applied to digitized spectra is now the standard method of analysis for radial velocities.

We define $s(n)$ to be the spectrum of a programme star, being some measure of the flux density at n bins, defined to be at equal intervals in $\ln \lambda$. Note that the measure of flux density, typically *counts* that are linearly related to flux density, has to be made relative to the local continuum level that is set to zero. This process of *rectification* of all spectra, including the template, is performed by fitting a least-squares cubic spline to selected points, or small regions, that are judged by the analyser of the data to be at the local continuum level. This rather subjective process is easy and accurate for early-type stars where there is much free continuum unaffected by spectral lines. But it becomes increasingly difficult as one moves towards later spectral types, and by mid-F-to-G-type stars, the procedure typically is to fit the least complicated curve to the 'tops' of the spectrum. It is not sensible to enter a cross-correlation process with a spectrum that shows substantial curvature in its continuum values, because that would lead to systematic errors in the location of the CCF on the x axis.

We also define $t(n)$ to be an equivalent spectrum of a *template* that has a *known radial velocity on the standard system*. That template can be the spectrum of a radial-velocity standard star, usually secured with the same instrument as the programme star, or it can be a synthetic spectral template calculated with the help of a model code for stellar atmospheres. Either way, it must be a good match to the spectral type of the programme star, within a few spectral subtypes. The only exception to this rule would be in a situation in which the spectral lines selected for the cross-correlation process were consistently presented over a wide range of spectral subtypes, such as the Fe I lines in the spectra of late-type stars, as used by CORAVEL. But the analyser must always be careful and must always try a range of templates.

Then the cross-correlation function $c(x)$ between the programme star and the template is defined as

$$c(x) = \int_{-\infty}^{\infty} s(n)t(n-x)\,dn \tag{3.6}$$

where $c(x)$ will have a maximum value when the template $t(n)$ has been shifted by $x = \ln[1 + (V/c)]$ to coincide with the stellar spectrum $s(n)$, where the velocity V is the *relative velocity* between the star and the template.

These observational data, $s(n)$ and $t(n)$, are composed of N discrete bins sampled at equal intervals in $\ln \lambda$, each with a measure of fractional flux density relative to the continuum level that is set to unity. If the continuum level is subtracted from the data in each spectrum, then we can form root-mean-square (RMS) uncertainties of that spectrum by

$$\sigma_s^2 = \left[\sum_{n=1}^{N} s(n)^2\right] \Big/ N; \qquad \sigma_t^2 = \left[\sum_{n=1}^{N} t(n)^2\right] \Big/ N \tag{3.7}$$

and the *normalized CCF* is then

$$c(x) = \frac{1}{N\sigma_s\sigma_t} \sum_{n=1}^{N} s(n)t(n-x) \tag{3.8}$$

The most efficient method to calculate $c(x)$ is via Fourier-transform techniques that make use of the fast-Fourier-transform (FFT) algorithm (see, e.g., ch. 2 and appendix C of Gray 1992). The reason for this speed is that convolution in the x domain becomes simple multiplication in the Fourier domain. Defining the discrete Fourier transforms of $s(n)$ and $t(n)$ by

$$S(k) = \sum_{n=1}^{N} s(n) \exp[2\pi ikn/N]; \qquad T(k) = \sum_{n=1}^{N} t(n) \exp[2\pi ikn/N] \tag{3.9}$$

then the Fourier transform of equation (3.8) is

$$C(k) = \frac{1}{N\sigma_s\sigma_t} S(k)T^*(k) \tag{3.10}$$

where T^* is the *complex conjugate* of T. Then the inverse transform of $C(k)$ is

$$c(x) = \frac{1}{N\sigma_s\sigma_t} \sum_{n=1}^{N} C(k) \exp[-2\pi ikx/N] \tag{3.11}$$

This Fourier process is of order $N \log N$ rather than N^2, so it can be performed more quickly. The applications of these Fourier techniques require the spectra to be as devoid as possible of discontinuities, so the sharp ends of the spectra have to be tapered by a cosine function, called

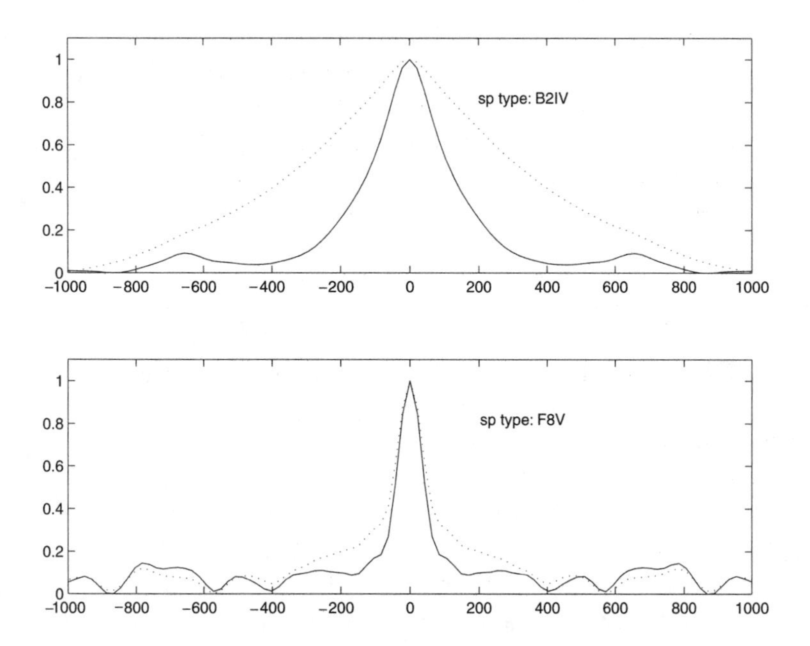

Fig. 3.8. Illustrations, via autocorrelation functions, of the effects of limiting the spectral range in a correlation. For the B2IV spectrum, inclusion of the broad H lines dramatically alters the appearance of the correlation function (dotted line) relative to that calculated from the use of the narrower He lines alone (solid line). The resultant measured velocity will be determined much less precisely when broad lines are included [see equation (3.13)]. For the F8V spectrum, the H lines are much narrower than in the B2IV spectrum, so that their effects, together with those of the broad G band of CH, are less dramatic but still significant.

apodizing with a cosine bell. Additional data bins (typically about 10% of the total data length) of value zero are added to the ends of the spectra so that the cross-correlation can be performed over a substantial range of relative velocities (± 1000 km s^{-1}) between the programme star and the template star.

Figures 3.8–3.14 illustrate varieties of spectra and CCFs for different combinations of template spectra and programme-star spectra and for double-lined spectroscopic binaries (SB2s) with similar and dissimilar components. These illustrations have been selected to demonstrate the effects of mismatching of spectra, the inclusion/exclusion of broad lines like H, and the remarkable ability of the cross-correlation process to extract the signature of a weak spectrum from a composite spectrum. Note that the cross-correlation of a template with itself, called an *autocorrelation*, results in a generally narrow CCF. The combined spectrum of a double-lined spectroscopic binary will, in general, produce a double-peaked CCF when it is cross-correlated with a template of a spectral type appropriate to that binary. Whether or not the two peaks will be well separated along the ln λ axis will depend upon the spectral resolution employed in the observation, the projected rotational velocities and the orbital velocities at the time of observation for the two stars in the binary, and the correct choices of spectral template and spectral lines within the template and programme spectrum.

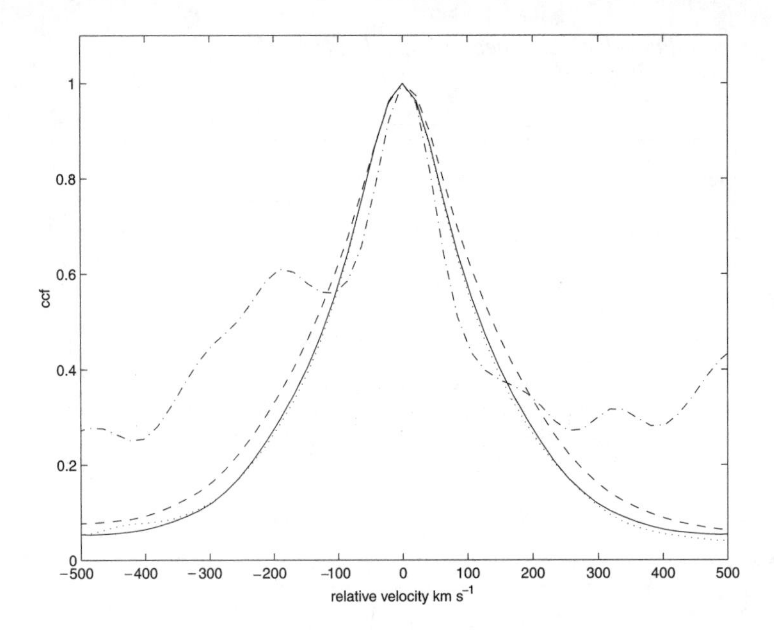

Fig. 3.9. CCFs for spectra of types O9V (dotted line), B5V (dash line), and A0V (dot-dash line) against a template spectrum of type B2V, together with the autocorrelation of the template (solid line). Note the near equality of the CCFs for the O9V and B2V spectra, whilst that for the B5V spectrum is somewhat broader, illustrating the effect of a partial mismatch of the spectra. That for the A0V spectrum is corrupted and unusable because of the substantial mismatch of the spectra.

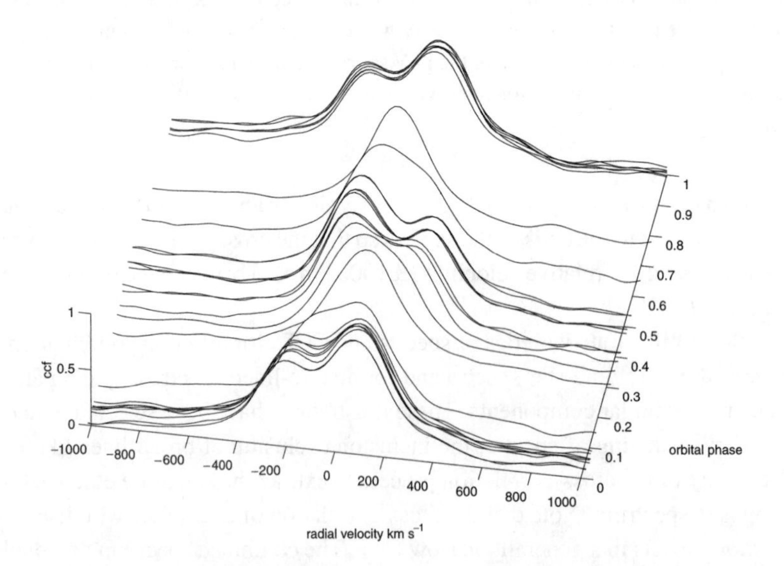

Fig. 3.10. A montage of CCFs for the O9.5V+O9.5V binary LZ Cep calculated from a set of spectra covering both quadrature phases (maximum positive velocity for the primary component is at phases 0.0 and 1.0). The selected template spectrum is for the star 10 Lac, also of spectral type O9.5V. Note that the CCFs are double-peaked at both quadratures, with clearly defined profiles, resulting in accurate determinations of radial velocities for both components. A few spectra nearer to the phases of conjunction (0.25 and 0.75 in the figure) show single-peak CCFs or strongly blended CCFs.

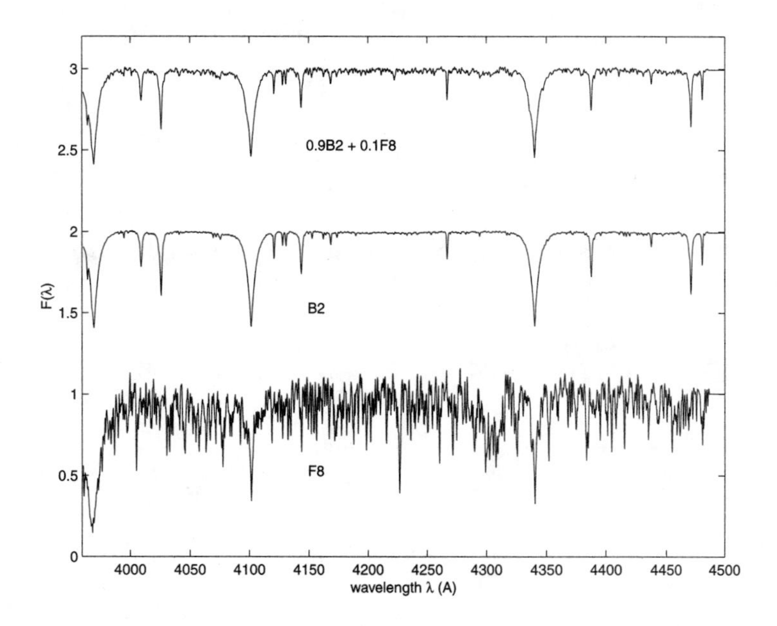

Fig. 3.11. A composite spectrum (top) for a simulated binary composed of a B2 star (middle) contributing 90% of the total flux density of the binary and an F8 star (bottom) contributing 10% of the total. The three spectra have been offset vertically for clarity. Note that the composite spectrum looks like that of the B2 star with some added noise; there is no obvious signature of the F8 star at this flux-density ratio. The B2-star spectrum has been red-shifted by 100 km s^{-1}, whilst the F8-star spectrum has been blue-shifted by 200 km s^{-1}.

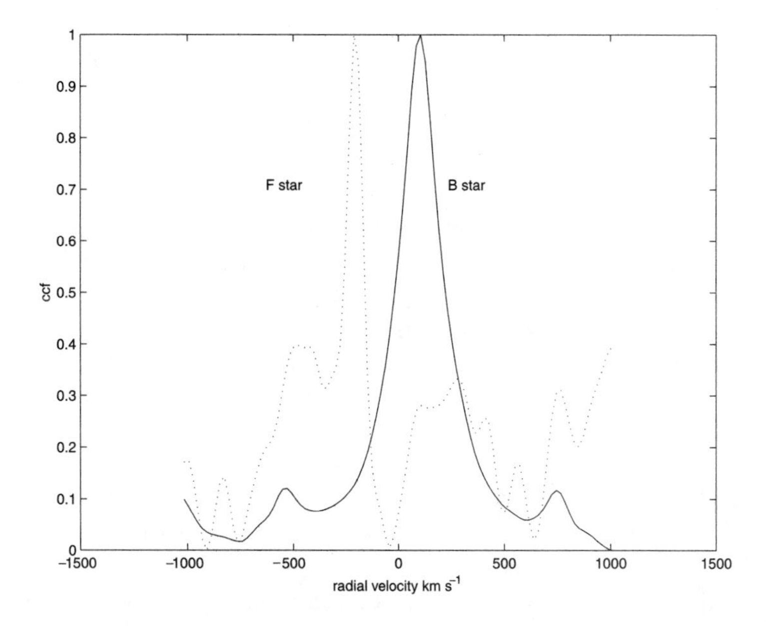

Fig. 3.12. CCFs between the composite spectrum for a B2+F8 binary and the B2 template (solid line) and the F8 template (dotted line). The velocities of the two components are recovered correctly. Note that in this simulation the effect of the B2 spectrum on the CCF generated with the F8 template is to raise the background level, but not sufficiently to distort the location and symmetric shape of the CCF peak.

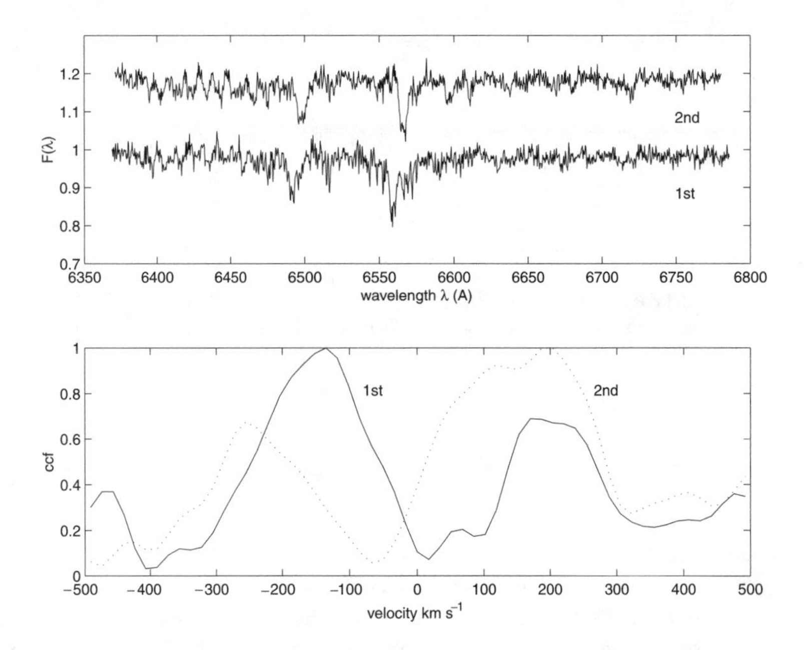

Fig. 3.13. Top: Two spectra from the mass-exchanging, eclipsing binary star V361 Lyr taken near the phases of first and second quadrature with the 4.2-m William Herschel Telescope and Cassegrain spectrograph (ISIS) and recorded on a CCD with a resolution of 0.38 Å px^{-1}. Apart from the Hα line, the presence of two sets of absorption lines in these composite spectra is not obvious to the eye. Bottom: CCFs from the foregoing two spectra relative to a G0V template spectrum clearly show two peaks corresponding to the two components of V361 Lyr seen at opposite sides of their barycentric orbits near the quadrature phases 0.25 (solid line) and 0.75 (dotted line). The wavelength region λ6480–6590 was excluded from the CCF calculation. Note that the CCF procedure readily identifies the secondary spectrum.

The question immediately arises as to which type of function will provide the best least-squares fit for these CCFs, so that the location and the *full-width at half-maximum* (FWHM) of the peak or peaks of the CCF can be determined accurately. *Gaussian profiles* offer an obvious choice, and these have been used extensively in analyses of data from binary stars. A least-squares procedure is invoked to determine the values of δ and μ in the expression

$$c(x) = c(\delta)\exp\left[\frac{-(x-\delta)^2}{2\mu^2}\right] \tag{3.12}$$

where $\delta = \ln[1 + (V/c)]$ gives the relative velocity between the programme star and the template, μ is the dispersion (or half of the FWHM of the Gaussian function), and $c(\delta) = c(x = \delta)$ is simply the scaling quantity to match the height of the peak of the CCF and normalize the profile to a maximum of unity. The half-FWHM is determined by two quantities: the intrinsic half-width τ of the spectral lines selected in the template spectrum and the observed half-width σ of the lines selected in the programme-star spectrum. Assuming that an analyst has made an excellent match between the template spectrum and the programme-star spectrum, then we can adopt the value of τ as the intrinsic half-width of the programme spectrum. Then the value

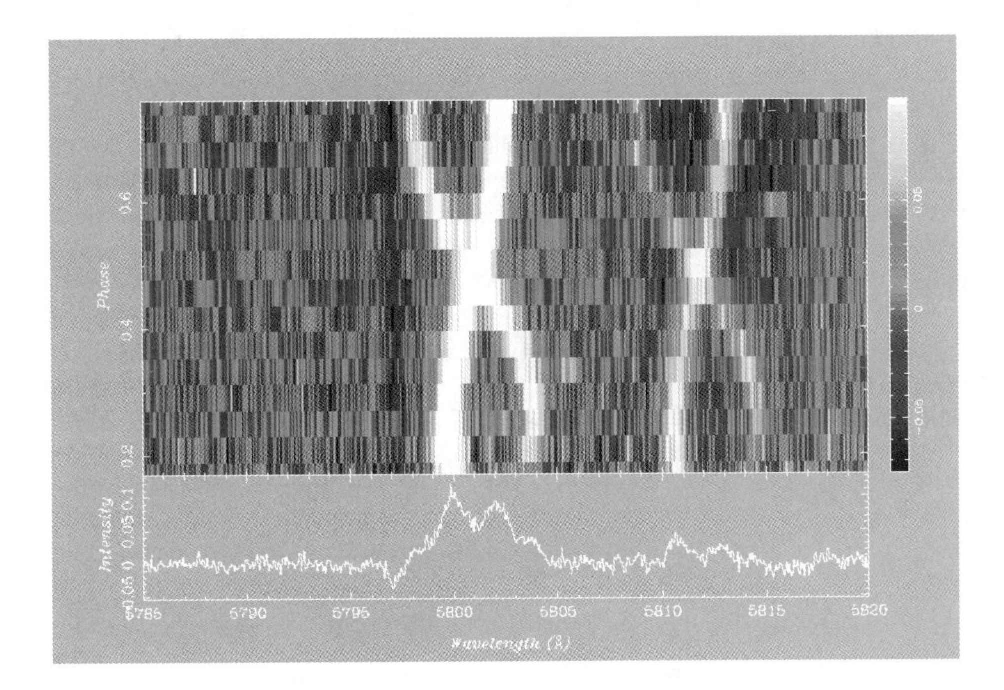

Fig. 3.14. Top: A grey-scale representation of the radial-velocity variations of the two components of the unusual spectroscopic binary system KV Vel (sdO+M) observed over most of the orbital cycle. The data were obtained with the 4-m Anglo-Australian Telescope and an échelle spectrograph, providing an effective resolution of about 0.1 Å. Bottom: The sum of all the spectra without any corrections for radial-velocity variations, showing the emission lines of C IV at λ5802 and λ5813 from the sdO star (as expected) as well as the same lines from the strongly heated facing hemisphere of the otherwise cool M star (not expected). Note that the high resolution of the spectra ensures that there are no difficulties with effects of spectral-line blending. Data from Hilditch et al. (1996).

of μ is given by the quadratic relationship (Tonry and Davis 1979)

$$\mu^2 = 2\tau^2 + \sigma^2 \tag{3.13}$$

This simple equation illustrates a point that was made earlier about the need for the spectral lines in the template and in the programme spectra all to have similar shapes and widths. If some mixture of broad and narrow lines is used, then the value of τ becomes greater, and the width of the CCF increases, thereby decreasing the accuracy with which a relative velocity can be determined. The value of σ is evidently related to the broadening of the spectral lines of the programme star beyond their intrinsic value for a slowly rotating and unperturbed star. The usual interpretation of σ is that the line broadening is due to substantial rotation of the programme star, which is to be expected in a close-binary-star system where the periods of axial rotation and orbital revolution are expected to be tidally locked, or at least very strongly linked (see Chapter 4). Accordingly, we can expect to find a quite simple relationship between measured values of μ and the *projected rotational velocity* $V_{\mathrm{rot}} \sin i$ of the programme star from these

CCFs. This expectation is borne out in practice, as demonstrated, for example, by Hill (1993), but it is important to recognize that the relationship between σ and $V_{\rm rot} \sin i$ depends also on the spectral types of the stars involved, as well as on the resolution of the spectrograph employed. The spectral lines in early-type stars, excluding the H lines, are intrinsically broader than those in late-type stars, so different relationships are required, as illustrated in Figure 3.15. The effects of the spectrograph resolution and other line-broadening processes are seen at low $V_{\rm rot} \sin i$ values, where the relationships become non-linear. Discussion of this topic of rotational velocities in binary stars will be continued in Section 5.5.9.

For stars that have $V_{\rm rot} \sin i < 50$ km s^{-1} it is often the case that a Gaussian profile will not provide a good representation of the derived digital CCF. The appropriate profile is the *Lorentzian* or *damping profile* that is found from classical line-broadening theory and from quantum theory. Figure 3.16 illustrates the differences between these Gaussian and Lorentzian

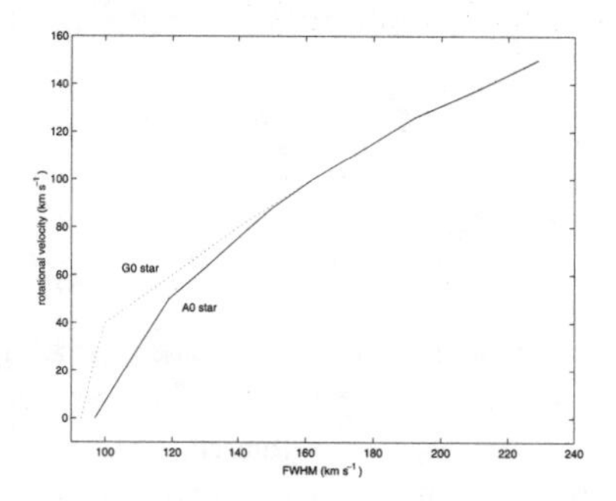

Fig. 3.15. Schematic illustration of the dependence of the measured FWHM from a CCF profile on rotational velocity. The dependence includes the resolution of the spectrograph employed and the intrinsic line widths in the stellar spectrum, narrower for late-type stars, broader for early-type stars.

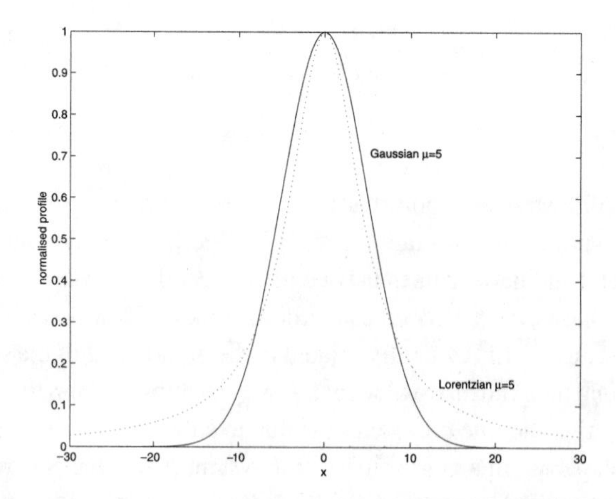

Fig. 3.16. Normalized Gaussian and Lorentzian profiles with the same dispersion of $\mu = 5$ units of the x axis.

profiles, with Lorentzians having sharper peaks than Gaussians, but with wider wings. The relevant formula is

$$c(x) = c(\delta)\,\frac{\mu}{\mu^2 + x^2} \tag{3.14}$$

where μ is half of the FWHM of the profile.

For about a decade, researchers using the cross-correlation method for radial velocities have used Gaussian profiles or, less frequently, Lorentzian profiles to fit functions to observed CCFs for binary stars and have found quite adequate fits to the data. However, there is increasing disquiet amongst these practitioners that the fitted functions do not achieve the quality of fit to the data that the data would seem to warrant, particularly when those data have better resolution and high S/N. An alternative procedure has been proposed by Hill (1993) and used by several investigators with very considerable success, particularly for binary systems composed of rapidly rotating O- and B-type stars. That procedure is to assume that the programme-star spectrum is the same as the (carefully selected) template and that it is broadened by some mechanism, such as more rapid rotation. Then we can consider the observed CCF to be a convolution of the autocorrelation of the template and the *rotational profile* discussed by Unsöld (1955) and by Gray (1992).

The rotational profile for a spherical star that is rotating rigidly at angular speed ω, with its rotation axis inclined at an angle i to the observer's line of sight, is given by the expression (Unsöld 1955; Gray 1992)

$$A(\Delta\lambda) = A(0)\frac{2(1-u)\left[1-\left(\frac{\Delta\lambda}{\Delta\lambda_{\rm max}}\right)^2\right]^{1/2} + \frac{1}{2}\pi\,u\left[1-\left(\frac{\Delta\lambda}{\Delta\lambda_{\rm max}}\right)^2\right]}{\pi\,\Delta\lambda_{\rm max}[1-(u/3)]} \tag{3.15}$$

where u is the standard linear limb-darkening coefficient (see Section 5.5.3), and the Doppler shift $\Delta\lambda = (\lambda r\omega \sin i)/c$. Here r is the perpendicular distance of a point on the projected disc of the stellar surface from its projected rotation axis, and when $r = R$, the radius of the star, $\Delta\lambda_{\rm max} = (\lambda R\omega \sin i)/c = (\lambda V_{\rm rot} \sin i)/c$. The quantity $A(0) = A(\delta\lambda = 0)$ is again used here to normalize the maximum of the profile to unity, as with the other profiles. This equation and its modifications for non-spherical stars are discussed further in Chapter 5. Figure 3.17 illustrates the form of the rotational profile for different values of the limb-darkening coefficient as a function of the dimensionless quantity $\Delta\lambda/\Delta\lambda_{\rm max}$.

An example of the convolution of this rotational profile with a narrower Lorentzian profile is given in Figure 3.18. It is clear that the resultant profile is different in shape from either Lorentzians or Gaussians. When such a rotational profile is convolved with the autocorrelation function from the selected template, it is found that the resultant digital profile matches very well the observed CCFs for binaries with rapidly rotating O- and B-type components.

The autocorrelation of the template will retain 'fine structure' relevant to the particular spectrum being studied, whilst the rotational profile broadens out the autocorrelation to that which is observed in a CCF. The analyser should experiment with different values of the input 'rotational velocity' for the rotational profile and determine which profile best fits the observed CCF for the particular programme star.

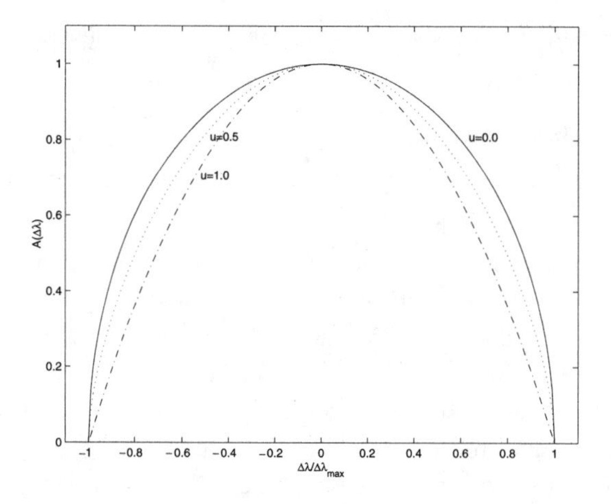

Fig. 3.17. Rotational profiles from equation (3.15) with the limb-darkening coefficients at $u = 0.0, 0.5$, and 1.0 corresponding to an apparent stellar disc that is uniformly illuminated, partially limb-darkened, and completely darkened at the limb, respectively.

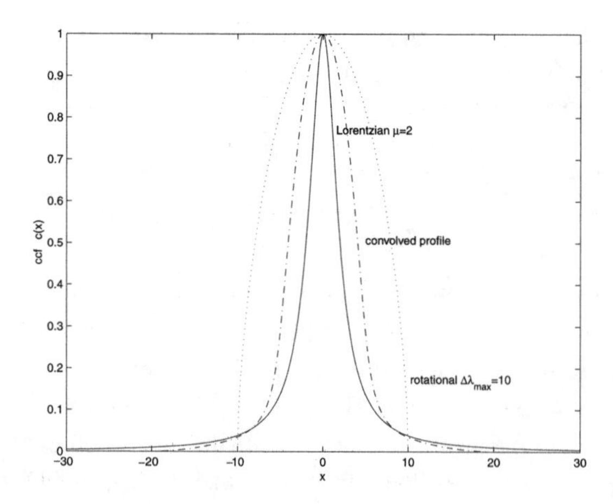

Fig. 3.18. Example profiles for a Lorentzian function (with dispersion $\mu = 2$ units in x), a rotational function (with $\Delta\lambda_{\max} = 10$ units in x and $u = 0.5$), and the resultant profile convolved from the Lorentzian and rotational functions.

There has been a substantial contribution to determination of orbits of O-type binaries by Stickland and colleagues that exploits the UV spectra in the IUE archive (see some 30 papers in *The Observatory* since 1987, most recently Stickland and Lloyd 1999). They have shown that careful application of the CCF techniques can provide accurate radial velocities because the UV spectra of such hot stars have many more spectral lines than the corresponding optical spectra.

3.2.6 Two-dimensional cross-correlation

The process of cross-correlating an observed spectrum against a well-chosen template has been shown to be most effective in determining accurate radial velocities for stars in binary systems. This is particularly the case both for SB2 systems where the two components have similar spectral types and for SB2 systems where the two stars are very dissimilar, as in the case of Algol systems. In this latter situation, Hill (1993) has shown that separate cross-correlation analyses with templates that are well suited to the dissimilar components can provide good radial velocities, even when the flux-density ratio between the two stars is quite extreme; for example, ratios as large as 10:1 are now routine, and a few Algol systems have yielded radial velocities for both components by this technique even when the flux ratio is 20:1.

Despite these successes, problems do occur with all the foregoing methods when the velocity separation between the components becomes small, typically less than about 50 km s^{-1}. It can be argued that such regions of velocity separation are not important for binaries in which the differences in velocities at the all-important quadratures are 300 km s^{-1} or greater, because the effects of spectral-line blending are too severe at small velocity separations, and the orbit solution may be compromised by systematic errors incurred from such attempted measurements. However, there are many binaries of interest in which such small velocity separations are all that are available, either because of the effects of orientation of the orbit to our line of sight and/or because the binary has a long orbital period (years) so that the true orbital velocities are small. For the types of binaries that are successfully studied by means of speckle interferometry to reveal visual orbits with periods of a few years, the ability to determine reliable radial velocities of both components becomes extremely important. Combination of the two techniques of observation can reveal the entire description of the two orbits and the masses of the component stars in such a binary system.

Appropriately, Zucker and Mazeh (1994) have introduced a generalization of the CCF procedure to include two templates simultaneously. This *two-dimensional cross-correlation* has been demonstrated by them to be extremely effective in determining the radial velocities of both stars when the velocity separation is quite small. Both one-dimensional and two-dimensional CCFs give the same answers at separations above 70–80 km s^{-1} for the typically employed spectral resolutions, but the two-dimensional technique can continue successful measurements down to separations of 20 km s^{-1} and less. The principle behind the technique is to replace the single template $t(n-x)$ in equation (3.8) by a combination of two templates with two different Doppler shifts, x_1 and x_2, and with a flux-density ratio l equal to that between the two stars in the observed binary. Then equation (3.8) becomes

$$c(x_1, x_2, l) = \frac{1}{N\sigma_s\sigma_t(x_1, x_2)} \sum_{n=1}^{N} s(n)[t_1(n-x_1) + lt_2(n-x_2)] \tag{3.16}$$

where the quantity $\sigma_t(x_1, x_2)$ is given by

$$\sigma_t^2(x_1, x_2) = \frac{1}{N} \sum_{n=1}^{N} [t_1(n-x_1) + lt_2(n-x_2)]^2 \tag{3.17}$$

The numerator in equation (3.16) can be rewritten into a form that is readily evaluated via FFT as

$$\sum_{n=1}^{N} s(n)t_1(n - x_1) + l \sum_{n=1}^{N} s(n)t_2(n - x_2) \tag{3.18}$$

The expression for $\sigma_t(x_1, x_2)$ can likewise be rewritten, for evaluation by FFT, as

$$\begin{aligned}\sigma_t^2(x_1, x_2) &= \frac{1}{N}\left[\sum_{n=1}^{N} t_1^2(n - x_1) + 2l \sum_{n=1}^{N} t_1(n - x_1)t_2(n - x_2) + l^2 \sum_{n=1}^{N} t_2^2(n - x_2)\right] \\ &= \sigma_{t_1}^2 + \frac{2l}{N} \sum_{n=1}^{N} t_1(n - x_1)t_2(n - x_2) + l^2\sigma_{t_2}^2 \end{aligned} \tag{3.19}$$

Thus the two-dimensional cross-correlation involves only three standard CCFs, one between the observed spectrum $s(n)$ and the template $t_1(n - x_1)$, a second between $s(n)$ and the template $t_2(n - x_2)$, and a third between the two templates. Zucker and Mazeh recommend that first estimates for the velocities can be pursued via standard one-dimensional CCFs, and then the additional work in their code TODCOR is restricted to a narrow range in the x_1, x_2 plane.

These three CCFs are, respectively,

$$\begin{aligned} c_1(x_1) &= \frac{1}{N\sigma_s\sigma_{t_1}} \sum_{n=1}^{N} s(n)t_1(n - x_1) \\ c_2(x_2) &= \frac{1}{N\sigma_s\sigma_{t_2}} \sum_{n=1}^{N} s(n)t_2(n - x_2) \\ c_{12}(x_2 - x_1) &= \frac{1}{N\sigma_{t_1}\sigma_{t_2}} \sum_{n=1}^{N} t_1(n - x_1)t_2(n - x_2) \end{aligned} \tag{3.20}$$

Then, for a known value of the flux-density ratio l, the two-dimensional CCF can be written

$$c(x_1, x_2, l) = \frac{c_1(x_1) + l'c_2(x_2)}{[1 + 2l'c_{12}(x_2 - x_1) + l'^2]^{1/2}} \tag{3.21}$$

where $l' = (\sigma_{t_2}/\sigma_{t_1})l$.

But if the value of l is not known with certainty, we can search for the 'best-fit' value of l that will maximize the correlation between the observed spectrum and the two selected templates. To do this, we differentiate the expression for $c(x_1, x_2, l)$ with respect to l and set it equal to zero; the resultant expression for l, as a function of the Doppler shifts (x_1, x_2), is given by

$$l_{\text{max-cc}}(x_1, x_2) = \left[\frac{\sigma_{t_1}}{\sigma_{t_2}}\right]\left[\frac{c_1(x_1)c_{12}(x_2 - x_1) - c_2(x_2)}{c_2(x_2)c_{12}(x_2 - x_1) - c_1(x_1)}\right] \tag{3.22}$$

One would expect that for a non-eclipsing binary system composed of constant, unperturbed stars, the flux-density ratio l would be constant. A first iteration through a set of data, or a best-quality subset of the data, should provide a value of $l_{\text{max-cc}}$ that could then be held fixed in a final solution. The CCF for the best-fit value of $l_{\text{max-cc}}$ is found to be, after some algebra,

$$c(x_1, x_2, l_{\text{max-cc}}) = \left[\frac{c_1^2(x_1) - 2c_1(x_1)c_2(x_2)c_{12}(x_2 - x_1) + c_2^2(x_2)}{1 - c_{12}^2(x_2 - x_1)}\right]^{1/2} \tag{3.23}$$

This two-dimensional cross-correlation procedure has been used very successfully by Zucker, Mazeh, Latham, and their research group in combined spectroscopic and interferometric studies of long-period binaries to establish complete descriptions of these systems.

3.2.7 Systematic errors in radial velocities

Systematic errors can be incurred in radial-velocity measurements from a variety of sources. The most obvious source is inadequate wavelength calibration due to a poor comparison spectrum. A poor comparison spectrum can result from incorrect integration times giving over-exposed or under-exposed emission lines, from a sparse distribution of lines through the selected wavelength range, and even from more fundamental matters like imprecise focussing, poor collimation of the source beam, and so forth. Assuming that the wavelength calibration has been performed accurately, the next possible sources of error lie in the selection of appropriate wavelength ranges for the cross-correlation process, in the correct spectral template, and in the choice of the most appropriate type of function to be used to match the derived CCF, as noted in the foregoing subsections.

Beyond these considerations, there remains one potentially difficult issue that we have not yet investigated thoroughly, namely, the *blending of the spectral lines* from the two components of a binary star. If the spectral resolution employed is not sufficient to separate the lines from the two components, then the measured positions of those lines will be systematically *dragged* from their true positions towards the systemic velocity of the binary. The result will be that the determinations of the semiamplitudes (K_1, K_2) of the two velocity curves will be systematically smaller than their true values. Because the evaluation of the two masses includes the term $(K_1 + K_2)^2 K_{1,2}$, it is obvious that serious under-estimates of the true masses will result.

Much has been written on the subject of the blending of pairs of spectral lines in the composite SB2 spectra. The first detailed investigation of pair blending was presented by Petrie, Andrews, and Scarfe (1967), where they demonstrated that the use of the then-new oscilloscope devices for accurate and more impersonal determination of spectral-line wavelengths in SB2s could result in significant systematic errors. At the spectral dispersion typically used in such studies at that time (15–30 Å mm^{-1}), corresponding to a resolution of 0.3–0.8 Å px^{-1}, lines were measured as being closer together than they actually were. For velocity differences between two stellar components of less than 200 km s^{-1}, systematic errors of at least 10% might result amongst the narrower He lines in B-type spectra, and significantly larger errors ($\approx$50%) might be incurred during measurements of the much broader and more diffuse H lines. Petrie et al. (1967) also noted that the more subjective method utilized in a visual-comparator/long-screw-micrometer system sometimes seemed to compensate or even over-compensate for the effects of blending, at least on the narrow He lines, but it was generally accepted by that time that velocity separations derived from H lines were always smaller than those from other lines and therefore were not trustworthy.

Pair blending was studied further on a theoretical basis by Tatum (1968), as well as by Batten and Fletcher (1971), particularly for close visual binaries observed spectroscopically near periastron. Hilditch (1973) studied such blending effects on all the He I, Si II, Mg II, and C II lines in the SB2, 57 Cygni, and demonstrated that the *diffuse-series* He I lines were more affected than the narrower *singlet-series* lines or the metallic lines (Si II, Mg II, C II). In addition, oppositely displaced pairs and triplets could result in serious blending problems in certain systems. All of those factors were placed on a more firm empirical foundation in a series of papers by Andersen and associates, a summary of which can be found in Andersen, Clausen, and Nordström (1980). Their conclusions were that the masses of the component stars were, on average, under-estimated by 10% when only *diffuse* He I lines were used, and by 40% when H lines were used, over the spectral range O8–A8.

Those systematic errors were found during studies of binaries with circular or mildly eccentric orbits, where the fraction of the orbital cycle during which velocity separations were less than 200 km s^{-1} was small, typically around ± 0.1 about each conjunction. The use of only narrow lines will ensure that systematic errors are not incurred, particularly at the crucial quadrature phases. However, in systems with highly eccentric orbits, velocity separations may remain quite small for very large fractions of an orbital cycle, thereby ensuring that the systematic effects of spectral-line blending will operate maximally even when care is taken to use only narrow lines.

The modern techniques employed in determining velocities by means of cross-correlation, as discussed in earlier sections, certainly can significantly mitigate the potential problems of pair blending. But such techniques are not perfect, and it is necessary for investigators to check that such systematic errors are absent from their results. With present-day computing power and digital spectra, it is now an easy task to perform all sorts of *spectral arithmetic* whereby we can add, subtract, multiply, and divide different proportions of the spectra of different stars. Accordingly, we can simulate the observed systems with sensible combinations of spectra and examine whether or not our CCF techniques can reproduce the velocities that are input to the simulation. But the best solution to all of these problems of line blending is to use high spectral resolution wherever possible. For example, échelle spectroscopy at a resolution of 0.1 Å px^{-1} is now quite readily attainable, thanks to the efficiency of modern CCD detectors.

One reason for mentioning the historical context of studies of pair blending is to warn against any simple comparison of sets of orbital elements for a given binary determined at different epochs. Such comparisons have been quite common for binary systems that exhibit secular changes due to various perturbations, as we discuss in the next chapter. As always, the investigator should be wary and should ensure that like is being compared with like.

There are additional possible sources of serious systematic errors affecting the measured velocities of the components of binary systems, all of which are consequences of the *proximity* of the two stars. Such stars may perhaps become non-spherical in shape, so that surface gravity will become a function of position across the stellar surface, leading to the phenomena of *gravity darkening* in each stellar atmosphere and the *ellipticity effect* on the observed light curve. That would also mean that the averaged centre of the integrated flux leaving each star might not be at the centre of gravity. The stars would mutually irradiate their facing hemispheres, resulting in higher temperatures in those parts of the stellar atmospheres than would be the case otherwise (the so-called *reflection effect*). Again, the integrated flux leaving each star will have a centre that may be displaced from the centre of gravity. The measured velocities for the stars may then have a component of the *axial rotation* included into the orbital motion, resulting in systematic errors. Both of these sources of systematic error can be serious in certain types of binaries. Calculation of corrections for these effects requires a full discussion of the shapes of the stars in *close* binary systems (see Chapter 4), as well as the properties of stellar atmospheres and their illumination, which are required to calculate the variations in total light from a binary as a function of orbital phase (see Chapter 5). An extreme example of the serious nature of such phenomena will be considered in Section 3.2.10 for the binary system KV Vel, composed of a hot sdO star and a cool M dwarf.

For the very strongly distorted stars found in the *contact* binaries, there are additional techniques for radial-velocity determination that prove to be more effective. Specifically, the *broadening-function* analysis of Rucinski (1992) and the *synthetic-line-profiles* analysis of CCFs from Hill (1993) both require discussion of the synthesis of light curves before their application. These are discussed more appropriately in Chapter 5 along with rotational velocities and non-Keplerian corrections. It is also possible for measured velocities to be systematically affected by the presence of gas streams and discs in interacting binaries and by the depth-of-formation effect in wind outflows from certain types of stars, such as Wolf-Rayet stars and O-type supergiants, amongst others. These latter issues are discussed in Chapter 7.

3.2.8 Tomographic separation of spectra

A set of spectra for a binary system normally will comprise about $m \sim 20$–30 observations covering a substantial range of the orbital cycle. Each observation will show a spectrum of the primary component at some projected orbital velocity, combined with a spectrum of the secondary, generally at a different velocity. If the individual velocities from all the spectra have been well determined from CCFs by means of the procedures discussed in the preceding sections, then it may be possible to Doppler-shift all the spectra to a rest frame and then separate the mean spectrum of the primary and the mean spectrum of the secondary from the combined total spectrum. That idea was the basis for applying the procedure of *tomography* to the spectra of binary stars. The technique was first used in this context of separation of spectra from the two components of a binary by Bagnuolo and Gies (1991), though the principles of tomography have been used in medical physics for many years (e.g., Barrett and Swindell 1981).

The resultant mean spectra for the two stars should then be of high quality, because they would be additions of all the m spectra in the sample, and the S/N ratio should be of the order $m^{1/2}$ greater than with a single observation. In addition, one could consider the procedure of radial-velocity analysis to be a more iterative one, with preliminary templates adopted from consideration of the spectral types of the stars in the binary, together with test runs to find the best separated peaks in the CCFs, leading to a first set of radial velocities for each component. These velocities could then be used in a tomography code to extract the best mean spectrum for each star. These mean spectra would serve as the templates in the next iteration through the original set of spectra to achieve more accurate radial velocities and more refined mean spectra via tomographic separation.

Tomographic separation of binary-star spectra is an iterative technique that relies on foreknowledge of the individual radial velocities of both stars at each time of observation and the relative brightnesses of the two stars. The starting point for the iteration may be to take an initial estimate of the spectrum of the dominant primary component, sp_i, and a flat continuum for the secondary, ss_i, or simply flat continua for both stars. Here, the subscript i is the ith pixel in each spectrum. These approximations to the spectra of the two components are then refined by applying corrections that are based upon the differences between the observed and the calculated combined spectra of the binary. These corrections are given by the equations:

$$
\begin{aligned}
\Delta sp_i &= \frac{\delta}{m}\sum_{k=1}^{m}\left[sb_{k,i} - \left(sp_i + ss_{i+s_k}\right)\right] \\
\Delta ss_i &= \frac{l\delta}{m}\sum_{k=1}^{m}\left[sb_{k,i-s_k} - \left(sp_{i-s_k} + ss_i\right)\right]
\end{aligned}
\tag{3.24}
$$

where Δsp_i and Δss_i are the corrections to the ith pixel of the primary and secondary spectra, respectively, δ is a number (0.3–0.8) that improves the speed of convergence of the iteration, m is the total number of combined spectra, $sb_{k,i}$ is the ith pixel of the kth combined spectrum shifted to the rest frame of the primary star, sp_i and ss_i are the ith pixels of the current approximations to the primary and secondary spectra, respectively, l is the flux-density ratio (secondary/primary), and s_k is the shift in radial velocity of the secondary spectrum relative to the primary spectrum in the kth combined spectrum.

For binaries composed of similar stars, the tomography technique gives excellent results, recovering both stellar spectra within about 50 iterations. The improved S/N in both extracted spectra ensures, at least, that improved spectral classifications can be made, but more generally, these extracted spectra can be analysed to provide spectral-line strengths and even detailed line profiles that can be compared directly with the predictions from model stellar atmospheres. Thus, these procedures open the way for determining stellar effective temperatures and surface gravities that are used later in analyses of eclipsing-binary light curves. These developments have not yet been exploited fully, though they remain very promising, particularly for binaries composed of O+O and O+B stars (e.g., Bagnuolo, Gies, and Wiggs 1992a; Bagnuolo et al. 1994; Hill et al. 1994; Thaller et al. 1995; Penny, Gies, and Bagnuolo 1997).

For binaries with dissimilar components, such as classical Algol types with a B- or A- type primary and a G- or K-type subgiant secondary, the situation is rather more complex. Here the flux-density ratio l is a strong function of spectral wavelength or pixel number i, because of the large difference in effective temperature between two such stars. There may also be a dependence of l on the orbital phase, or k, if the system is an eclipsing binary and/or has strongly non-spherical stars (the ellipticity effect, see Chapter 5). So $l \equiv l_{k,i}$. Accordingly, equations (3.24) need to be modified (Maxted 1994; Maxted et al. 1994a,b) to take account of the wavelength dependence of l:

$$
\begin{aligned}
\Delta sp_i &= \frac{\delta\overline{lp_{k,i}}}{m}\sum_{k=1}^{m}\left[sp_{k,i} - \left(lp_{k,i}sp_i + ls_{k,i}ss_{i+s_k}\right)\right] \\
\Delta ss_i &= \frac{\delta\overline{ls_{k,i}}}{m}\sum_{k=1}^{m}\left[sp_{k,i-s_k} - \left(lp_{k,i}sp_{i-s_k} + ls_{k,i}ss_i\right)\right]
\end{aligned}
\tag{3.25}
$$

where $\overline{lp_{k,i}} = 1/(1 + l_{k,i})$ and $\overline{ls_{k,i}} = l_{k,i}/(1 + l_{k,i})$ are the fractional mean flux densities of the primary and secondary stars averaged over the wavelength region that is being analysed.

The foregoing procedure certainly extracts a high-quality spectrum for the dominant component of the binary, but the results for the secondary component are much less satisfactory, as one would expect for flux-density ratios of 10:1 or more. Again, these procedures are really in their infancy as compared with, say, the development of techniques for cross-correlation. Yet much can be learned from their application, particularly with regard to extraction of spectra from barely visible secondary components in the mass-exchanging Algol systems. It is these stars that may well display evidence of extensive mass loss and may reveal anomalous chemical abundances on their surfaces.

For the technique of tomography to be really successful, we need spectra of high S/N that are well distributed around the orbital cycle, so that many 'viewing angles' of the resultant combined spectrum will be available for analysis. The same remark applies to the procedure described by Simon and Sturm (1994) as the *disentangling* of binary-star spectra. Here, the analyser attempts to find the optimum set of radial velocities for the two components of a binary from a set of spectra, together with the extracted mean spectrum of each star – that is, cross-correlation and tomography being operated simultaneously. Because such observed spectra are so similar and may be close to being linear combinations of others, the set of spectra is singular, and the technique of *singular-value decomposition* must be employed in order to achieve a solution to the problem (e.g., Press et al. 1992). Sturm and Simon (1994) and Simon, Sturm, and Fiedler (1994) applied the procedure to sets of spectra for the O+O binaries DH Cep and Y Cyg and found that they needed to start their solutions from preliminary estimates of the radial velocities determined by standard cross-correlation procedures, as well as a flux-density ratio that was known accurately from photometry and the ratios of absorption-line strengths. For the technique of disentangling to be effective, it seems that the demands on the set of original spectra must be quite stringent. The S/N ratios for the spectra should all be very similar and of good quality, and all phases of the orbital cycle need to be sampled equally well. The severe constraints concerning allocation of observing time for many of the world's telescopes are such that observers are unlikely to be able to reach the second of those objectives. So the sequential use, and possible iteration, of the techniques of cross-correlation and tomography will continue for some time to be the most effective form of analysis. Hadrava (1995) has presented an alternative technique based on Fourier transforms for separating the two spectra from sets of composite spectra of binary systems.

3.2.9 Orbital periods from radial velocities

The general problem of finding periodicities amongst sets of data has troubled the physical sciences for a long time. Several techniques have been developed, and they essentially fall into two categories: those working in the *time domain* and those working in the *frequency domain*. The time-domain procedures usually are non-parametric methods, that is, they do not prescribe the form of the periodic variation that is being searched for within a set of data. They involve exploring a range of possible periods, first by placing all the data points into *phase* order at each selected period and then looking for a minimum in the dispersion of the data points about some mean curve. These are the *phase-dispersion-minimization* (PDM) techniques, which can be particularly effective for variations that display abrupt changes, such as those seen in the light curves of binary stars around the eclipses. The two most common methods are the PDM technique itself, presented by Stellingwerf (1978), and the string-length method, discussed by Dworetsky (1983); both of which are incorporated into the code PERIOD by Dhillon and Privett (1995). For smoother variations, like those seen in radial-velocity curves for binary stars, it is often better to use the frequency-domain techniques, as exemplified by taking Fourier transforms of the data string and examining the *power spectrum* or *periodogram* of the data, as explained in *Numerical Recipes* (Press et al. 1992).

For spectroscopic binaries, sets of radial-velocity data usually are acquired via either of two modes of observation, namely, by investigation of photometrically variable binaries whose orbital periods are already well known from photometry (see Chapter 5), or as part of a general survey of some stars where the fraction of binaries in the sample is being sought.

In the former case, the known *ephemeris*, that is, the orbital period coupled with a time of primary-eclipse, for example, drives the planning of the observational programme. Typically, a simple linear ephemeris of the form

$$T_{\text{pr.min}} = T_0 + n_{\text{cyc}} P \tag{3.26}$$

is used, where successive times of primary-eclipse minima, $T_{\text{pr.min}}$, are calculated from a well-defined eclipse minimum T_0 for integer values of the number of orbital cycles n_{cyc} with orbital period P. Clearly, successive times of other orbital phases also can be calculated with equation (3.26) by using non-integer values of n_{cyc}. The expected times of maximum and minimum radial velocities in the orbital cycle of the binary (the *quadrature* phases) can be calculated in advance, and the observations can be concentrated around those phase intervals. Other orbital phases can be studied as well to investigate evidence for mass transfer between components or eclipse effects.

In the latter case, the radial-velocity data are acquired in a more random fashion, as dictated by the allocated telescope time, the weather, and the number of stars being investigated. There are obvious procedures that need to be incorporated into such a programme, such as ensuring that several observations are secured within one night on each star in order to find out if the orbital period is less than 1 day. But the general result from such surveys is that the data set for each star is a sample of radial velocities secured over unequally spaced intervals of time that may span anything from several nights to several years. Such a data set can prove to be very difficult to interpret: There is no certainty that any maxima and minima have been observed, the distribution of observations around the orbital cycle may not be optimum, and the fact that astronomical observations are generally restricted to the night-time fraction of the 24-hour day ensures that many possible spurious periodicities may be found. However, a saving grace for spectroscopic binaries is that radial-velocity curves generally are reasonable approximations to cosine curves (see Chapter 2) and therefore can readily be represented by sine and cosine series. Accordingly, one of the most effective methods for determination of the orbital periods of spectroscopic binaries from unequally sampled data is to use the *periodogram* analysis based on Fourier techniques.

The list of papers and monographs on the subject of period-finding techniques is extensive, and it is not relevant here to discuss all the procedures that have been developed. We shall briefly consider, as an example of current practice in astronomy, the *Lomb-Scargle periodogram* (Lomb 1976; Scargle 1982) as presented by Horne and Baliunas (1986) and Press and Rybicki (1989), which is particularly effective for determining periodicities in sets of radial-velocity data that are obtained over unequally spaced intervals of time.

For a set of radial velocities V_i obtained at times t_i, with $i = 1, \ldots, N_{\text{tot}}$, we compute the *mean* and the *variance* of the data set, or the *signal*, by, respectively,

$$\overline{V} \equiv \frac{1}{N_{\text{tot}}} \sum_{i=1}^{N_{\text{tot}}} V_i; \qquad \sigma^2 \equiv \frac{1}{N_{\text{tot}} - 1} \sum_{i=1}^{N_{\text{tot}}} (V_i - \overline{V})^2 \tag{3.27}$$

For a range of frequencies f, we also compute a time offset τ for each value of the angular frequency, $\omega \equiv 2\pi f$, by the equation

$$\tan(2\omega\tau) = \frac{\sum_{j=1}^{N_{\text{tot}}} \sin 2\omega t_j}{\sum_{j=1}^{N_{\text{tot}}} \cos 2\omega t_j} \tag{3.28}$$

Then the Lomb-Scargle *normalized periodogram* provides a measure of the spectral power in the signal as a function of ω and is defined by

$$P(\omega) \equiv \frac{1}{2\sigma^2} \left\{ \frac{\left[\sum_j (V_j - \overline{V}) \cos \omega(t_j - \tau)\right]^2}{\sum_j \cos^2 \omega(t_j - \tau)} + \frac{\left[\sum_j (V_j - \overline{V}) \sin \omega(t_j - \tau)\right]^2}{\sum_j \sin^2 \omega(t_j - \tau)} \right\} \tag{3.29}$$

Note that the power $P(\omega)$ is a positive quantity, because only squared terms are included. Note also that the periodogram is normalized by the variance of the data that are used to calculate equation (3.29) not by the uncertainty in an individual observation. This correct normalization ensures that the power $P(\omega)$ provides a measure of the probability that a peak at some frequency ω will be real. Horne and Baliunas (1986) discuss this issue at some length and provide examples of how effective the Lomb-Scargle periodogram can be in detecting a period within a signal even when the total number of observations is as low as 12.

It is obvious that the range of frequencies over which a search for any periodicity in the signal is made should be limited to sensible values. The *lowest* frequency (longest period) that can be detected will be equal to the reciprocal of the length of the whole data string, namely, $f_L = 1/(t_{N_{\text{tot}}} - t_1)$. The *highest* frequency, for equally spaced data, is formally given by the *Nyquist frequency*, $f_N = 1/(2\Delta t)$, where Δt is the *sampling interval* in the data string. For unequally spaced data, it is common practice to estimate a pseudo-Nyquist frequency from the minimum value of Δt. A little careful thinking by the analyst about the range of frequencies to be searched, $f_L < f < f_N$, will save a lot of computing time.

Chapter 2 of Gray's text *The Observation and Analysis of Stellar Photospheres* (1992) provides a very helpful discussion of the properties of Fourier transforms and the transforms of simple functions. For our immediate purposes, note that the Fourier transform of a cosine/sine function of frequency f is an even/odd pair of δ functions at frequencies $\pm f$. The radial-velocity curve for a binary orbit is an approximate cosine curve, so that the Fourier transform of data representing such a curve will be a pair of δ functions. The power $P(\omega)$ is the square of the Fourier transform of the data, so that we can expect a pure cosine curve to reveal a periodogram that has one δ-function peak at $\pm\omega = \omega_{\rm true} = 2\pi f_{\rm true}$. In reality, all observational data have associated *noise* from observational errors, and noise is signal with a wide range of frequencies. If low-level noise is present in a signal that has a true frequency $\omega_{\rm true}$, then the resultant periodogram displays low-level 'spikes' at a wide range of frequencies, and the δ function from the signal is broadened. Increase the noise, and it will eventually swamp the signal. If there are multiple frequencies present in the signal, then multiple broadened δ functions will occur in the periodogram. In addition, the times of observation will introduce further peaks in the periodogram that will depend on the details of the sampling of the signal. If the sampling is performed at equal intervals of time, then the contamination of the pure data signal will be at its most severe; if the sampling is uneven, then the effects will be reduced, but never eliminated.

The question then arises as to how the analyser can distinguish between the peaks in the periodogram caused by the purely periodic signal and those caused by the sampling. The first part of the answer is to compute another periodogram, via equation (3.29), but with the data values $V_j - \overline{V}$ all set to unity and the variance set to unity. Then we obtain the periodogram called the *data window* that simply depends upon the manner in which the sampling of the signal was performed and does not depend upon the signal itself. The data window will reveal a sequence of peaks at frequencies ω_w. The second part of the answer is then straightfoward: Compare the two periodograms, that of the signal and that of the window, either by a formal cross-correlation or simply by comparing by eye the two plotted periodograms. The signal periodogram will display peaks at $\pm\omega_{si}$ $(i = 1, \ldots, n)$ corresponding to the n periodicities in the signal itself, together with *sidelobes*, or *aliases*, at $\pm(\omega_{si} \pm \omega_{wj})$, where ω_{wj} $(j = 1, \ldots, m)$ are the frequencies of the m peaks in the data window.

As a useful example, Figures 3.19 and 3.20 illustrate these aspects of periodogram analyses with real radial-velocity data for the star HR1165 taken from Jarad, Hilditch, and Skillen (1989). Note that the most typical sidelobe encountered in astronomical observations is strongly evident in this example, that of the sidereal day. The foregoing equation relating frequencies in the signal and window functions explains the most commonly known relationship in astronomy between a true period P_t and spurious periods P_s, namely,

$$\frac{1}{P_s} = \frac{1}{P_t} \pm 1.0027 \tag{3.30}$$

where the constant is the reciprocal of the sidereal day expressed in solar days.

Once a period for a set of radial velocities has been determined, it is clearly necessary to ascertain how well a function with that period fits the original data. A simple procedure is to fold the data, by means of a linear ephemeris with the determined period, and see if the data do represent some clearly defined function like a sine or cosine. If that is the case, then the residuals from such a sine-wave fit to the data should be examined for further evidence of periodicities. This procedure is described as *pre-whitening* the data, because it is removing a sine-wave variation and may lead to the revised data being simple noise with a wide range of frequencies, or *white noise*. Such proved to be the case for HR1165 (Figure 3.21), and the resultant semiamplitude of the velocity curve, assuming zero eccentricity, was $K_1 = 15\,\mathrm{km\,s^{-1}}$.

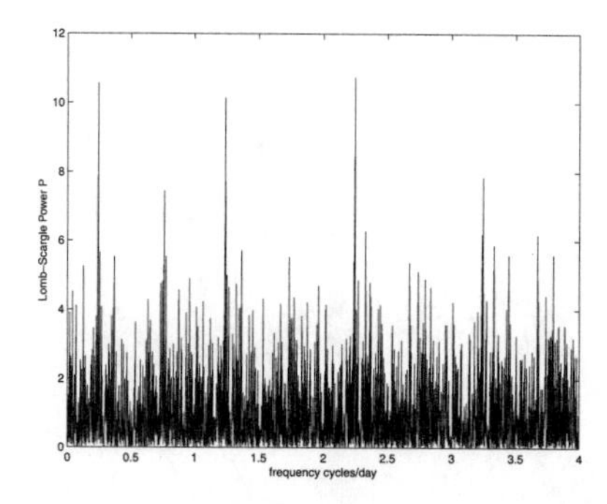

Fig. 3.19. The Lomb-Scargle periodogram (for frequencies between 0.0 and 4.0 cycles per day) of radial velocities for the star HR1165 reported by Jarad et al. (1989) and calculated with the code PERIOD from Dhillon and Privett (1995). Note that several large peaks occur, the three strongest being separated by one cycle per day. The peak at the frequency of 0.24184 cycle per day gives the true period, and the others are aliases caused by the observational sampling, as confirmed by the periodogram of the data window (Figure 3.20), for when the data are pre-whitened by this frequency, the periodogram is reduced to noise only (Figure 3.21).

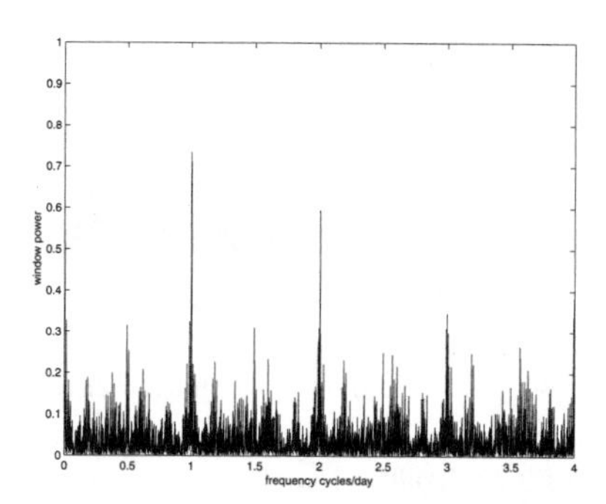

Fig. 3.20. The periodogram of the data window for the observations of HR1165 calculated with PERIOD, as noted in Figure 3.19. Note the obvious regularity of the peaks spaced at intervals of 0.5 cycle per day and the consequent simple explanation of all the peaks in Figure 3.19 as the peak and the sidelobes of the true signal frequency.

With a period of 4.135 days, it is reasonable to conclude that such a variation in radial velocities for this Be star is due to orbital motion about a low-mass companion being viewed at some low inclination. But this evidence is not conclusive, and one must be careful about assigning any observed radial-velocity variation to orbital motion; a low-amplitude variation could be due to modes of oscillation on the surface of the star.

A second example of the power of the periodogram technique is seen in the radial-velocity data for the known eclipsing binary star DM Per obtained by Hilditch et al. (1986) and Hilditch, Hill, and Khalesseh (1992). That system was known to have an orbital period of 2.7277 days from photometric observations. But when the radial-velocity data from the primary component

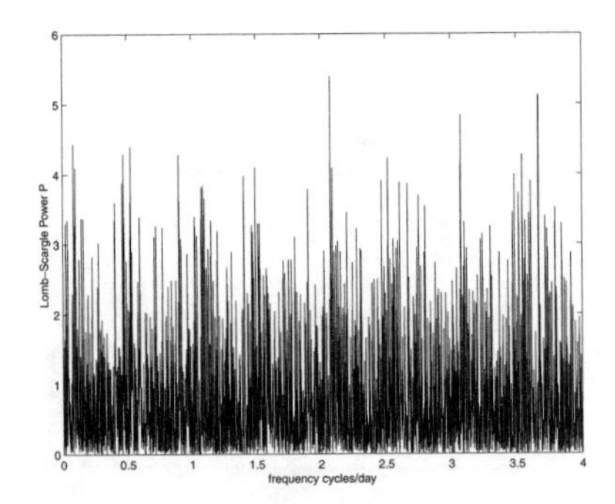

Fig. 3.21. The periodogram (from PERIOD) of the original data on HR1165 after it has been pre-whitened by a sine curve with the determined period of 4.1349 days, showing that there is now little evidence of any other significant periodicity in these residuals.

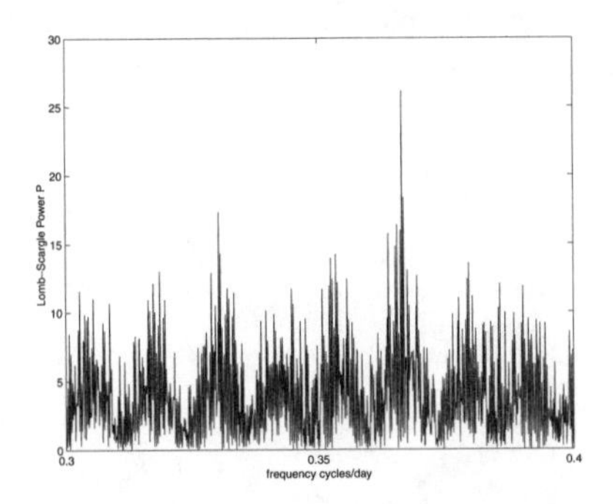

Fig. 3.22. The Lomb-Scargle periodogram (from PERIOD) of the radial velocities of the primary component of DM Per for the frequency range that encompasses the known photometric period of this eclipsing binary. The photometric period of 2.7277 days ($f = 1/2.7277 = 0.3666$ cycle/day) is unambiguously recovered.

were plotted against the orbital phase on the basis of the well-defined photometric ephemeris, it was found that the radial-velocity curve displayed a scatter of data points that was about six times greater than what would have been expected on the basis of the quality of the spectrograms employed. In addition, it was clear that data secured within one or two orbital cycles of the eclipsing-binary period were much more consistent. The possibility that the eclipsing binary might be in orbit about a third body had to be considered, and a periodogram analysis of the data was conducted. Figure 3.22 illustrates the periodogram of the data around the frequency range that includes the known eclipsing-binary period, and Figure 3.23 shows the corresponding data window. The 2.7277-day period is unambiguously recovered, and is not contaminated by any additional sidelobes from the data window. When the short-period orbital motion is subtracted from the original data, which cover a time interval of nearly 7000 days, the periodogram at low frequencies (Figure 3.24) reveals two closely spaced peaks around 100 days. One peak is the alias of the other, caused by the first sidelobe in the transform of the data window, as illustrated

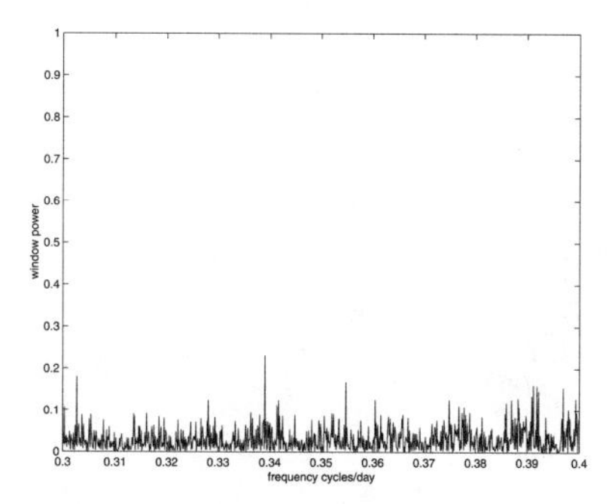

Fig. 3.23. The data window (from PERIOD) of the times of observation of radial velocities of the primary component of DM Per for the frequency range that encompasses the known photometric period of this eclipsing binary. Note that the data window does not contaminate the peak corresponding to the short-period orbit.

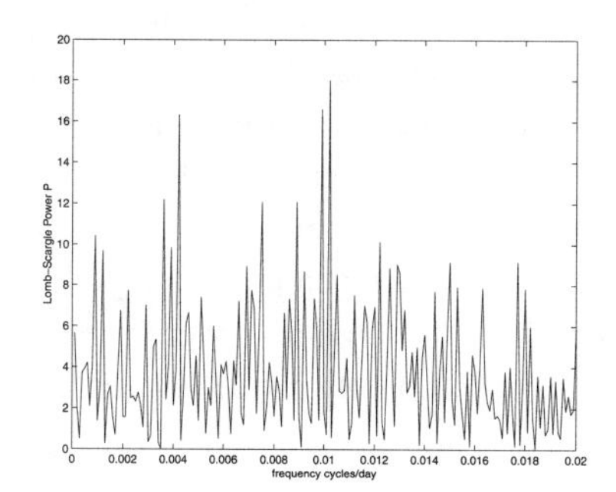

Fig. 3.24. The Lomb-Scargle periodogram (from PERIOD) of the radial velocities of the primary component of DM Per after the short-period orbital motion has been subtracted from the data. The two peaks near $f = 0.01$ cycle/day are clearly the strongest, and one is the alias of the other. Subsequent orbital analyses showed that the higher-frequency peak at $f = 0.0102$ cycle/day is the preferred solution.

in Figure 3.25. Subsequent iterations of the solutions by Hilditch et al. (1992) showed that the 98-day period was the correct value for the orbit of the eclipsing binary about the third body.

3.2.10 Spectroscopic orbital elements

In the preceding subsections we have considered the details of observation and reduction procedures for securing stellar spectra, and with particular reference to binary stars we have investigated the most effective methods to determine radial velocities. The periodogram technique for establishing preliminary values for orbital periods from radial-velocity data has been illustrated, and we have reached the final goal of this section, namely, to define the orbital elements for a spectroscopic binary from a set of such velocities.

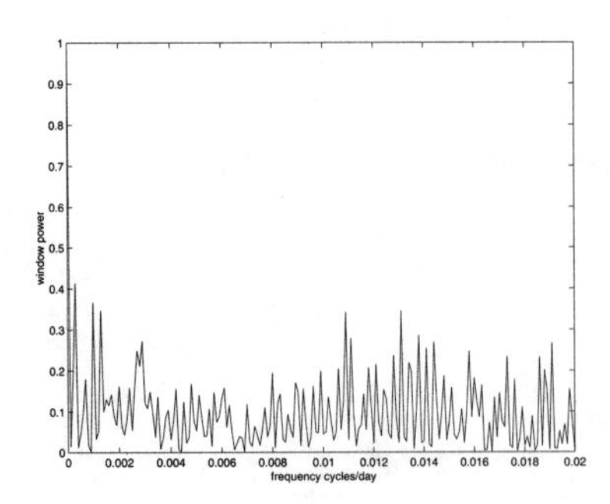

Fig. 3.25. The data window (from PERIOD) of the times of observation of radial velocities of the primary component of DM Per for the frequency range 0.0–0.02 cycle/day showing only one significant peak corresponding to the length of the data string, 6989 days, or $f = 0.00014$ cycle/day.

The earliest procedures developed for such work were predominantly graphical, and the method of Lehmann-Filhés (1894) is well described in Aitken's text *The Binary Stars* (1935, ch. 6), together with summary discussions of other graphical methods; see also Petrie (1962). Although those methods are now of historical interest only, it is recommended that one example of such a method be worked out, purely as an enjoyable exercise in manual graph plotting and geometry, particularly so when some good data are employed and when the analyser is not required to repeat the effort too often! Such methods were used to provide preliminary values for the orbital elements that were subsequently revised by means of an iterative least-squares solution for a set of equations to yield the final values for those elements. In present-day practice, it is sufficient to make simple estimates of the orbital elements visually from a graph plot of velocities versus orbital phase, utilizing, where necessary, some example radial-velocity curves like those given in Figures 2.6–2.10 to estimate values for e and ω. These estimates can then be entered as starting values into an iterative code that will provide the final solution for that set of data. The following discussion presents two methods for iterative solutions that differ only in their use of alternative reference times.

The n observations of radial velocities V at times t are to be used to determine the spectroscopic orbital elements $(K, e, \omega, T, P, \gamma)$ for a particular binary system. The relevant equation, from Chapter 2, is

$$V = K[\cos(\theta + \omega) + e \cos \omega] + \gamma \tag{3.31}$$

and we can write, schematically, that $V = F(K, e, \omega, T, P, \gamma)$, because T and P enter the equation for V via the true anomaly θ. The values for these orbital elements that best represent the n observations of (V, t) are found by pursuing an iterative linear least-squares solution based on the foregoing relationship, a method known as *differential corrections*, for obvious reasons to be outlined next.

If a small change ΔV in V is produced by small changes ΔK, Δe, and so forth, in the orbital elements, or our estimates of those elements, then we can expand equation (3.31) as a first-order Taylor series about a preliminary solution and write, to first order,

$$\Delta V = \frac{\partial F}{\partial K}\Delta K + \frac{\partial F}{\partial e}\Delta e + \frac{\partial F}{\partial \omega}\Delta\omega + \frac{\partial F}{\partial T}\Delta T + \frac{\partial F}{\partial P}\Delta P + \frac{\partial F}{\partial \gamma}\Delta\gamma \tag{3.32}$$

Because the function F is analytic, equation (3.31), the partial derivatives in equation (3.32) can be derived to yield the following result, first shown by Lehmann-Filhés (1908):

$$\begin{aligned}\Delta V = {} & [\cos(\theta+\omega) + e\cos\omega]\Delta K + K\left[\cos\omega - \frac{\sin(\theta+\omega)\sin\theta\,(2+e\cos\theta)}{(1-e^2)}\right]\Delta e \\ & - K[\sin(\theta+\omega) + e\sin\omega]\Delta\omega + \left[\sin(\theta+\omega)(1+e\cos\theta)^2\frac{2\pi K}{P(1-e^2)^{3/2}}\right]\Delta T \\ & + \left[\sin(\theta+\omega)(1+e\cos\theta)^2\frac{2\pi K(t-T)}{P^2(1-e^2)^{3/2}}\right]\Delta P + \Delta\gamma \end{aligned} \tag{3.33}$$

In this equation, all the coefficients of the differential-correction terms ($\Delta K, \Delta e, \ldots$) can be calculated from the preliminary estimates for the orbital elements. We can identify ΔV with $(O - C)$, the difference between the observed (O) velocity and the calculated (C) velocity at the time of observation t, with the calculated velocity coming from equation (3.31), again using the preliminary estimates for the elements. Note that to calculate the radial velocity from equation (3.31) at some observed time t, it is necessary to solve Kepler's equation for the value of the eccentric anomaly E and then calculate the true anomaly θ from the standard relationship given in Table 2.1, taking care to assign the result to the correct quadrant in $0 \le \theta \le 2\pi$.

Then the only unknowns in equation (3.33) are the differential-correction terms themselves. But the foregoing equation can be written for each observation of (V, t), and so we have a set of n simultaneous equations for n observations that can be solved for the 6 differential-correction terms provided the set is sensibly over-determinate, with n substantially greater than 6. A linear least-squares solution can be performed to find the values of ΔK, and so forth, that will best match the quantities $(\Delta V, t)$ with the input preliminary values for the orbital elements. A first iteration yields revised values ($K_1 = K_0 + \Delta K$, etc.) that can be entered for a second iteration, and so on. Provided that the initial estimates are reasonable, with the observations sampling most orbital phases, the procedure will converge rapidly. Such least-squares solutions used to be performed by hand, and so it is not surprising to see that in the older research papers the preliminary elements were very carefully estimated from the graphical procedure, and the number of iterations was limited, typically to two. Present-day computing power makes the iteration process almost trivial.

As always, there are caveats. If the only source for the orbital period is the set of radial velocities themselves, then clearly the ΔP term must be retained in the least-squares solution. If the binary displays eclipses, or other photometric variations that are more sharply defined than the velocity variations, then such observations may provide a more accurate orbital period that can then be fixed in the preceding solution process.

If the orbital eccentricity is substantial ($e > 0.15$), so that the longitude of periastron ω and the time of periastron passage T are well defined, the preceding Lehmann-Filhés method will provide reasonable results. For values of $e < 0.15$, solutions via that method become less stable because of the increasing indeterminacy of the ω term as $e \to 0$. To obviate this problem, Sterne (1941) introduced a new method of analysis in which the time of periastron passage T is replaced by the *time of zero mean longitude* T_0, more readily understood as the *time of ascending-node passage*. Sterne defined *mean longitude* L by $L = 2\pi(t - T)/P + \omega = \eta + \omega$, where η is the *mean anomaly*, or the fractional orbital phase expressed in radian measure. The advantage of using T_0 rather than T is that the time of ascending-node passage, which is the *time of maximum positive velocity*, is *always* defined, regardless of the orbital eccentricity. It is perhaps unfortunate that Sterne's method has not been adopted more widely, because it is applicable to orbits of any eccentricity, and it can be simplified, for orbits with small eccentricity, to *Sterne's simple method*.

When T and ΔT are replaced by T_0 and ΔT_0, respectively, the differential-correction equation (3.33) is revised to the following expression, which differs only by one term in the coefficient of $\Delta\omega$:

$$\begin{aligned}\Delta V = {} & [\cos(\theta+\omega) + e\cos\omega]\Delta K + K\left[\cos\omega - \frac{\sin(\theta+\omega)\sin\theta(2+e\cos\theta)}{(1-e^2)}\right]\Delta e \\ & - K\left[\sin(\theta+\omega) + e\sin\omega - \frac{\sin(\theta+\omega)(1+e\cos\theta)^2}{(1-e^2)^{3/2}}\right]\Delta\omega \\ & + \left[\sin(\theta+\omega)(1+e\cos\theta)^2\frac{2\pi K}{P(1-e^2)^{3/2}}\right]\Delta T_0 \\ & + \left[\sin(\theta+\omega)(1+e\cos\theta)^2\frac{2\pi K(t-T_0)}{P^2(1-e^2)^{3/2}}\right]\Delta P + \Delta\gamma \end{aligned} \tag{3.34}$$

Sterne provided a modification of equation (3.34) for the case where $e \to 0$, namely,

$$\begin{aligned}\Delta V = {} & \cos L \Delta K + Ke\cos\omega\cos 2L + Ke\sin\omega\sin 2L \\ & + \frac{2\pi K\sin L}{P}\Delta T_0 + \frac{2\pi K(t-T_0)\sin L}{P^2}\Delta P + \Delta\gamma \end{aligned} \tag{3.35}$$

where we can make the simple approximation that $L = (\eta+\omega) \simeq (\theta+\omega)$. Andersen (1983) has demonstrated that this simplified method is reliable only for $e < 0.03$ and will lead to systematic errors if it is used on orbits of higher eccentricity. When orbits are circular, as is often the case amongst *close* binaries, then many of the foregoing complications disappear, and one can solve for the simpler parameters, K, γ, T_0, usually with the orbital period fixed at a value defined by photometric or pulse-timing data.

In Figures 3.26–3.29 we present examples of radial-velocity data of high quality that have been derived from different observing and analysis techniques. All the data sets share a common theme, namely, that the quadrature phases, or maxima and minima in the velocity curves, have been well defined by observations repeated on different nights, thereby ensuring that independent observations have confirmed the values of radial velocities that have been derived. Note the precision with which the semiamplitudes have been determined (typically, uncertainties of $\pm 1\%$ or less), thereby ensuring that the minimum masses of the component stars have been determined to uncertainties of less than $\pm 3\%$. Such accuracies are not easily achieved. They require particular attention to ensuring that sufficient observations have been obtained and to ensuring that rigorous procedures of data reduction and analysis have been followed.

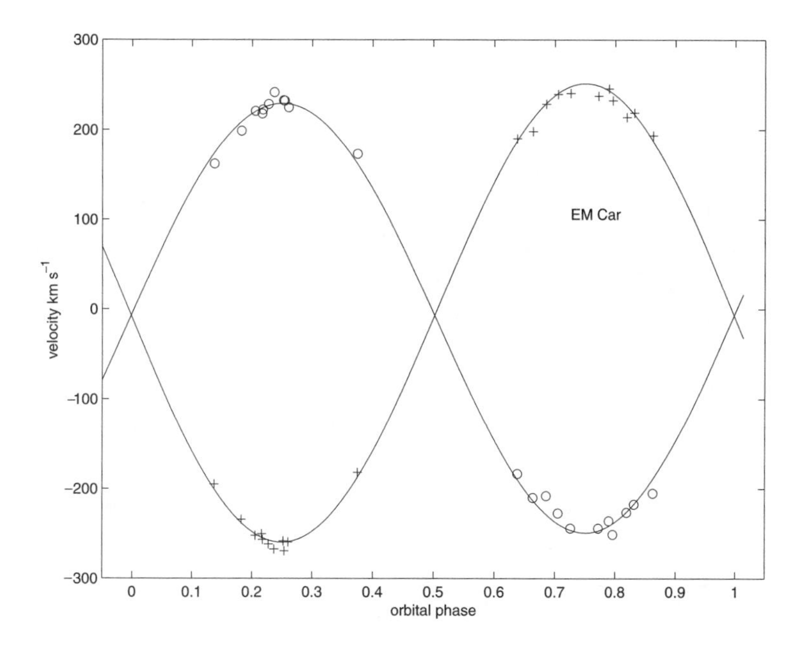

Fig. 3.26. The radial velocities for both components of the eclipsing binary star EM Car (O8V+O8V) plotted against orbital phase, together with the final velocity curves determined via the foregoing techniques. The data and the orbit solution are taken from Andersen and Clausen (1989). The original spectra were recorded by the 1.5-m telescope at the European Southern Observatory on fine-grain photographic emulsion at a reciprocal dispersion of 20 Å mm^{-1}, giving an effective resolution of about 0.7 Å. Velocities were determined by means of an oscilloscope comparator from a set of carefully selected absorption lines that did not suffer from spectral-line blending. The orbital elements are $K_1 = 239.1 \pm 1.7$ km s^{-1}, $K_2 = 255.4 \pm 1.5$ km s^{-1}, $\gamma = -7.0 \pm 1.5$ km s^{-1}, $e = 0.012$, $\omega = 6°$, $P = 3.4142765$ days, and T (primary minimum) = HJD2445038.8001, adopted from the related photometric studies.

3.3 Orbits from sources of pulsed radiation

3.3.1 Radio-pulsar binaries

Observation of pulsars at radio wavelengths is effectively a form of photometry, measuring a received flux density of radiation as a function of time, usually at several wavelengths simultaneously. Radio astronomers often make use of multi-channel receivers with perhaps 64 channels, each having band passes of 1–5 MHz and distributed over approximately 300 MHz. The practical difficulties for observations of radio pulsars lie principally in the fact that the pulsed signals are weak in comparison with the background signal due to the Milky Way galaxy, typically 10^{-4} of the background noise. For all but the brightest pulsars, the pulses lie well within the noise level, and consequently considerable care must be exercised in extracting meaningful results from the raw data.

A number of factors can assist with the determination of the pulsed signal and its periodicity. Firstly, the remarkable regularity of the pulses ensures that the application of Fourier transforms

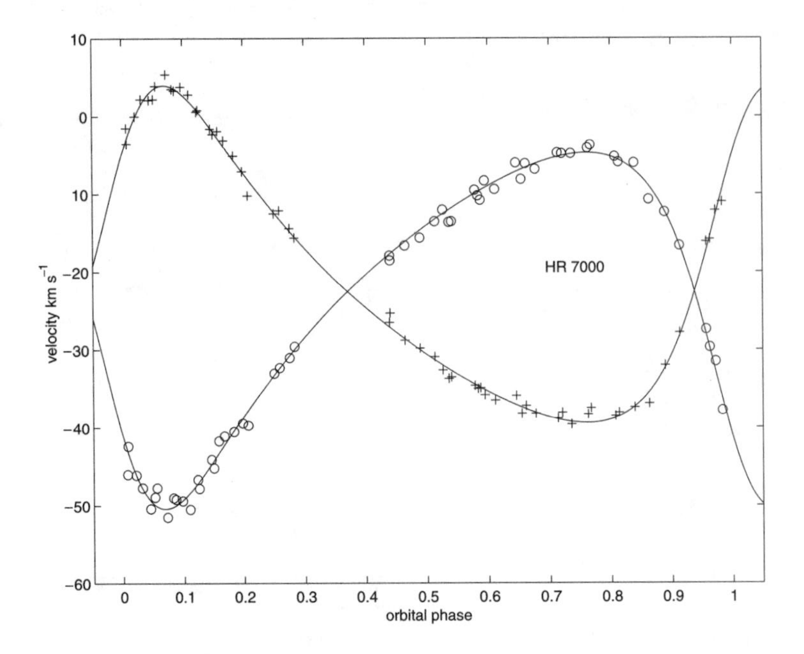

Fig. 3.27. The radial velocities for both components of the spectroscopic binary HR7000 (F4V+F6V) plotted against orbital phase, together with the final velocity curves. The data and the orbit solution are taken from Griffin et al. (1997). The radial velocities were determined from the CORAVEL radial-velocity spectrometer on the Swiss 1-m telescope at Haute Provence Observatory. The orbital elements are $P = 39.526 \pm 0.004$ days, $K_1 = 21.68 \pm 0.15\,\mathrm{km\,s^{-1}}$, $K_2 = 22.85 \pm 0.19\,\mathrm{km\,s^{-1}}$, $\gamma = -22.52 \pm 0.08\,\mathrm{km\,s^{-1}}$, $e = 0.372 \pm 0.005$, $\omega = 306.5° \pm 0.9°$, and T (periastron) = HJD2449257.66 ± 0.08.

can identify a periodic signal buried in noise. As a simple example, Figure 3.30 illustrates the ability of periodogram analysis to extract such a periodic signal from within noisy data. Although the artificial data set is only 100 seconds long, with idealized pulses, a Lomb-Scargle periodogram can identify the true period of 1.35 seconds, or a frequency of 0.7407 cycle s^{-1}, even from data in which the noise amplitude is three times the amplitude of the pulses. In practice, real pulsars emit pulses with quite variable amplitudes and asymmetric profiles, but Figure 3.30 indicates that with long data strings, typically 20,000 seconds (≈6 hours) of observation in a single session, a periodogram analysis can reveal an otherwise hidden pulsar signal.

Secondly, it was found from the early studies of the brighter pulsars that the mean shape of the pulse profile for a given pulsar was remarkably consistent, despite the well-known fact that the overall flux density of each individual pulse may well be quite variable. As a consequence, a preliminary estimate of the pulse period is sufficient to allow data to be added together in phase bins to define a mean pulse profile. This digital profile can be cross-correlated with the data string to refine determinations of each pulse arrival time and hence improve the pulse period and the mean profile shape. A few iterations are sufficient to yield a pulse period, with all data put into phase order. Typical accuracies are at microsecond levels. Thirdly,

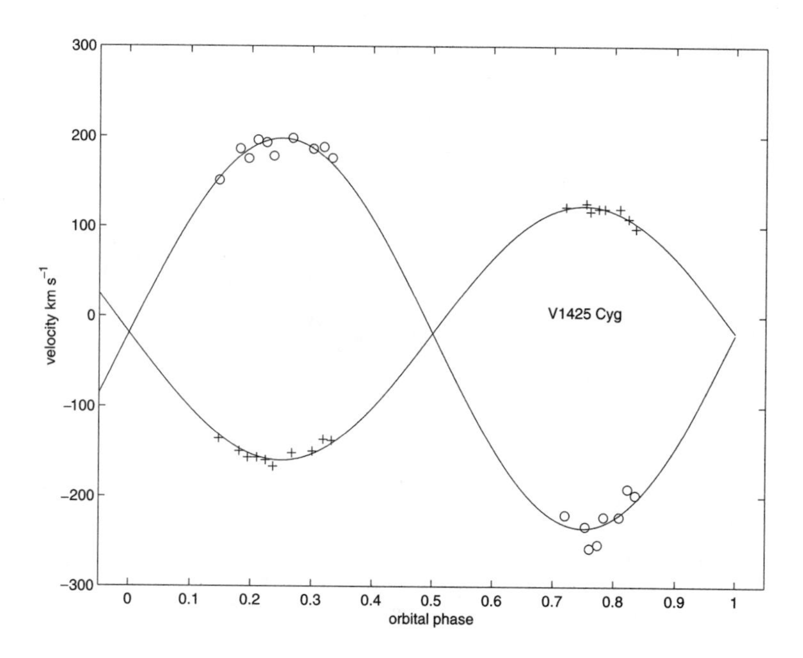

Fig. 3.28. The radial velocities for both components of the eclipsing binary V1425 Cyg (B5V+B9V) plotted against orbital phase, together with the final velocity curves. The data and the orbital elements are taken from Hill and Khalesseh (1993). The original spectra were recorded by the 1.2-m telescope at the Dominion Astrophysical Observatory and a diode array at a reciprocal dispersion of 20 Å mm^{-1} and effective resolution of about 0.7 Å. Velocities were determined iteratively from initial cross-correlation of the spectra against a standard star, followed by tomographic reconstruction of the mean primary- and secondary-star spectra, and the use of those spectra to yield revised velocities and hence improved reconstructed mean spectra. The orbital elements are $K_1 = 142.3 \pm 1.8\,\mathrm{km\,s^{-1}}$, $K_2 = 221.7 \pm 2.2\,\mathrm{km\,s^{-1}}$, and $\gamma = -19.1 \pm 1.8\,\mathrm{km\,s^{-1}}$ in a circular orbit, with the photometric observations of Lee (1989) providing the ephemeris of T (primary minimum) = (HJD2445969.0590 ± 0.0005) + (1.2523879 ± 0.0000008).

the passage of radio photons through the interstellar medium ensures a dispersion in pulse arrival times at the observer that is frequency-dependent. Accordingly, data from different channels can be analysed independently at first, and then added together, after making the frequency-dependent time shifts, to improve the overall determination of the pulse period and the interstellar-dispersion measure. Further details of the procedures involved in surveys for radio pulsars can be found in the text *Pulsar Astronomy*, by Lyne and Graham-Smith (1990).

Refinements to this preliminary value for the pulse period for a given pulsar have to be conducted in concert with improvements in determining the position of the pulsar on the sky (right ascension and declination), any proper motion components, and any possible variability of the pulse period. For this work to be carried out to the accuracy demanded by the data, we require knowledge of the position of the Earth relative to the barycentre of the Solar System to greater accuracy than a simple two-body representation of the Earth's orbital motion. The position of the Earth at a given time has to be established on the basis of a many-body

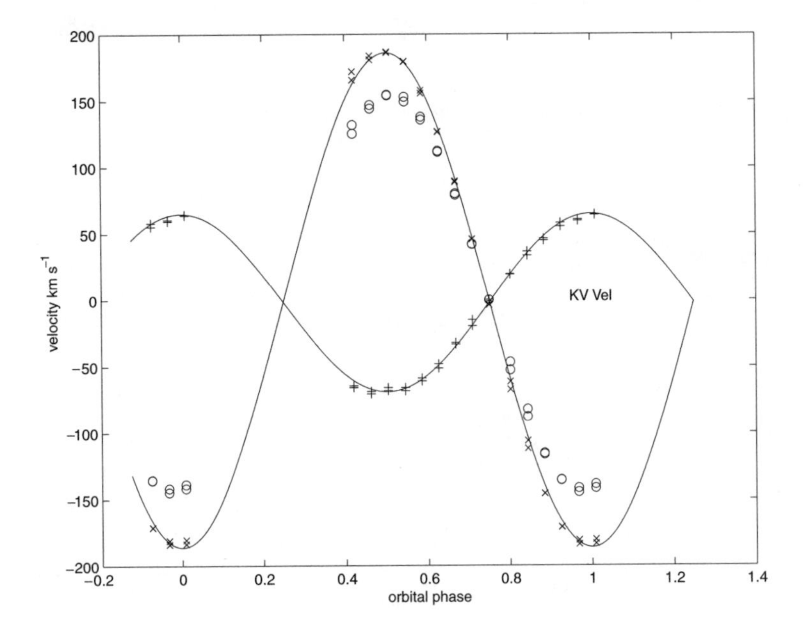

Fig. 3.29. The radial velocities for both components of the non-eclipsing binary KV Vel plotted against orbital phase (+, primary component; o, secondary component), together with fitted velocity curves, the corrected velocities x, and the velocity curve for the secondary component. This system provides an extreme example of systematic under-estimation of the semiamplitude of a velocity curve caused by the *reflection effect*, discussed in Chapter 5. Data from Hilditch et al. (1996).

representation of the planetary system, including the gravitational effects of all the planets when long-term (years) secular variations in pulse periods are being investigated. As noted in Section 2.10, the relativistic clock correction also must be included. We can then secure improved values for the *pulse period* τ or the *pulse frequency* f, its first derivative $\dot{f} = df/dt$, and the other positional parameters, in the following manner.

The pulse frequency can be expanded as a Taylor series, with

$$f = f_0 + \dot{f}(t - T_0) + \frac{1}{2}\ddot{f}(t - T_0)^2 + \cdots \tag{3.36}$$

where f_0 is the frequency at time $t = T_0$. For a *fixed* $f = 1/\tau$, the *pulse phase* is $\phi = (1/\tau)(t - T_0)$, but, in general, the pulse phase is defined by

$$\phi = \int_{T_0}^{t} f\,dt \tag{3.37}$$

Thus, the pulse phase at a time t is

$$\phi = \phi_0 + f_0(t - T_0) + \frac{1}{2}\dot{f}(t - T_0)^2 + \cdots \tag{3.38}$$

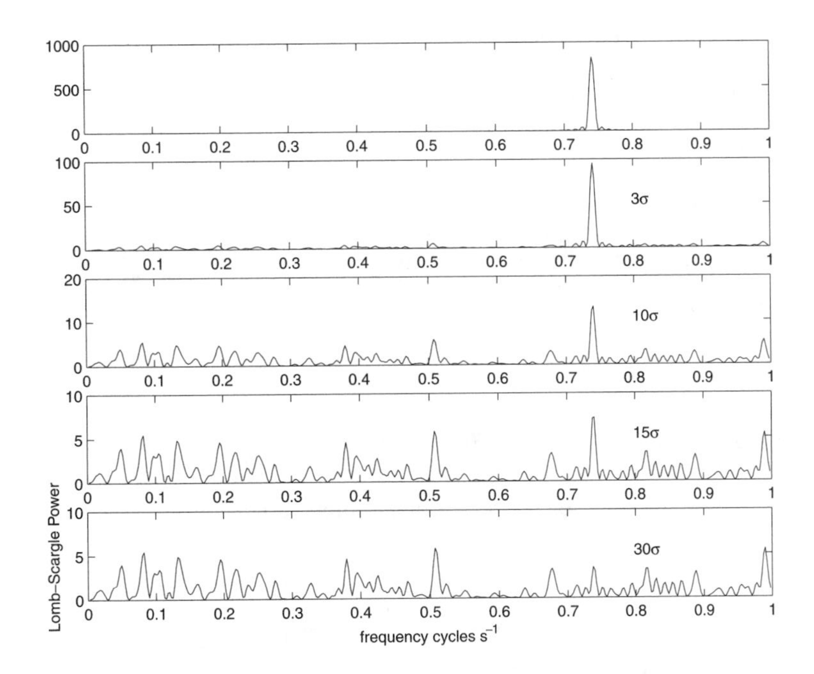

Fig. 3.30. A set of Lomb-Scargle (L-S) periodograms calculated from an artificial and idealized data set of symmetric pulses of unit height, with Gaussian half-widths of 0.03 second and a pulse period of 1.35 seconds, observed over an interval of 100 seconds. The top panel shows the L-S periodogram for the input perfect data, and the lower panels illustrate how the L-S power decreases dramatically as the amount of random noise N added to the input data increases from 3 to 30 times the standard deviation σ of the input data set, where $\sigma = 0.194$ in this example. When $N = 15 \times \sigma$, the peaks of the input data are only one-third of the noise amplitude, and yet the L-S periodogram still identifies the correct periodicity in the data. When $N = 30 \times \sigma$, and the noise amplitude is about six times the pulse amplitude, the periodicity has become ambiguous.

Equation (3.38) can be used to calculate predicted phases using initial values for f_0. If there is an error in f_0, then the residuals between the observed and calculated pulse phases will show a linear change with time t; if $\dot{f}$ is significant, then the residuals will show a parabolic deviation away from the constant-period line. For nearby pulsars, where the proper motion may be measurable, these timing residuals will demonstrate an *annual* sine curve of linearly increasing amplitude as the pulsar moves away from its initial position in right ascension (α) and declination (δ). Accordingly, for a single pulsar, an 8-parameter fit of differential corrections is required to match the residuals between the theoretical model and the observations of pulse phases, namely, Δf_0, $\Delta \dot{f}$, $\ddot{f}$, $\Delta\alpha$, $\Delta\delta$, μ_α, μ_δ, $\Delta\phi_0$. When the pulsar is a member of a binary system, then orbital motion about the barycentre of the binary system is included, and hence 14 parameters may be required. The light-travel time across the projected orbit, and also the frequency change due to the Doppler effect caused by the motion of the pulsar in its orbit, will cause changes to the pulse arrival times that will be additional to the 8 parameters cited.

The light-travel time (see Section 2.8) is

$$z/c = (r/c)\sin(\theta + \omega)\sin i \tag{3.39}$$

leading to a phase lag of

$$\Delta\phi_L = -fz/c \tag{3.40}$$

The frequency shift due to the Doppler effect is

$$\Delta f_D/f = \dot{z}/c \tag{3.41}$$

and it leads to a phase shift of

$$\Delta\phi_D = \int_{T_0}^{t} \Delta f_D\, dt = \frac{\dot{z}}{c} f_0(t - T_0) + O(\dot{z}\dot{f}) + \cdots = \frac{\dot{z}}{c} f_0 \frac{z}{c} \tag{3.42}$$

to first order. Hence, the additional term for binary motion in the pulse-phase equation (3.38) is

$$\Delta\phi_B = \Delta\phi_D + \Delta\phi_L = \frac{z}{c}\left[f_0\frac{\dot{z}}{c} - f_0 - \dot{f}(t - T_0)\right] \tag{3.43}$$

To justify the first-order approximation, note that for the well-known pulsar binary PSR1913+16, $\dot{z}/c \approx 0.001$, so that $\Delta\phi_D \approx 0.001\Delta\phi_L$. Nevertheless, $\Delta\phi_D$ is a measurable quantity for radio pulsars and can be used to obtain preliminary values for orbital elements.

3.3.2 Pulsar orbital elements

Although observations of a radio pulsar from a single observing session can provide a useful determination of the pulse period, because pulse periods typically range from 1.5 milliseconds to 4 seconds, it is necessary to conduct many repeat observations over time scales of days, weeks, and perhaps years in order to search for evidence of binary motion. Accordingly, it is necessary to consider the likelihood that the pulse period may be changing with time, perhaps because of the well-known loss of rotational angular momentum by the pulsar, or because of sudden jumps or glitches in pulse phase. These monotonic increases in pulse period typically amount to values of order 10^{-5} s yr^{-1} and hence are not problematic when establishing *preliminary* values for orbital elements. The proper motion of the pulsar across the sky also may be significant, reflecting substantial space motions, but here the effect is a purely *annual* one of linearly increasing amplitude.

The effect of the binary motion of a pulsar on the measured pulse periods determined from separate, short (5-minute) sets of observations is much larger, amounting perhaps to values as large as 8×10^{-5} seconds from day to day, depending on the particular orbital characteristics. Such changes in pulse period, or frequency, can be used, via equation (3.41), to establish the *radial-velocity curve* for the pulsar in its barycentric orbit, provided that the sets of observations have monitored the orbital cycle sufficiently. That velocity curve can then be analysed by the methods described in Section 3.2 to derive preliminary values for orbital elements. Subsequent analyses, utilizing all the pulse-arrival-time data, can be used to refine the values for these elements, together with period changes due to other causes, and to determine proper-motion components.

Manchester and Peters (1972) and Manchester, Taylor, and Van (1974) showed how pulse-arrival-time data could be used to establish accurate proper-motion vectors from observations secured over several years. The pulse-phase equation (3.38), can be used to calculate the expected integer phases of arrival of pulses from initial values of τ or f, and perhaps $\dot{f}$. The *observed* pulse phases will, in general, be somewhat different because of errors in the assumed parameters and because of additional terms due to proper motion and to binary motion. It is appropriate to consider a least-squares analysis of the data, in a manner similar to that used for the velocity curves of spectroscopic binaries. Taking $\Delta\phi_0$, Δf, $\Delta\dot{f}$, $\Delta\alpha$, and $\Delta\delta$ to be the corrections needed for the adopted zero point, frequency, frequency derivative, right ascension, and declination, we can write the appropriate residuals equation for each observation to be solved by a least-squares procedure as

$$R(t) \equiv \frac{(\phi_{\text{obs}} - \phi_{\text{calc}})}{-f} = -\Delta\phi_0 - \Delta f(t - T_0) - \frac{1}{2}\Delta\dot{f}(t - T_0)^2 + A\Delta\alpha + B\Delta\delta \tag{3.44}$$

where the coefficients A and B are given by

$$A = \frac{r_E}{c}\cos\delta_E\cos\delta\sin(\alpha - \alpha_E) \tag{3.45}$$

$$B = \frac{r_E}{c}[\cos\delta_E\sin\delta\cos(\alpha - \alpha_E) - \sin\delta_E\cos\delta] \tag{3.46}$$

and the residual $R(t)$ is given in seconds of time. Here the (α_E, δ_E) are the coordinates of the Earth, and the (α, δ) are the assumed coordinates of the pulsar at the same epoch, relative to the barycentre of the Solar System. These A and B coefficients have periods of one year, and so data spanning several years are necessary to establish the proper-motion components.

For a radio pulsar in a binary system, the $R(t)$ residuals will also display variations of the form discussed in Sections 2.8 and 3.3.1. To establish the necessary residuals equation for determining the orbital parameters, we follow the same procedure as discussed in Section 3.2.10 for spectroscopic binaries and consider a first-order Taylor-series expansion of the pulse-phase terms relevant for binary motion, equation (3.43). However, in this case we need to consider only the dominant term in that equation, namely, $-f_0 z/c$, because we are looking for small improvements to the preliminary values for the orbital elements.

The term z/c in equation (3.39) can be expressed as

$$z/c = x[(\cos E - e)\sin\omega + (1 - e^2)^{1/2}\sin E\cos\omega] \tag{3.47}$$

where $x = a_p \sin i/c$, and a_p is the semimajor axis of the barycentric orbit of the pulsar. Then $F = -f_0 z/c$, and we can write

$$\Delta F = \frac{(\phi_{\text{obs}} - \phi_{\text{calc}})}{-f_0} = \frac{\partial F}{\partial x}\Delta x + \frac{\partial F}{\partial e}\Delta e + \frac{\partial F}{\partial \omega}\Delta\omega + \frac{\partial F}{\partial E}\Delta E \tag{3.48}$$

We can use Kepler's equation, $E - e\sin E = 2\pi(t - T_0)/P$, to rewrite the ΔE expression in terms of the orbital elements e, P, and T_0 and achieve the final result:

$$R(t) = \frac{(\phi_{\rm obs} - \phi_{\rm calc})}{-f_0} = [(\cos E - e)\sin\omega + (1-e^2)^{1/2}\sin E\cos\omega]\Delta x$$
$$-\left[x\sin E\frac{\sin\omega\sin E - (1-e^2)^{1/2}\cos\omega\cos E}{(1-e\cos E)} + x\sin\omega + \frac{xe\cos\omega\sin E}{(1-e^2)^{1/2}}\right]\Delta e$$
$$+x[\cos\omega(\cos E - e) - (1-e^2)^{1/2}\sin\omega\sin E]\Delta\omega$$
$$+x\left[\frac{\sin\omega\sin E - (1-e^2)^{1/2}\cos\omega\sin E}{(1-e\cos E)}\right]2\pi\,(t-T_0)\frac{\Delta P}{P^2}$$
$$+x\left[\frac{\sin\omega\sin E - (1-e^2)^{1/2}\cos\omega\sin E}{(1-e\cos E)}\right]2\pi\frac{\Delta T_0}{P} + \sin E\,\Delta g \tag{3.49}$$

Blandford and Teukolsky (1976) have presented a rigorous analysis of the pulse-arrival-time problem for pulsar binaries developed in the framework of general relativity. Their final equation for improving a preliminary set of values for orbital elements via the foregoing differential-correction procedure is the same as that given in equation (3.49), in which all terms are of Newtonian origin except for the additional term on the last line, $\sin E\Delta g$, which is of relativistic origin. The quantity

$$g = \frac{2\pi a_p^2 e}{c^2 P}\left[2 + \frac{m_p}{m_c}\right] \tag{3.50}$$

is the sum of the second-order Doppler effect and the gravitational redshift and is expressed in terms of the masses of the two components of the binary, m_p for the pulsar and m_c for its companion. Because the coefficient for Δg depends on the orbital phase in a manner similar to that for the coefficients for Δx and $\Delta\omega$, it is usual practice to set $\Delta g = 0$ for a set of data that span only a year or so. To evaluate Δg, it is necessary to accumulate data over a decade or more, so that the orbital elements and their possible secular variations can be decoupled from the effects of gravitational radiation. Exactly that decoupling has been achieved for the pulsar binary PSR1913+16 from observations spanning 10 years (Weisberg and Taylor 1984).

For secular variations in the orbital elements to be established, equation (3.49) can be modified straightforwardly by making the substitutions

$$\Delta x \to \Delta x + t\Delta\dot{x}; \qquad \Delta\omega \to \Delta\omega + t\Delta\dot{\omega} \tag{3.51}$$
$$\Delta e \to \Delta e + t\Delta\dot{e}; \qquad \Delta P \to \Delta P + t\Delta\dot{P} \tag{3.52}$$

but a discussion of secular variations due to all the possible sources of perturbations acting on binary-star orbits is left until Chapter 4.

The orbital parameters determined for binary systems that contain radio pulsars are the most precise known. Typical uncertainties for semimajor axes are $\pm 0.1\%$, and eccentricities and longitudes of periastron, for example, are known to six or seven significant figures. Pulse periods are defined to 1 part in 10^9, and period derivatives to 1 part in 10^4. Taylor, Manchester, and Lyne (1993) have provided a catalogue of 558 known radio pulsars, of which 24 are in binary systems. As examples of the data obtained, Figures 3.31 and 3.32 show the velocity

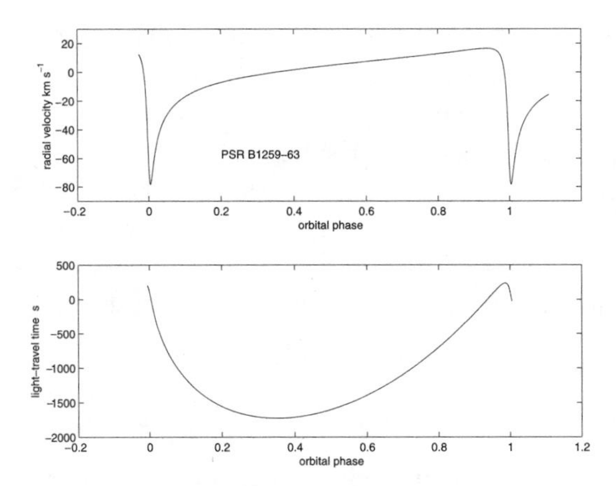

Fig. 3.31. The inferred velocity curve and light-travel-time curve for the radio pulsar + Be-star binary system PSRB1259-63, as determined by Johnston et al. (1994) from detailed radio timing data over 5 years. Values for the orbital elements are determined very precisely with such data: e.g., $a_p \sin i = 1295.98 \pm 0.01$ light-seconds, $P_b = 1236.79 \pm 0.01$ days, $P_p = 47.762053919 \pm 0.000000004$ milliseconds. The companion Be star is about seven times more massive than the pulsar, and its expected velocity curve would have an amplitude of only 10 km s^{-1}, which would be difficult to determine for a B star with emission lines.

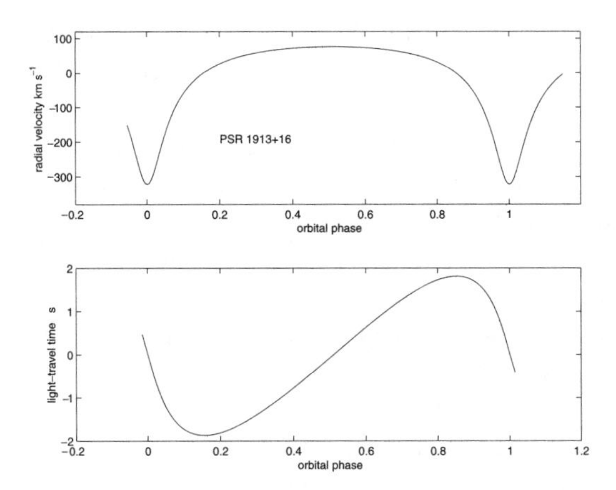

Fig. 3.32. The inferred velocity curve and light-travel-time curve for the radio-pulsar binary system PSR1913+16, as determined by Taylor et al. (1976) from detailed radio timing data over 1 year. Values for the orbital elements are determined very precisely with such data: (e.g., $a_p \sin i = 2.3371 \pm 0.0001$ light-seconds, $P_b = 27906.980 \pm 0.002$ seconds, $P_p = 59.029995272 \pm 0.000000005$ milliseconds). From a longer series of data that establishes the apsidal motion in this system (see Chapter 4), the total mass of this pulsar binary system is found to be 2.82 $M_{\odot}$.

curves and light-travel-time curves for the pulsar systems PSRB1259-63 and PSR1913+16. These two systems are extremely important for understanding the evolution of binary stars. PSRB1259-63 is in a highly eccentric, long-period orbit about a main-sequence Be star, and it may be a rare example of a system in the transition stage between the first supernova event for a massive binary (producing the pulsar) and the later mass-exchanging stage of a massive x-ray binary. PSR1913+16 was the first pulsar binary to be discovered. It has a short orbital period and a high eccentricity and has been used to examine the predictions of general relativity with regard to apsidal motion and gravitational radiation, with great success (see Chapter 4).

3.3.3 X-ray-pulsar binaries

The radio pulsars have narrow pulse profiles that occupy only small fractions (≈ 0.03) of their pulse periods, and their mean profiles have remarkably consistent shapes. By contrast, the x-ray pulsars exhibit pulse profiles that extend over about 50% of the pulse period and seem to depend on the x-ray luminosity, the overall source luminosity, and the observed energy band (or wavelength range) as well. When the first x-ray pulsars were discovered, the sensitivities of the x-ray detectors were quite low, and it was necessary to use coarse phase binning in order to achieve a reasonable signal. Typically, only 10–20 phase bins were used per pulse period, and as a consequence, the accuracy achieved in the timing of pulse arrivals was significantly poorer than for the radio pulsars. Our x-ray detectors have been improved remarkably in terms of sensitivity, and it is now standard to see phase binning at 1% of the pulse period. The improved S/N also demonstrates the remarkable variation in pulse shape as a function of x-ray outburst luminosity and wavelength. White et al. (1995) provided a list of the 32 known x-ray binaries; they show a very wide range of pulse periods, from 69 milliseconds to 835 seconds. There have also been detailed investigations of the *quasi-periodic oscillations* exhibited by x-ray pulsars (where the time resolution is required to be about 1 millisecond), with instruments like the Rossi X-ray Timing Explorer. Such studies are exploring the region of the inner accretion disc around a central neutron star and are independent of the orbital motion.

The pulse-arrival-time data reflect the light-travel time across the projected barycentric orbit of an x-ray pulsar, in exactly the same manner as discussed for radio pulsars, but it is generally the case that the frequency shifts due to the Doppler effect are not measurable because of the physical attributes of the x-ray pulses. Accordingly, the equations from the preceding section are simplified to consider only the light-travel time. In addition, many of the x-ray binaries of short orbital period have circular orbits, so that the terms involving eccentricity and longitude of periastron can be neglected. Just as for radio pulsars, the x-ray pulsars display substantial changes in pulse period because of rotational angular-momentum loss (*spin-down*) or gain (*spin-up*) via mass transfer in these semidetached systems, and that must be taken into account when analysing the pulse arrival times.

Accordingly, the equation usually adopted for interpreting the pulse arrival times from x-ray pulsars is

$$t_n = t_0 + nP_{p0} + \frac{1}{2}n^2 P_{p0}\dot{P}_p + z/c \qquad (3.53)$$

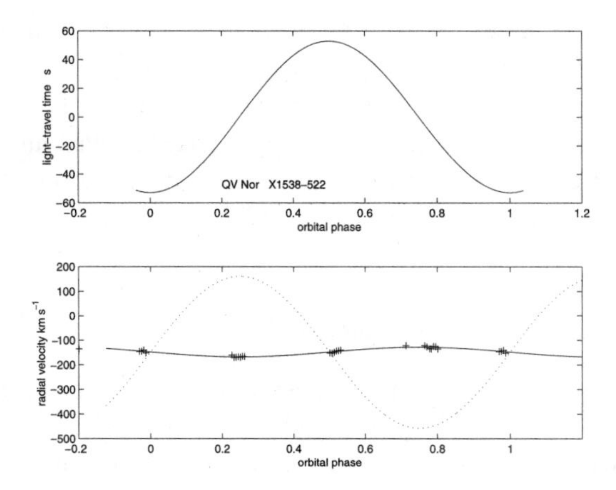

Fig. 3.33. The light-travel-time curve (top) and velocity curves (bottom) for the x-ray-pulsar binary system QV Nor, as determined by Makishima et al. (1987) and Reynolds et al. (1992) from x-ray pulse-timing data and optical spectroscopic data, respectively. The + signs are the observed radial-velocity data for the B-type supergiant, and the solid line is the orbit fit to those data. The dotted line is the inferred velocity curve for the pulsar derived from the model fit to the observed light-travel-time data.

where t_n is the arrival time of, for example, the peak of the nth pulse, t_0 is the adopted epoch at which the pulse period is P_{p0}, the third term allows for secular changes in the pulse period, and the last term is the light-travel time, given by equation (3.39). As noted in Section 2.8, the x-ray pulse-timing data can be combined with spectroscopic observations of the normal companion star, which can readily be observed in the ultraviolet, optical, or near-infrared region of the spectrum. The x-ray observations can define precise values for the orbital period and the orbit shape, so that the number of free parameters to be solved for in radial-velocity data for the companion can be reduced to a minimum. The combined data will provide a complete description of the orbit of the x-ray binary, together with minimum masses for both components. The orbital inclination to the line of sight can be determined from photometric observations, to be discussed in Chapter 5. As an example of the combination of x-ray pulse-timing data with spectroscopic data to determine the radial velocity of the companion, Figure 3.33 illustrates the light-travel time across the projected barycentric orbit of a pulsar, together with the final velocity curves for the two components in the HMXB, QV Nor. The radial-velocity data for the B-type supergiant in QV Nor are plotted together with the final orbit fit to those data. But it seems to be common practice in papers reporting x-ray pulse-timing observations for the data to be plotted on graphs only and not published in tabular form, even though the total number of pulse times used is typically 30–100. This is regrettable, because the analysis cannot easily be repeated by other researchers. The final model fit to the pulse-timing data provides values of the parameters for the pulsar's orbit, from which a velocity curve has been calculated and plotted in Figure 3.33. Note the very large difference in semiamplitudes of the two velocity curves, reflecting a mass ratio of 16:1 for the B star relative to the pulsar.

A new development in these studies of x-ray binaries is the recommendation by Wilson and Terrell (1998) that the pulse-timing data from the x-ray source, including the durations of any x-ray eclipses, be combined with the radial-velocity data from the companion to provide a *simultaneous* solution for all the data. They argue that the resultant unified descriptions of the orbital characteristics of such x-ray binaries are much improved, with the system Vela X-1 : GP Vel being analysed as one example.

3.3.4 Binaries containing pulsating stars

The true *pulsating* stars, those whose photospheres are observed to be experiencing radial or non-radial oscillations with regular periods, are also to be found in binary systems. Szabados (1992) has shown that the frequency of binary stars amongst the classical Cepheid variables is at least 50%, a value shared with other normal stars. Here the observed pulsation periods are to be found in the range 2–20 days, and the orbital periods from 300 days to several thousand days. Such long orbital periods reflect the fact that the orbits are large, sensibly measured in astronomical units, which is necessary to allow a star to evolve to a supergiant size and pass through the Cepheid instability strip in the HR diagram. Evans et al. (1998) have established reliable masses for Cepheid variables by combining optical spectroscopy of the pulsator with ultraviolet spectroscopy of the blue companion star.

Other pulsators are also known to be present amongst binary components, such as β Cephei stars and δ Scuti stars, with pulsation periods of hours, pulsating white-dwarf stars, with pulsation periods of minutes, and the pulsating sdB stars (EC14026 stars), recently discovered by the Edinburgh-Cape blue-object survey (Stobie et al. 1997; Kilkenny et al. 1997). These latter objects pulsate with at least two dominant oscillation periods (typically 2–3 minutes), and in some cases they have many periodicities in evidence. One of these sdB stars, PG1336-018, has recently been found to be a member of a short-period ($P = 2.4$ hours) eclipsing binary system (see Figure 5.15), thereby opening the possibility of determining directly the mass, radius, and luminosity for these unusual objects (Kilkenny et al. 1998).

Within our current context of using pulse-timing data to establish orbital elements, one may wonder if these pulsating stars can provide the necessary light-travel-time information. There are two critical factors: Firstly, the pulsation period must be a small fraction of the orbital period, so that the pulsational data can provide an adequate sampling rate of the orbital motion. Secondly, the ratio of the uncertainty ΔP_p in the pulsation period to the projected barycentric orbit size $a \sin i/c$ must be sufficiently small. For the radio pulsars, $\Delta P_p/a \sin i/c \approx 10^{-9}$, so that light-travel-time effects are easy to observe, and pulse frequency changes due to the Doppler effect are also observable, as we have seen. For the x-ray pulsars, $\Delta P_p/a \sin i/c \approx 10^{-2}$, and light-travel-time effects are still easily measurable. For the known Cepheids in binary systems, we find ratios of 10^{-4} to 10^{-2}. In principle, the periodic advances and delays in the times of maximum/minimum light from Cepheids in binaries, due to their orbital motion, should be discernible. But there are practical difficulties, because of the long orbital and pulsation periods, of securing enough data with sufficient phase coverage to determine the light-travel times. Nevertheless, such data have been acquired, for example, for the Cepheid binary AW Per by Vinko (1993), who showed that use of the photometric times of maximum light of the Cepheid combined with use of its spectroscopic radial velocities significantly improved the determination of its barycentric orbit.

The prototype pulsating star of early-B spectral class, β Cephei, has a long history of spectroscopic and photometric observations, in part because of its brightness and in part because of the early discovery of its variability, with a period of 4.5 hours, by Frost (1906). An analysis by Pigulski and Boratyn (1992) of all the published data showed that the known deviations of the pulsation cycle from a linear ephemeris were sensibly interpreted as the light-travel-time effect across the projected orbit of β Cephei about its companion, discovered via speckle interferometry by Gezari, Labeyrie, and Stachnik (1972). Its orbital period has been found to be 91.6 years, and, fortuitously, its orbit is nearly edge-on, with $i = 86.7°$. Thus, the criterion for detection of the light-travel time is well satisfied by this system, at about 10^{-5}.

For the pulsating sdB star in PG1336-018, the current uncertainty in the dominant pulsation periods is about 0.3 second, and the light-travel time across the orbit is about the same, $a \sin i / c \approx 0.3$ second. Continued photometry at very high S/N and very-high time resolution will be required to discern the effects of orbital motion, as well as the multiple intrinsic periodicities of this star.

3.4 Orbits from astrometric observations: Resolved binaries

3.4.1 Visual, speckle, and direct interferometric observations

Light from a distant star approaches the top of the Earth's atmosphere with the characteristics of a plane wave-front. During its passage through the successively more dense zones of the atmosphere, the wave-front is refracted by the different densities of the air masses that it encounters, with the result that the wave-front becomes tilted relative to the direct line of sight, and also it may become somewhat non-planar, or even 'corrugated'. Successive wave-fronts may not, in general, follow the same process, because of the natural turbulence of the atmosphere, caused by several factors. Thermal motion of the air occurring within a few tens of metres above the observing site is a major cause, and mass motions of large volumes of air at many levels above the observer also contribute. The rapidly developing technology of *adaptive optics* is demonstrating how high-speed control of optical surfaces can compensate for the tilted, non-planar wave-fronts and provide near-diffraction-limited images of stellar sources.

The stellar image seen through a small telescope appears somewhat fuzzy, the so-called *seeing disc*, and may exhibit image motion by amounts of at least the diameter of the image if the moving air masses are larger than the diameter of the telescope. The so-called correlation length, or Fried's parameter, of the Earth's atmosphere is wavelength- and seeing-dependent, being of order 10 cm for visible wavelengths and 1 m for the infrared region. The visual observer, with a small telescope, sees the overall effect as a dancing of the stellar image. *Scintillation* is the name given to the rapid fluctuations across the receiver aperture, which are seen with the unaided eye as the familiar twinkling of starlight. The direct visual observer, using a long-focus telescope with a large scale at the focal plane, will see a resolved binary very clearly in those moments of excellent seeing, whilst a longer-term recording device, like a photographic plate or a CCD, will integrate the image and lose those moments of extra clarity and resolution. As a result, photographs or CCD frames of resolved binaries have proved to be extremely effective means for measuring separations above 2–3 arcsec, whereas a visual observer utilizing a micrometer could take advantage of repeated brief intervals of excellent seeing to measure separations between 0.2 and 2 arcsec. Typical accuracies achieved are about ± 0.01 arcsec in separation, and about $\pm 1°$ in position angle. More details of the observational techniques applicable to direct images of resolved binaries can be found, for example, in van der Kamp (1967) and Heintz (1978).

With the use of larger telescopes, the image motion is mostly removed, except in very poor, turbulent conditions, whilst the fuzzy image seen at high magnification provides a visual impression of a boiling, convective, speckled pattern. When recorded with integration times of milliseconds, the speckle pattern is frozen and can be resolved into a distribution of light and dark patches. These are caused by interference induced by phase fluctuations on the incoming wave-front due to atmospheric turbulence. It was shown by Labeyrie (1970) that the speckle pattern contains substantial spatial information about the diffraction-limited resolved image of a star. For a resolved binary star, the speckle pattern can provide a determination of angular separation ρ, position angle ϕ, and the brightness difference between the components. *Speckle interferometry* therefore requires high-speed recording of the speckle images via a single large (2–4-m) telescope. The faintness limit for direct speckle observations is currently around magnitude 8 for recordings obtained with an intensified CCD camera on a 4-m telescope. Successful detections of binaries are made when projected separations are in the range $0.035 < \rho < 1.5$ arcsec, with magnitude differences between components of $\Delta m < 3$ mag (Mason et al. 1998). The lower limit is set by the diffraction limit of the telescope employed, and the upper limit is determined by the size of the turbulence cells in the atmosphere. Essentially, speckle patterns of stars will be very similar when viewed through the same 'isoplanatic patch', amounting to about 2–3 arcsec on the sky, that is, all separations of interest to the speckle technique.

The earliest speckle observations of binary stars, conducted during the 1970s, utilized image intensifiers, cameras, 100-ft rolls of photographic film, and exposure times of 0.02 second to record the speckle images seen through a 4-m telescope (e.g., McAlister 1977). The technology has progressed to intensified CCD cameras with digitally recorded integrations of typically 0.015 second (e.g., Hartkopf 1992) that can be analysed directly, rather than after further photographic processing, which loses the flux-density information contained within the images. A typical 'observation' consists of a sequence of short integrations lasting for about 1 minute.

Extraction of the astrometric data of (ρ, ϕ) and the magnitude difference between the components is now achieved by following the *directed-vector autocorrelation* (DVA) algorithm developed by Bagnuolo et al. (1992b). Recall from Section 3.2.5 that a one-dimensional autocorrelation function can be written

$$a(x) = \int_{-\infty}^{\infty} s(n)s(n-x)\,dn \tag{3.54}$$

A speckle pattern is a two-dimensional image on the sky (x, y), so that the autocorrelation function is

$$a_{\mathrm{sp}}(x, y) = \int_{-\infty}^{\infty} \int_{-\infty}^{\infty} s(n, m) s(n - x, m - y) \, dn \, dm \tag{3.55}$$

The speckle pattern of a single star will yield an autocorrelation function (ACF) of an Airy disc, a single three-dimensional peak in (x, y, ACF) space when all the speckles correlate exactly with themselves. Such an analysis could be performed, for example, by using the foregoing one-dimensional equation for a range of offsets, $\pm\Delta y$, to provide a sequence of ACFs across the (x, y) plane, a two-dimensional histogram of (x, y, ACF). If the observed object is a binary system, the speckle patterns of both stars will be superposed at the appropriate (ρ, ϕ), and the two patterns will be very similar in structure for separations less than about 2–3 arcsec. An autocorrelation of a binary speckle pattern will yield three peaks: the central peak, as before, and two smaller peaks. Envisage performing one-dimensional ACFs from left to right across the image. A first peak will occur when the speckles from the right component correlate with the those of the left component. The large central peak will occur when all the speckles correlate with themselves exactly. A third peak will occur when the speckles from the left component correlate with those of the right component. If no attempt is made to distinguish between the brighter and fainter components at this stage of analysis, then the two subsidiary peaks will be the same, and there will be a 180° ambiguity about the position angle ϕ for the binary, which can lead to incorrect interpretation of the data and hence wrong orbital periods, eccentricities, and so forth. The DVA technique of Bagnuolo et al. (1992b) introduces a discrimination based on the brightness differences between speckles. If the pixel flux densities f_i and f_j are such that $f_i \geq f_j$, then the two-dimensional histogram is incremented at position $[(x_i - x_j), (y_i - y_j)]$; if $f_i \leq f_j$, the increment is placed at position $[(x_j - x_i), (y_j - y_i)]$. Thus, direction, brighter to fainter, is preserved, and the true orientation will result in a 'principal' peak larger than the smaller 'ghost' peak. Examples of such two-dimensional histograms are given by Bagnuolo et al. (1992b). Calibrations of separation and position angle can be achieved by observing a known single star through a double slit placed in the converging beam of the telescope. The autocorrelation of the speckle images will provide a set of peaks with positions and orientations defined by the telescope optics, the geometry of the double slit, and the observed wavelength defined by a narrow-band filter.

Numerous alternative techniques have been proposed for reconstructing diffraction-limited images from speckle interferograms, and Weigelt and Wirnitzer (1983) have provided a list of references to them. The procedure most commonly used for resolved binaries has certainly been the preceding method developed by Bagnuolo et al. (1992b), but there has also been significant use of the *speckle-masking* method first proposed by Weigelt (1977). Here the speckle images are processed to yield a direct image of the source, rather than an autocorrelation, and as a result there is no ambiguity about the position angle of the binary. The procedure involves a substantial understanding of Fourier transforms and correlation theory and has been fully described by Weigelt and Wirnitzer (1983) and Lohmann, Weigelt, and Wirnitzer (1983); here, an outline will suffice.

There are n speckle interferograms in a typical 'observation', each providing an intensity distribution $I_n(\vec{x})$, where $\vec{x}$ is a position vector in the speckle image. The intensity distribution in each image is related to that of the source, $S(\vec{x})$, via the equation

$$I_n(\vec{x}) = S(\vec{x}) * P_n(\vec{x}) \tag{3.56}$$

where $P_n(\vec{x})$ is the point-spread function (PSF) of the atmosphere above the telescope at the instant of the nth observation. The $*$ sign denotes convolution of the two quantities S and P_n. These individual PSFs are not known, and so the determination of S is not straightforward; we need to use the sequence of data I_n in an observation to find an adequate description of the P_n that will permit a determination of S. The procedure involved in *speckle masking* is to offset a speckle interferogram from itself by an amount that is given by a *masking vector* $\vec{m}$ and then compute a third-order correlation function (TCF) defined by

$$T_{\text{image}}(\vec{x}) = \left\langle \int \int [I_n(\vec{x'})I_n(\vec{x'} - \vec{m})]I_n(\vec{x'} + \vec{x})\, d\vec{x'} \right\rangle \tag{3.57}$$

Here the angle brackets denote an average over the n images of the observation. As a result, it is not necessary to determine the individual PSFs, only the ensemble average.

For work on resolved binary stars, the masking vector $\vec{m}$ is made equal to the separation vector of the binary, which can be determined from a preliminary analysis, for example, via autocorrelation. By calculating the foregoing TCF, T_{image}, for the range of $\vec{x}$ that covers the image field, an estimate of the ensemble average PSF is found, because the displaced and overlapped images are convolved to yield only those parts of the images that are shared by both, namely, the ensemble average PSF. There are complications, because some of the overlaps are spurious, and Lohmann et al. (1983) have described how these can be subtracted from the TCF to provide the third-order correlation function of the source intensity distribution S, namely, $T_{\text{source}} = [S(\vec{x})S(\vec{x} - \vec{m})] \otimes S(\vec{x})$, where the circled cross denotes cross-correlation. The final image of the source, $S(\vec{x})$, can be determined from T_{source} if $\vec{m}$ has been correctly chosen, so that $S(\vec{x})S(\vec{x} - \vec{m}) \approx \delta(\vec{x})$, where $\delta(\vec{x})$ is the delta function. Then, because $\delta(\vec{x}) \otimes S(\vec{x}) = S(\vec{x})$, the brightness distribution of the source is extracted from the n speckle images. The orientation of the binary is preserved, as well as the brightness difference between the two components. Examples of the use of these techniques have been reported, for example, by Leinert et al. (1990).

The 1990s witnessed the breakthrough in direct, multi-element optical interferometry that had been promised for so long. The Mark III Optical Interferometer installed at the Mt. Wilson Observatory, as described by Shao et al. (1988), has led the way towards realizing the goal of determining the orbits of resolved binaries down to angular separations of the order of 10–20 milliarcseconds, with uncertainties of only $\pm 1\%$. These new observational techniques reported, for example, by Armstrong et al. (1992a,b) and Hummel et al. (1993, 1994, 1998), are extending the lower limit of speckle interferometry and resolving the orbits of spectroscopic binary stars for the first time in a systematic manner that will place our knowledge of binary orbits, distances, and masses on a much firmer foundation. These developments will result in a major improvement in the stellar distance scale. The review paper by Armstrong et al. (1995) provides a summary of the historical development of astronomical optical interferometry, explains the major technical difficulties in performing optical interferometry, and highlights the many instruments that are currently being developed to exploit these new abilities in the control of optical systems. Specifically for studies of resolvable binary stars, we noted in Chapter 1 the new Navy Prototype Optical Interferometer (NPOI) and the imminent arrival of the CHARA and SUSI interferometric arrays, and others are expected to be operational within the next few years. As for the earlier discussion of speckle interferometry, we shall summarize the measurements made in long-baseline optical interferometry to determine the angular separations and position angles for resolved binary stars.

The Mark III Interferometer, the NPOI, and the forthcoming CHARA and SUSI instruments are all multi-element arrays of siderostats. The baselines between any two of these moveable mirrors can have orientations of north–south (N–S), east–west (E–W), or various intermediate options. For each pair of mirrors, the instrument acts like a Michelson interferometer, such that the wave-fronts from a single on-axis source to both mirrors can be brought to a focal plane where they create interference fringes. A quantitative measure of whether or not the light and dark bands of the interference pattern are readily seen is provided by the *visibility* V, introduced by Michelson, of

$$V = \frac{I_{\max} - I_{\min}}{I_{\max} + I_{\min}} \tag{3.58}$$

where the I values are the observed flux densities at the maxima and the adjacent minima of the fringe pattern. It can be shown (e.g., Hecht 1992) that the expression for the variation of V across the focal plane for a star with a uniformly illuminated disc of angular diameter θ_i is

$$V_i = 2\left|\frac{J_1(\pi B\theta_i/\overline{\lambda_0})}{(\pi B\theta_i/\overline{\lambda_0})}\right| \tag{3.59}$$

where B is the separation between the two mirrors, the *baseline* of the interferometer, $\overline{\lambda_0}$ is the mean wavelength of the (narrow) pass band used for the observations, and J_1 is the Bessel function of the first kind.

Because the two stars in the binary are independent sources of light, the phases of their incoming wave-fronts are incoherent, and so the flux densities within each fringe pattern are simply additive. Accordingly, maximum visibility will be achieved in the combined signal when the two patterns overlap completely, and minimum visibility will occur when the two patterns are completely out of phase, when the baseline length has changed by an amount $\Delta d = \overline{\lambda_0}/2\theta_b$ (Hecht 1992).

Hanbury-Brown (1974) discussed the properties of Michelson and intensity interferometers and provided an expression for the *squared visibility*, V^2, appropriate for a binary star with components of angular diameter θ_1 and θ_2 with a brightness ratio β and separated by an angular distance θ_b, namely,

$$V^2 = \frac{C}{(1+\beta)^2}\left[\beta^2 V_1^2 + V_2^2 + 2\beta|V_1||V_2|\cos(2\pi\theta_b)B\cos\alpha/\overline{\lambda_0}\right] \tag{3.60}$$

Here the individual visibilities of the two components are $V_{1,2}$, as given by equation (3.59), and the angle α is that between the position angle of the binary on the sky and the direction of the baseline of the interferometer projected onto the sky at the time of observation. The quantity C is a normalization constant determined by observations of known single stars that

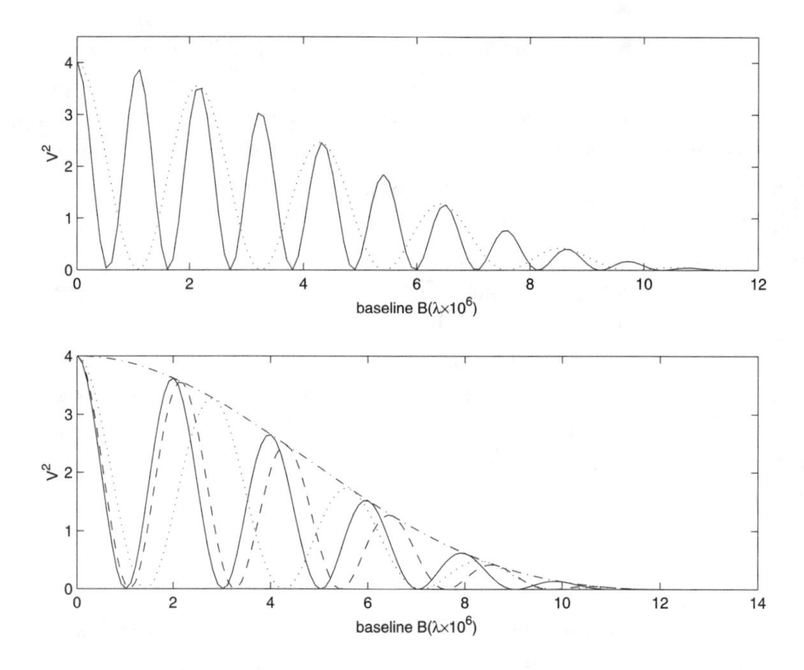

Fig. 3.34. Top: The squared visibility, equation (3.60), for a binary star plotted as a function of the baseline length B. When the angular separation between the two stars is doubled, the spacings between successive maxima and minima are halved. Bottom: The squared visibility for a fixed value of θ_b and a range of values for the angle α between the position angle of the binary and the baseline projected onto the sky ($\alpha = 0°$, $22.5°$, $45°$, and $90°$ for the solid, dash, dotted, and dot-dash lines).

are unresolved at the resolution employed. Note that in all these equations, the quantity $B/\overline{\lambda_0}$ appears. So the baseline length is measured in units of the wavelength used for that observation; hence multiple baselines are possible for a given spatial arrangement of the interferometer by employing a range of narrow-band filters. Note that the range of possible values of the visibility for a single star is between 0 and 1. Hanbury-Brown's expression, equation (3.60), for a binary star can range between 0 and 4. However, in practice, the contrast between fringes is less than the idealized picture, and typically the maximum value of V^2 for a single star is about 0.75. Observations of unresolved single stars are used to calibrate the instrument and to normalize the maximum values of V^2 for binaries to unity (Figure 3.34).

Interferometric observations of binary stars are interleaved with those of known single stars at accurately established positions in order to monitor the atmospheric perturbations and to calibrate the baseline. Two arms of the interferometer are used, oriented, for example, N–S and E–W, with baselines of fixed spatial length at the instrument. Then the rotation of the Earth during a night of observation provides a slow continuing change in baseline length projected onto the sky at the position of the binary, together with equivalent changes in the orientations of the N–S and E–W arms relative to the position angle of the binary at the time of observation. The equations that govern these projections are derived via straightforward spherical trigonometry using the (y, x) coordinates, or $(\Delta\alpha \cos\delta, \Delta\delta)$ given by equation (2.62). The projections

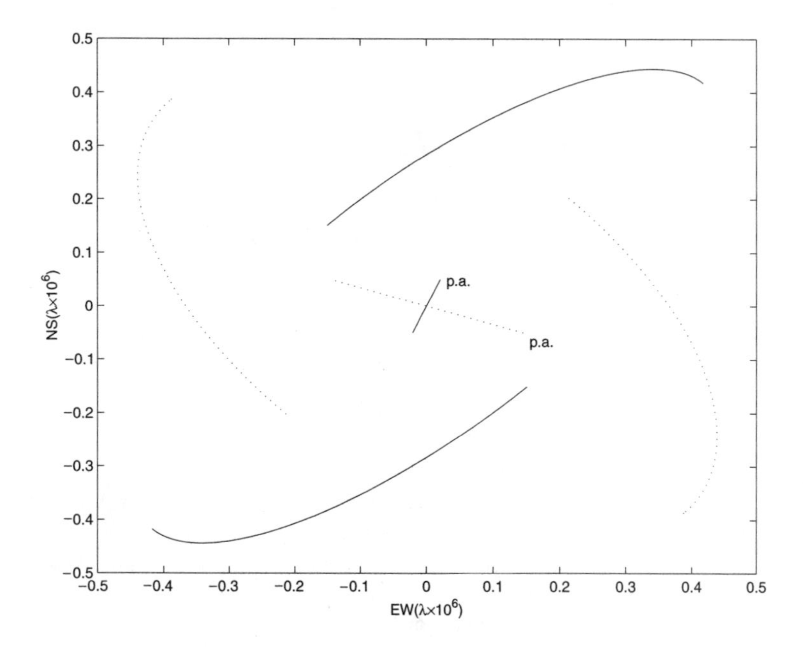

Fig. 3.35. A schematic representation of the observations of a resolved binary star by an interferometer. Because of the Earth's rotation, the lengths of the N–S and E–W baselines projected onto the sky at the binary star are slowly and continuously changing. Hence the elliptical arcs (solid lines) for the two components of the binary at a given separation and position angle from one night of observations. The dotted lines illustrate the projections when the binary was observed on a different night. The baseline length B is measured in millions of wavelengths, so that observations at different wavelengths effectively change the value of B.

NS and EW of the binary's (x, y) coordinates onto the N–S and E–W baselines are given by

$$\mathrm{NS} = x\cos(\phi - \delta)\cos(ha) + y\sin(ha) \tag{3.61}$$

$$\mathrm{EW} = x\cos(\phi - \delta)\sin(ha) + y\cos(ha) \tag{3.62}$$

where the binary star at declination δ is observed at an hour angle ha from an interferometer at latitude ϕ on the Earth's surface. Figure 3.35 provides a schematic illustration of the varying projections of the binary-star separation onto N–S and E–W baselines as the Earth rotates. The changing baseline then samples the squared visibility function, V^2, providing n observations at n times during a night. The V^2 time plots, Figure 3.36, for the different arms of the interferometer and for different wavelengths of observation can then yield the angular separation and position angle of the binary on that night. For the binary stars observed thus far with a direct interferometer, the orbital periods have been greater than 100 days, so the amount of binary orbital motion during a night has been negligible.

In summarizing this subsection, we see that these three independent methods of observation for resolvable binary stars yield the same quantities: time, angular separation, and position

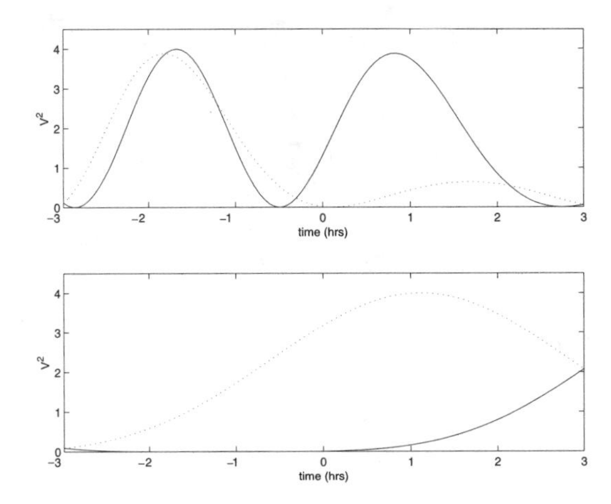

Fig. 3.36. An example of the variation of the squared visibility V^2 for a binary star as a function of time of observation, using the E–W (solid lines) and N–S (dotted lines) arms of an interferometer. The two panels refer to two separate nights of observations and are for the same orientations of the binary as in Figure 3.35. Other examples of V^2–time plots can be found, for example, in the papers by Armstrong et al. (1992a,b) and Hummel et al. (1993, 1994, 1998).

angle, or (t, x, y). So all the sets of data on resolved binaries can be analysed with the same standard procedures, as described in the next subsection.

3.4.2 Orbital elements from resolved binaries

The visual binary stars were the first double star systems to be discovered, as noted in Chapter 1. Methods for determination of the orbits of such binaries have been developed over more than 150 years, and many of the star systems that have been discovered have not yet completed a single orbit since their identification. A very valuable treatise on the subject is *Double Stars*, by Heintz (1978), who considers the historical development of such studies, including the methods used to determine orbital parameters from data sets of angular separation and position angle that cover only a fraction of the whole orbit. Of course, there are those double stars with orbital periods of just a few years, and they have been studied extensively, but the majority of the visual binaries that have been discovered have long orbital periods, simply because the linear separations between components must be large in order to allow angular resolution of at least a few tenths of an arcsecond at nearby stellar distances.

Figure 3.37 illustrates a set of observations for a resolved binary, with the data covering more than one complete orbit. Such plots of orbital motion projected onto the sky are intuitively obvious and appealing to the eye. But because these are graphs of one measured projection, x, versus the other, y, with their associated uncertainties of measurement, they compare unfavourably with the data from spectroscopic binaries, where one measured quantity, *radial velocity*, is plotted against the times of observation, and this latter quantity is known

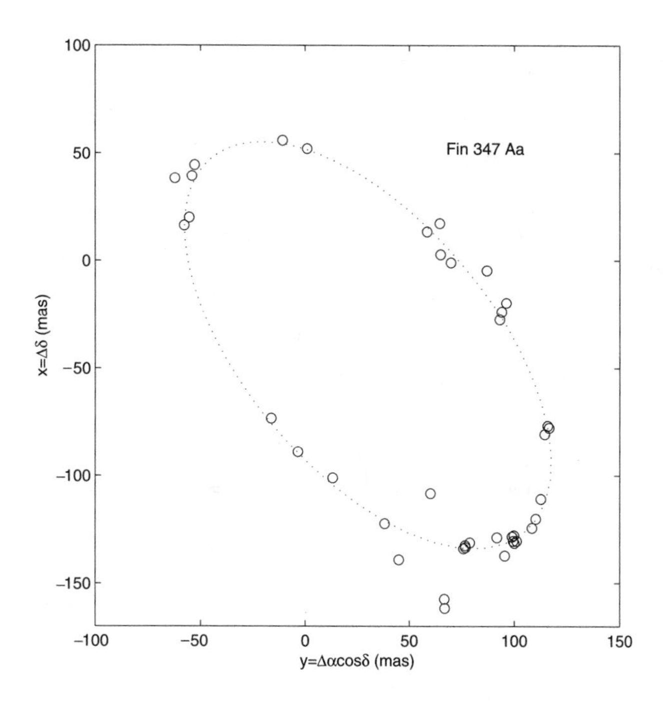

Fig. 3.37. Speckle-interferometer observations of the projected orbit of the double-lined spectroscopic binary Fin 347Aa, made by Mason et al. (1996). The dotted line is the final model orbit calculated from these speckle data. Note that the scales are measured in milliarcseconds (mas). (The orientation of this figure is north up and east to the right.)

with much greater precision than any other. It has been standard practice in visual-binary work to make separate graphs of (t, x) and (t, y) in order to ensure that the two measured quantities do provide mutually consistent results and permit two determinations of the orbital period. Such *time–displacement curves* are very useful aids in the search for the correct orbital parameters for a particular binary system. Van der Kamp (1967) describes how they can be used to provide graphical estimates of orbital period and eccentricity, which can be valuable starting points for the orbital analysis, particularly when the data are not the most accurate. Some examples of such curves are given in Figures 3.38–3.41 for a range of values of (e, i, ω, Ω).

For the more recent data sets, such as speckle and direct interferometric observations covering complete binary orbits in a few years, the analysis is quite straightforward and makes use of the equations introduced in Chapter 2, specifically the Thiele-Innes constants, equations (2.63). Using preliminary estimates of (P, T, e), determined from projected orbits (x_i, y_i) and/or from $[(t_i, x_i), (t_i, y_i)]$ curves, Kepler's equation is solved iteratively for each of the n observations, as in the case of spectroscopic observations. Then the *elliptical rectangular coordinates* X_i and Y_i are calculated for each observation. With n observations, where n well exceeds the four unknown Thiele-Innes constants (A, B, F, G), a least-squares solution of the n equations

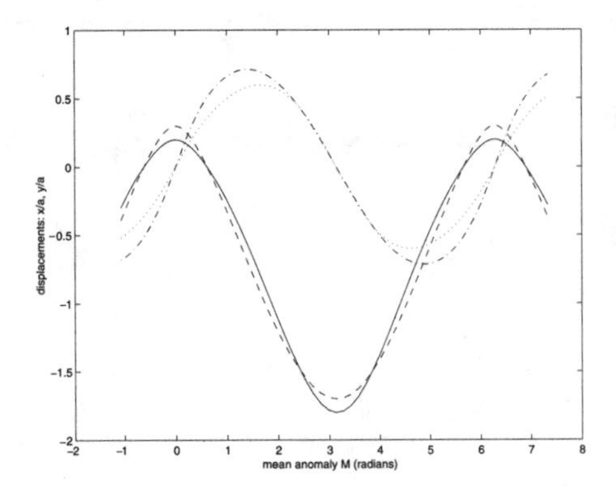

Fig. 3.38. Time–displacement curves for orbits of eccentricity $e = 0.5$ [(t, x), solid line; (t, y), dotted line] and $e = 0.7$ [(t, x), dash line; (t, y), dash-dot line], with $i = \omega = \Omega = 0°$.

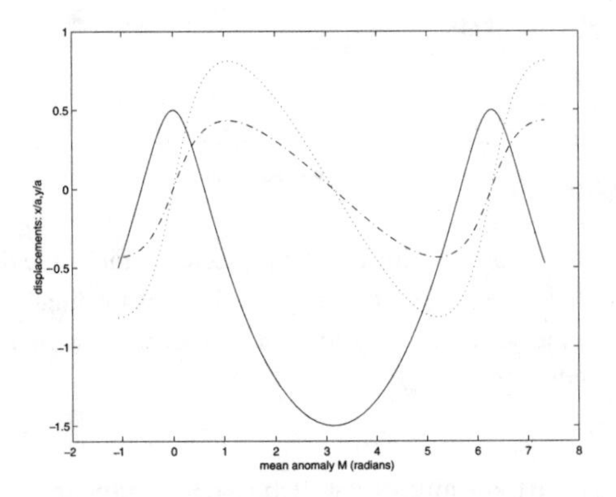

Fig. 3.39. Time–displacement curves for orbits with $e = 0.5$ and $i = 20°$ [(t, x), solid line; (t, y), dotted line] and $i = 60°$ [(t, x), dash line; (t, y), dash-dot line], with $\omega = \Omega = 0°$.

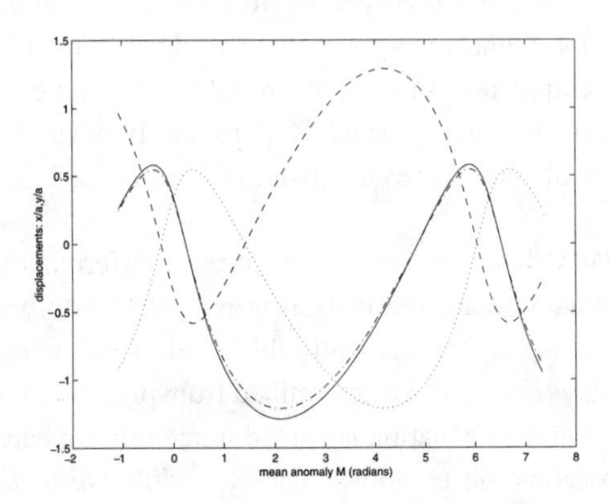

Fig. 3.40. Time–displacement curves for orbits with $e = 0.5$, $i = 20°$, and $\omega = 45°$ [(t, x), solid line; (t, y), dotted line] and $\omega = 135°$ [(t, x), dash line; (t, y), dash-dot line], with $\Omega = 0°$.

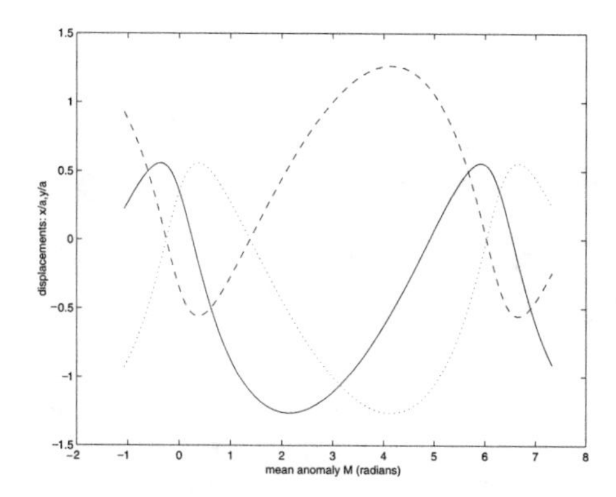

Fig. 3.41. Time–displacement curves for orbits with $e = 0.5$, $i = 20°$, and $\Omega = 45°$ [(t, x), solid line; (t, y), dotted line] and $\Omega = 135°$ [(t, x), dash line; (t, y), dash-dot line], with $\omega = 0°$.

[equations (2.64)], for those n observations will provide values for the four Thiele-Innes constants. A grid search is usually required in these solutions in order to refine the values of P, T, and e in particular. As for spectroscopic orbits, the calculation of orbital phases or mean anomalies is the first step in the analysis, and discordant values of P, T, and e will give incorrect values of the eccentric anomalies E_i and hence incorrect values of (A, B, F, G) and the corresponding orbital elements.

Hartkopf, McAlister, and Franz (1989) discussed the significant complications that can arise when the data set includes observational data from different types of studies, such as visual observations performed with telescopes of different apertures, as well as data from speckle and direct interferometry. The interferometric methods are generally much more accurate. But a long series of visual data may define an orbital period very accurately, even though the individual values of (ρ, ϕ) may be 10 times less accurate than modern speckle data. Figure 3.42 provides an illustrative example of the combination of visual observations with more accurate speckle observations of the spectroscopic-visual triple system HD202908. The visual data extend over 116 years and help to define the orbital period of 78.5 years, and the speckle data cover 19 years around the most recent periastron date (1987), where the angular separation decreased to only about 40 milliarcseconds.

For direct interferometry, where many observations per night provide values for the visibility as a function of time at several different wavelengths, Hummel et al. (1993) described a procedure for utilizing all the individual observations, rather than adopting mean values of (ρ, ϕ) for each night. The preliminary determination of the orbit is performed in the manner described earlier, with these nightly average values. Then theoretical visibility functions are calculated from the preliminary orbital elements, the position of the binary on the sky, and the known parameters of the interferometer. These theoretical functions are then fitted by a least-squares procedure directly to all of the observational data, including those nights of observation where the observed visibilities do not pass through successive maxima and minima (e.g., Figure 3.36). Figure 3.43 shows an excellent example of the quality of present-day orbit

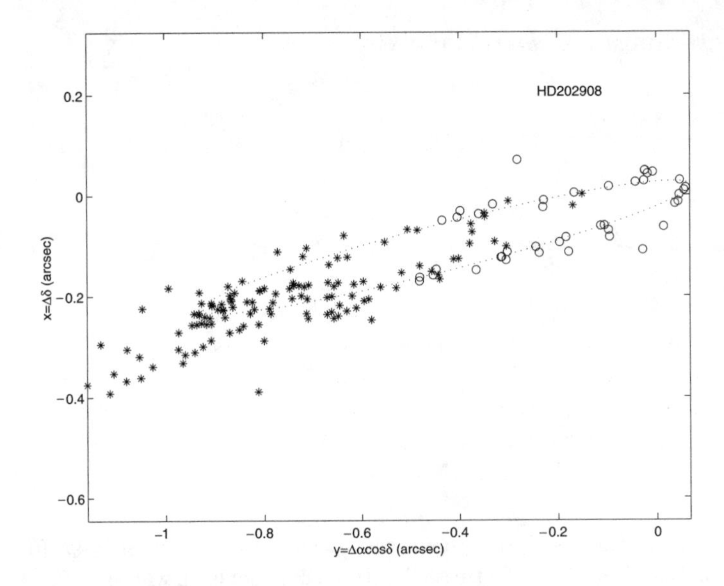

Fig. 3.42. Visual (∗) and speckle-interferometer (○) observations of the projected orbit for the spectroscopic-visual triple system HD202908, made by Fekel et al. (1997). The dotted line is the final model orbit calculated from these visual and speckle data. (The orientation of this figure is north up and east to the right.)

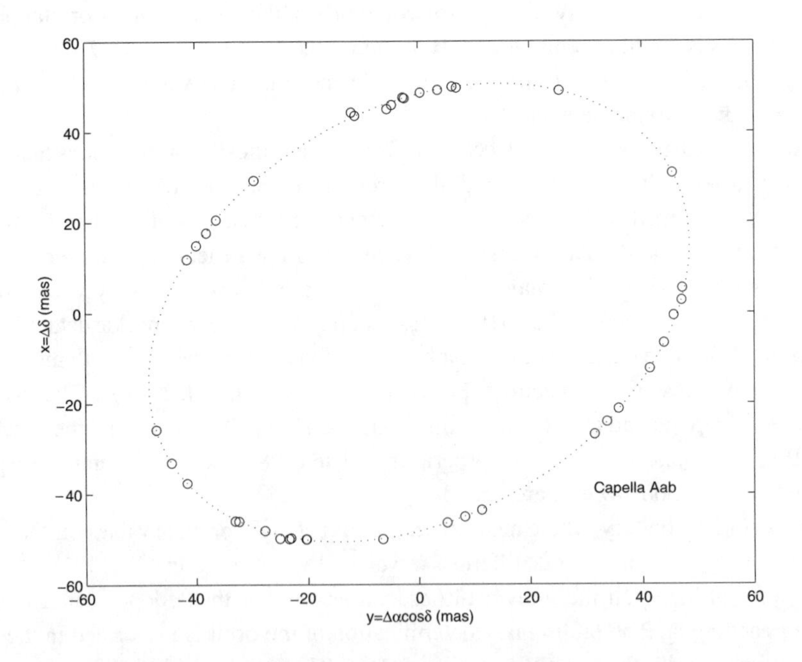

Fig. 3.43. Observations of the projected orbit of the double-lined spectroscopic binary star Capella, made by Hummel et al. (1994) with the Mark III long-baseline interferometer. In combination with published spectroscopy (Barlow et al. 1993) the masses of the two stars were determined with uncertainties of ±2% and the radii with uncertainties of ±2–4%, and evidence was presented for the presence of cool spots on the surface of component Ab, of spectral type G1III. (The orientation of this figure is north up and east to the right.)

determinations by interferometry, which can now exceed the accuracy attained by the best spectroscopic observations.

3.5 Orbits from polarimetric observations

Our ability to measure accurately the state of polarization of optical radiation from astronomical sources developed rapidly in the 1970s. *Photopolarimetric observations* of various binary stars illustrated a variety of weak polarization phenomena that were phase-locked with the binary orbits. Rudy and Kemp (1978) provided a first description, and Brown, McLean, and Emslie (1978) (BME) presented a general discussion of polarization by electron (Thomson) scattering in optically thin stellar envelopes of arbitrary shape and illuminated by any number of point sources. They applied their theoretical model to the interpretation of data from binary stars and demonstrated that reliable determinations of the orbital inclination i and the longitude of the ascending node Ω could be made from consideration of the variations with orbital phase of the *Stokes parameters* used to describe the state of polarization. The theory was extended to consideration of binary stars with eccentric orbits by Brown et al. (1982), and to the effects of occultations within binary systems by stars of finite size by Drissen et al. (1986a), Robert et al. (1990), and St. Louis et al. (1993). Thus photopolarimetry has developed into a powerful technique that can be used to determine (i, Ω) values for binaries independently of other procedures such as the light curves of eclipsing binaries or spectroscopic orbits. Here we illustrate the main results for determining (i, Ω).

Of course, the technique requires a measurable polarimetric signal, and the degree of polarization is typically quite small, $p < 1\%$. To achieve the necessary S/N for following the phase-locked *variations* in such a small signal, it has been necessary to use broad-band *UBVRI* filters in photopolarimeters. Measurable polarization is mainly limited to binaries containing at least one hot component that emits sufficient UV photons to ionize hydrogen and provide free electrons for Thomson scattering of that radiation to take place, or to binaries where there is a ready supply of free electrons, such as the hot ionized winds from O stars and Wolf-Rayet (WR) stars. The most successful applications of the BME theory, with the previously mentioned extensions, have been made by the Montreal group in studying the WR+O-star binaries. Here, the O-star photons are scattered off the electrons in the WR wind, with a concentration of that scattering taking place on the hemisphere of the WR outflow that faces the O star. As the binary rotates, the observer sees that part of the system at different orientations and consequently observes phase-locked variations in the degree of polarization p and its phase angle θ that can be described by the *Stokes parameters* Q and U, defined as

$$Q = p\cos(2\theta); \qquad U = p\sin(2\theta) \tag{3.63}$$

The observed variations of Q and U with orbital phase ϕ are sensibly represented by the following Fourier series, with $\lambda = 2\pi\phi$:

$$Q(\lambda) = q_0 + q_1\cos(\lambda) + q_2\sin(\lambda) + q_3\cos(2\lambda) + q_4\sin(2\lambda) \tag{3.64}$$

$$U(\lambda) = u_0 + u_1\cos(\lambda) + u_2\sin(\lambda) + u_3\cos(2\lambda) + u_4\sin(2\lambda) \tag{3.65}$$

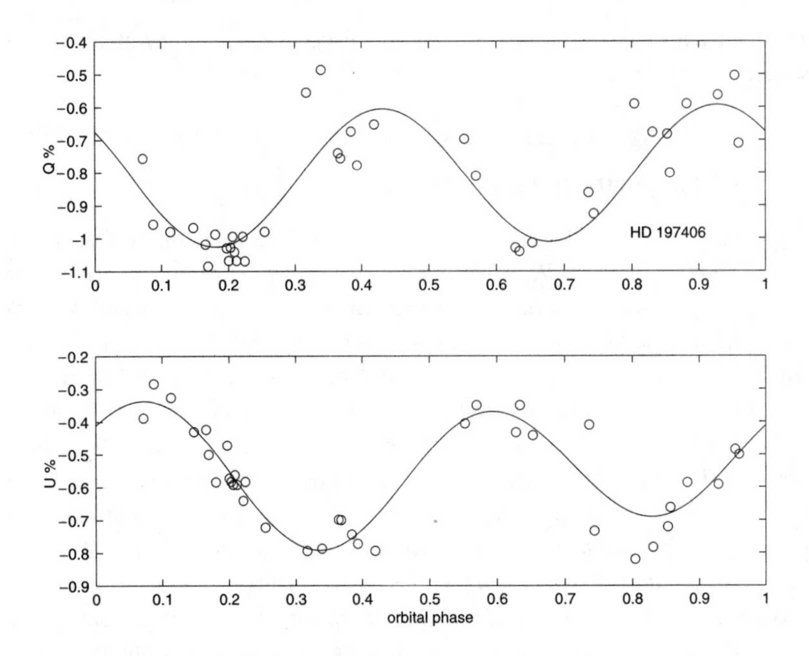

Fig. 3.44. Photopolarimetric observations of the WN7 binary system HD197406, from Drissen et al. (1986a,b), showing the Stokes (Q, U) parameters (o) plotted against orbital phase ϕ. The solid lines through the data are Fourier sine/cosine series, including terms up to the second harmonic $(2 \times 2\pi\phi)$.

as illustrated in Figure 3.44 for the WR binary HD197406. These values of $Q(\lambda)$ and $U(\lambda)$, smoothed by the Fourier-series representation, can be plotted in the (Q, U) plane. These variations of Q and U during the orbital motion trace out a Lissajous-type figure (as described by the BME theory) whose shape and orientation will depend on the orientation of the binary to the observer's line of sight, the angles (i, Ω), and the distribution of the density of scatterers around the binary. An example of such photopolarimetric data is given in Figure 3.45 for the WR binary HD197406.

The quantities Q_+ and U_+ can be derived from the data, defined as

$$Q_+(\lambda) = \frac{1}{2}[Q(\lambda) + Q(\lambda + \pi)] \tag{3.66}$$

$$U_+(\lambda) = \frac{1}{2}[U(\lambda) + U(\lambda + \pi)] \tag{3.67}$$

so that only the second harmonics contribute. This closed elliptical locus in the (Q, U) plane is that produced by a binary with its scatterers being distributed symmetrically about the orbit plane; the locus is executed twice per binary orbit.

As shown by BME, the geometry and the Fourier coefficients of the (Q_+, U_+) locus provide two alternative methods for determining (i, Ω). The eccentricity of the (Q_+, U_+) ellipse provides a determination of the orbital inclination i, and its orientation in the (Q, U) plane determines Ω relative to the observer's reference frame for the polarimetry. (It is usual practice for

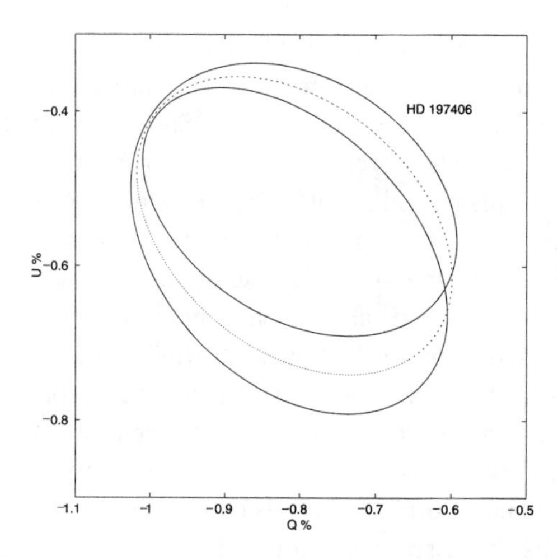

Fig. 3.45. The smoothed (Q, U) variations with orbital phase for the binary system HD197406 plotted in the (Q, U) plane. The Lissajous-type figure is typical of that seen in many binary systems and reflects the orientation of the orbit as well as the distribution of scatterers around the system. The dotted line is the (Q_+, U_+) locus defined by the second-harmonic terms only and is executed twice over the binary orbit.

the phase angle of the polarization to be measured relative to the north point.) The eccentricity of the ellipse is

$$e = \sin^2 i/(2 - \sin^2 i) \tag{3.68}$$

and its major axis lies at an angle Ω to the Q axis. So the ellipse becomes circular for $i = 0°$, and linear for $i = 90°$.

Alternatively, one can use algebraic formulae that include only the second-harmonic Fourier coefficients $(q_i, u_i; i = 3, 4)$ of equations (3.64) and (3.65) to determine (i, Ω):

$$\begin{aligned} \left(\frac{1-\cos i}{1+\cos i}\right)^4 &= \frac{(u_3+q_4)^2+(u_4-q_3)^2}{(u_4+q_3)^2+(u_3-q_4)^2} \\ \tan\Omega &= \frac{D-T}{B-C} = \frac{B+C}{D+T} \end{aligned} \tag{3.69}$$

where

$$\begin{aligned} T &= \frac{u_3+q_4}{(1+\cos^2 i - 2\cos i)} \\ B &= \frac{u_4-q_3}{(1+\cos^2 i - 2\cos i)} \\ C &= \frac{u_4+q_3}{(1+\cos^2 i + 2\cos i)} \\ D &= \frac{q_4-u_3}{(1+\cos^2 i + 2\cos i)} \end{aligned} \tag{3.70}$$

Equations (3.69) and (3.70) are from a paper by Drissen et al. (1986a) that corrects a small misprint in BME in the equation for i and provides an improved method for Ω. Again, there is ambiguity in the value of Ω, with Ω and $\Omega + 180°$ being equivalent. Standard convention is to accept the value $\Omega \leq 180°$. The effect of interstellar polarization is simply to displace the (Q, U) locus in the (Q, U) plane, without affecting its shape.

This polarimetric technique for finding (i, Ω) is particularly helpful in the context of WR+O-star binaries, because the light curves of such systems, the usual sources for i, are frequently distorted by the effects of photometric absorption along varying path lengths through the circumstellar envelope and winds of the WR star, the wind of the O star, and the collisional interactions of the two winds in such binaries. The representation of the polarimetric observations by a smoothed variation described by Fourier coefficients can allow systematic errors to be incurred in the determination of the orientation angles. One way to minimize such errors is to ensure that the entire range of orbital phases is covered equally well, but variability caused by changing distributions of the scatterers in the binary can add complications. The orbital periods of WR+O-star systems typically are days to tens of days, so that complete coverage of the orbital cycle is a difficult task for any single observing site on the Earth's surface.

An alternative procedure, again based on the BME theory, has been introduced by Harries and Howarth (1996a) and found to be very effective. The position angle θ of the polarization that is intrinsic to the binary system is always perpendicular to the plane containing the line of centres of the two stars and the observer's line of sight. The angle θ changes as the binary rotates, and it provides information on the binary orbit projected onto the sky that is similar to the description used for resolved binaries. Harries and Howarth have shown that the following equation provides the necessary link between the observable quantities and the projected orbit on (x, y):

$$\theta = \frac{1}{2} \arctan(U/Q) = \arctan(y/x) + \pi/2 \tag{3.71}$$

This is illustrated in Figure 3.46.

Recall that (x, y) are related to the orbital elements $(P, T, e, i, \omega, \Omega)$, so that a complete specification of the projected orbit is possible.

The method requires the interstellar polarization vector to be subtracted vectorially before the analysis, but it has the advantage that it is not sensitive to global changes in the density distribution of scatterers, the overall rate of mass loss that is observed to occur in these hot, rapidly evolving systems. As for the (Q, U)-plane approach, it is necessary to omit from the analysis those orbital phases where occultations or 'core eclipses' are taking place, unless a good prescription is available for taking proper account of occultations of parts of the WR outflow.

Improvements in CCD technology in the 1990s have permitted high-quality spectropolarimetry to be performed on these types of binary systems. Continuum regions of the spectrum that are free from the potential distortions of the polarization signal within the broad emission lines of WR stars can be identified and used to determine the continuum polarization (p, θ) at many different wavelengths. Such results from the interacting WR+O-star binary CQ Cephei (Harries and Hilditch 1997a), for example, are in very good agreement with the broad-band

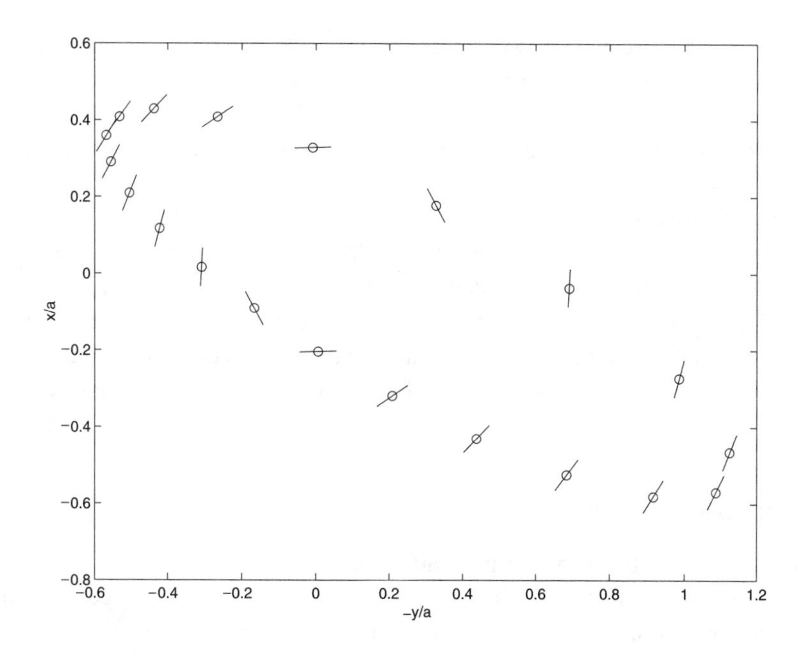

Fig. 3.46. A schematic polarimetric orbit derived from the Stokes parameters Q and U with orbital elements of $e = 0.4$, $\omega = 45°$, $\Omega = 60°$, and $i = 75°$ and projected onto the plane of the sky (y, x) in the same manner as observations of astrometric binaries (north up, and east to the left). The vector through each 'observation' point denotes the orientation of the angle θ defined in the text.

photopolarimetry of Drissen et al. (1986b) and Piirola and Linnaluoto (1989). The advantage of spectropolarimetry lies in the fact that the observations provide values for p and θ across the spectral lines as well as the continuum. In WR stars, the broad emission lines are generally formed outside the continuum thermalization radius, with those from less ionized species being formed farthest away. The emission-line flux is therefore unpolarized, and observations confirm substantial reductions in the value of p through these spectral lines. The interstellar component can therefore be separated from the intrinsic contribution to (q_0, u_0) directly.

This *spectropolarimetric-orbit* technique is proving to be of particular importance for the *symbiotic stars*. Recall from Chapter 1 that such binary systems are composed of red-giant stars (with substantial outflowing winds) and small, hot companion stars. For many years, the presence of unidentified but very strong emission lines at λ6825 and λ7082 in the spectra of symbiotic stars had defied explanation. A solution proposed by Schmid (1989) was that they were due to O VI λ1032 and λ1038 photons from the hot star being *Raman-scattered* off H atoms in the red-giant wind and producing the observed emission lines. Spectropolarimetric surveys of symbiotic stars (Schmid and Schild 1994; Harries and Howarth 1996b) have established the global properties and detailed secular variations of these lines amongst many systems, whilst theoretical models for the production of these lines (Schmid 1992; Harries and Howarth 1997) are proving to be very successful. During those surveys, it has been found that the strong polarization ($p \approx$ 5–15%) across these emission lines shows reasonable variations coupled to the binary orbit in those binaries whose periods have been determined from radial-velocity work. In particular, the polarization vector in the blue wings of the two emission lines has been found to be perpendicular to the binary line of centres (Schmid and Schild 1994). This part of the line profile is formed predominantly in that region of the red-giant outflow that is seen to be blue-shifted by the source of the O VI photons (Harries and Howarth 1997). The geometry that produces the variations of the polarization vector that is phase-locked to the binary orbit is the same as that noted earlier with reference to the WR+O-star systems, and the general model introduced by Harries and Howarth (1996a) was first used by them to establish the binary orbit of the symbiotic star SY Mus. This new development for studying symbiotic systems is a major breakthrough in the search for reliably determined orbital parameters that can provide realistic determinations of the masses of the evolved stars in these strongly interacting systems.

3.6 Combined solutions

Simultaneous solutions for data sets from different methods of observation are now being pursued to yield improved descriptions of binaries of various types. Typical combinations of observations are the use of radial velocities with pulse-timing data and the use of radial velocities with astrometric or polarimetric data, as well as addition of x-ray eclipse durations or complete light curves to those combinations. Morbey (1975, 1992) has championed this combined approach for spectroscopic/astrometric binaries, and Wilson (1979) and Wilson and Terrell (1998) have pursued combinations of radial velocities with general photometric variations (light curves) and with x-ray timing data, respectively.

In all cases it is necessary to establish *preliminary* values for orbital elements by some structured approach, such as generating *grids* of possible radial-velocity curves, pulse-timing curves, time–displacement curves, or light curves that may be close to those observed for the particular binary under investigation. For radial-velocity and pulse-timing curves, such grids can be limited to a few values of e and ω, as illustrated earlier in this chapter, and semiamplitudes can readily be estimated directly from the plotted data. The differential-correction procedures described in Sections 3.2.10 and 3.3.2 provide very fast and robust solutions subject to well-constrained values of orbital periods and epochs (e.g., the time of maximum positive velocity). For astrometric/polarimetric data, grids may have to be more extensive, because e, i, ω, and Ω are involved. Beyond such initial estimates, the analyst can pursue techniques like downhill simplex schemes, genetic algorithms, or differential corrections if the prior grid search is fine enough. Entire books have been written on such numerical procedures, and I would simply recommend that the texts *Data Reduction and Error Analysis for the Physical Sciences*, second edition (Bevington and Robinson 1992), and *Numerical Recipes*, second edition (Press et al. 1992), be consulted for complete descriptions, algorithms, and subroutines that can be adopted.

When combining such data into a final least-squares optimization procedure, it is necessary to make all of the (observed−calculated) residuals $R(t)$ dimensionless by dividing them by the appropriate *variance* of the data set. For example, in the now frequent investigations of spectroscopically resolved binaries, the adopted least-squares procedures aim to minimize the value of χ^2 for the three-dimensional solutions of the projected-orbit data (ρ_j, ϕ_i) and the radial-velocity data V_k, where χ^2 is defined by

$$\chi^2 = \sum_i w_i(\Delta\phi_i/\sigma_\phi)^2 + \sum_j w_j(\Delta\rho_j/\sigma_\rho)^2 + \sum_k w_k(\Delta V_k/\sigma_V)^2 \tag{3.72}$$

The weighting factors w are required to reflect the differing accuracies of the input data, and considerable care must be exercised in deciding what values to assign to them. Hartkopf et al. (1989) have provided a substantial discussion of these issues in relation to combinations of astrometric data from different sources: visual, speckle, interferometric, and radial-velocity data.

4 Perturbations, the Roche model, and mass exchange/loss

4.1 Introduction

If binary stars simply executed their orbits according to Newtonian theory for point masses, then interest in their properties would have waned long ago, save for the need to improve determinations of stellar masses. The universe is rather more exciting, however, and at least the close binary stars (as defined in Chapter 1) display all manner of perturbations and interactions that guarantee that they will continue to provide an abundance of astrophysical phenomena that will require explanation. In this chapter we consider a sequence of progressively greater departures from the point-mass, spherical-star model that we used in Chapters 2 and 3.

We consider, firstly, a theory of mild perturbations, or deviations from the idealized spherical shape for a star, which theory can fully explain the observed phenomena of *apsidal motion*, the *circularization* of orbits, and the *synchronization* of stellar axial-rotation periods and orbital periods. The stars in such binary systems become *tidally locked*, such that two stellar hemispheres face each other, and two are permanently averted. The logical extension of these perturbations is to the *Roche model* for binary stars that is applicable to tidally locked systems in circular orbits. Here the stars can be virtually spherical in shape when their radii (R) are small relative to their separation $(a)(R/a < 0.10)$, and they appear no different from those in the earlier point-mass theory. But the Roche model also permits stars to become seriously distorted from spherical shape, with $R/a > 0.20$, far beyond the limitations of the earlier perturbation theory. The Roche-model prescription has been used very successfully for study of all types of interacting binary stars: the non-spherical but detached stars that produce binary light curves, classified as EB type; the EW type, in which the stars reach contact or overcontact states (see Figure 1.2); the mass-exchanging Algols, cataclysmic variables, and x-ray binaries. Lastly, in the context of this Roche model we shall consider the consequences of *exchanges of mass* and *exchanges of angular momentum* between the components of interacting binaries, as well as *mass losses* and *angular-momentum losses* from binary systems, and how those events affect orbital periods, eccentricities, and separations between components. These latter phenomena play crucial roles in the *evolution* of binary stars and explain many of the behaviours exhibited by the interacting and unusual binary systems that are found in the Milky Way galaxy and other nearby galaxies.

There are also binary systems for which the Newtonian theory of gravitation cannot provide an adequate description, such as those containing neutron stars and black holes. Here, the general theory of relativity is the required prescription for gravitation, because the concentration of the mass of a neutron star (typically 1.4 $M_{\odot}$) into a small volume (typical radius about 10 km) can so distort space-time, as can the singularity of a black hole. In the present context, we can regard such general-relativity effects as simple perturbations upon standard Newtonian theory.

Finally, there is the well-known fact that some 20% of binary-star systems have been found to be physical members of triple-star or multiple-star systems (Abt 1983). Any additional point-mass source of gravitation in a system will provide a perturbation of the two-body motion of the binary, and its size obviously will depend on the mass of the perturber and its distance from the binary.

4.2 General perturbations

A little more than two centuries ago Lagrange completed his work on the mathematical theory of perturbations of the two-body motion of a planet around the Sun caused by the presence of other bodies. The theory can be formulated in terms of a *perturbing gravitational potential* S, such that the total gravitational potential experienced by the planet is $\Phi = \Phi_0 + S$, where Φ_0 is the gravitational potential of the isolated point-mass two-body system in which the planet would move in an unchanging, closed, elliptical orbit as discussed in Chapter 2. Beyond the requirement of the theory that $S \ll \Phi_0$, so that the perturbations will be small in comparison with the main control on the orbital motion set by Φ_0, the source of S is quite general and is therefore applicable to many situations encountered in observed systems of binary stars. The perturbing potential S can be caused by one or both stars in the binary being non-spherical in shape, or by one or both having such extreme gravitational potentials that general-relativity corrections to the Newtonian theory are required, or by membership of the binary in a triple or multiple system, or even by various combinations of all three potential contributions.

Much effort has been devoted to applying Lagrange's theory in studying the motions of the planets about the Sun and the motions of satellites about planets. Various formulations have been used, and the reader is referred to classical works on the subject of celestial mechanics, such as those by Smart (1953) and Brouwer and Clemence (1961), for more details. A helpful summary description of the derivation of *Lagrange's planetary equations* is given by Roy (1978) in his text *Orbital Motion*, and later we shall use one form of those equations. This set of *planetary equations* describes how the orbital elements that define the two-body motion will be forced to change with time as a direct consequence of the presence of a perturbing potential S, which is sometimes also called a *disturbing function*. The orbital elements will no longer be constants of the motion, but will become functions of time. The instantaneous values of the elements a, e, i, ω, Ω, T, and P at a given time t are called the *osculating elements* of the ellipse. *Osculating* means 'just touching' or 'kissing', or, in this context, it means that these values for the orbital elements describe the two-body orbital motion only at this instant t. One orbital period later, $t + P$, say, the shape of the orbit will have changed slightly, and the two loci will be just touching at various points around their orbit paths.

One form of *Lagrange's planetary equations*, taken from Roy's text (1978), is as follows:

$$
\begin{aligned}
\frac{da}{dt} &= \frac{2}{na}\frac{\partial S}{\partial \chi} \\
\frac{de}{dt} &= \frac{1}{na^2 e}\left[(1-e^2)\frac{\partial S}{\partial \chi} - (1-e^2)^{1/2}\frac{\partial S}{\partial \omega}\right] \\
\frac{d\chi}{dt} &= -\frac{(1-e^2)}{na^2 e}\frac{\partial S}{\partial e} - \frac{2}{na}\frac{\partial S}{\partial a} \\
\frac{d\Omega}{dt} &= \frac{1}{na^2(1-e^2)^{1/2}\sin i}\frac{\partial S}{\partial i} \\
\frac{d\omega}{dt} &= \frac{(1-e^2)^{1/2}}{na^2 e}\frac{\partial S}{\partial e} - \frac{\cot i}{na^2(1-e^2)^{1/2}}\frac{\partial S}{\partial i} \\
\frac{di}{dt} &= \frac{1}{na^2(1-e^2)^{1/2}}\left[\cot i\frac{\partial S}{\partial \omega} - \operatorname{cosec} i\frac{\partial S}{\partial \Omega}\right]
\end{aligned}
\tag{4.1}
$$

where $n^2 a^3 = G(m_1 + m_2)$ and $\chi = -nT$, n being the *mean daily motion*.

It is clear from these *planetary equations* that the functional form of S will determine which of the orbital elements will change with time. From an observational perspective, the two most important causes of perturbations of two-body point-mass motion are the non-spherical shapes of the stars in a close binary and the presence of *a third body* in the gravitational system. We shall consider these two types of perturbations separately, though recognizing that both can occur in some systems.

4.2.1 Apsidal motion: Observations

The binary stars for which perturbations are significant encompass all those systems that have stars that cannot be considered to be point masses, that is, all close binaries. Tidal torques are set up on each star by the action of its companion, with the result that the stellar-rotation axes become aligned perpendicular to the orbital plane of the binary, the axial-rotation periods become equal to the orbital period, and the departures from spherical shape for both stars are made symmetric with respect to the orbital plane. The enforced synchronous rotation of each star also ensures some rotational flattening, with the polar radius smaller than the equatorial value for each star. The perturbation that is seen most readily by observational astronomy is *apsidal motion*, the precession of the orbit in its own plane, described by the differential term $d\omega/dt$. The observed apsidal rotation periods, the time taken for an orbit to precess through $360°$, range from a few years to a few centuries and hence have been quite well defined for a number of close binaries. Eventually, these tidal torques will also circularize the orbit, making $e \to 0$, in addition to synchronizing rotation/orbit periods and aligning axes, but all these effects take place on much longer time scales than a few apsidal periods. We have to study these longer-term effects via a statistical approach involving the studies of many binary systems with a range of ages so that we can determine observational constraints on the time scales as functions of orbital period or separation. We shall consider some aspects of these theoretical

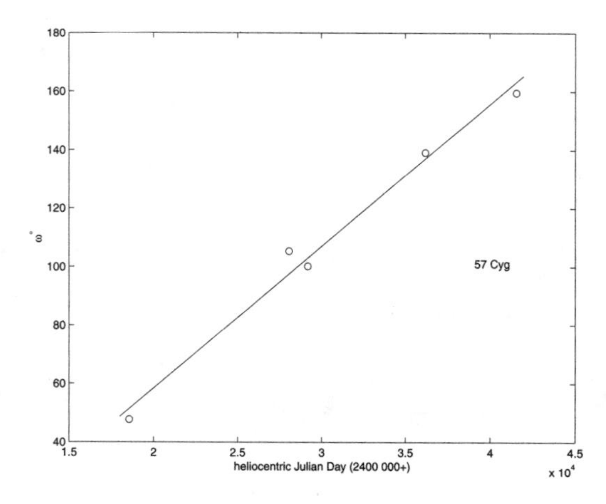

Fig. 4.1. Determination of the longitude of periastron ω for the spectroscopic binary 57 Cyg from five discrete sets of radial-velocity data obtained between 1910 and 1972.

expectations and observational data on orbit circularization, and so forth, later in this chapter. But because there was observational evidence for the existence of apsidal motion before an adequate theoretical explanation had been developed, we shall first consider the observational consequences of $\omega \equiv \omega(t)$ before providing a summary of the theoretical models. As noted in Chapter 1, apsidal motion provides a critical test for the interior structure of a star, its radial-density distribution, as well as a test of general relativity.

Discrete sets of radial-velocity data or pulse-timing data can provide determinations of the orbital elements, as discussed earlier. Provided that such data sets are obtained over time scales that are short compared with significant values for the differentials $d\omega/dt$, and so forth, then sensible results will be obtained. A sequence of such discrete data sets, therefore, can reveal apsidal motion quite readily, and several apsidal-motion systems have been discovered in this manner. An example is the non-eclipsing, spectroscopic binary 57 Cyg, for which discrete sets of radial velocities have been compiled over 60 years. Figure 4.1 illustrates the observed changes in the value of the longitude of periastron, ω, over that time, resulting in a determination of the *apsidal period* U of $U = 203 \pm 4$ years (Hilditch 1973).

A second observational manifestation of apsidal motion is found from study of the times of eclipse minima for eclipsing binary stars with eccentric orbits. If apsidal motion is present, then the observed times of minima will not display a simple linear dependence on the orbital period, because the time intervals between successive primary or secondary eclipses will differ from the time intervals between successive times of periastron passage. We can reasonably call the average time interval between successive eclipses of the same type (primary, secondary) or the time interval between successive nodal passages the *sidereal period* $P_{\rm sid}$ of the binary. The time interval between successive times of periastron passage is known as the *anomalistic period* P, because we measure the various *anomalies* (true, mean, eccentric) from the line of apsides through periastron. The situation is illustrated in Figure 4.2 for the relative orbit of a primary star about a secondary star that is fixed at the focus O of the relative orbit. Here the

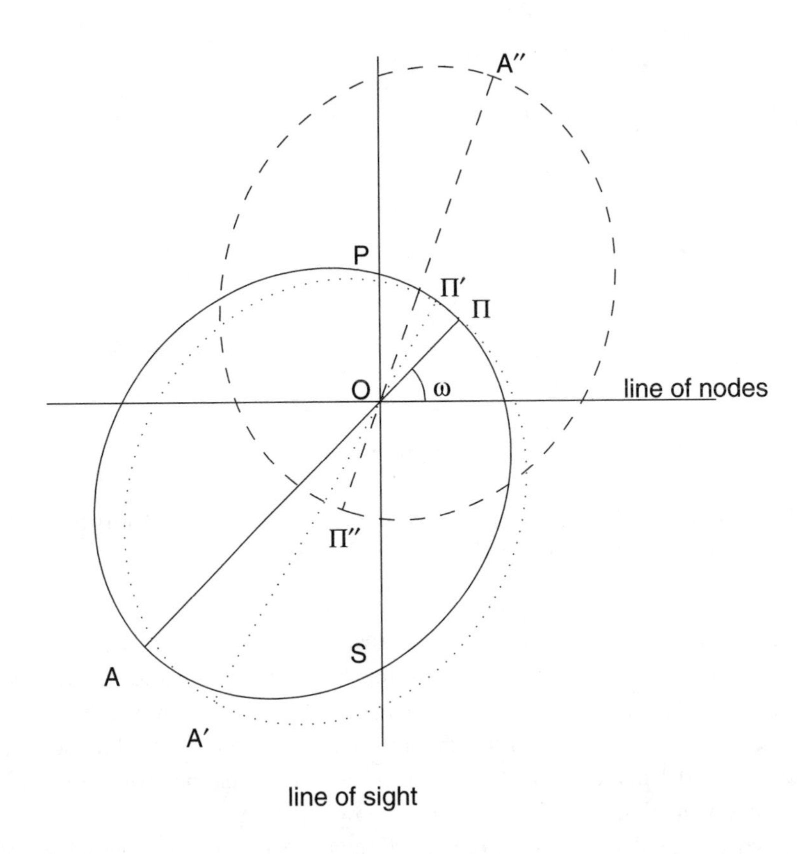

Fig. 4.2. Schematic illustration of the rotation of an orbit in its own plane: *apsidal motion*. In this illustration, three discrete orbits are shown for $\omega = 45°$ (solid line), $\omega = 60°$ (dotted line), and $\omega = 250°$ (dash line). The major axes are labelled ΠA, $\Pi' A'$, and $\Pi'' A''$, respectively.

primary star is eclipsed at position P by the secondary, giving rise to primary eclipse, whilst the secondary is eclipsed when the primary is at position S. The eclipses occur always along the observer's line of sight to the binary, and if the orbit is precessing, then the observations of minima will not occur at even time intervals, and the phase in the orbit of secondary eclipse will move relative to mid-primary eclipse, which is usually taken to define phase 0.0.

From Figure 4.2 it is clear that *primary* eclipses will occur when $\theta + \omega - \pi/2 = 0$, where the true anomaly θ is the angle ΠOP. This is the case when the orbit plane is in the line of sight, so that the orbital inclination is $i = 90°$. For inclined eccentric orbits, $\theta + \omega$ will be close to $\pi/2$, but not equal to it. At some initial epoch t_0, when the sidereal period is P_0, the true anomaly at mid-eclipse will be θ_0, and $\omega = \omega_0$. At a later epoch t_n, n reference periods (P_0) later, $\theta_n \neq \theta_0$ when mid-eclipse occurs because of the effects of apsidal motion. Unfortunately, observational data do not directly provide measures of the true anomaly θ_n, but only the time of mid-eclipse, which can be expressed in terms of a fractional orbital phase, $(t_n - t_0)/P_0$, say, or its equivalent in radian measure, called the *mean anomaly*, $\eta = 2\pi(t_n - t_0)/P_0$. Before the days of digital computers it was common practice to make approximate representations of functions that were difficult to evaluate by means of series expansions. There exists a set of such equations, called the *equations of the centre*, that relate the three anomalies (θ, E, η) to one another by means of series expansions in powers of the orbital eccentricity e (e.g., Smart 1953). The equation relating the observable η to θ is, in general form,

$$\eta = \theta + 2\sum_{k=1}^{\infty} \frac{(-1)^k}{k} e^k \frac{\left[1 + k(1 - e^2)^{1/2}\right]}{\left[1 + (1 - e^2)^{1/2}\right]^k} \sin k\theta \tag{4.2}$$

So at the nth eclipse we have

$$\eta_n - \eta_0 = \eta_n = \frac{2\pi(t_n - t_0)}{P_0} = \theta_n - \theta_0 - 2e\sin\theta_n + 2e\sin\theta_0 + O(e^2, e^3, \ldots) \tag{4.3}$$

where we can define $\eta_0 = 0$ at $t = t_0$. Because $\theta = \pi/2 - \omega$, we can write

$$(t_n - t_0) = (\omega_0 - \omega_n)\frac{P_0}{2\pi} - \frac{eP_0}{\pi}\cos\omega_n + \frac{eP_0}{\pi}\cos\omega_0 + O(e^2, e^3, \ldots) \tag{4.4}$$

Hence the observed displacements of the times t_n of primary-eclipse minima, relative to a linear ephemeris based on the initial sidereal period P_0 and the initial epoch t_0, will consist of the following: the first term, which will be linear if $\omega(t)$ is a linear function of time; the second term, which will contribute a cosine variation for $\omega = \omega(t)$; and the additional terms that will complicate the overall waveform beyond a simple cosine variation if the orbital eccentricity is significant, so that e^2 and higher-order terms will be substantial contributors. This form of variation is precisely that which is seen in the binary systems whose precessions have been monitored for one or more apsidal periods. Figure 4.3 is a good example of such data for the eclipsing binary V526 Sgr. Secondary eclipses occur when $(\theta + \omega - 3\pi/2) = 0$, so the foregoing equation applies with the cosine term, and so forth, having the *opposite signs*. The deviations of times of eclipse minima for primary and secondary eclipses from a linear ephemeris have the same form, but the waveform terms are of opposite sign when the cause of those deviations is apsidal motion. This is a particular signature of apsidal motion that can be used to rule out other causes of apparent period changes, such as orbital motion about a third body where both sets of eclipses show the same variation, or the effects of mass exchange/loss.

It is clear from the foregoing equation that binary systems with substantial orbital eccentricity will show larger deviations of eclipse times for a given apsidal period than will those with more nearly circular orbits (obviously, for circular orbits, when all points in the orbit are at periastron, there cannot be any apsidal motion). Most known apsidal-motion binaries have $e > 0.1$, and so it is not possible to ignore the higher-order terms involving e^2, and so forth. The preceding simple theory would have to become more rigorous if it was to provide a correct interpretation of the observational data. Extensions of the series expansion were made up to terms in e^3 by Todoran (1972), and the problem was extensively revisited by Giménez and Garcia-Pelayo (1983), who included terms up to e^5 and orbits at arbitrary inclinations. Most recently, Lacy (1992) has considered the problem afresh, utilizing iterative numerical techniques to secure exact solutions that provide accurate representations of the observations and yield improved apsidal periods and orbital eccentricities.

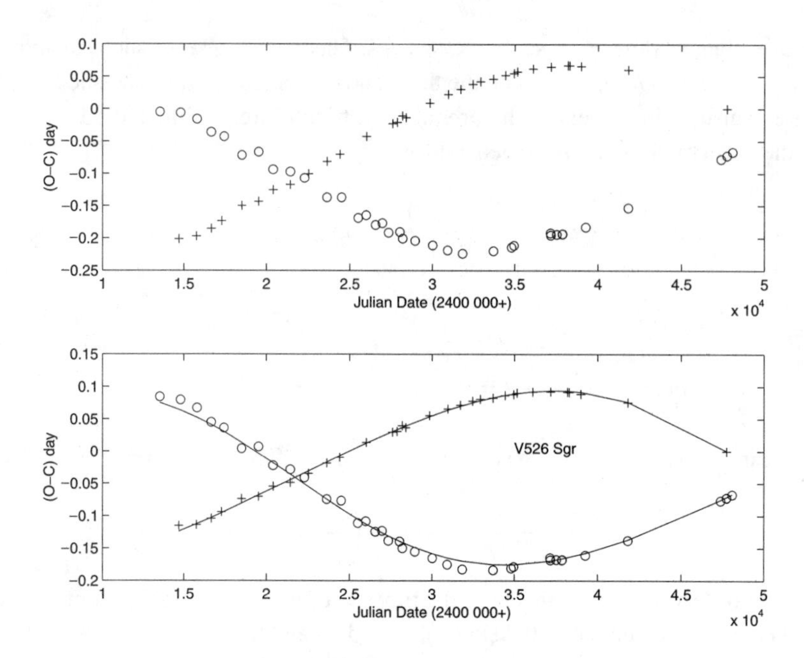

Fig. 4.3. Top: An ephemeris curve for the eclipsing binary V526 Sgr using an initial estimate only of the sidereal period. Note the linearly increasing departure from such an initial ephemeris and the skew-symmetric variations superposed (crosses, primary eclipses; circles, secondary eclipses). Bottom: The revised ephemeris curve with the final value of the average sidereal period and the final fit to the observations according to Lacy's method. *Source*: Data from Lacy (1993).

Lacy's method is based on the assumption that the line of apsides advances at a constant rate that can be expressed as

$$\omega(t) = \omega_0 + \dot{\omega}\epsilon \tag{4.5}$$

where ω_0 is the longitude of periastron at time t_0 expressed in radians, $\dot{\omega}$ is the rate of change of ω in units of radians per sidereal period, and ϵ is the *epoch* of the observation of a primary or secondary eclipse in units of the sidereal period:

$$\epsilon = (t - t_0)/P_{\rm sid} \tag{4.6}$$

Rather than adopting some initial sidereal period P_0, as done earlier, Lacy's method uses an *average sidereal period* $P_{\rm sid}$, defined by the *average time between eclipses* of the same type over the set of times of eclipses that are being studied. So the relationship between the sidereal and anomalistic periods is simply

$$P_{\rm sid} = P(1 - \dot{\omega}/2\pi) \tag{4.7}$$

The actual times of eclipses will deviate from a linear ephemeris based on the average sidereal period in a waveform manner that will be skew-symmetric by an amount that will depend on the eccentricity of the orbit. The expected or calculated (C) time of mid-eclipse is given by

$$C = \text{round}(\epsilon) P_{\text{sid}} + t_0 \tag{4.8}$$

where round (ϵ) is the value of the epoch rounded to the nearest whole number. The foregoing equation must be applied separately to the primary eclipses and the secondary eclipses, with the reference time t_0 referring to the times of mid-primary and mid-secondary eclipses. Hence the deviations, observed – calculated ($O - C$), can be calculated and plotted as in Figure 4.3.

Because $\theta_p = \pi/2 - \omega$ at primary eclipses and $\theta_s = 3\pi/2 - \omega$ at secondary eclipses when the orbital inclination $i = 90°$, the two waveforms have opposite signs but are of the same shape, as illustrated in Figure 4.3. When $i \neq 90°$, these values of $\theta_{p,s}$ are somewhat different from the preceding. The eclipse minima in eccentric-orbit binaries do not occur exactly at the conjunctions, but at the time when the projected separation between the centres of the two stars, δ, is at a minimum, where

$$\delta = \frac{(1 - e^2)}{(1 + e\cos\theta_{p,s})}[1 - \sin^2 i \sin^2(\theta_{p,s} + \omega)]^{1/2} \tag{4.9}$$

and the semimajor axis of the relative orbit is adopted as the unit of length (see Chapter 5). In general, this equation must be solved to find the values of $\theta_{p,s}$ that will minimize δ. An estimate of ω is required first, and this can be obtained from the observed time of the minimum together with the initial estimates of ω_0 and $\dot{\omega}$. A numerical method due to Brent, as described by Press et al. (1992), can be used to find the appropriate values of $\theta_{p,s}$. Once these values of $\theta_{p,s}$ have been determined, we can follow the usual route to establish accurate values for the mean anomalies via the set

$$\begin{aligned} E_{p,s} &= 2\tan^{-1}\left[\left(\frac{1-e}{1+e}\right)^{1/2} \tan\frac{\theta_{p,s}}{2}\right] \\ \eta_{p,s} &= E_{p,s} - e\sin E_{p,s} \\ \Delta t_{p,s} &= \eta_{p,s} P/2\pi \end{aligned} \tag{4.10}$$

This procedure obviates the need to use series expansions between θ and η.

The foregoing procedure can also be used to determine the time of periastron passage, T_0, that occurs immediately before the first reference time of minimum, t_0, that is, when $\omega = \omega_0$, with $T_0 = t_0 - \Delta t_0$. Then the *predicted* times for all the observed eclipse minima can be calculated from the set

$$\begin{aligned} N &= \text{round}[(C - T_0)/P] \\ T &= T_0 + NP + \Delta t \end{aligned} \tag{4.11}$$

Several iterations of this set of equations are required to arrive at the best solution for e, ω_0, $\dot{\omega}$, P, and t_0, with the value of the inclination i supplied from an analysis of the light curve of the eclipsing binary. Both Lacy and Holmgren, Hill, and Scarfe (1995) have described the implementation of a non-linear least-squares algorithm for the analysis based on the foregoing set of equations. Figures 4.3 and 4.4 illustrate modern observations of the times of primary

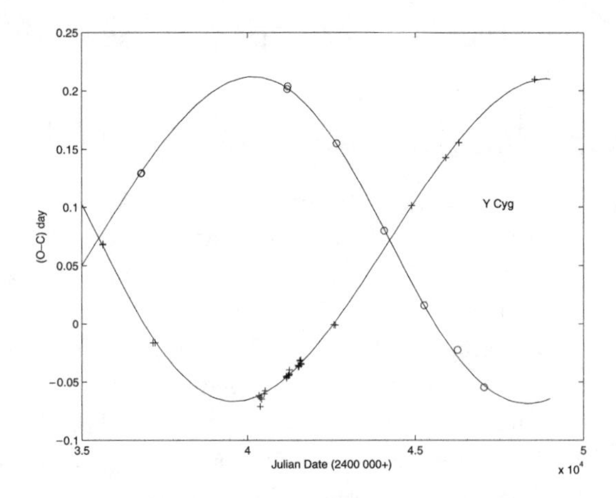

Fig. 4.4. Accurate ephemeris curves of the eclipsing binary Y Cyg determined from photoelectrically determined times of minima (crosses, primary eclipses; circles, secondary eclipses), together with the final fit to the data determined by Holmgren et al. (1995).

and secondary minima collated to form these *ephemeris curves*, together with the final model fits to the observations. Note that $\dot{\omega}$ and e are determined to uncertainties of better than $\pm 1\%$ from the times of minima determined photoelectrically over about 20 years.

4.2.2 Apsidal motion: Theory

There are three sources of perturbing gravitational potential S in an isolated close binary system. These are the general-relativity correction to Newtonian gravitational theory (which we shall consider later in this section), the *rotational potential* created by a body spinning on its axis, and the *tidal potential* exerted on each star by its companion.

For the majority of close binary systems that have been studied sufficiently accurately, it has been found that the axial-rotation periods are equal to the orbital periods, such that stars rotate *synchronously*. Thus equatorial rotation speeds for these stars of radius R and period P are

$$V_{\rm rot} = 2\pi R/P \approx 50(R/{\rm R}_{\odot})/P({\rm day})\ {\rm km\ s^{-1}} \tag{4.12}$$

and any main-sequence star in a typical close binary will be rotating rapidly. Such rapid rotation leads to flattening of the polar regions of the star and to equatorial extension.

A particle on the surface of a star of radius r at co-latitude ϕ' will experience a centripetal acceleration $\vec{\omega} \times (\vec{\omega} \times \vec{r})$ of magnitude $\omega^2 r \sin\phi' = \omega^2 \rho$ directed towards the axis of rotation (the vector $\vec{\omega}$) and perpendicular to that axis (Figure 4.5). That centripetal acceleration $\omega^2 \rho$ corresponds to a *rotational potential*

$$\Phi_{\rm rot} = -\frac{1}{2}\omega^2 \rho^2 = -\frac{1}{2}\omega^2 r^2 \sin^2\phi' \tag{4.13}$$

which contributes to the total potential experienced by the star.

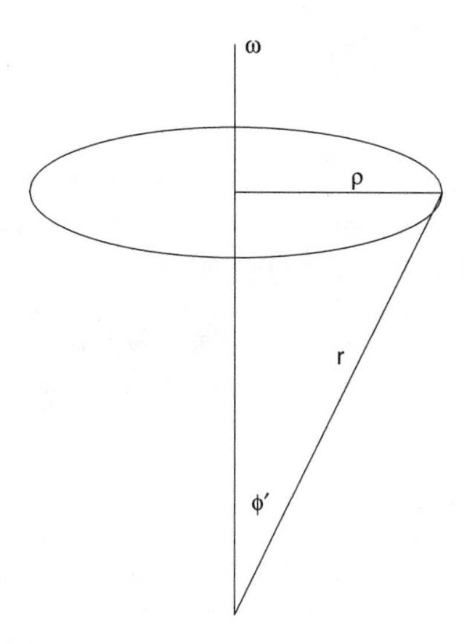

Fig. 4.5. A particle at position (r, ϕ') on the surface of a star of radius r rotating at an angular velocity $\vec{\omega}$.

The standard theory of gravitational potentials says that a general expression can be written for the potential over the surface of a rotationally symmetric but non-spherical body (i.e., a spheroid) in terms of Legendre polynomials $P_n(\cos\phi)$ of the form

$$\Phi(r) = -\frac{Gm}{r}\left[1 + \sum_{n=2}^{\infty} J_n \left(\frac{R_{\rm eq}}{r}\right)^n P_n(\cos\phi)\right] \tag{4.14}$$

where the body has mass m and equatorial radius $R_{\rm eq}$, and r denotes the position on the surface at latitude ϕ where $\phi' = \pi/2 - \phi$. The most important Legendre polynomial is the $n = 2$ term, where

$$P_2(\cos\phi) = \frac{1}{2}(3\cos^2\phi - 1) \tag{4.15}$$

Note that the latitudinal dependence enters as a $\cos^2\phi$ term ensuring symmetry about the equatorial plane. The J_n are numerical coefficients that need to be determined from other theory or experiment, and they express the amplitudes of the P_n terms.

In his classical paper on apsidal motion in binary systems, Sterne (1939) represented the total potential experienced on the surface of the primary of mass m_1 at position (r, ϕ) as

$$\Phi_t = \Phi_p + \Phi_{\rm rot} + \Phi_d \tag{4.16}$$

where

$$\Phi_p = -\frac{Gm_1}{r} - \frac{J_2 P_2(\cos\phi)}{r^3} \tag{4.17}$$

is the potential at (r, ϕ) due to the non-spherical shape of the primary, $\Phi_{\rm rot}$ is the rotational potential given by equation (4.13), and

$$\Phi_d = -\frac{Gm_2}{d^3} r^2 P_2(\cos\phi) \tag{4.18}$$

is the distorting, or tidal, potential created by the secondary of mass m_2 when it is at a distance d from the centre of the primary component. In this latter term, the secondary is taken to be a point mass. The same type of expression can be written for the secondary component, taking the primary to be a point mass for calculation of the tidal potential. The J_2 terms can be expressed in terms of the distribution of the density with radial distance within the star concerned, determined from the theory of stellar structure. The perturbing potential for the primary is then given by

$$S_p = -\frac{J_2 P_2(\cos\phi)}{r^3} + \Phi_{\text{rot}} + \Phi_d \tag{4.19}$$

and a similar equation can be written for the secondary, S_s. The total perturbing potential is $S = S_p + S_s$, and these expressions can be used in Lagrange's planetary equation for $d\omega/dt$ to predict the apsidal rotation rate for a given binary system. Sterne's final expression for the apsidal period U for a binary system composed of stars of masses $m_{1,2}$ and mean radii $\bar{R}_{1,2}$, in a *relative* orbit of semimajor axis a, eccentricity e, and orbital period P, is

$$\begin{aligned} \frac{P}{U} &= \frac{k_{12}\bar{R}_1^5}{a^5}\left[\frac{m_2}{m_1}[15f(e) + \gamma_1 g(e)] + \gamma_1 g(e)\right] \\ &\quad + \frac{k_{22}\bar{R}_2^5}{a^5}\left[\frac{m_1}{m_2}[15f(e) + \gamma_2 g(e)] + \gamma_2 g(e)\right] \end{aligned} \tag{4.20}$$

The terms $f(e)$ and $g(e)$ are given by

$$\begin{aligned} f(e) &= 1 + \frac{13}{2}e^2 + \frac{181}{8}e^4 + \cdots \\ g(e) &= 1 + 2e^2 + 3e^4 + \cdots = (1 - e^2)^{-2} \end{aligned} \tag{4.21}$$

and $\gamma_{1,2} = (\omega_{1,2}/\omega_K)^2$ are the squared ratios of the actual angular speeds of the stars to the synchronous or Keplerian value ω_K.

The $k_{12,22}$ are referred to as the *apsidal constants* for the primary and secondary components, respectively. Note that with the exception of $k_{12,22}$, *all* the terms in equation (4.21) are determinable from a combination of spectroscopic and photometric observations giving radial-velocity curves for both components and complete light curves. The two major terms involving $k_{12,22}$ are not separable, however, and so observations can provide only a weighted mean value of $k_2 = w_1 k_{12} + w_2 k_{22}$. There are several close binaries in which the components are very similar stars, with nearly equal masses and radii, and synchronous rotation (e.g., Y Cyg), and so the resultant value $k_2 \approx (k_{12} + k_{22})/2$. From the beginnings of the first numerical models of stellar structure (e.g., Schwarzschild 1958), comparisons have been made between values of k_2 derived from observations of eclipsing binaries with apsidal motion and values calculated from the theoretical models. The observed values were always less than the predicted values by factors of 2–3, and that problem persisted until Giménez and Garcia-Pelayo (1982) showed that the issue could be resolved by a combination of more accurate observational data, better theoretical models, and, critically, an appreciation that the value of the apsidal constant k_2 was a strong function of evolutionary age across the main-sequence band in an HR diagram. The discrepancy arose because most observed systems were distributed across the main sequence at various ages, and the early models had been calculated for the zero-age main sequence. Giménez and Claret (1992) have discussed the major improvements that have been made in both observation and theory and have indicated where some discrepancies still lie, even for well-studied binaries. Continuing refinements in our models of stellar structure, incorporating more realistic representations of mass loss caused by stellar winds, some convective-core overshooting, and stellar rotation (Claret 1995), have further improved the agreement between theory and observation. A good example of such results concerns the eclipsing binary system EK Cep, for which Claret, Giménez, and Martin (1995) have reported excellent agreement between the many observations of this system and the latest stellar models and have demonstrated that the binary is composed of a main-sequence primary of mass $2.02 \pm 0.01\ M_\odot$ at the beginning of its hydrogen-burning phase, with an age of 2×10^7 years, and a pre-main-sequence secondary of mass $1.12 \pm 0.01\ M_\odot$.

Despite all the excellent progress in this area, there remain a few binary systems for which the observed rates of apsidal motion are abnormally small, much less than would be expected from theory. The two most famous systems are AS Cam and DI Her, where the observed rates are less than one-third of the rates expected from the combination of the foregoing classical quadrupole (tidal) effect and the general-relativity effect to be discussed later (Guinan and Maloney 1985; Maloney, Guinan, and Boyd 1989). These discrepancies, amongst others, have led to a major re-examination of the classical theory of apsidal motion caused by tidal distortions (Quataert, Kumar, and Ao 1996) and to investigations of whether or not inclined spin-orbit coupling (Company, Portilla, and Giménez 1988) or third bodies in these discrepant systems might lead to reduced apsidal rotation rates (Khaliullin, Khodykin, and Zakharov 1991; Khodykin and Vedeneyev, 1997).

The classical theory is based on the approximation that a star adjusts its equilibrium shape instantaneously in response to the changing tidal distortion caused by a companion in an eccentric orbit. This is the so-called *equilibrium tide*, the hydrostatic adjustment of the structure of the star to the perturbing potential. But there exists, also, the so-called *dynamical tide*, which is a collective term to describe a number of physical phenomena that include free modes of oscillation of the star (p, f, and g modes) and the dissipation of energy in a fluid by means of viscous effects in turbulence, by convection, and by radiative damping, where compressed, heated gases will radiate more energy than in a less compressed state and hence will lose energy. Sterne's original formulation of the problem adopted the limiting case in which the orbital period was much longer than any free-oscillation periods. Quataert et al. (1996) included a full set of Legendre-polynomial terms to describe the shape of each star, and they have shown that the classical apsidal-motion formula, equation (4.20), does give very accurate results for most binary systems because the periods of the low-order, quadrupole p, f, and g modes usually are smaller than the duration of periastron passage by a factor of 10 or more. They have shown that such dynamical effects are not important in the two systems AS Cam and DI Her and that other mechanisms must be sought to explain the observations. However, it has also been shown that these dynamical tides can be extremely important in certain systems if the ratio of the duration of periastron passage to mode periods is nearer to unity, and even periastron recession is possible. Smeyers, Willems, and Van Hoolst (1998) have recently investigated the possibility that dynamical tides can be *resonant* with a free-oscillation mode for a star. Such resonant dynamical tides would be important and could operate in the same sense as the classical formula (apsidal advance) or in the opposite sense (apsidal recession). These theoretical advances in understanding the tidal interactions in close binaries, together with the possibilities of stellar-rotation axes not being aligned perpendicular to the orbit plane, can be coupled with recent major advances in securing high-resolution échelle spectroscopy to open new investigations of stellar interactions in binaries.

For many binary systems composed of ordinary main-sequence stars, the observed apsidal rotation rates are substantial, at several degrees per year, and their agreement with Sterne's formulation is very good. The additional contribution of the *general-relativity apsidal motion* is usually smaller by a factor of 10, because the separation between the stars is quite large. However, in circumstances where the stars are small, but quite massive, and are at small separations, the relativity effect can become dominant. Such is the case for binaries composed of one or two neutron stars, the radio pulsars, in short-period orbits.

General relativity shows that in a relative orbit of semimajor axis a and eccentricity e, the gravitational force per unit mass of the orbiter at a distance d from its companion is somewhat different from the standard Newtonian value and is given by

$$-\frac{G(m_1+m_2)}{d^2} - \frac{3G^2(m_1+m_2)^2 a(1-e^2)}{c^2 d^4} \tag{4.22}$$

The second term is the perturbing force, that is, the gradient of the perturbing potential S that is directed purely radially:

$$\frac{\partial S}{\partial d} = -\frac{3G^2(m_1+m_2)^2 a(1-e^2)}{c^2 d^4} \tag{4.23}$$

Note that this perturbing force depends on the separation d as d^{-4}, so that its 'range' is much shorter than the usual d^{-2} dependence, but neutron stars have $m \approx 1.4\ \mathrm{M_\odot}$, and the mass term is squared.

The effect of S on the binary orbit can be assessed by using Lagrange's planetary equation for $\dot{\omega}$. Here $\partial S/\partial i = 0$, so that

$$\dot{\omega} = \frac{(1-e^2)^{1/2}}{na^2}\frac{1}{e}\frac{\partial S}{\partial e} \tag{4.24}$$

We can write

$$\frac{\partial S}{\partial e} = \frac{\partial S}{\partial d}\frac{\partial d}{\partial e} \tag{4.25}$$

and ask for a useful expression to give $\partial d/\partial e$. Sterne (1939) showed that the *secular* dependence of d on e is given by

$$\left(\frac{a}{d}\right)^3 = 1 + \frac{3}{2}e^2 + \frac{15}{8}e^4 + \cdots (+\text{ periodic terms}) \tag{4.26}$$

Then

$$\frac{\partial}{\partial e}\left(\frac{a^3}{d^3}\right) = \frac{\partial}{\partial e}\left[1 + \frac{3}{2}e^2 + \frac{15}{8}e^4 + \cdots\right] = 3e(1-e^2)^{-5/2} = \frac{\partial}{\partial d}\left(\frac{a^3}{d^3}\right)\frac{\partial d}{\partial e} \tag{4.27}$$

So

$$\frac{\partial d}{\partial e} = \frac{3e(1-e^2)^{-5/2}}{\frac{\partial}{\partial d}\left(\frac{a^3}{d^3}\right)} = \frac{3e(1-e^2)^{-5/2}}{a^3(-3d^{-4})} \tag{4.28}$$

and

$$\frac{\partial S}{\partial e} = \frac{3G^2(m_1+m_2)^2 e(1-e^2)^{-3/2}}{c^2a^2} \tag{4.29}$$

Then the formula for $\dot{\omega}$, equation (4.24), becomes

$$\dot{\omega} = \frac{(1-e^2)^{1/2}}{na^2e}\frac{3G^2(m_1+m_2)^2 e(1-e^2)^{-3/2}}{c^2a^2} \tag{4.30}$$

and, with $n^2a^3 = G(m_1+m_2)$, we obtain the much-quoted result for relativistic apsidal advance:

$$\dot{\omega}_{\text{rel}} = \frac{6\pi G(m_1+m_2)}{Pc^2a(1-e^2)} \tag{4.31}$$

For binary systems like PSR1913+16, where one radio pulsar is observable and the companion is not detected at any frequency, this expression for the observed apsidal motion due to relativity effects can be combined with that for the mass function $f(m)$ of the system to yield the total mass of the system. That is, because $a = a_1(m_1+m_2)/m_2$, then

$$\dot{\omega} = \frac{6\pi G m_2}{Pc^2a_1(1-e^2)} \tag{4.32}$$

Also, the mass function is

$$f(m) = \frac{m_2^3\sin^3 i}{(m_1+m_2)^2} \tag{4.33}$$

The quantities $f(m)$ and $(a_1 \sin i)$ are determinable directly from analysis of the pulse-timing data. The quantity $\dot{\omega}$ is determinable from discrete sets of pulse-timing data repeated over a number of years, in the same manner as for radial-velocity data from ordinary binaries. So we can write

$$\dot{\omega}(a_1\sin i) = \frac{6\pi G(m_2\sin i)}{Pc^2(1-e^2)} \tag{4.34}$$

and we can determine the quantity $(m_2 \sin i)$ because all other quantities are known. Hence we obtain the *total mass* (m_1+m_2) from

$$(m_1+m_2) = \frac{(m_2\sin i)^{3/2}}{[f(m)]^{1/2}} \tag{4.35}$$

For the particular system PSR1913+16, the observed apsidal rotation rate is $\dot{\omega}_{\rm rel} = 4.2$ degrees per year, and the total mass is found to be 2.82 $M_\odot$, or twice the mass of a standard neutron star. This particular binary seems to be composed of an active radio pulsar with a radio-quiet neutron star as companion.

4.2.3 Triple systems

It is recognized that about 20% of binary systems are members of triple or multiple systems. The binary system will move in an orbit about the barycentre of the triple or multiple system, and the most obvious manifestations of that motion will be changes in the systemic velocity γ of the binary and the light-travel-time effect across the projected wider orbit. The observed triple systems typically are composed of a close binary (A-B) and a much more distant companion (C). These more common triple systems have orbital characteristics like Algol, where the close binary (A-B) will have a nearly circular orbit ($e \approx 0.01$) and a period P_1 of only a few days, and the wide binary (AB-C) may have an eccentric orbit with a period P_3 of hundreds of days. There are a few triple systems in which the close A-B system has an eccentric orbit. With substantial values of P_3, it is inevitable that variations in the systemic velocity of the A-B pair will be quite small. For example, a binary composed of two 5-$M_\odot$ stars with a 1-$M_\odot$ companion in a $P_3 = 200$-day orbit will display a variation in γ velocity of semiamplitude $K_3 = 16$ km s^{-1} provided that the orbits are co-planar and in the observer's line of sight. For more random orientations, the effects will be smaller. Such variations in systemic velocity have been observed; examples include Algol itself, where the A-B orbit has a period of 2.8 days, and the AB-C orbit has $P_3 = 1.86$ years and semiamplitude $K_3 = 12$ km s^{-1} (Hill et al. 1971, 1993; Tomkin and Lambert 1978), and DM Per, where the close orbit has $P_1 = 2.7$ days, and the wide orbit has $P_3 = 98$ days, with $K_3 = 26.5$ km s^{-1} (Hilditch et al. 1992) (see Chapter 3). If the radial-velocity data for a binary are obtained in short discrete intervals of time, separated by longer intervals, then such γ-velocity variations will be very apparent. But in more typical sets of radial-velocity data, which are obtained over substantial lengths of time at a rate of only a few spectra per night, the radial-velocity variations due to the close orbit are convolved with those due to the wide orbit, and it can be very difficult to disentangle the two sets of variations. Searches via a Lomb-Scargle periodogram analysis of all the available radial-velocity data may provide unambiguous determinations of both periods (A-B and AB-C), together with numerous aliases. But independent photometric observations of eclipses, or even ellipsoidal variations, in the A-B orbit will constrain the range of possible periods. Additionally, a subset of the radial-velocity data obtained over an interval of time that is short compared with any likely value of P_3 will provide an initial estimate of the A-B orbit. That can be subtracted from the observed velocities, and the resultant velocity differences can be searched, via periodogram analysis, for evidence of the AB-C orbit.

If the A-B pair is an eclipsing system, or if one of its components emits pulsed radiation, then it will be possible to observe the effects of the light-travel time across the orbit, around the barycentre of the triple system, of the A-B pair that is projected into our line of sight. The times of eclipse minima from an eclipsing A-B pair will deviate from a linear ephemeris according to

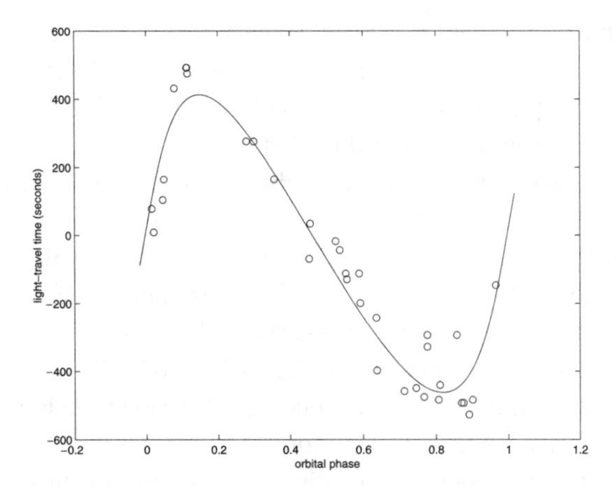

Fig. 4.6. Deviations of observed times of eclipses from a linear ephemeris for the eclipsing binary IU Aur. The data, plotted as open circles, are from Mayer (1987). The line is the expected light-travel time across the P_3 orbit, with $e_3 = 0.54$, $\omega_3 = 5°$, and $a_3 = 1.47$ AU, and it is tilted at $i = 45°$ to the line of sight, which represents approximately the observed deviations.

the light-travel-time equation given in Chapter 2, equation (2.57). The value of the semimajor axis a is given by Kepler's third law, with the quantity $M = Gm_3^3/(m_1+m_2+m_3)^2$ appropriate for the orbit of the body of mass (m_1+m_2) about the barycentre of the $(m_1+m_2+m_3)$ system. Note that the times of both the primary and secondary eclipse minima will deviate in the same manner from a linear ephemeris when the cause of those deviations is orbital motion about a third body, unlike the case already discussed for apsidal motion, where the deviations have opposite signs for primary and secondary eclipses. An example is given in Figure 4.6 for the eclipsing system IU Aur, which moves in an orbit of 294 days about a third body.

Note that the theoretical curve in Figure 4.6 provides only an approximate representation of the data and does not fully match the amplitude of the curve, although the shape is mainly correct. The reason for this discrepancy lies with the *perturbations* on the motion of the A-B pair caused by the presence of the companion.

A *third body* in orbit about a binary system can perturb the orbital characteristics of the binary significantly, provided that the third body is sufficiently massive and close to the binary system. There is no a priori reason why the various orbits of these stars in multiple systems should be co-planar, and, indeed, we can expect a random distribution of orbital orientations, at least initially. Whether or not the cumulative effects of perturbations will make multiple systems become more co-planar is open to question. The particular characteristics of a triple system will induce perturbations on *all* of the orbital elements of the A-B pair and the AB-C pair in amounts that will depend on those characteristics. Thus we observe some triple systems with apsidal motion, others with orbital-inclination changes, and others with all elements varying with time. The observed apsidal motion can be a combination of the third-body-induced effect and the tidal effects between the close pair in an eccentric orbit, and they will not necessarily act in concert. Although apsidal motion always advances ($\dot{\omega}$ is positive), the particular variations of e, i, ω, and Ω within each system may result in the net observed motion being much smaller than would be expected on the basis of tidal effects alone. This explanation seems to be the one that is most favoured to explain the systems like DI Her and AS Cam that display anomalously slow apsidal motion, as noted in the preceding section.

One of the most obvious consequences of perturbations is a change in the orbital inclination of the A-B binary if it is an eclipsing system. The depth of an eclipse is particularly sensitive to the value of the orbital inclination, which is reflected in the uncertainty with which such values are determined, typically less than $\pm 1°$ (see Chapter 5). An example is the *former* eclipsing binary system SS Lac. This system was recorded as a normal eclipsing binary with an eccentric orbit of period $P_1 = 14.4$ days and clearly defined eclipses of depth 0.5 mag when observed in the first half of the twentieth century. At some time around 1940–1960 the eclipses stopped occurring and have not been detected since that time. Tomasella and Munari (1998) have determined a spectroscopic orbit for the system that confirms the earlier photometric determinations of the quantity $e \cos \omega$ from the orbital phase of secondary eclipse (see Chapter 5). Thus the perturbing potential is contributing to a change in the orbital inclination, observed to be $di/dt = 0.13 \pm 0.01$ degrees per year, though it is not changing the significant eccentricity of $e = 0.12$, nor the longitude of periastron, $\omega = 332°$. A positive identification of the existence of a third body in SS Lac has not yet been made.

A second example is the eclipsing system IU Aur ($P_1 = 1.8$ days), which has been studied repeatedly over nearly two decades by Mayer, Drechsel, and others; see Mayer and Drechsel (1987) for a review, and Drechsel et al. (1994) for an update. The third component in the system has been identified from three independent sets of observations: a well-defined light-travel-time effect with a period of 294 days; the requirement for a third-light contribution of 22% to obtain solutions to the observed light curves; and the variation with time of the depths of the eclipses, increasing by more than 0.2 mag over 20 years and reaching their maximum depth with total eclipses at $i = 90°$ around 1990. A recent spectroscopic investigation by Harries, Hilditch, and Hill (1998) made substantial revisions to the velocity semiamplitudes, the mass ratio, and the total masses and dimensions, but did not alter the description of the system as a semidetached binary. Harries et al. (1998) were unable to detect a third component in their spectra, despite the 22% third light, and concluded that it could be hidden in the cross-correlation peak at low velocity relative to the binary's systemic velocity. A third body has been resolved by using speckle interferometry (Mason et al. 1998) and by *Hipparcos*, with an angular separation of only 0.14 arcsec and a magnitude difference relative to the binary of $\Delta V = 1.36$ mag, which is in excellent agreement with the derived third-light contribution of 22%. However, with a distance to the binary about 1600 pc, this angular separation corresponds to a linear separation of about 222 AU and an orbital period of 3300 years, far longer than the light-travel-time period of 294 days. A likely explanation is that the perturber of the binary orbit is another, much closer companion that can contribute little light to the total brightness of the system.

Rather than speculating, we may ask if such perturbations can provide quantitative information about the mass and the orbit of the perturbing third body in such systems. Early theoretical work on the three-body problem, with all components being considered to be point masses, developed from studies by Brown (1936a–c, 1937). Brown's work was a modification of the 'lunar theory' in celestial mechanics, namely, the dynamical theory to follow the motion of the

Moon in the triple system Moon + Earth + Sun. Various approximations were made to reduce the complexity of the equations, one of them being that the orbit of component C remained an unperturbed ellipse. A description of that theory and its subsequent developments can be found in a book by Kopal (1978), where several sets of perturbation equations for all six elements of the close-pair orbit are presented. Certain very useful conclusions can be drawn from that theory regarding the influence of the third body on the close-binary orbit, and they can be divided into three sets of perturbations. These are the *short-period*, *long-period*, and *apse-node*, or *secular*, perturbations, which have periods of P_1, P_3, and of order (P_3^2/P_1), respectively. The magnitudes of these perturbations Δp relative to their respective orbital parameters p are $\Delta p/p \approx (P_1/P_3)^2$, (P_1/P_3), and unity, respectively. With the aforementioned typical values for P_1 and P_3, the short-period perturbations clearly are far too small to be observable with current technology. The long-period perturbations are small but may be significant in some systems. But the rotation of the line of apsides, the *apse* perturbation, and that of the nodal line, the *node* perturbation, are both expected to be substantial and to move in opposite directions. Whilst the apse advances, the node recedes, and they do so at comparable rates. Thus the sum $\omega + \Omega$ tends to remain fairly constant according to the Brown-Kopal theory. The remaining elements of the close-pair orbit may also show secular variations, rather than periodic ones, although the semimajor axis of the close-pair orbit has been shown to be secularly constant.

As an example of the type of variations to be expected on the basis of the Brown-Kopal theory, Figure 4.7 illustrates the secular variations for some of the close-pair elements that

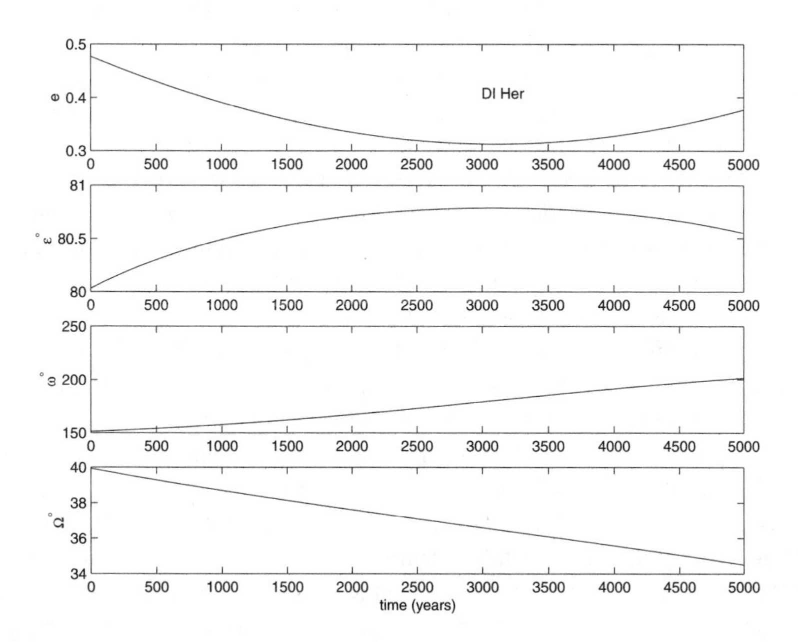

Fig. 4.7. An illustrative example of secular variations of the orbital elements of a close binary (DI Her) caused by the presence of a third body. These curves were generated by a simple numerical integration of the perturbation equations from the Brown-Kopal theory.

might be expected for the eclipsing binary DI Her if it were part of a triple system. Recall that DI Her displays an anomalously slow apsidal rotation rate, much slower than would be expected from normal tidal effects within a binary. One possible explanation is that the perturbations of the elements of DI Her all contribute to slowing the apsidal advance. Figure 4.7 shows the secular variations in e, ϵ, ω, and Ω calculated according to equations 6.153–6.156 of Kopal (1978). Here ϵ is the angle of inclination between the close-pair orbit and the wide orbit, with the wide orbit being considered to be unperturbed. The mass of the third body was taken to be 1.0 $M_\odot$, with a circular orbit of period 10.5 years, and with $\epsilon = 80°$ initially. It is not clear how far Kopal's equations for the secular perturbations per wide-orbit period (P_3) can be extrapolated. But Figure 4.7 does illustrate that significant variations can be expected. Khaliullin et al. (1991) have performed a more rigorous numerical integration of the appropriate Lagrange equations, rather than adopting approximate analytic solutions as in the Brown-Kopal theory, and specifically they have explored the range of possible values for the third-body mass and orbit that would be compatible with the available observational data on DI Her and that might explain the anomalously slow apsidal motion. They have succeeded in achieving that explanation with entirely reasonable values for the wide-orbit period (a few years) and the mass of component C (0.1–1 $M_\odot$). The lack of appropriate observational data on DI Her secured over a long time interval (a century or so) means that more specific conclusions cannot be drawn.

An alternative theory of perturbed-triple-system motions was developed by Harrington (1968, 1969), who also included perturbations of the wide orbit. That theory was made more directly applicable to observed triple systems by Söderhjelm (1975, 1982). In the 1975 paper, Söderhjelm derived equations for the variations with time of the orbital parameters of the close pair that were periodic on the P_3 time scale (the long-period perturbations), together with an equivalent set of equations for the longer-term secular perturbations. Those equations were derived with the eccentricity of the close orbit $e_1 \equiv 0$ and therefore have restricted validity. In the 1982 paper, the long-period-perturbation equations and the secular-perturbation equations were given in full with no restriction on the value of e_1. Söderhjelm also demonstrated, via numerical integrations of the Lagrange equations, that his analytical formulae could provide a correct description of the expected perturbations provided that the ratio of orbital periods P_3/P_1 was greater than about 50.

The *long-period perturbations* are generally found to be quite small, with the orbital elements of the close pair varying by 1–2% at most during the orbital cycle about the triple-system barycentre. As an example, Figure 4.8 provides an illustration of the periodic variations of the elements e_1 and ω_1 of a hypothetical close pair during its P_3 orbit, when its orbital elements are equal to those of IU Aur, but with $e_1 = 0.5$. Such variations would not be determinable from discrete sets of radial-velocity data secured at intervals around the P_3 orbit. However, Söderhjelm's 1982 illustration for the system ξ UMa does indicate periodic variations in e of ± 0.02, and in ω of $\pm 2°$. The most important long-period perturbation would seem to be that on the mean anomaly in the close orbit, the δl_1 term (equation 5) in the 1982 paper. As noted by Söderhjelm (1975) and by Mayer (1990), the addition of this dynamical term to the geometrical light-travel-time effect increases the amplitude of the theoretical curve used to match the $(O - C)$ variations of times of minima about a linear ephemeris, with the result that

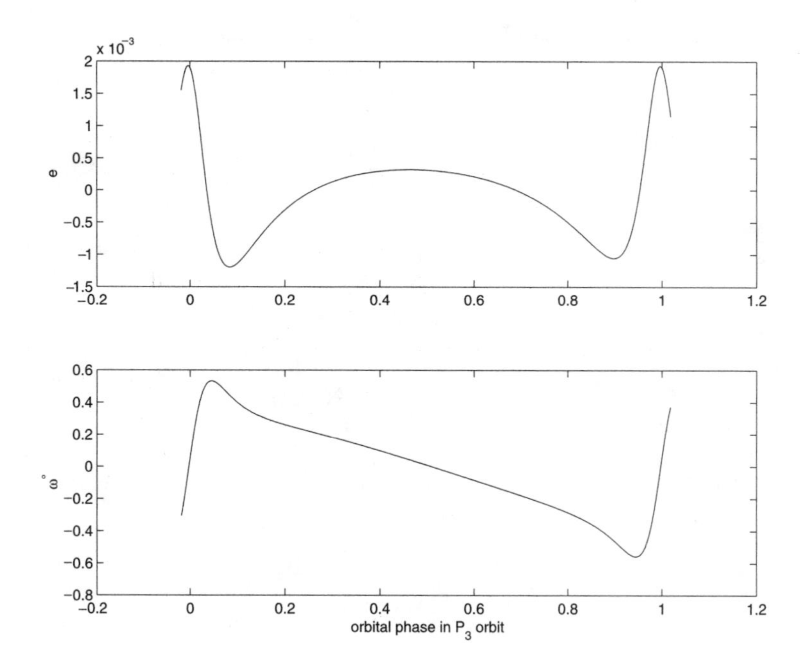

Fig. 4.8. The long-period perturbations of the orbital elements e and ω for the close orbit of a hypothetical triple system, where $e = 0.5$ and $\omega = 0°$ initially, and the masses and orbital periods are the same as in IU Aur. Perturbations calculated according to formulae from Söderhjelm (1982). Note the vertical scales for e and ω.

a lower mass is inferred for the perturbing third body. Figure 4.9 illustrates the same data set as in Figure 4.6, but the final theoretical curve fits the data very well when the two effects are combined. In this illustration, co-planar orbits are adopted, rather than being at 45° to each other, and the mass of the third body is reduced from 18 $M_\odot$ to 8 $M_\odot$.

The *secular* or *apse-node* perturbations are significantly more important. Here the variations in e_1, i_1, ω_1, and Ω_1 can be substantial, $\Delta p/p \approx 0.1$–1, as discussed and illustrated by Söderhjelm (1982). Variations in e_1 and ω_1 may be cyclic, with sufficient amplitudes to ensure that for close pairs with short orbital periods, tidal interactions may be enhanced significantly. Such additional effects could reduce the time scales (millions of years) over which it is expected that close-binary orbits will become circularized. Over time scales of a few decades or centuries we find that the orbit precesses about the total-angular-momentum vector for the triple system, with the effect of changing the orientation of the eclipsing-binary orbit relative to the observer's line of sight. The observed orbital inclination i_1 of the close pair is expected to change cyclically as a result of these secular perturbations. A secular regression of Ω_1 occurs in a nodal period, given by Söderhjelm (1975), of

$$P_{\text{node}} = \frac{4(1+q_1+q_3)P_3^2\left(1-e_3^2\right)^2(1+q_1)^2}{3\left\{q_1 \cos i_{I1}\left[q_3(1+q_1+q_3)\left(1-e_1^2\right)\right]^{1/2} + (1+q_1)^2 q_3 \cos i_{I3}\left(1-e_3^2\right)^{1/2}\right\}\cos j} \tag{4.36}$$

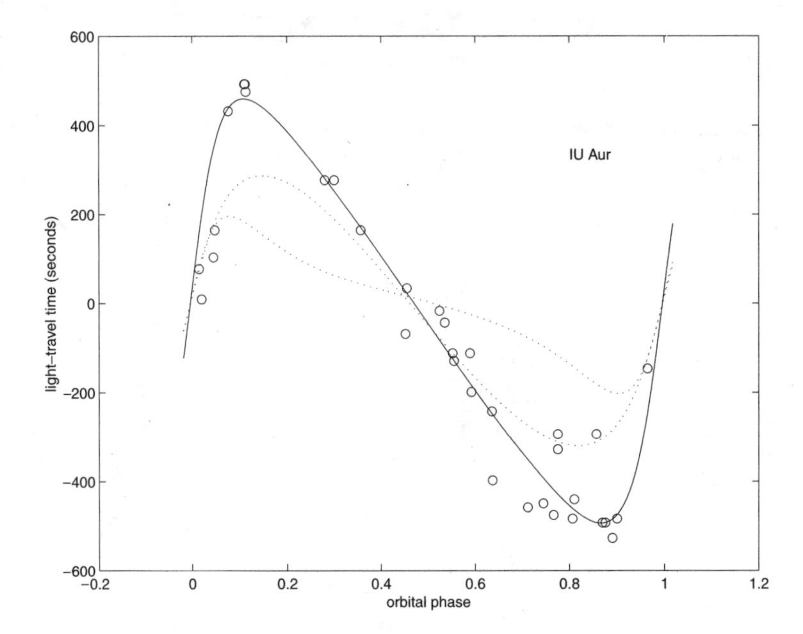

Fig. 4.9. Deviations of observed times of eclipses from a linear ephemeris for the eclipsing binary IU Aur. The data, plotted as open circles, are from Mayer (1987). The solid line is the total of the geometrical light-travel-time effect (shown by a dotted line) and the dynamical perturbation (shown by a more complex dotted line) for the representative model noted in the text.

which, for the model for IU Aur, gives $P_{\text{node}} \approx 293$ years. This changing aspect of the orbit relative to the observer ensures a change in the observed value of the orbital inclination i_1, as illustrated in Figure 4.10. The observed orbital inclination is related to two other angles in the figure by means of an equation from standard spherical astronomy applied to the spherical triangle *ABC*:

$$\cos i_1 = \cos I \cos i_{I1} - \sin I \sin i_{I1} \cos \Omega_1 \tag{4.37}$$

where $C\hat{A}B = I$ is the angle between the invariable plane of the triple system and the observer's plane of the sky, $A\hat{C}B = i_{I1}$ is the angle between the angular-momentum vector of the close orbit and that of the invariable plane, and $AC = \Omega$ is the longitude of the ascending node of the close orbit as measured from the intersection of the two fundamental planes in the problem, namely, the invariable plane and the observer's tangent plane of the sky. The same kind of equation can be written for the wide orbit, so that i_{I3} will be the angle between the angular-momentum vector of the wide orbit and that of the invariable plane, and i_3 will be the inclination of the wide orbit to the observer's plane of the sky. Then the angle $j = i_{I1} + i_{I3}$ is the inclination between the close-orbit and wide-orbit planes.

As an example, in Figure 4.11 the four determinations of the observed close-orbit inclination to the line of sight in the system IU Aur are plotted as functions of time. Also shown is the expected secular perturbation of i_1 caused by the third body in the system, which suggests that the observable range for i_1 could extend over $\pm 30°$. Clearly, the available data hardly constrain

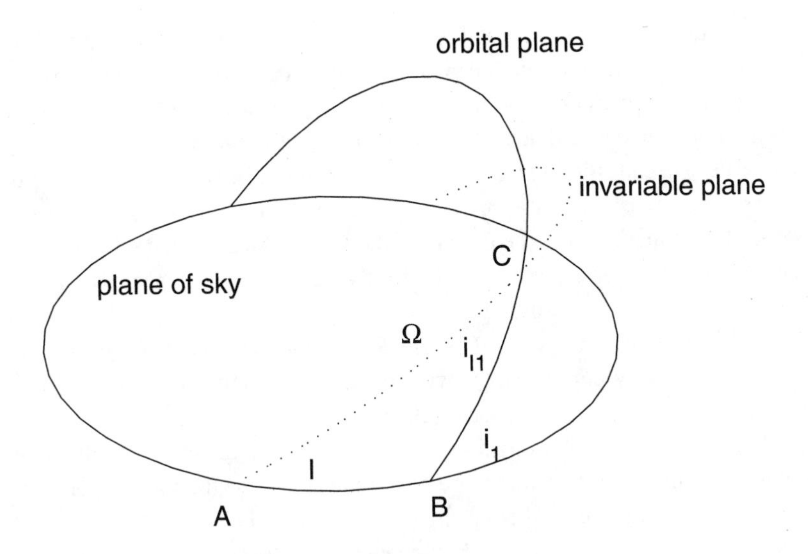

Fig. 4.10. The orientation of an orbit to the invariable plane of the triple system and the observer's plane of the sky. Both of these latter planes are unchanging, whilst both the close-orbit plane and the wide-orbit plane are liable to perturbations.

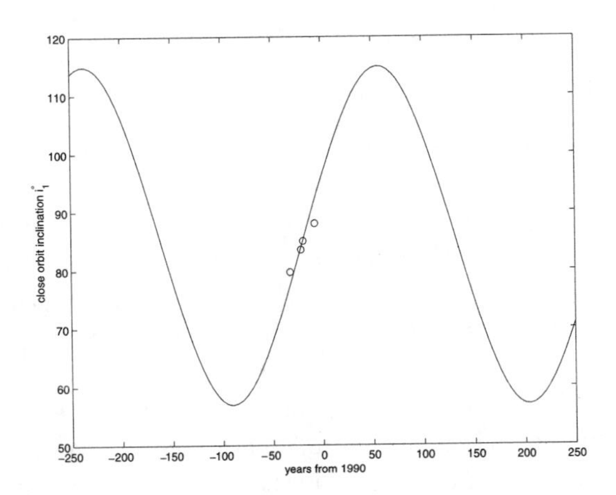

Fig. 4.11. Four determinations of the observed close-orbit inclination i_1 of IU Aur to the observer's plane of the sky are shown as open circles. Data from Drechsel et al. (1994). The line shows the expected changes in i_1 caused by regression of the node in a period $P_{\mathrm{node}} = 293$ years.

the range of possible third-body orbits that could match these data! Even in combination with the light-travel-time data, the ranges of acceptable third-body orbits and masses remain too large.

As we have already seen, the apsidal motion is readily observed, because its effects are immediately apparent in both photometric and spectroscopic data. The complication for the investigator of a triple system may lie in separating the two contributions to the total observed apsidal motion: that of the tidal effects in a close pair and that of the perturbing effects from the third body on the motions of the close pair. The few anomalous cases of apsidal motion may well be explainable by appeal to the properties of triple systems, but gathering the observational evidence that will prove such an interpretation will be difficult because of the time scales (decades to centuries) involved. It might be argued, therefore, that these investigations of the dynamical perturbations within triple systems are of only theoretical interest. When *visual* binaries were the only resolved binary systems, and they had long orbital periods and wide separations, these perturbations were not readily apparent, save in exceptional cases. Now that several short-period eclipsing binaries in triple systems have been discovered and investigated, it is clear that the combination of repeated acquisitions of light curves and radial-velocity curves can provide evidence of light-travel-time effects, orbital-inclination changes, and systemic-velocity changes. But it seems that such data cannot yet constrain the possible orbit solutions to a sensibly small range. However, now that speckle interferometry and direct interferometry are beginning to resolve the shorter-period triple systems that include spectroscopic and eclipsing binaries, the nodal perturbations should become measurable directly, thereby adding another independent observational constraint on possible configurations. The shorter-period, closer, triple systems are expected to exhibit perturbations on shorter time scales as well, so that investigations of real-time perturbations may be possible.

4.2.4 Circularization and synchronization

The same theory of orbital perturbations caused by tidal forces between the components of close binaries has been used by Zahn (1975, 1977, 1978, 1989) and Zahn and Bouchet (1989) to investigate the time scales over which binary orbits can be expected to become circularized. Those papers have also considered the issue of the synchronization of stellar axial-rotation periods with orbital periods. A useful review of the adequacy of tidal theory has been presented by Zahn (1992), including references to other theoretical and observational work in the field.

Zahn's discussion (1977) of tidal friction in close binary stars establishes sets of equations for the secular evolution of the semimajor axis a, the orbital eccentricity e, and the product of the moment of inertia I and the axial-rotation angular velocity ω_r. The problem is divided into two parts, one dealing with the equilibrium tide, and the other with the dynamical tide. For both tides, it is necessary also to separate stars into two classes: those that have convective envelopes (CE) and those that have radiative envelopes (RE). For the *equilibrium tide*, the most efficient dissipation mechanism, in the sense of the shortest time scale, is the turbulent friction that occurs in the convective envelopes of late-type stars. The friction time scale for a star of mass m, radius R, and luminosity L is estimated to be

$$t_f \sim (mR^2/L)^{1/3} \tag{4.38}$$

which is about 1 year for CE stars. The relevant secular equations for the effects of the equilibrium tide are, with corrections noted by Zahn (1978),

$$\frac{\dot{a}}{a} = -12\frac{k_2}{t_f}q(1+q)\left(\frac{R}{a}\right)^8\left[\left(1-\frac{\omega_r}{\omega_K}\right)+e^2\left(23-\frac{27\omega_r}{2\omega_K}\right)+O(e^4)\right] \tag{4.39}$$

$$\frac{\dot{e}}{e} = -\frac{3}{2}\frac{k_2}{t_f}q(1+q)\left(\frac{R}{a}\right)^8\left[\left(18-\frac{11\omega_r}{\omega_K}\right)+O(e^2)\right] \tag{4.40}$$

$$\frac{d}{dt}(I\omega_r) = \frac{6k_2}{t_f}q^2 mR^2\left(\frac{R}{a}\right)^6\left[(\omega_K-\omega_r)+e^2\left(\frac{27\omega_K}{2}-\frac{15\omega_r}{2}\right)+O(e^4)\right] \tag{4.41}$$

In these equations, terms are given for only one component of the binary, so that they must be calculated for each star with the appropriate values of the stellar radius R, moment of inertia I, apsidal constant k_2, angular velocity ω_r, and friction time t_f, together with the values of a and q for the binary system and the orbital (or Keplerian) angular velocity ω_K. Note that in the formula for the rate of change of a, $\dot{a} < 0$ if the orbital angular velocity exceeds the rotational angular velocity. That is, a tide that is *lagging behind* the motion of the companion will cause the orbit to decay. If, however, $\omega_r > \omega_K$, then the tide will lead and cause the orbit to become larger. The evolution of e is rather more complex, because the terms involving ω_r and ω_K have various numerical constants involved, and the terms can become positive or negative depending on the ratios of these angular speeds.

The corresponding formulae that provide estimates of the *synchronization* and *circularization* time scales for CE stars are

$$t_{\text{sync}} = \frac{1}{6q^2k_2}\left(\frac{mR^2}{L}\right)^{1/3}\frac{I}{mR^2}\left(\frac{a}{R}\right)^6 \approx 10^4\left[\frac{1+q}{2q}\right]^2 P^4 \text{ years} \tag{4.42}$$

$$t_{\text{circ}} = \frac{1}{84q(1+q)k_2}\left(\frac{mR^2}{L}\right)^{1/3}\left(\frac{a}{R}\right)^8 \approx 10^6 q^{-1}\left[\frac{1+q}{2}\right]^{5/3} P^{16/3} \text{ years} \tag{4.43}$$

where the orbital period P is in days. Thus, for a mass ratio $q = 1$ and $P = 1$ day, we obtain $t_{\text{sync}} \approx 10^4$ years, and $t_{\text{circ}} \approx 10^6$ years. But note the strong dependence on the orbital period. For $P = 10$ days, $t_{\text{sync}} \approx 10^8$ years, and $t_{\text{circ}} \approx 10^{11}$ years. Accordingly, the theory predicts that observations will show a *cut-off period* below which orbits will be circularized well within the lifetimes of the stars on the main sequence, and above which the binaries will retain their orbital eccentricity unless acted upon by other disturbances. Normal stellar evolution will cause the stars to increase in size across the main sequence, and much more so during the expansion to the red-giant phase. Thus, tidal perturbations will increase in effectiveness as the ratios of the stellar radii to their mutual separation increase. Furthermore, stellar evolution may well lead to exchanges of mass between components and to mass losses from the systems, which effects will act on the orbital parameters rather more quickly than will tidal perturbations.

The predicted switch from circular orbits to elliptical orbits for a sample of stars occurs at a cut-off period of around 8 days, according to most theoretical estimates for stars with convective envelopes, the exact value depending on the details of the dissipation mechanisms acting in the stellar envelopes.

For RE stars, the dissipation mechanisms acting on the equilibrium tide are not sufficient to cause significant secular changes in their orbits on stellar-evolution time scales. However, radiative damping is the dominant dissipative mechanism acting on the *dynamical* tide in early-type stars, and the time scales for synchronization and circularization are given by Zahn (1977) as

$$t_{\rm sync} = \frac{1}{5 \times 2^{5/3}} \left(\frac{R^3}{Gm} \right)^{1/2} \left(\frac{I}{mR^2} \right) \frac{1}{q^2(1+q)^{5/6}} \frac{1}{E_2} \left(\frac{a}{R} \right)^{17/2} \tag{4.44}$$

$$t_{\rm circ} = \frac{2}{21} \left(\frac{R^3}{Gm} \right)^{1/2} \frac{1}{q(1+q)^{11/6}} \frac{1}{E_2} \left(\frac{a}{R} \right)^{21/2} \tag{4.45}$$

where a star of mass m, radius R, and moment of inertia I is in a binary system of mass ratio q and separation a. The tidal-torque constant E_2 is determined from stellar-structure theory and is provided by Zahn (1975) in tabular form as a function of stellar mass. For a binary system composed of two 10-$M_\odot$ stars with radii of about 5 $R_\odot$, $E_2 \sim 10^{-6}$, and $I/mR^2 \sim 10^{-1}$ (Zahn 1975). For an orbital period of 5 days, $a = 33.44$ $R_\odot$, and $t_{\rm sync} \sim 7 \times 10^6$ years, whilst $t_{\rm circ} \sim 2 \times 10^9$ years. These time scales are very considerably longer than those for CE stars, a result that is confirmed in a statistical sense by observations to be noted later on the distributions of orbital eccentricities for binaries composed of RE and CE stars. Note, in particular, the very strong dependence of $t_{\rm sync}$ and $t_{\rm circ}$ on the ratio (a/R), the separation divided by the stellar radius. For RE stars in wide binaries, these time scales are very much larger than their evolution times; for short-period binaries, with $(R/a) \sim 0.25$, rather than $(R/a) = 0.15$ as in the foregoing example, the time scales are reduced by a factor of about 100, so that we can reasonably expect such systems to exhibit synchronous rotation (as is observed in most cases), but circular orbits only for those systems with the shortest periods (again, as is observed).

At the conference Binaries as Tracers of Stellar Formation, with proceedings edited by Duquennoy and Mayor (1992), much time was devoted to consideration of the observational data and theoretical modelling of the tidal synchronization and circularization processes. Also discussed were the roles of circumstellar discs and circumbinary discs in the pre-main-sequence stages of binary-star evolution, with the findings demonstrating that positive eccentricity evolution can be expected during that phase. Many of the observational data presented for binaries in the general field, as well as in clusters, demonstrated that there was encouraging agreement with the tidal theory. As examples, Figure 4.12 illustrates the distribution of orbital eccentricity as a function of the logarithm of the orbital period for several samples of binary stars. The sample of field binaries composed of G dwarf stars illustrates the results very well for stars with convective envelopes. There seem to be no *eccentric-orbit systems* with periods less than about 8 days; above $P \approx 8$ days, the range of values for e increases, and there seem to be no binaries with *circular orbits*. The pre-main-sequence binary sample shows similar behaviour, in agreement with the theory of Zahn and Bouchet (1989) that for systems with short orbital periods, tidal circularization occurs before the main-sequence stage is reached. These findings can be contrasted with those for the O-star binaries, on the one hand, and the field G-K dwarfs and giants observed by Griffin in his long-running programme, on the other. For the radiative-envelope O stars, we see that stars with orbital periods of about $P < 2$ days have only circular orbits, though both circular and eccentric orbits occur for stars with $2 < P < 20$ days. Above that value, the eccentricity distribution is similar to that seen in the low-mass systems. For the mixture of late-type dwarfs and giants, circular orbits can occur over a wide range of

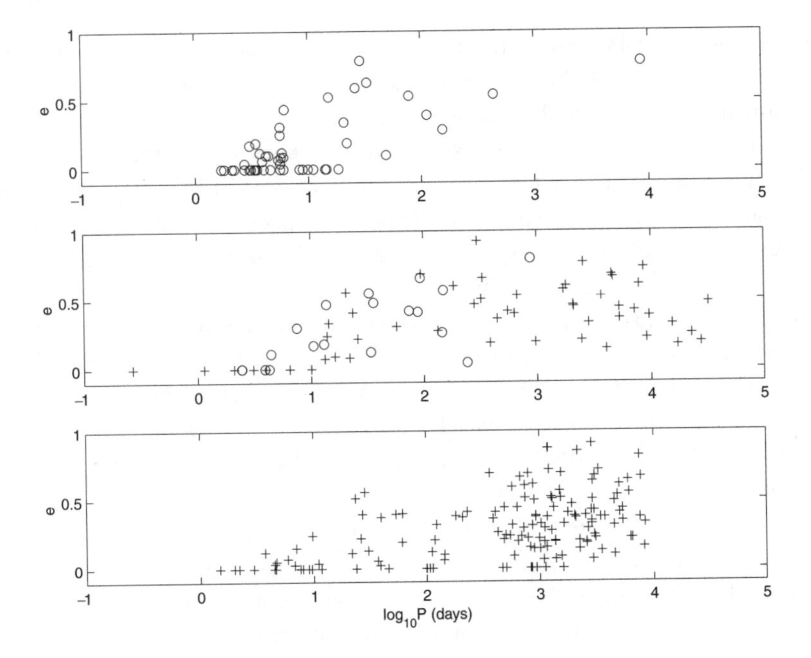

Fig. 4.12. Observed distributions of orbital eccentricity (e) as a function of orbital period ($\log_{10} P$). Top: O-star binaries from the sample compiled by Mason et al. (1998). Middle: Pre-main-sequence stars (circles) compiled by Mathieu (1992) and field G dwarf stars reported by Duquennoy and Mayor (1991). Bottom: Field late-type stars reported by Griffin in his *Observatory* series of papers (*no. 1–141*).

periods, up to 1000 days, perhaps reflecting the fact that these systems have enough room for their components to evolve as single stars to giant dimensions, allowing tidal perturbations to become important. Thus, any short-period binary star that has an eccentric orbit is likely to be a young system, unless it is attended by companions that perturb its otherwise two-body motion.

The agreement between the observed e–P distributions and the expectations from tidal theory is very encouraging. With time scales for circularization of 10^6 to 10^{11} years, similar to those for stellar evolution, we have to rely on such statistical studies. However, it is worthwhile noting that the theory of Zahn and Bouchet predicts quite rapid evolution in e and P values around the ages of 10^4 to 10^5 years for low-mass CE stars in short-period binaries. Continuing work on pre-main-sequence binaries may well reveal further statistical evidence for such perturbations of these orbital elements.

As noted earlier, the theoretical expectations for the *synchronization* of rotational and orbital periods typically are less than the circularization time scales by a factor of 10^2 to 10^3. Accordingly, we can expect to find most close binary systems in such a state of synchronism, and that has indeed been the case. Again, there are specific predictions from tidal theory about the ratio of rotational speeds to orbital speeds at various stages in the pre-main-sequence evolution of stars, and these *non-synchronism* predictions need to be investigated observationally. An interesting example is the eclipsing-binary system TY CrA (Casey et al. 1998), where the main-sequence primary (3.2 $M_\odot$) executes a circular orbit in 2.88 days around a pre-main-sequence secondary (1.6 $M_\odot$); the two stars are rotating synchronously, and the age of the system is estimated at 3×10^6 years. However, it does have to be borne in mind that tidal dissipation may not be the only physical effect operating during these times. For low-mass CE stars there may be substantial magnetic fields generated as a result of quite rapid rotation. If the complex magnetic-field structures on such stars possess open field lines, as does the Sun, then we might expect the outflowing and ionized stellar wind from such stars to interact with the open field lines and control their geometry. The open field lines would exert a braking torque on the star, thereby causing it to rotate more slowly. If the star was a member of a binary system, then the tidal perturbations might enforce synchronous rotation, and angular-momentum losses due to the escaping stellar-wind particles spiralling along open field lines would come from the binary orbital motion as well as the stellar-rotation motion. Such *magnetic braking* does seem to be operating in a number of different types of binary systems, amongst them being the RS CVn binaries, the contact binaries, and the cataclysmic variables. The higher-mass RE stars do not suffer such complications, and our understanding should profit from detailed studies of the early evolutionary stages of such stars.

There are a few relatively unevolved binary systems in which the rotational velocities of the stars have been determined to be significantly non-synchronous. An example is TZ For (Andersen et al. 1991), where the more massive component is now a G8III, 2.05-$M_\odot$ star orbiting an F7III, 1.95-$M_\odot$ companion in a 75.7-day circular orbit. The more massive star rotates synchronously, whilst the secondary rotates at several times the synchronous value. The system Capella has similar properties, and both are explained in terms of the more massive star having sufficient room to expand to the red-giant stage without hindrance, with the relative size of the giant star ultimately ensuring circularization of the orbit as well as its own synchronism. The originally RE secondary has not yet reached the synchronous stage. It is appropriate to record the recommendation from Casey et al. (1998) that the rotational evolution of the stars should be considered in parallel with their normal evolution to enhance the quality of the comparisons that can be made against theoretical models.

For the more evolved binary systems that have undergone mass-transfer events between their components, we see the whole range of rotational properties, from synchronism to slow rotators and to extremely rapid rotators, such as the millisecond pulsars. The physical explanations for these different systems require assessments of the relative contributions of the tidal effects discussed earlier and the consequences of the transfer of mass and angular momentum from one star to its companion via accretion streams, columns, discs, and stellar winds, as well as consideration of losses of mass and angular momentum from the binary system. Some examples are discussed at the end of this chapter.

4.3 The Roche model for binary stars

In the preceding section, Legendre polynomials were used to describe small departures from sphericity for the stars in close binaries, and it was found that the resultant tidal theory successfully explains the distribution of orbital eccentricities and the time scales for circularization and synchronization observed in detached close binaries. If all binaries were formed as well-detached systems and remained that way, so that the stars could be considered to be essentially spherical, with minor tidal bulges, then the use of Legendre polynomials to describe their shapes would have been more extensive. But some binaries seem to be formed as exceedingly close systems, with the two stars on or near the zero-age main sequence (ZAMS) in the HR diagram, and the stellar surfaces nearly touching, as seen, for example, in V701 Sco (Bell and Malcolm 1987), where the light curve demonstrates that both stars must be decidedly non-spherical. In any event, as noted in Chapter 1, stars change their characteristics with time, with a star's radius increasing very substantially at different stages of its evolution. Consequently, any close binary will ultimately develop into one with stars that have extreme tidal distortions, with roughly ellipsoidal shapes, at least at some stage of its evolution. Therefore, it is necessary to introduce an alternative formulation for the shapes of stars that can provide a quantitative description of spheres as well as tidally distorted, nearly ellipsoidal shapes. That formulation is the Roche model, which is based on consideration of the total gravitational potential in a system of two point masses that move in circular orbits about their barycentre. That is, the tidal forces will have produced their inevitable result: stars that are tidally locked in synchronous rotation and are moving in circular orbits.

The Roche model for binary stars is named in honour of the nineteenth-century French mathematician Edouard Roche, who investigated the mathematics of the *restricted three-body problem*. An analytic solution for the general motion of three masses governed by their mutual gravitational field could not be found. In the quest for such a solution, mathematicians pursued solutions of the restricted problem, where two masses move in circular orbits about their common centre of mass, whilst the third body is of infinitesimal mass and moves in the gravitational field of the other two massive bodies. For a given value of the total gravitational potential Φ experienced by the third body, it was found that a three-dimensional hypothetical surface could be constructed around the two point masses that would represent the region of space where the motion of the third particle would be zero relative to a coordinate system that would rotate uniformly with the two masses in their circular orbits. For a range of values of Φ, these *surfaces of zero velocity* close to each mass point were found to be nested around each mass point, with the more distant surfaces enclosing both masses. The application to binary stars is readily apparent, particularly when it is appreciated that these zero-velocity surfaces are identical with *surfaces of constant gravitational potential* in a two-body system. The surface of a star is an equipotential surface, and so a description of these surfaces provides a means of quantifying the shapes of stellar surfaces in a binary system.

We consider the calculation of the location of the equipotential surfaces surrounding two mass points that are in orbit about their common centre of mass. Circular orbits are assumed, so that a reference frame with its origin at the centre of mass of the more massive star, of mass m_1, rotates at constant angular speed ω with the binary system. The less massive star, of mass m_2, is at a distance $a \equiv 1$ from the origin. In this rotating frame, the two stars are fixed in position at (0, 0, 0) and (1, 0, 0) respectively, as shown in Figure 4.13.

The gravitational potential Φ experienced at the position $P(x, y, z)$ is the sum of the two point-mass potentials and the rotational potential, namely,

$$\Phi = -\frac{Gm_1}{r_1} - \frac{Gm_2}{r_2} - \frac{\omega^2}{2}\left[\left(x - \frac{m_2}{(m_1+m_2)}\right)^2 + y^2\right] \tag{4.46}$$

where $r_1 = (x^2 + y^2 + z^2)^{1/2}$, and $r_2 = [(x-1)^2 + y^2 + z^2]^{1/2}$.

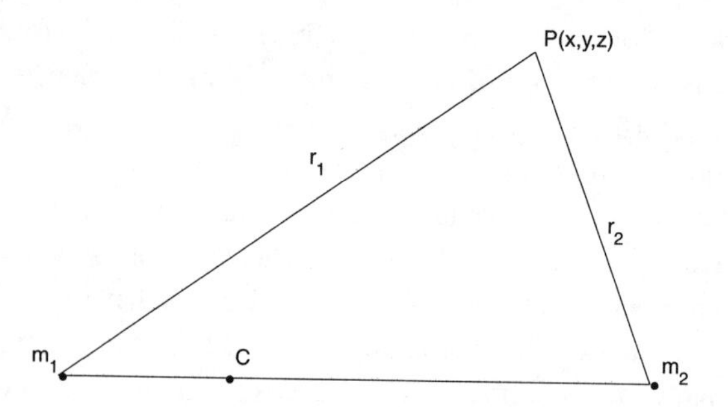

Fig. 4.13. Two masses, $m_{1,2}$, with their common centre of mass at C, are separated by unit distance. The coordinate system is centred at the mass m_1, with the x axis directed towards the mass m_2, the y axis in the orbital plane, and the z axis perpendicular to the orbital plane. The point $P(x, y, z)$ is at a distance r_1 from m_1, and r_2 from m_2.

With $a \equiv 1$ and synchronous rotation in a circular orbit, we have

$$\omega^2 = \left(\frac{2\pi}{P}\right)^2 = \frac{G(m_1 + m_2)}{a^3} = G(m_1 + m_2) \tag{4.47}$$

If we define the quantity $\Phi_n = -2\Phi/G(m_1 + m_2)$ and the mass ratio $q = m_2/m_1$ $(0 < q \leq 1)$, then

$$\Phi_n = \frac{2}{(1+q)r_1} + \frac{2q}{(1+q)r_2} + \left(x - \frac{q}{(1+q)}\right)^2 + y^2 \tag{4.48}$$

The quantity Φ_n is a normalized potential, and it can be calculated for any location (x, y, z) around the two mass points. Hence, the *surfaces of constant potential*, or *equipotential surfaces*, within the rotating frame can be found, all expressed in terms of the separation a that has been adopted as the unit of distance, and evidently dependent upon the mass ratio q. Figure 4.14 illustrates a section of those surfaces in the orbital plane of the binary system at $z = 0$ for a mass ratio of $q = 0.4$. Note that the shapes of the surfaces near to each mass point, $m_{1,2}$, are circles in this (x, y) cross section, and the same applies in the (y, z) and (x, z) planes. Stars in well-detached binaries have spherical shapes. As one moves farther away from the two stellar centres, the surfaces become more distorted, particularly along the *line of centres* that joins the two masses, though the averted hemispheres remain approximately spherical. Eventually the independent surfaces surrounding the two mass points will touch each other, at the location labelled L_1 in the figure. This point is referred to as the *inner Lagrangian point*. The two surfaces that just touch at $L_1$are referred to as the *Roche limits* for the two components of the binary system. They define two three-dimensional limiting volumes, usually called the *Roche lobes*. The reason that these volumes are limits is simply that they define the maximum volume that a star can occupy in a binary system and still have all its constituents (atoms, ions, etc.) under its own gravitational control. The actual physical size of each limiting volume is defined primarily by the separation a, and to a lesser extent by the mass ratio q. When the mass ratio is unity, the two Roche limiting volumes are equal in size. As q decreases from unity, the relative

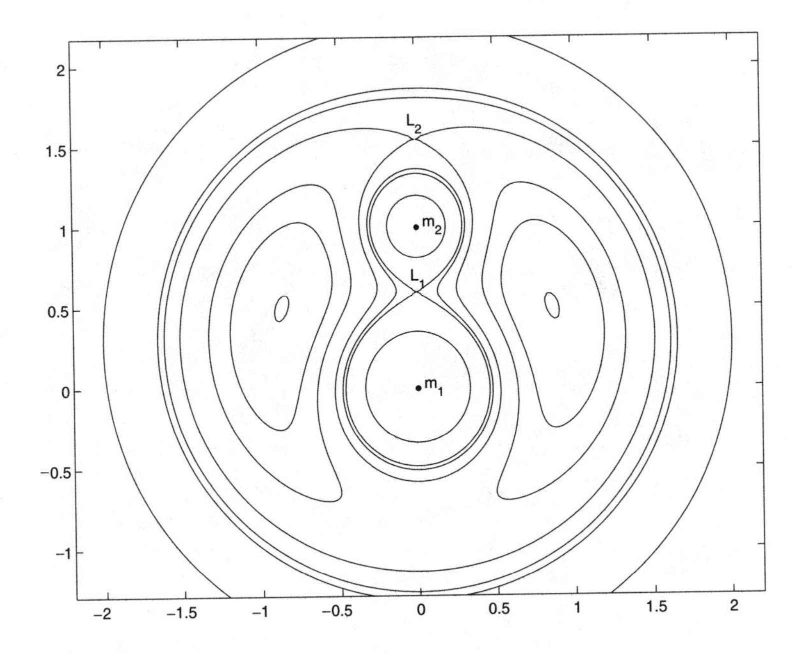

Fig. 4.14. A cross section in the orbital or (x, y) plane of the surfaces of constant normalized potential Φ_n for a binary system with a mass ratio $q = 0.4$. The values of Φ_n are, from the curves closest to the two mass points $m_{1,2}$ and moving outwards: 5.0, 3.9075 (at L_1), 3.8, 3.559 (at L_2), 3.2, 3.0, 2.8 (near $L_{4,5}$, the local potential maxima), and then increasing again at larger distances. The $L_{4,5}$ points are inside the smallest oval contours that form equilateral-triangle points with the line of centres through the masses at $m_{1,2}$.

size of the Roche lobe of the more massive star increases, whilst that of the less massive star decreases.

In Figure 4.14, the equipotential surface that lies immediately outside the figure-8 Roche limit illustrates the shape taken on by a *contact* binary system. In these systems, the two stars are in physical contact by means of a narrow neck in the region of L_1, and they are surrounded by a common envelope occupying the region between the Roche limit and the equipotential surface appropriate for the surface of the system. Beyond that region, a second equipotential surface is shown; it surrounds the entire binary and displays an intersection point at L_2. This is the *outer Lagrangian point*, through which matter can escape most easily from the gravitational field of the binary. There is an additional intersection, not shown in Figure 4.14, on the opposite side of the two masses from L_2, known as the L_3 point, which offers an additional escape route through a zone at a higher potential than L_2. At greater distances from the two masses lie the potential maxima, labelled L_4 and L_5, which occupy positions that make equilateral triangles with the two masses.

As an aid to appreciating some of these issues, Figure 4.15 provides a mesh-surface representation of the value of the normalized potential surrounding the two mass points and seen in the orbital (x, y) plane. The two deep potential wells for the two stars are obvious, and the L_1 point lies at the col between the two stars. As each star fills more of its available volume, its

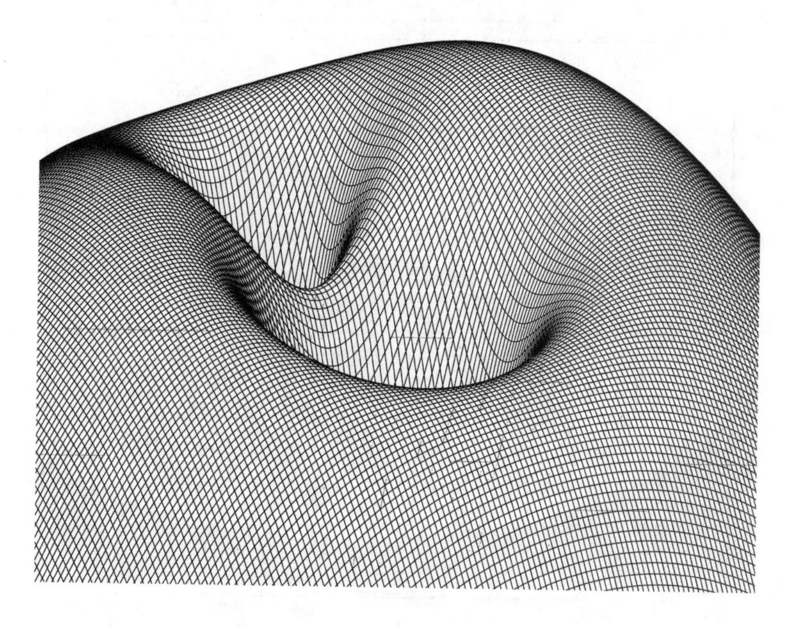

Fig. 4.15. A mesh-surface representation of the value of the normalized potential Φ_n in the orbital plane of a binary system. The figure illustrates the two potential wells of the stars, as well as the inner Lagrangian point L_1 and the two potential maxima at $L_{4,5}$ at the equilateral-triangle points relative to the line joining the two centres of mass. The L_2 point lies on the line of centres at the back of the figure.

surface approaches the top of its potential well. Once a stellar surface reaches its Roche lobe, matter can spill over the L_1 col more easily than anywhere else and fall down the potential well of the companion star. The potential barrier in all other directions is larger, and therefore matter would require more energy to overcome it than in the L_1 direction. It must be appreciated that Figure 4.15 is a representation of the potential seen in the (x, y) plane. Similar cross sections can be generated in the (y, z) and (x, z) planes. The overall three-dimensional effect is that the L_1 point acts as a nozzle through which matter can travel more easily than anywhere else from one component of the binary to its companion. The same description can be applied to the L_2 point and the ejection of matter from the binary system through the lowest part of the potential barrier.

In summary, the *shape* of a star in a binary system will be defined by the equipotential surface that is appropriate for that star. For well-detached binary systems, where R/a is approximately 0.1 or less, the stars are spherical because the equipotential surfaces close to each mass point are that shape. For binary systems in which the stars are relatively closer to each other, $R/a \approx 0.2$–0.3, the stars are non-spherical and ultimately can fill their respective Roche lobes. An important quantity, particularly with respect to binary-star evolution, is the volume of the Roche lobe for a star, and hence its *effective radius*, r_L. The effective radius is also called the *volume radius* because it is the radius of the sphere of the same volume as the Roche lobe. The most extensive tabulation of the details of Roche geometry, specifically for binary-star studies, has been provided by Mochnacki (1984), where values for r_L are given to five significant figures as functions of the mass ratio q, together with other useful quantities. Eggleton

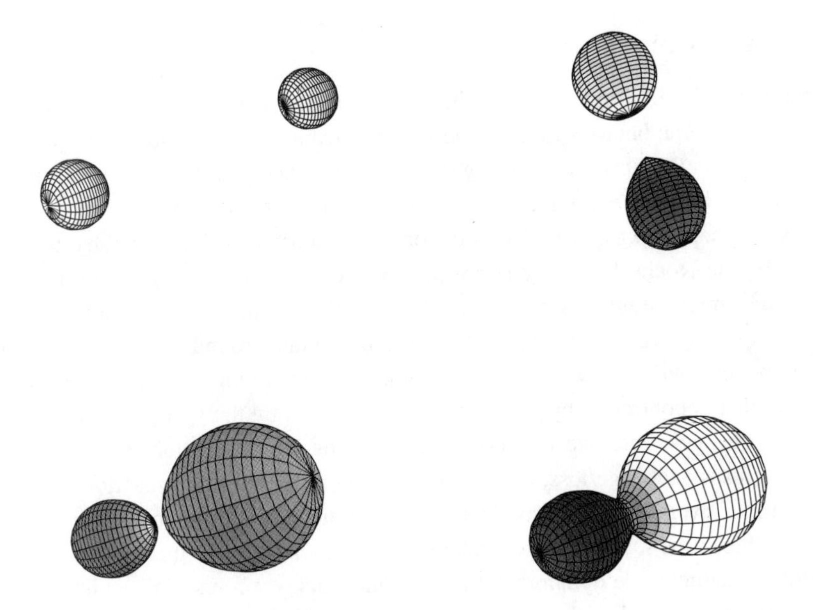

Fig. 4.16. Roche-model surfaces for four different binary-star systems seen at a variety of orientations. Clockwise from top left: a detached system with $q = 0.85$; a semidetached system with $q = 0.3$; a contact system with $q = 0.4$ and both components surrounded by a common envelope; a semidetached system with $q = 0.25$.

(1983) has presented a convenient formula for r_L that is accurate to about $\pm 1\%$ for all values of q, from zero to infinity, and is now widely used, namely,

$$r_L = \frac{0.49q^{2/3}}{0.69q^{2/3} + \ln(1 + q^{1/3})} \tag{4.49}$$

For the lower-mass star, put $q = m_2/m_1$, so that $q \leq 1$, to evaluate $r_{L,2}$. For the higher-mass star, put $q = m_1/m_2$, so that $q \geq 1$, to evaluate $r_{L,1}$. The actual effective radius for a star at the Roche lobe is then $R_L = r_L a$, with a being the separation, or the semimajor axis of the relative orbit.

The illustrations in Chapter 1 of detached, semidetached, and contact binaries and the cataclysmic variable were calculated from this Roche model. Figure 4.16 provides further illustrations of the shapes of stars according to this model. The details of the equations used to generate these models will be given in Chapter 5, where we shall consider the photometric properties of binary stars.

For the case of asynchronous rotation of the two stars, it is standard practice to multiply the rotational-potential term in equation (4.46) by a quantity F^2, where F is the ratio of the axial-rotation speed to the synchronous-rotation speed; see Wilson (1979) for a discussion of the development of this aspect, including references. When stars do rotate asynchronously, their associated equipotentials are distinct, and equation (4.46) must be applied with each star in turn being regarded as star 1 in the equation.

4.4 Mass exchange/loss

A star can reach the limiting surface of its Roche lobe by a variety of routes, and they may act in concert in some binary systems. The star may expand through normal stellar-evolution processes to fill its Roche volume during the main-sequence phase (case A), during the transition to the much larger red-giant phase (case B), or during the post-red-giant or supergiant phase (case C), depending on the size of the orbit or the orbital period, as noted in Chapter 1. Alternatively, the Roche lobe may shrink down onto the surface of the star as a result of orbital-angular-momentum loss by means of magnetic braking or gravitational-wave radiation. Because of the existence of this upper limiting volume around a star in a binary system, binary components will lose or gain mass and angular momentum to or from their companions in a manner that is controlled by the Roche geometry and by the speed with which the stars react to such changes. Later we shall consider the changes in Roche-lobe sizes due to various processes of mass exchange/loss, but firstly we must note the stellar reaction times that are convolved in the overall mechanisms of mass and angular-momentum exchange and loss.

There are three time scales that are fundamentally important in the theory of stellar structure and evolution, namely, the dynamical, thermal, and nuclear time scales, which describe the reaction times of a star to changes in its equilibrium state. The dynamical time scale corresponds to the time required for a star to react to departures from hydrostatic equilibrium and is

$$t_{\mathrm{dyn}} = (2R^3/Gm)^{1/2} \approx 40\left[\left(\frac{R}{R_\odot}\right)^3 \frac{M_\odot}{m}\right]^{1/2} \text{ minutes} \tag{4.50}$$

The star reacts quickly to departures from hydrostatic equilibrium, and it does so adiabatically. The thermal time scale measures reactions to departures from thermal equilibrium and is

$$t_{\mathrm{th}} = Gm^2/RL = (3.0 \times 10^7)\left(\frac{m}{M_\odot}\right)^2 \frac{R_\odot}{R}\frac{L_\odot}{L} \text{ years} \tag{4.51}$$

whilst the nuclear time scale, characteristic of the main-sequence lifetime, is

$$t_{\mathrm{nuc}} = (7 \times 10^9)\frac{m}{M_\odot}\frac{L_\odot}{L} \text{ years} \tag{4.52}$$

Of particular importance here is the dependence of stellar radius on changes in mass, expressed, for example, by a term $\zeta = d \ln R/d \ln m$, which can be calculated from stellar models for each of the above time scales, as well as for the Roche geometry in a given binary system, from, for example, equation (4.49) and $R_L = r_L a$. Hjellming and Webbink (1987) have described how these terms ζ_{dyn}, ζ_{th}, and ζ_{nuc} can be compared with ζ_L, corresponding to the changing size of the Roche lobe as the mass ratio of the binary changes. If $\zeta_L > \zeta_{\mathrm{dyn}}$, then mass transfer will occur on a dynamical time scale, because the star will not be able to readjust its hydrostatic equilibrium quickly enough to keep pace with the Roche-lobe changes. If $\zeta_{\mathrm{dyn}} > \zeta_L > \zeta_{\mathrm{th}}$, then hydrostatic equilibrium can be restored easily, but the slower restoration of thermal equilibrium will not be able to keep pace with the shrinkage of the Roche lobe as the star loses mass. Thus mass transfer from the loser will take place on a much more gentle thermal time scale. If $\zeta_{\mathrm{dyn,th}} > \zeta_L$, then the star will lose mass only because of stellar-evolution processes. The time scales for Roche-lobe shrinkage due to magnetic braking, in systems where one or both components have substantial magnetic activity and stellar winds, encompass the range t_{th}–t_{nuc}, depending on the orbital period at the ZAMS stage; those for gravitational-wave radiation are comparable to t_{nuc}.

Theoretical models for mass-transfer/loss processes show that mass-exchange/loss rates of 10^{-5} to 10^{-4} $M_\odot$ per year can be expected in dynamical-time-scale events, which may last 10^4 to 10^5 years. The probability of catching a binary in such a rapid mass-transfer process is exceedingly low, however, because of the short duration of the event relative to the stellar-evolution time scale. In such events, the positive feedback from the mass-exchange process forcing the Roche lobe down onto the surface of the mass-losing star will ensure its continuation through to the time when the mass ratio of the system will have been completely reversed and the Roche-lobe sizes will increase again. These facets of the process will be demonstrated later in discussions of conservative and non-conservative mass transfer. For thermal-time-scale events, mass-transfer rates are 10^{-7} to 10^{-6} $M_\odot$ per year, similar to those found observationally for radiatively driven, stellar-wind mass losses from O-type stars. On nuclear time scales, the rates are still lower, at 10^{-11} to 10^{-8} $M_\odot$ per year, which are rates typical also of mass losses due to winds from red-giant stars and due to magnetic braking amongst solar-type stars.

Mass transfer between components of binary systems, or mass loss from binary systems, can occur by several different routes, each resulting in different rates of change of the binary orbital period. The orbital period is the best-defined quantity for any binary system, and hence changes in its value can be detected fairly readily, provided that sufficient observational effort is devoted to monitoring those binaries where mass-exchange processes are likely to be occurring. The common signatures of mass transfer/loss are variable emission lines in spectra, asymmetric and variable light curves, substantial and/or variable x-ray- and/or radio-flux densities, although this is not to say that such variations are due exclusively to mass exchanges. If such signatures identify on-going mass transfer, the variations in the orbital period can provide an independent measure of the rate of mass exchange or mass loss. Discussion of the interpretation of these signatures is deferred until Chapter 7, on techniques of surface imaging. This section examines what can be expected in the way of orbital-period changes due to various modes of mass transfer and mass loss, giving an indication of whether or not we can expect to determine observationally the mass-transfer rates in various binary systems.

4.4.1 Conservative mass transfer

The term *conservative mass transfer* describes the simplest case, namely, that all the mass lost by one component is gained by its companion, and so the total mass of the binary is conserved, together with the total orbital angular momentum J_{orb}. Strictly speaking, the total angular momentum of the binary is composed of the sum of the orbital angular momentum for the two mass points, as given in Chapter 2, and the rotational, or spin, angular momentum of each star. In practice, we find that the total spin angular momentum is no more than 1–2% of J_{orb} in most cases, and so it can be ignored, but only just.

Thus we have constant total mass, $m_1 + m_2 = M_{\text{tot}}$, and $dm_1 = -dm_2$. Also,

$$J_{\text{orb}} = \left[\frac{G m_1^2 m_2^2 a(1 - e^2)}{M_{\text{tot}}} \right]^{1/2} \tag{4.53}$$

so that we can write $a = \text{constant}/(m_1 m_2)^2$. Kepler's third law states that $G M_{\text{tot}} = 4\pi^2 a^3/P^2$. If we define m_{1i}, m_{2i}, and P_i as the *initial* values for the two masses and the orbital period before any mass-exchange process, and m_1, m_2, and P are those values after mass exchange, then

$$\frac{P}{P_i} = \left[\frac{m_{1i} m_{2i}}{m_1 m_2} \right]^3 \tag{4.54}$$

This equation can be differentiated with respect to time to give

$$\frac{\dot{P}}{P_i} = 3 \left[\frac{m_{1i} m_{2i}}{m_1 m_2} \right]^2 \left[-\frac{\dot{m}_1}{m_1^2 m_2} - \frac{\dot{m}_2}{m_1 m_2^2} \right] m_{1i} m_{2i} \tag{4.55}$$

But $\dot{m}_1 = -\dot{m}_2$ for conservative mass transfer, so that

$$\begin{aligned} \frac{\dot{P}}{P_i} &= 3 \left[\frac{m_{1i} m_{2i}}{m_1 m_2} \right]^3 \left[\frac{m_1 - m_2}{m_1 m_2} \right] \dot{m}_1 \\ &= 3 \frac{P}{P_i} \left[\frac{m_1 - m_2}{m_1 m_2} \right] \dot{m}_1 \\ \frac{\dot{P}}{P} &= \frac{3 \dot{m}_1 (m_1 - m_2)}{m_1 m_2} \end{aligned} \tag{4.56}$$

Thus $\dot{P}$ depends on $\dot{m}_1$, as we would reasonably expect. But note the bracketed term $(m_1 - m_2)$, which can change sign as one star loses mass whilst the other gains mass. If the initially more massive star transfers mass to the initially less massive star, then $\dot{m}_1 < 0$, $\dot{P} < 0$, and the orbital period decreases, together with the absolute size of the orbit, in accordance with Kepler's third law. This period decrease will continue until $m_1 = m_2$ and the orbit size has reached a minimum. If the star of mass m_1 continues to lose mass after $m_1 < m_2$, then $\dot{P} > 0$, and both P and a will increase again. The absolute sizes of the Roche lobes, R_L, will depend primarily on the semimajor axis of the relative orbit a, and only secondarily on the value of the mass ratio q. This is readily apparent from $R_L = r_L a$ and equation (4.49).

As an example, consider a 5+2-$M_\odot$ binary with an orbital period of 10 days and a conservative-mass-transfer rate of 10^{-7} $M_\odot$ per year. Equation (4.56) gives a value of $\dot{P}/P = 9 \times 10^{-6}$ per year, so that the change in the binary orbital period over an interval of 10 years will be $\Delta P = 0.00018$ day $= 15.5$ seconds. Such a change will be detectable from a sequence of well-determined times of eclipse minima, because orbital periods typically are determined to an accuracy of 1 part in 10^7, as shown in Chapter 5.

4.4.2 Non-conservative mass transfer: Mass loss

The mode of mass transfer that can be expected to occur most often in nature is that in which only some of the mass and some of the angular momentum are transferred from the loser to the gainer, with the rest being lost from the binary system. Several such *non-conservative* mechanisms seem possible, chief amongst them being: (1) mass transfer/loss caused by a stellar wind, regardless of whether or not the loser fills its Roche lobe, (2) events of Roche-lobe overflow (RLOF), which are rapid, occurring, for example, on a dynamical time scale, and (3) sudden catastrophic mass loss, such as from a nova or a supernova event on one of the components of the binary. Angular-momentum loss (AML) due to a weak stellar wind from a magnetically active star in a binary will be effective for reducing the orbital period, but will not significantly alter the mass of the system, whilst gravitational-wave radiation (GWR) will also reduce the orbital period, but on a longer time scale.

The most simple representation for mass loss due to a wind is to consider a spherically symmetric wind that does not interact with the companion star. Although the model is simplistic, it is a useful starting point. Here we can write $\dot{m}_1 < 0$, $\dot{m}_2 = 0$ (it does not matter which star is the loser), and we note that the linear velocity of the loser in its binary orbit remains constant. The orbital angular momentum changes only because there is mass loss. Kepler's third law is

$$P^2 = \frac{4\pi^2 a^3}{G(m_1 + m_2)} \tag{4.57}$$

which can be differentiated to give the rate of change of the orbital period:

$$\begin{aligned} 2P\dot{P} &= \frac{4\pi^2}{G}\left[\frac{3a^2\dot{a}}{(m_1+m_2)} - \frac{a^3\dot{m}_1}{(m_1+m_2)^2}\right] \\ \frac{\dot{P}}{P} &= \frac{1}{2(m_1+m_2)}\left[\frac{3\dot{a}(m_1+m_2)}{a} - \dot{m}_1\right] \end{aligned} \tag{4.58}$$

To find an expression for $\dot{a}/a$, note that the requirement for a constant linear velocity for the loser is that $a_1 2\pi/P = \text{constant}$. Combined with the usual relationship $a_1 = am_2/(m_1 + m_2)$, we have

$$\frac{am_2}{(m_1+m_2)}\,\frac{2\pi G^{1/2}(m_1+m_2)^{1/2}}{2\pi a^{3/2}} = \text{constant} \tag{4.59}$$

Thus $a(m_1 + m_2) = \text{constant}$, so that $\dot{a}(m_1 + m_2) + a\dot{m}_1 = 0$, and hence we obtain the final simple equation

$$\frac{\dot{P}}{P} = \frac{-2\dot{m}_1}{(m_1+m_2)} \tag{4.60}$$

Thus for $\dot{m}_1 < 0$ caused by a stellar wind, such that $a_1 2\pi/P = \text{constant}$, the binary period P must *increase*, regardless of which star is losing mass. For typical wind-driven mass-loss rates of 10^{-7} $M_\odot$ per year in a massive 20+10-$M_\odot$ binary with $P = 10$ days, equation (4.60) shows that $\Delta P \sim 7 \times 10^{-7}$ days over 10 years. It would be necessary to monitor such a system for decades to see positive evidence for period changes due to wind-driven mass loss, an unfortunate situation considering that the theory of radiatively driven winds in hot stars would no doubt benefit from an independent determination of mass-loss rates in O stars. For binaries containing Wolf-Rayet stars, where the wind-driven mass-loss rates are one or two orders of magnitude larger, the period changes that have been observed can reasonably be interpreted as due to wind-driven mass loss.

A more rigorous examination of the general problem of wind-driven mass transfer and RLOF has been conducted by Tout and Hall (1991), with the reasonable assumption for closely interacting binaries that the orbital eccentricity is zero. If the donor is losing mass at a rate of $\dot{m}_1$, some of that mass will be accreted onto the gainer at a rate $\dot{m}_2$, and the rest, amounting to $\dot{m} = \dot{m}_1 + \dot{m}_2$, will be lost from the system. Note that both $\dot{m} < 0$ and $\dot{m}_1 < 0$, whilst $\dot{m}_2 > 0$. The orbital angular momentum that is lost by the binary because of the escaping matter will be the sum of the fraction of orbital angular momentum lost by the mass loser and that contributed by any other mechanisms, say KJ, so that we can write

$$\dot{J} = \dot{m} a_1^2 2\pi / P + KJ \tag{4.61}$$

which, because $J = 2\pi a^2 m_1 m_2 / P(m_1 + m_2)$, and $a_1/a = m_2/(m_1 + m_2)$, can be rewritten as

$$\frac{\dot{J}}{J} = \frac{m_2}{m_1} \frac{\dot{m}}{(m_1 + m_2)} + K \tag{4.62}$$

The alternative form of the equation for J, given in Chapter 2, is

$$J = \frac{G^{1/2} m_1 m_2 a^{1/2}}{(m_1 + m_2)^{1/2}} \tag{4.63}$$

which yields, upon differentiation,

$$\frac{\dot{J}}{J} = \frac{\dot{m}_1}{m_1} + \frac{\dot{m}_2}{m_2} + \frac{\dot{a}}{2a} - \frac{\dot{m}}{2(m_1 + m_2)} \tag{4.64}$$

Likewise, Kepler's third law can be differentiated to give

$$\frac{\dot{P}}{P} = \frac{3\dot{a}}{2a} - \frac{\dot{m}}{2(m_1 + m_2)} \tag{4.65}$$

These three equations can be combined to provide the final equation relating the derivative of the period to the rate of mass transfer and loss:

$$\frac{\dot{P}}{P} = -\frac{2\dot{m}}{(m_1 + m_2)} - \frac{3\dot{m}_2(m_1 - m_2)}{m_1 m_2} + 3K \tag{4.66}$$

This equation sensibly gives our previous results for conservative mass transfer via RLOF, equation (4.56), when $\dot{m} = K = 0$ and $\dot{m}_1 = -\dot{m}_2$. It also provides the simple wind-driven mass-loss equation, equation (4.60), when $\dot{m}_2 = K = 0$ and $\dot{m} = \dot{m}_1$, and it allows the introduction of an additional angular-momentum-loss mechanism, via the K term, which might be via magnetic braking or by gravitational-wave radiation.

For magnetic braking, van't Veer and Maceroni (1992) have thoroughly reviewed and cited much of the earlier theoretical and observational work, both on single stars in open clusters and on binary systems composed of solar-type stars, and have adopted

$$K = xm(kR)^2 \alpha \omega^\beta \tag{4.67}$$

for a star of mass m, radius of gyration (kR), and angular-rotation speed ω. The x term is included to allow for the possibility that the star may not be regarded as a rigidly rotating body (when $x = 1$), and, in general, $x \leq 1$. The time dependence of the rate of rotation is taken to follow the form of the Weber-Davis (1967) law, $\dot{\omega} \propto -\omega^3$, which leads to the Skumanich (1972) relationship $\omega \propto t^{-1/2}$ determined observationally from studies of the rotational properties of solar-type stars in open clusters with a range of ages; hence, $\beta = 3$ and $\alpha < 0$ in this case. But subsequent studies of many open clusters, particularly the younger ones, have shown that a value of $\beta = 3$ does not hold for all rotational speeds, but is limited to speeds of $v_{\mathrm{rot}} < 10 \times V_{\mathrm{rot}\odot}$. A review paper by Collier Cameron, Jianke, and Mestel (1991) discusses the comparisons between observation and theory, noting that while the Weber-Davis law holds for low rotational speeds, when basic dynamo theory expects a linear dependence between a star's magnetic field B and its angular speed, $B \propto \omega$, the same cannot be expected at higher angular speeds. There seems to be substantial uncertainty about the value of the exponent $\beta \leq 3$, with several mechanisms contributing to a reduction of the power-law dependence of $\dot{\omega} \propto \omega^\beta$, such as centrifugal driving of the stellar wind and the possibility of dynamo saturation. Accordingly, van't Veer and Maceroni have considered a wide range of values for β in attempting to match theoretical models to the observed bimodal distributions of orbital periods for close binary stars containing solar-type stars. They have concluded that the magnetic-braking relationship, $\omega \propto f(t)$, should exhibit a low value of $\beta \sim 0$–0.5 at high rotational velocities, $200 > v_{\mathrm{rot}} > 20$ km s^{-1}, with a rapid transition to the standard $\beta = 3$ relationship below $V_{\mathrm{rot}} = 10$ km s^{-1}. The expression derived by Guinan and Bradstreet (1988) for the rate of period change amongst detached binaries of solar type is based on the same representation, but uses an earlier calibration of the Skumanich relationship that, as shown earlier, is strictly applicable only to slowly rotating stars.

The rate of angular-momentum loss due to gravitational-wave radiation (GWR) is given by the standard expression, from Landau and Lifshitz (1962), of

$$\dot{J}_{\mathrm{GWR}} = \frac{-32G^{7/3}}{5c^5}(2\pi)^{7/3}\frac{(m_1 m_2)^2}{(m_1+m_2)^{2/3}P^{7/3}} \tag{4.68}$$

This equation can be modified to be combined with equation (4.66) in the form

$$\frac{\dot{J}_{\mathrm{GWR}}}{J} = K = \frac{-32G^3}{5c^5}\frac{m_1 m_2(m_1+m_2)}{a^4} \tag{4.69}$$

It is possible that non-conservative mass transfer may occur via another route, namely, RLOF through L_1 to the secondary star, followed by ejection from the binary system through L_2, which is located on the side of the secondary that is farthest from the primary component. Such a mechanism could operate during the rapid phase of RLOF, when the secondary could not react sufficiently quickly to accommodate all of the inflowing matter from the primary, or perhaps in contact systems. It would lead to a decrease in the orbital period, in part because of the conservative flow from the more massive star to the less massive star, ensuring that $\dot{P} < 0$ even if only a small fraction of that flow was accreted, and in part because of the loss of angular momentum from the system because of the matter escaping from L_2. From equations (4.64) and (4.65), we can write

$$\frac{\dot{P}}{P} = 3\left[\frac{\dot{J}}{J} + \frac{\dot{m}}{(m_1 + m_2)} - \frac{\dot{m}_1}{m_1} - \frac{\dot{m}_2}{m_2}\right] \tag{4.70}$$

With $\dot{m} = \dot{m}_1 + \dot{m}_2$, let us assume for simplicity that the secondary does not accept any of the transferred mass. Then $\dot{m}_2 = 0$, and $\dot{m} = \dot{m}_1$. The angular-momentum loss through L_2 can be written as $\dot{J} = \dot{m}d^2 2\pi/P$, where d is the distance of the L_2 point from the centre of mass of the binary system. Using Kepler's third law to replace the period P in the expression for $\dot{J}$, and the standard expression for J in equation (4.63), we can then rewrite the expression for the change in orbital period as

$$\frac{\dot{P}}{P} = 3\dot{m}\left[\frac{(m_1 + m_2)}{m_1 m_2}\frac{d^2}{a^2} - \frac{m_2}{m_1(m_1 + m_2)}\right] \tag{4.71}$$

Because $\dot{m} < 0$, and all the terms inside the square brackets are positive, with the first term larger than the second, it follows that $\dot{P} < 0$, and the orbital period must decrease during such an event.

The preceding formulations have assumed circular orbits, which is reasonable for most closely interacting binaries because they will already have passed through the stages of synchronization and circularization before reaching the events of mass transfer and loss. But the third mode of mass transfer/loss, identified at the beginning of this subsection, is the supernova event, which might make a binary orbit eccentric again, or even disrupt the binary completely. In such an event, we can adopt $\dot{m}_1 < 0$ for the star undergoing the supernova explosion, and $\dot{m}_2 = 0$, because it is unlikely that the companion will accrete any significant amount of the matter that is ejected at such high speeds ($V \sim 10^4$ km s^{-1}) from the supernova. Then the total energy of the binary, $C = -Gm_1m_2/2a$, provides a general expression for the rate of change of the size of the relative orbit:

$$\frac{\dot{a}}{a} = \frac{\dot{m}_1}{m_1} - \frac{\dot{C}}{C} \tag{4.72}$$

Likewise, the equation for the total orbital angular momentum, $J^2 = Gm_1^2m_2^2a(1 - e^2)/(m_1 + m_2)$, provides a general expression for the rate of change of the orbital eccentricity:

$$\frac{\dot{e}e}{(1 - e^2)} = \frac{\dot{a}}{2a} - \frac{\dot{J}}{J} + \frac{\dot{m}_1}{2m_1}\left[\frac{(m_1 + 2m_2)}{(m_1 + m_2)}\right] \tag{4.73}$$

Thus the orbital parameters a and e depend on $\dot{m}_1$ and on the consequent $\dot{C}$ and $\dot{J}$. In a supernova (SN) event, we can expect an increase in e, because equation (4.73) shows that for orbital-angular-momentum loss $\dot{J}/J < 0$, the term $-\dot{J}/J > 0$, and if it is greater than $\dot{a}/a < 0$ plus the term involving $\dot{m}_1 < 0$, then $\dot{e} > 0$. It clearly is not easy to make generalizations about the consequences of SN events in binaries. The explosion itself need not be symmetrical, and the surviving core may receive a kick velocity in any direction. The star on which the SN event takes place is likely to be the less massive component in the system, because the binary should have undergone a case-B mass-exchange/loss process to yield a pre-SN star with a helium core, as noted in Chapter 1. With the explosion occurring on the less massive star, the binary stands a greater chance of continuing. But to calculate the sequence of events for a particular binary would require us to know the amount of mass lost, together with the amounts of energy and angular momentum that were removed from the system. Verbunt (1993) has discussed earlier work on this problem, using the assumption that each star in a binary retains the same orbital velocity immediately after the explosion that it had just before the explosion, that is, that the explosion is instantaneous and perfectly symmetrical. The resultant expression for the eccentricity of the orbit after the SN event is

$$e = \frac{\Delta m}{(m_1 + m_2 - \Delta m)} \tag{4.74}$$

where Δm is the amount of mass lost during the SN event. If, for example, 90% of the mass of the SN progenitor is lost from the binary, then we obtain $e = 0.9q/(1 + 0.1q)$, where $q \leq 1$, and hence the eccentricity lies in the range $0.09 \leq e \leq 0.82$ for $0.1 \leq q \leq 1.0$. Thus we can expect that an SN event occurring on a star in a binary system will at least make the binary orbit quite eccentric again, and may even disrupt the binary system completely. The fact that we observe binary systems in which remnant cores of Type-II SNe, namely, neutron stars, are components at least shows that some systems survive these catastrophic events. The system PSRB1259-63, which contains a radio pulsar and a Be star, has an orbital eccentricity of $e = 0.87$ and is therefore a good example of this mechanism.

Shara et al. (1986) have applied the same general theory to the particular case of a nova explosion on one component of an interacting binary. They have shown that the mass loss resulting from the ejection of the nova shell dominates over all other considerations and ensures that the separation and hence the orbital period of the binary must increase and halt the exchange of mass by RLOF from the red-dwarf star to its white-dwarf companion. Only in the case of extreme mass ratios, $q \leq 0.01$, does the effect of any frictional-angular-momentum loss (due to the binary orbiting in an ejected resistive medium) dominate over the mass loss and cause orbital shrinkage. This problem has been discussed further by Livio, Govarie, and Ritter (1991).

4.5 Observed changes in orbital periods

Having provided a summary of the period changes to be expected from the various possible modes of mass transfer and loss, it is appropriate to illustrate the value of these theoretical considerations by applications to observed binary systems of different types in which mass-transfer/loss processes are known to be taking place. At the same time, we must be aware that other mechanisms for orbital-period changes may be operating in some systems, specifically, magnetic-activity cycles in rapidly rotating solar-type stars, and exchanges of angular momentum between the stars' rotations and their mutual orbit due to tidal interactions when synchronization and circularization may have been disturbed by evolution processes.

It is standard and sensible practice to define a linear *ephemeris* for a source of variable brightness of the form

$$T_{\text{calc}} = T_{\text{ref}} + \epsilon P \tag{4.75}$$

where T_{calc} is the calculated time of a maximum or a minimum brightness level, T_{ref} is a reference time of the same type, and ϵ is the number of cycles in the period P, as defined by equation (4.6). For eclipsing binary stars, T_{calc} and T_{ref} refer to the times of mid-primary or mid-secondary eclipses. We can compare the observed times of mid-primary or mid-secondary eclipses, T_{obs} – which can be determined to an accuracy of a few seconds in favourable cases (± 0.0001 day) – with those calculated from a *linear* ephemeris, T_{calc}, such as that given earlier, to form an *observed–calculated* $(O - C)$ *curve*, or an *ephemeris curve*, where

$$(O - C) = T_{obs} - T_{calc} = \Delta T(\epsilon) = T_{obs} - T_{ref} - \epsilon P \tag{4.76}$$

From a sequence of such times of minima extending over many years, or even decades in some cases, it is possible to determine whether or not the orbital period has remained constant. Values of $(O - C)$ that scatter about a straight line on a graph of $(O - C)$ versus time clearly show a constant orbital period, defined to an accuracy that is determined both by the precision of the individual times of minima and by the length of time over which the times of minima have been determined. The straight line will have zero slope only if the value for the period used in equation (4.76) is the correct one; an incorrect, but constant, value for the period will only introduce a tilt to the $(O - C)$ curve because of the accumulation of small differences, $\Delta T(\epsilon)$, per orbital cycle extending over the set of times of minima. Times of minima determined from photoelectric or CCD photometry typically are accurate to a few seconds (± 0.0001 day), whilst those from photographic photometry and visual estimates may well be less accurate by a factor of 10 or more. But such photographic or visual estimates can be exceedingly useful when their use can extend the time axis of the ephemeris curve by a factor of 5 or 10.

But, in general, we can expect that the ephemeris curve for a binary will be more complex, and we have seen earlier in this chapter how we can recognize the signatures of apsidal motion in $(O - C)$ data and of light-travel time across barycentric orbits in multiple systems. A very useful and general prescription for interpreting ephemeris curves is that given by Kalimeris, Rovithis-Livaniou, and Rovithis (1994), which will be summarized here.

If we write that an observed time of minimum, $T_{obs}(\epsilon)$, is given by

$$T_{obs}(\epsilon) = T_{calc}(\epsilon) + \Delta T(\epsilon) \tag{4.77}$$

where $T_{calc}(\epsilon)$ is calculated from a linear ephemeris with an adopted period P_{le}, so that

$$T_{calc}(\epsilon) = T_{ref} + \epsilon P_{le} \tag{4.78}$$

then the instantaneous value of the orbital period at cycle ϵ is simply

$$P(\epsilon) = T_{obs}(\epsilon) - T_{obs}(\epsilon - 1) = P_{le} + \Delta T(\epsilon) - \Delta T(\epsilon - 1) \tag{4.79}$$

If we then represent the differences $\Delta T(\epsilon)$ by a general polynomial form

$$\Delta T(\epsilon) = \sum_{j=0}^{n} c_j \epsilon^j \tag{4.80}$$

we find that

$$P(\epsilon) = P_{\text{le}} + \sum_{j=0}^{n} c_j \epsilon^j - \sum_{j=0}^{n} c_j (\epsilon - 1)^j \tag{4.81}$$

and hence that

$$\begin{aligned} \frac{dP}{d\epsilon} &= \sum_{j=0}^{n} c_j j \epsilon^{j-1} - \sum_{j=0}^{n} c_j j (\epsilon - 1)^{j-1} \\ &= \sum_{j=0}^{n-1} c_{j+1} (j+1) \epsilon^j - \sum_{j=0}^{n-1} c_{j+1} (j+1)(\epsilon - 1)^j \end{aligned} \tag{4.82}$$

Becuase $\epsilon = (t - T_{\text{ref}})/P_{\text{le}}$, it follows that $d\epsilon/dt = 1/P_{\text{le}}$, and hence we can calculate the rate of change of the orbital period at any instant by

$$\frac{dP}{dt} = \dot{P} = \frac{dP}{d\epsilon}\frac{d\epsilon}{dt} = \frac{1}{P_{\text{le}}}\frac{dP}{d\epsilon} \tag{4.83}$$

The simplest example, beyond a linear ephemeris, is to consider a quadratic form for $\Delta T(\epsilon) = c_2\epsilon^2 + c_1\epsilon + c_0$. Inserting this expression into equation (4.82) provides the result that $dP/d\epsilon = 2c_2$. Thus, an ephemeris curve that displays a quadratic form is demonstrating that the orbital period of a binary system is changing at a constant rate with respect to cycle number ϵ, and the coefficient of the ϵ^2 term is $c_2 = P_{\text{le}}\dot{P}/2$. The c_2 coefficient may be determined from the ephemeris curve by a least-squares solution provided that the quadratic ephemeris is deemed to represent the observations correctly. Hence, in this simple case, we obtain a direct measure of $\dot{P}$ that can be inserted into equation (4.56), for example, to derive the mass-transfer rate between the components in the conservative case.

The most extreme examples of such quadratic ephemeris curves are those displayed by the interacting binaries β Lyr and SV Cen, where the orbital periods are found to be changing rapidly. In a comprehensive review of the well-studied semidetached system β Lyr, Harmanec and Scholz (1993) have shown that a single quadratic ephemeris can provide an excellent match to the radial-velocity data and to the times of minima extending back over more than two centuries to 1784. The orbital period of 12.9 days has been *increasing* at a mean rate of 18.9 s yr^{-1} over the past 100 years, which corresponds to a mass-transfer rate, via equation (4.56), of about 2×10^{-5} $M_\odot$ per year from the less massive B6–8II star to the more massive Be star. De Greve and Linnell (1994) have shown that the system has evolved non-conservatively through an interval of rapid mass exchange and loss to its current configuration in which the initial mass ratio has been reversed; now the mass loser is the less massive star, which is transferring mass conservatively to the more massive gainer, albeit still at a high rate, according to these observations. The large disc that is inferred from other observations to hide the mass-gaining Be star is further evidence for the considerable rate of accretion of mass.

The B1V+B6.6III binary SV Cen has a much shorter orbital period than β Lyr, at $P = 1.66$ days, and has been shown to be in a contact state. The orbital period has been observed to have been *decreasing* over the past century at a remarkably linear rate, with

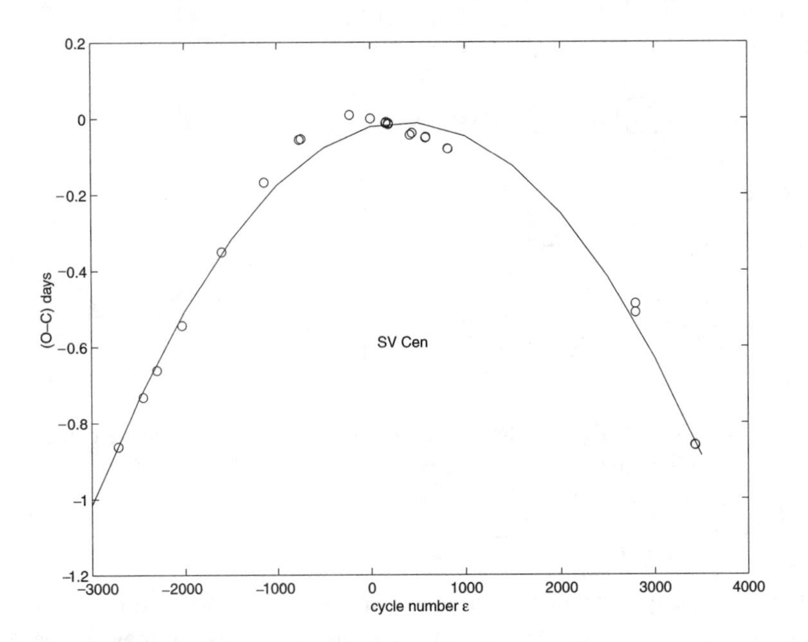

Fig. 4.17. The ephemeris curve for the interacting binary SV Cen, established from times of eclipse minima (circles) determined from photoelectric and CCD photometry over some 30 years, displays an approximate quadratic form, as shown by the solid line. The coefficient of the ϵ^2 term is $c_2 = -(8.9 \pm 0.2) \times 10^{-8}$, giving a value of $\dot{P}/P = -(2.36 \pm 0.05) \times 10^{-5}$ per year from this set of data, which is close to the mean value given by Drechsel et al. (1982).

superposed quasi-sinusoidal deviations, with a mean value of $\dot{P}/P = -2.15 \times 10^{-5}$ per year, rather greater than that for β Lyr, and of opposite sign. Figure 4.17 illustrates the ephemeris curve for SV Cen determined from photoelectric and CCD photometry over the past 30 years, showing the quadratic nature of the main variation, together with additional smaller deviations. The sign and the rate of the period change suggest, for a conservative-mass-exchange process, that mass is being transferred from the more massive star to its less massive companion at a variable and high rate of about 1–4$\times 10^{-4}$ $M_\odot$ per year.

The only sensible conclusion is that mass and angular momentum are being lost from the system. Drechsel et al. (1982) have used a formula from a review paper of Kruszewski (1966) (his equation 208) relevant to outflow from the outer Lagrangian point L_2 and calculated by considering the motions of particles according to the restricted three-body problem. They have concluded that the mass-loss rate is $\dot{m} \sim 5 \times 10^{-5}$ $M_\odot$ per year. If use is made instead of the more simple formula in equation (4.71), then $\dot{m} \sim 4 \times 10^{-6}$ $M_\odot$ per year when the masses of the two components are taken from Rucinski et al. (1992) as 8.60 $M_\odot$ and 6.05 $M_\odot$. Drechsel et al. (1982) have discussed substantial evidence from ultraviolet spectra for outflow of matter from the binary at speeds of 1000 km s^{-1}.

An intriguing recent discovery concerns the eclipsing-binary system V361 Lyr, which has an orbital period of 0.31 day, that is, in the main part of the distribution of orbital periods for contact binaries of the W-UMa-type. Yet the light curve demonstrates decisively that the system is not in contact, and its asymmetric shape can be interpreted as being due to an accretion hot spot on the facing hemisphere of the secondary star (Andronov and Richter 1987). Together with accurate light curves from Kaluzny (1990, 1991), Hilditch et al. (1997) used their spectroscopic data to provide a complete description of the system that involved the techniques of light-curve synthesis (see Chapter 5) and eclipse mapping (see Chapter 7). The 1.26-$M_\odot$ primary component fills its Roche lobe and is transferring mass conservatively at a rate of about 2×10^{-7} $M_\odot$ per year to the 0.87-$M_\odot$ secondary that fills only about 60% of its limiting volume. The mass-transfer stream from the L_1 point impinges directly onto the secondary at an angle to the line of centres that is entirely consistent with hydrodynamical calculations and produces an accretion hot spot that is about 5000 K hotter than its surrounding photosphere. The published times of primary minima determined photoelectrically extend over only 17 years so far, and so the deviations from a linear ephemeris are only now becoming significant, as shown in Figure 4.18. This ephemeris curve helps to confirm the interpretation of all the spectroscopic and photometric data, rather than define the mass-transfer rate in an absolute sense.

The more common kind of ephemeris curve does not display simple linear or quadratic behaviour, but is composed of subregions that can be well represented by straight lines, quadratics,

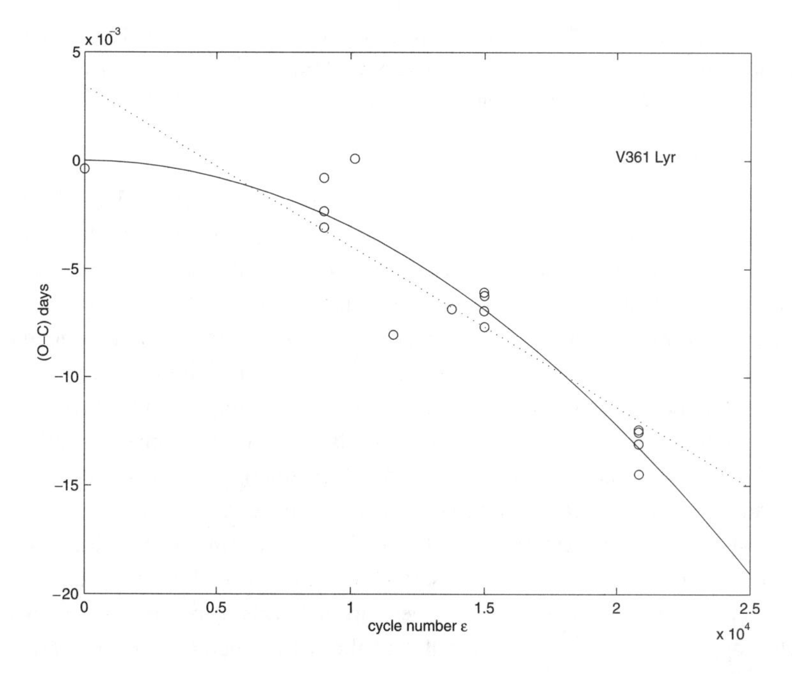

Fig. 4.18. The ephemeris curve for the mass-exchanging binary V361 Lyr, established from times of eclipse minima (circles) determined from photoelectric and CCD photometry over about 17 years, displays an approximate quadratic form, as shown by the solid line. A linear ephemeris, shown by the dotted line, gives a value of χ^2 that is about 35% greater than the quadratic representation. The coefficient of the ϵ^2 term is $c_2 = -(3.05 \pm 1.03) \times 10^{-11}$, giving a value of $\dot{P}/P = -(2.3 \pm 0.9) \times 10^{-7}$ per year from these data.

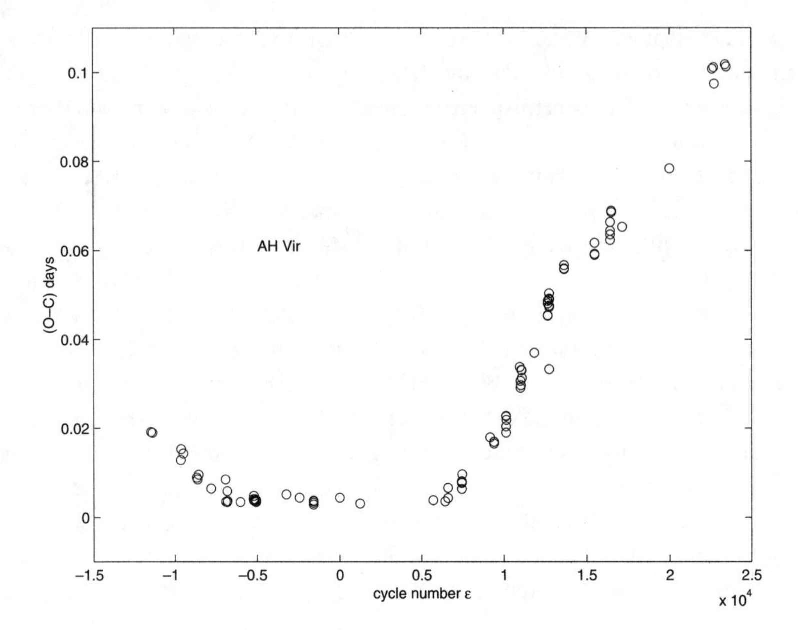

Fig. 4.19. The ephemeris curve for the contact binary AH Vir established from times of minima determined via photoelectric observations only, and collated by Demircan et al. (1991). The $(O - C)$ values have been calculated relative to a linear ephemeris.

or higher-order polynomials that are piecewise continuous. Kalimeris et al. (1994) have illustrated such ephemeris curves for several W-UMa-type contact binaries and have shown how the foregoing prescription can be applied to a sequence of subregions. In order to avoid the discontinuities in the values of $\dot{P}$ that are inherent in a piecewise-continuous representation of the ephemeris curve, Kalimeris et al. (1994) have used spline interpolation to join the various subregions. It is possible to consider the direct use of a least-squares cubic spline to fit the entire ephemeris curve, rather than dividing it into separate regions, and thereby determine directly the instantaneous rate of change of the orbital period from numerical differentiation. But the application of splines requires a well-defined ephemeris curve without substantial gaps in the data string, something that is not readily achieved observationally.

Figure 4.19 shows the ephemeris curve for the binary system AH Vir, drawn from times of minima determined via photoelectric observations and collated by Demircan, Derman, and Akalin (1991) from many earlier studies. There are intervals of nearly constant period, and intervals that approximate to sine waves, or at least alternating decreases and increases in orbital period. It is typical of many ephemeris curves for short-period binaries that contain at least one component that is a rapidly rotating late-type star with a convective envelope and hence can be expected to be magnetically active. Such stars are to be found in Algol systems, RS CVn systems, contact systems, and cataclysmic variables, as demonstrated by Hall (1989, 1991). A second example of this type of ephemeris curve is shown in the data on the cataclysmic variable UX UMa, published by Rubenstein, Patterson, and Africano (1991) and plotted in Figure 4.20.

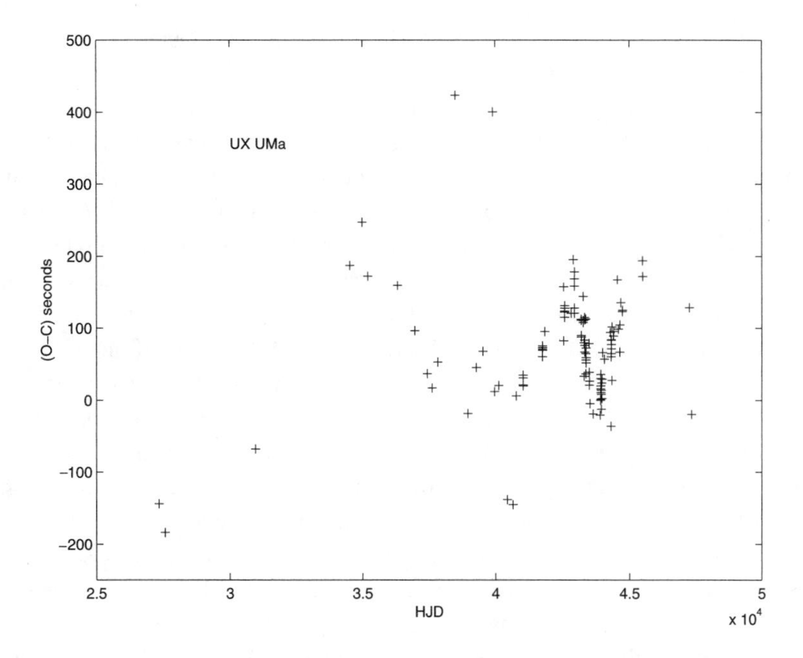

Fig. 4.20. The ephemeris curve for the cataclysmic variable UX UMa established from times of minima determined via averaged photographic data and direct photoelectric observations published by Rubenstein et al. (1991). The $(O - C)$ values have been calculated relative to a linear ephemeris. Most of the photoelectric timings have uncertainties of a few seconds, no greater than the size of the symbols used in the figure.

Such an ephemeris curve usually cannot be represented by a single function, and there seems always to be more scatter about a low-order mean curve than would be expected on the basis of the known quality of the observations. The overall evidence suggests that there is cyclic behaviour in the $(O - C)$ residuals that is not exactly periodic. The typical range of indicated periods is years to decades, and the amplitudes of the cycles are about 0.03 day. Such cycles should not reasonably represent orbital motion about a third body, because the inferred masses are usually very low, unless the orbits are highly inclined to each other, and in any event third-body motion would require strictly periodic variations that are not observed, except in a few cases such as SS Ari (Demircan and Selam 1993). An alternative mechanism for producing cyclic variations in the orbital period has been proposed by Applegate (1992), whereby a magnetically active star undergoing its standard magnetic-activity cycle would have a variable J_2 coefficient [equation (4.18)] and hence variable tidal effects on the orbit of the binary, thus producing variations in the orbital period of the system. This *Applegate mechanism* seems to be promising as a single explanation for the morphology of ephemeris curves for many types of binaries, and it seems potentially powerful as an aid to our understanding of the dynamo activity in rapidly rotating stars. Thus the arguments made by Kalimeris et al. (1994) that ephemeris curves ought to be analysed very carefully indeed, taking note of all apparent deviations from a mean smooth curve, should be considered very seriously. Most recently, Donati (1999) has shown that the RS CVn binary HR1099 exhibits an orbital-period change that is cyclic, with a period of 18 ± 2 years. He suggests that the Applegate mechanism, as modified by the work of Lanza, Rodonó, and Rosner (1998), provides the most sensible interpretation of these results.

The x-ray-binary systems, of both high mass and low mass, have the emission of their x-radiation powered by accretion of matter onto the neutron star. That matter is transferred to the neutron star from the companion either by means of RLOF through L_1 or by means of a stellar wind from a detached O-type companion. Accordingly, one might expect that orbital-period changes would be observable in such systems, in accordance with the mass-exchange formulae presented earlier. Amongst the HMXBs, only two systems (SMC X-1 and Cen X-3) among the 24 systems listed by White et al. (1995) show confirmed orbital-period changes as determined from accurate x-ray-pulse arrival times. Both show decreases in orbital periods, with values of $\dot{P}/P \sim -(2\text{–}3) \times 10^{-6}$ per year that are well represented by quadratic ephemeris curves (Kelley et al. 1983; Levine et al. 1993). If such period decreases were due to a conservative RLOF, then equation (4.56) would suggest that the mass-transfer rate was $\dot{m} \sim -1 \times 10^{-6}\ \mathrm{M}_\odot$ per year from the O–B supergiant to the neutron-star companion. Such a rate is about two orders of magnitude too large if the x-ray flux from these systems is to be explained by accretion of matter onto a compact source (see Chapter 7 and the Eddington luminosity). If the O–B star loses mass because of a substantial stellar wind that does not interact significantly with the companion, then equation (4.60) shows that the orbital period will increase, although at only a marginally detectable rate. To explain the well-established decreases in these orbital periods, we must consider some of the alternative mechanisms that have been mentioned earlier. Both Kelley et al. (1983) and Levine et al. (1993) have discussed these matters *in extenso* and have shown that tidal torques between the large O–B star in each system and its companion neutron star may provide much of the solution if it can be established that the O–B star is rotating at less than the synchronous rate, equation (4.39). Reynolds et al. (1993) have provided some evidence, in the case of SMC X-1, that the optical primary is indeed rotating at less than the synchronous rate, at $\omega_r \sim 0.75\omega_K$, but more extensive and more precise work is needed here.

To illustrate the point, consider equation (4.39) for the case of a circular orbit, as in SMC X-1. From Kepler's third law for constant total mass of the binary, we can write

$$\frac{\dot{a}}{a} = \frac{2}{3}\frac{\dot{P}}{P} \tag{4.84}$$

and hence equation (4.39) becomes

$$\frac{\dot{P}}{P} = -18\frac{k_2}{t_f}q(1+q)\left(\frac{R}{a}\right)^8\left(\frac{\omega_K - \omega_r}{\omega_K}\right) \tag{4.85}$$

From Reynolds et al. (1993), $q = 0.091$, $a = 25.3\ \mathrm{R}_\odot$, $R \sim 17\ \mathrm{R}_\odot$, and $\omega_r \sim 0.75\omega_K$. With the observed value of $\dot{P}/P = -3.36 \times 10^{-6}$ per year, we obtain a value for the ratio $t_f/k_2 = 5525$ years. The evolution age of the O–B star seems to be about 10^7 years, so that the value of the apsidal constant, according to the models of Claret and Giménez (1992), is $k_2 \sim 10^{-3}$ for an approximate SMC chemical composition. Hence, to explain the observed rate of period decrease, the current value of the friction time would have to be $t_f \sim 5$ years, comparable to the values estimated for the lower-mass, convective-envelope stars like the Sun.

A possible alternative scheme to explain the orbital decay would be to postulate RLOF through L_1, followed by ejection of the stream of matter out through L_2, with only some fractional interaction with the neutron star. Then equation (4.71) would also allow a period decrease to be expected. Kelley et al. (1983) considered that possibility, but tentatively discounted it, owing to lack of positive evidence for substantial column densities ($\sim 10^{26}$ cm^{-2}) in the x-ray data.

Amongst the LMXBs, White et al. (1995) have listed only four systems for which orbital-period changes have been determined. Two have decreasing periods, one increasing, and one is apparently variable in sign. None of them fits the straightforward picture of simple mass transfer from one component to the other. Perhaps the low-mass companions to the neutron stars are magnetically active stars like those in Algol, RS CVn, and CV systems, and perhaps these have more control over the orbital-period changes than do the causes postulated by simpler arguments involving exchanges and losses of mass and angular momentum.

In summary of this section, the observational data do not fully support the simple expectation of orbital-period changes depending solely on exchanges and losses of mass and angular momentum. There are some systems for which consistent interpretations of all available data, including period changes, do support direct RLOF and/or mass loss. But for the majority of the binary stars for which the observational evidence provides conclusive proof of mass-transfer processes, the observed orbital-period changes seem just as likely to be governed by magnetic-activity cycles or by tidal torques between axial-rotation rates and orbital rates as by mass transfer/loss.

5 Photometry and polarimetry: Stellar sizes and shapes

5.1 Introduction

Measurement of the brightness of astronomical sources is a fundamental part of all astronomy. The first detections of variations in brightness levels led to the beginning of the recognition of the existence of eclipsing binary stars, thanks to the work of Goodricke (1783), as noted in Chapter 1. Since that time, systematic searches for photometrically variable stars have been conducted by visual, photographic, and photoelectric techniques, and many eclipsing, or photometrically variable, binary stars have been discovered in the process, again as discussed in general terms in Chapter 1. It is therefore no accident that the observational technique of photometry has been the mainstay of studies of binary stars in general, at least in part because photometry can be conducted with very modest equipment and relatively small telescopes. Recent developments in CCD technology that have made it possible to record digital images of small areas of the sky with remarkable photometric precision have further enhanced the value of telescopes of modest apertures in conducting front-line research – for example, the recent discoveries of substantial numbers of eclipsing binaries in the nearby galaxy M31 by Kaluzny et al. (1998, 1999) and Stanek et al. (1998, 1999), who used a telescope of 1.3-m aperture. The current prospects for setting up a global network of programmable robotic telescopes to conduct continuous photometric monitoring of variable sources, including binary stars, are very encouraging, and that will herald a new era in our understanding of the properties of binary stars – magnetic-activity cycles, outbursts, changes in accretion structures, for example.

In addition, these same CCDs make it possible to perform spectroscopic or spectrophotometric studies of binary stars at a sampling rate that is becoming comparable to that achieved hitherto only via photometry. As one example, Baptista et al. (1998a,b) have used low-resolution spectra secured with the Faint Object Spectrograph on the Hubble Space Telescope, with integration times of 5 seconds and intervals between exposures of only 0.1 second, to investigate the properties of the accretion disc in the cataclysmic variable UX UMa. Appropriately, in this chapter we discuss the techniques and applications of multi-colour photometry and of spectrophotometry for determining the shapes and sizes of the stars in binaries, and the distributions of stellar-surface brightnesses, orbital inclinations, as well as the photometric evidence for the existence of accretion structures within binaries. We shall also discuss the major importance of polarimetry and spectropolarimetry, particularly for determining orbital inclinations and stellar sizes in Wolf-Rayet binaries. In Chapter 6 we shall summarize the current status of our knowledge of the masses and absolute dimensions of stars in binaries, as determined from application of the techniques learned in Chapters 3–5. But we shall defer until Chapter 7 discussion of the techniques used to determine maps of the surface brightnesses of stars and of accretion discs, streams, and columns, because they usually follow on from a prior, or at least initial, investigation of a binary via more straightforward photometric procedures.

5.2 Flux densities, magnitudes, and colour indices

It is an unfortunate fact that the astronomical literature has, in the past at least, made use of terms like 'luminosity' and 'intensity' in somewhat vague and ill-defined ways. Within the field of photometry of eclipsing binaries, for example, the term 'intensity' has been used to describe (1) the amount of radiation from the binary as a whole that is received by the observer, (2) the amount of radiation emerging from a particular element of a stellar surface in a particular direction, and (3) the logarithm of received x-radiation in a particular photon energy range. Unnecessary confusion has reigned for far too long, and I hope that this section will rectify that situation.

The *luminosity* L of a source is defined to be the total amount of radiant energy emitted over all wavelengths per unit time in all directions. So the most appropriate units for L are simply joules per second (J s^{-1}), or watts (W), or solar units ($L_\odot$). How bright an astronomical source *appears* to be depends upon the luminosity of the source and upon its distance d from us. An observer can make quantitative measurements, using various detectors, of the *received* radiant energy per unit time and per unit area. The observer would hope to be able to express such measurements in units of joules per second per square metre (J s^{-1} m^{-2}), or watts per square metre (W m^{-2}). Photometric detectors do not provide measurements of received radiant energy in such useful units directly, but have to be *calibrated* in some manner against some standard sources, and a summary of such calibrations will be given later. The measured *apparent brightness* is more correctly called *flux density* f, and ideally has units of watts per square metre (W m^{-2}); it is often simply called *flux*. The link between flux density, luminosity, and distance is given by the inverse-square law for the propagation of radiation, so that $f = L/4\pi d^2$ W m^{-2}. Because electromagnetic radiation spans so many orders of magnitude of frequency or wavelength, it is not possible to design a single type of telescope and detector that can measure all incident radiation from a source. Accordingly, our measured flux density is limited by some filter, or pass band, whose properties are determined by the particular combination of telescope optics (or beam convergers), filters, and detectors, and therefore the units of flux density most commonly encountered are watts per square metre per nanometre (W m^{-2} nm^{-1}) or ergs per square centimetre per second per angstrom (erg cm^{-2} s^{-1} $Å^{-1}$).

Before we consider these matters further, it is necessary to take note of some of the historical aspects of the development of photometry in astronomy that gave rise to some unusual definitions. But one must accept the necessity of getting used to these terms, because they are so embedded into all of astronomy that they will not go away, and in any event, many of them are really useful. A cursory look at the history will explain why these terms are still in use.

Hipparchos, in the second century B.C., divided the visible stars into six classes, or *magnitudes*, according to their apparent brightness as seen by the naked eye. The first class contained the brightest stars (magnitude 1), and the sixth class contained the faintest stars still visible to the naked eye (magnitude 6). These terms have been continued in use to the present day. In the late 1700s, W. Herschel used one of his telescopes, with a variable-size diaphragm inserted into it, to estimate *by eye* the *relative brightnesses* of stars. By reducing the aperture size of the telescope via the diaphragm, he could make stars appear fainter, because his *detector area* was decreased. For example, consider a field of view containing two stars of unequal brightnesses. Then record the radius r_1 of the diaphragm needed to make the fainter star just disappear; then record the radius r_2 needed to make the brighter star just disappear. The ratio of the two areas, $\pi r_1^2/\pi r_2^2$, gives the ratio of the apparent brightnesses of the two stars. With this technique, Herschel found that the average first-magnitude star was about 100 times brighter than the average sixth-magnitude star. Note that throughout these experiments, the *detector* is the human eye, which has a limited response to electromagnetic radiation, with only a small range of wavelengths constituting *visible* radiation, or light.

In the 1800s it was found that the response of the human eye to the brightness of light was not linear. If the flux densities of three sources are in the proportions 1:10:100, the *brightness difference* between the first and the second sources is perceived to be equal to the difference between the second and the third sources. Equal flux-density ratios, namely, 1:10, in each case correspond to *equal apparent brightness differences* for the eye. The human perception of brightness is logarithmic.

At the time of those developments, the only available detector for starlight was the human eye, aided by telescopes. So in 1856, N. Pogson suggested that a more formal categorization of magnitudes be accepted. With a difference of 5 magnitudes in apparent brightness corresponding to a ratio of 100 in flux density, Pogson proposed that a difference of 1 magnitude should be the fifth root of 100, namely, $(100)^{1/5} \approx 2.512$. So if two stars have *apparent magnitudes* m_1 and m_2, then the difference $m_1 - m_2$ corresponds to a ratio of received flux densities, f_1/f_2, via the equation

$$m_1 - m_2 = -2.5 \log_{10}\left(\frac{f_1}{f_2}\right) \tag{5.1}$$

The negative sign is included purely to make the brighter stars with larger values of f have smaller values of magnitude m. This definition agrees exactly with Pogson's suggestion. For $m_1 - m_2 = -1$ mag, we obtain $f_1/f_2 = 2.512$, to three decimal places. Note that there is no explicit zero level built into the magnitude scale, so we can write

$$m_\lambda = -2.5 \log_{10} f_\lambda + C_\lambda \tag{5.2}$$

where the λ subscript indicates that these values will depend upon the wavelength of observation, and the quantity C_λ is a constant that is arbitrarily defined by reference to a set of standard stars.

When the introduction of photographic plates revolutionized optical astronomy in the latter part of the nineteenth century, it was sensible to continue the magnitude system because photographic emulsion also responds almost logarithmically to received radiation. This system of recording apparent brightness in terms of the dimensionless logarithmic quantity called apparent magnitude m has been used ever since in optical astronomy and has been carried over into infrared astronomy as well. Radio astronomy, ultraviolet astronomy, and x-ray astronomy usually use flux density for measurements of received radiation. Flux densities are now also used in optical and infrared astronomy because photoelectric devices and CCDs respond linearly to received radiation. It so happens that the SI units of flux density, watts per square metre, are inconveniently large for most astronomical sources. So flux densities are expressed in janskys (Jy), where 1 Jy $= 10^{-26}$ W m^{-2}, or even millijanskys.

Thus a magnitude system provides measures of *relative* flux density in dimensionless units. A magnitude system is defined by the values of magnitudes assigned to a set of standard stars located around the sky whose constancy in brightness has been established over many repeated observations (tens to hundreds). A magnitude system usually involves measurements of flux density at several different wavelength regions, or pass bands, which are defined by coloured-glass or interference filters, together with the response of the detector, which is also wavelength-sensitive. Provided that the pass bands of the magnitude system have well-defined lower and upper wavelength limits, such a set of magnitudes can be determined from photoelectric or CCD photometry to very high precision, about ±0.2–0.5%, *independently* of any calibration of the set in terms of an absolute flux density in standard SI units. Accordingly, if a set of photometric observations of a particular binary star, for example, is presented in terms of magnitudes defined by a specific system, then those values can be compared with other data obtained at any other time, without the need for any absolute flux-density calibration. But one must be very careful, when comparing the apparent magnitudes of stars obtained at different times, say, to ensure that the magnitudes are defined in the same way. From photographic photometry, there are the *photographic blue* (m_{pb}) and *photovisual* (m_{pv}) magnitudes. From photoelectric photometry, there are the *UBVRI* magnitudes and the *uvby* and DDO systems, for example, each defined, in terms of their respective pass bands and selected standard stars, with lesser or greater precision. Although some of these magnitudes are similar, such as the visual (V), yellow (y), and m_{pv} pass bands, the transformations between any two can possibly be non-linear, and great care must be exercised in making any comparisons. This text is not the place to delve into such complications, and the reader is referred to useful texts like *Astronomical Photometry* by Henden and Kaitchuk (1982) for further details.

The absolute calibration of a magnitude at a given wavelength λ involves determination of the constant C_λ in equation (5.2). That determination, in units of watts per square metre per nanometre (W m^{-2} nm^{-1}) (i.e., flux density per unit wavelength interval in nanometres), has been fraught with difficulties. In brief, these difficulties have concerned the lack of a known constant astronomical source at a precisely known distance and luminosity, which has meant that only laboratory calibration sources could be used, and they could not be observed through the Earth's atmosphere in the same manner as stars and would not have the same spectral-energy distributions as most stars. However, much effort has been devoted to solving this problem, and Gray (1992) has presented a helpful discussion of the recent calibrations. In terms of the V magnitude, as defined in the *UBVRI* system discussed in Section 5.3.1, a star of spectral type A0, with colour index $(B - V) = 0.0$ and of magnitude $V = 0.0$, has a flux density of $f_\lambda = 3.54 \times 10^{-12}$ W m^{-2} Å^{-1} = 990 photons s^{-1} cm^{-2} Å^{-1} at a wavelength of $\lambda = 5556$ Å.

The links among the apparent magnitude of a star defined within certain wavelength limits and its distance and luminosity are made via another term called *absolute magnitude* that has the same wavelength limits. Thus, for V magnitudes, the absolute visual magnitude M_V is defined to be the apparent magnitude of a star as seen from a distance of 10 parsecs (pc). From the inverse-square law, the ratio of received flux density at distance d, $f(d)$, to that at distance 10 pc, $f(10)$, is, in the absence of any interstellar extinction,

$$\frac{f(d)}{f(10)} = \left[\frac{10\,\text{pc}}{d}\right]^2 \tag{5.3}$$

Then the difference in magnitudes at d and at 10 pc is

$$V - M_V = -2.5\log_{10}\left[\frac{f(d)}{f(10)}\right] = 5\log_{10}\left[\frac{d}{10\,\text{pc}}\right] \tag{5.4}$$

or

$$V - M_V = 5\log_{10}[d] - 5 \tag{5.5}$$

provided that the distance d is given in parsecs.

Because we do not observe the entire electromagnetic spectrum of an astronomical source easily, if at all, it is necessary to determine corrections to a derived absolute visual magnitude that will represent integration of the entire spectrum. This quantity is called the *bolometric correction*, BC, and is related to the *bolometric absolute magnitude* M_{bol} by the simple relationship

$$M_{\text{bol}} = M_V + \text{BC} \tag{5.6}$$

where BC is a negative quantity. It is calculated from theoretical models for the emergent spectrum from stars of different temperatures and has been checked and largely confirmed by means of photometric observations in the ultraviolet, visible, and infrared parts of the spectrum for a number of standard stars. Being a magnitude quantity, the zero point for BC is again arbitrary and has been defined so that BC is close to zero for stars that emit most of their radiation in the visible part of the spectrum. Thus BC ~ -0.1 for stars with effective temperatures of $T \sim 7000$ K, where most of the radiation is emitted in the visible part of the spectrum. It becomes more negative for both hotter and cooler stars, reaching values of BC ~ -3.8 for the hottest and coolest stars, most of whose radiation is emitted in the ultraviolet and infrared parts of the spectrum, respectively.

The connection between the absolute bolometric magnitude of a star $M_{\text{bol}*}$ and its luminosity L_* is usually given relative to the Sun by

$$M_{\text{bol}*} - M_{\text{bol}\odot} = -2.5\log_{10}\left[\frac{L_*}{L_\odot}\right] \tag{5.7}$$

In addition to comparing the flux densities of different sources at a given wavelength, or through a specific pass band, we can compare flux densities received at different wavelengths (λ_1, λ_2) from a single source and form a *colour index*, CI, given by the equation

$$\text{CI} = -2.5\log_{10}\left[\frac{f_{\lambda_1}}{f_{\lambda_2}}\right] + C_{\lambda_1} - C_{\lambda_2} \tag{5.8}$$

Such a colour index provides a crude measure of the slope of the spectral-energy distribution emitted by the source, for it is simply the ratio of the received flux densities at two wavelengths converted into logarithmic form. Thus a carefully selected colour index can provide an estimate of the temperature of the source, and we shall consider these issues later. The term *colour index* is used throughout optical and infrared photometry because it was developed in parallel with the magnitude systems, but it has been extended into ultraviolet photometry, and the idea appears in x-ray photometry as the *hardness ratio*, or *hard colour* and *soft colour*. These x-ray quantities are ratios of flux densities received in different pass bands, so that the hardness ratio typically measures the proportion of hard x-rays, with energies above about 2.5 keV, to soft x-rays, with energies below 2.5 keV. For example, the hard colour used by Hasinger and van der Klis (1989) is the ratio of flux densities in the 6–20-keV band and the 4.6–6-keV band, and the soft colour is the ratio in the 3–4.5-keV band and the 1–3-keV band. These bounds on the x-ray energy ranges seem to be dictated by the particular experiment on a given x-ray satellite, rather than some adopted standard values that are found amongst the optical and infrared photometric systems.

Having defined some of the basic elements of photometry, we can now consider in outline the advantages and disadvantages of some of the various photometric systems that have been developed for stellar astronomy.

5.3 Photometric systems and spectrophotometry

Multi-colour photometry is low-resolution spectrophotometry that can be used to establish the effective temperatures of stars and to provide indicators of the abundances of heavy elements, or the *metallicity*, and of surface gravity. The types of systems selected for studies of different binary stars will depend on numerous factors, including the *speed* of the photometry required for time resolution of the variations in flux density due to eclipses and other phenomena, the apparent brightness of the star and the precision of the photometry required, the difficulty of separating the spectral-energy distributions of the two stars in the binary, and the effects of interstellar extinction. For example, photometry of a faint cataclysmic variable that contains a non-synchronously rotating white dwarf may require integration times of about 1 second to achieve the time resolution necessary to separate the white-dwarf oscillations from the other variations of the binary. The observations that discovered such phenomena were done in so-called *white light*, without any filters and using the entire spectral response of a photomultiplier in order to achieve a sufficient level of signal to noise. Alternatively, an eclipsing binary with a period of tens of days that is seen through significant interstellar extinction will require precise multi-colour photometry tied to a standard system in order to combine the data from the different nights accurately, establish the temperatures of the two stars as distinguished from the reddening effects of interstellar dust, and follow the exact course of the long eclipse phases so that precise sizes and shapes of the stars can be determined.

The brief discussion of the calculation of signal-to-noise (S/N) ratios given in Chapter 3, and specifically equations (3.1) and (3.2), can be used to estimate S/N values for a proposed set of observations. For photometers, in contrast to spectrographs, the total number of glass–air interfaces may be smaller and there may not be a grating involved, except in certain multi-channel photometers. Pass-band filters are needed, and these allow only about 50–70% of the incident light through; so it is possible that the overall efficiency of the telescope-photometer will be about 35%, perhaps only 10% greater than that for a modern spectrograph. But in photometry, all the light is imaged onto a few pixels, rather than being spread out into a spectrum across many pixels, and so we can expect to achieve much higher S/N values for a given integration time in photometric observations than in spectroscopic/spectrophotometric observations. For worthwhile photometry on binary stars, the observer should be aiming for S/N values greater than 100. If it is anticipated that serious eclipse mapping will be performed on the data (see Chapter 7), then ideally one needs S/N $\sim$ 250–500. Such data with uncertainties of 0.4–0.2% $\simeq$ 0.004–0.002 mag can be achieved with careful attention to detail in photoelectric or CCD photometry. Photographic photometry provides uncertainties of order 10%, unless several observations are combined to reduce errors to about 5%. Visual photometry is less accurate again, but it can be extremely useful in providing, for example, advance warning of outbursts in cataclysmic variables, as monitored by amateur astronomy networks in the United States, New Zealand, Germany, Japan, the United Kingdom, and elsewhere.

We have already considered some of the spectral-classification criteria for ordinary stars (Section 3.2.2), many of which were identified in the early part of the twentieth century as a direct result of empirical findings, as well as predictions from the developing theories of atomic physics and stellar atmospheres. Many of those criteria have been tacitly employed in the development of photometric systems, such as the dependence of the slope of the Paschen continuum on the effective temperature of the star, the dependence of the strength of the Balmer discontinuity on both temperature and luminosity, the sensitivity of the Balmer H lines to changes in surface gravity at a given temperature, and the sensitivity of the CN band to both luminosity and chemical composition. The observed crowding of large numbers of Fe lines around the 4000–4200-Å region has also been used as a *metallicity* indicator (e.g., Keenan 1963; Strömgren 1963).

A given pass band j in a photometric system can be characterized by a response curve $S_j(\lambda)$ that accounts for the spectral responses of the filter, the detector, and the optics of the telescope and the photometer. The lower and upper wavelength limits for the response curve are λ_1 and λ_2, respectively. The corresponding observed magnitude m_j for a star with a spectral-energy distribution $E(\lambda)$, measured in flux densities, as modified by its interaction with the interstellar medium, is given by

$$m_j = -2.5 \log_{10} \left[\int_{\lambda_1}^{\lambda_2} E(\lambda) S_j(\lambda)\, d\lambda \right] + C_j \tag{5.9}$$

where the constant C_j is a zero-point value chosen to satisfy some convention about the magnitude system.

A well-defined photometric system, with its n pass bands, has all its $S_j(\lambda)$ $(j = 1, \ldots, n)$ response curves defined almost entirely by the specified filters and is therefore essentially independent of the particular telescope + photometer combination that is employed for the observations. The *effective wavelength* λ_{ej} for each filter is defined by

$$\lambda_{ej} = \frac{\int_{\lambda_1}^{\lambda_2} \lambda E(\lambda) S_j(\lambda)\, d\lambda}{\int_{\lambda_1}^{\lambda_2} E(\lambda) S_j(\lambda)\, d\lambda} \tag{5.10}$$

whilst the *mean* or *constant-energy* wavelength λ_{cj} is defined by

$$\lambda_{cj} = \frac{\int_{\lambda_1}^{\lambda_2} \lambda S_j(\lambda)\,d\lambda}{\int_{\lambda_1}^{\lambda_2} S_j(\lambda)\,d\lambda} \tag{5.11}$$

Clearly, λ_{ej} depends on the spectral-energy distribution of the star observed, and so it will vary from star to star. In the case of eclipsing binaries composed of stars with very different temperatures, the value of λ_{ej} can vary significantly through the duration of an eclipse, because the observed energy distribution will change substantially. This potentially serious problem is obviated more or less completely when the pass bands are *narrow*, with $\lambda_2 - \lambda_1 = \Delta\lambda \lesssim 100$ Å, and the photometry is effectively *monochromatic*. For the *intermediate-band* photometric systems, with $\Delta\lambda \sim 300$ Å, the problem is solved by carefully selecting the locations of the pass bands to be in regions of smoothly varying continua, and not near spectral discontinuities or major absorption/emission lines. For *wide-band* photometric systems, with $\Delta\lambda \sim 1000$ Å, the potential for changes in λ_{ej} is substantial, particularly during eclipses of hot/cool stars by their cool/hot companions. The observations themselves will look consistent and accurate, but attempts to match them with a theoretical model calculated at a fixed wavelength may well be difficult.

One may wonder why the use of broad-band photometry has persisted for so long. The justification lies in the following explanation: Assuming that $E(\lambda)$ can be expanded as a Taylor series about λ_{cj} in powers of λ in the interval λ_1–λ_2, we obtain

$$E(\lambda) = E(\lambda_{cj}) + (\lambda - \lambda_{cj})E'(\lambda_{cj}) + \cdots \tag{5.12}$$

so that

$$\int_{\lambda_1}^{\lambda_2} E(\lambda)S_j(\lambda)\,d\lambda = E(\lambda_{cj}) \int_{\lambda_1}^{\lambda_2} S_j(\lambda)\,d\lambda + E'(\lambda_{cj}) \int_{\lambda_1}^{\lambda_2} (\lambda - \lambda_{cj})S_j(\lambda)\,d\lambda + \cdots \tag{5.13}$$

Provided that $E(\lambda)$ can be described with sufficient accuracy in the interval λ_1–λ_2 by the series truncated after the first derivative, then the foregoing definition of the constant-energy wavelength λ_{cj} allows the coefficient of the $E'(\lambda_{cj})$ term in equation (5.13) to be zero. Then we obtain

$$m_j = -2.5 \log_{10} \left[E(\lambda_{cj}) \int_{\lambda_1}^{\lambda_2} S_j(\lambda)\,d\lambda \right] + C_j \tag{5.14}$$

Thus, wide-band photometry can be justified when the sections of the spectral-energy distribution that are measured by the selected filters have no more than simple linear slopes across each wavelength interval. Such is the case for several regions of stellar spectra when viewed at low spectral resolution, so that the details of many small spectral lines are lost. But major spectral features, such as molecular bands or series discontinuities, for example, cannot be described by simple linear slopes, and so they should be avoided.

Attempts to describe the spectral-energy distribution of a star by means of wide-band photometric observations at three to four wavelengths clearly will be quite crude. Changing to intermediate-band filters that sample specific features at four to six wavelengths can be a marked improvement, whilst narrow-band filters provide essentially *monochromatic* flux densities that can be matched more readily by theoretical models.

The efficiency of CCDs now allows time-resolved spectroscopy and spectrophotometry to be carried out, at least on larger telescopes with apertures of 2–4 m. The technique of spectroscopy usually means slit spectroscopy, where the width of the entrance slit to the spectrograph is matched to the resolution of the detector to achieve maximum spectral resolution, typically at less than 1 Å. The normal slit widths used in astronomy project to about 1 arcsec on the sky, somewhat less than the typical seeing disc for a point source, unless the site is very good and special steps are taken to optimize image quality. Under standard observing conditions, not all of the light from a stellar source will pass through the slit to be recorded. Thus photometry from slit spectroscopy is not possible. Some estimate of these so-called *slit losses* can be made if the orientation of the long axis of the entrance slit on the sky is chosen to include another nearby star that may not be intrinsically variable, though this practice opens the possibility of systematic errors caused by atmospheric dispersion when the slit is not placed parallel to the direction to the zenith through the star position at the time of observation. But to achieve photometric accuracy, it is clearly necessary to open up the entrance slit to about 10 arcsec in order to allow all of the light through to the spectrometer and the detector. The spectral resolution is thereby degraded, and it has been common practice in spectrophotometry to use a spectral resolution of about 10 Å. Spectrophotometry provides the greatest information about the detailed energy distribution of a source, whereas the wider each pass band becomes, the more difficult it is to say anything about $E(\lambda)$. The question of which type of observation to carry out on a particular binary system is clearly a matter of optimizing wavelength resolution and time resolution. For a given size of telescope and efficiency of detector, an intermediate-band photometric system, say, will provide much better time resolution than will spectrophotometry, albeit at some loss of information about the detailed energy distribution. Ground-based telescopes with apertures of 4 m or more, as well as the Hubble Space Telescope, can now provide time-resolved spectra or spectrophotometry that can challenge the short integration times achievable with 1-m telescopes with CCDs and filter photometers. But the papers published thus far still attest to the enormous value of pursuing multi-colour photometry.

The remaining parts of this section are concerned with brief reference notes on the more commonly used photometric systems. These systems are all defined empirically on the basis of the magnitudes and colour indices for a set of standard stars observed through a particular set of filters. We note the sources for the sets of standard stars, the intrinsic colour relationships, the interstellar reddening corrections, and the absolute-magnitude calibrations relative to colour indices and spectral types for each system. We also note recent calibrations of these systems in terms of stellar effective temperatures, surface gravities, and chemical compositions.

5.3.1 The *UBVRI* system

The original *UBV* photometric system (for ultraviolet, blue, visual) was developed by Johnson and Morgan (1953) to complement the visual classification of stellar spectra in the MKK system (Morgan, Keenan, and Kellman 1943); see Morgan and Keenan (1973) for a review of stellar spectral classification. It is a wide-band system ($\Delta\lambda \sim 1000$ Å), which means that the central wavelengths and the shapes of the pass bands are defined in terms of the convolution of the spectral responses of the filters, the detector employed, and the optics of the telescope. In its original definition, only telescopes with all-reflection optics and located at high-altitude observing sites were employed. The best detector at that time was the 1P21 photomultiplier with an S4 photocathode, which has a predominantly blue spectral response. The wide pass bands ensured that adequate signals were recorded with modest telescopes for stars as faint as the tenth magnitude.

The *UBV* system was extended by Johnson and colleagues into the near infrared with the introduction of the *RI* pass bands (red, infrared) when the S1 photocathode was developed sufficiently for the faint signals from astronomical sources. The original set of standard stars that are distributed over the sky includes mainly stars of naked-eye brightness, and the present-day set is listed in the *Astronomical Almanac*, published annually. Over the past 40 years or so there have been important developments in photomultipliers with wider spectral responses, such as the extended S20 and the GaAs photocathodes that allow observations to be made in all of the *UBVRI* pass bands extending from 330 nm to 900 nm, and more recently there has been the introduction of CCDs with a variety of spectral responses. Whilst all *UBV* observations have been tied to the original photometric system, the definitions of the *RI* pass bands have been changed from those originally used by Johnson to those introduced by Kron and Cousins when the S20 and GaAs photomultipliers were developed. Thus one must be careful when comparing colour indices $[(V-R), (V-I)]$ of stars obtained at earlier times to determine whether those indices are Johnson-system values, usually written $[(V-R)_J, (V-I)_J]$, or Kron-Cousins-system values, usually written $[(V-R)_C, (V-I)_C]$. Cousins (1980) provided transformation equations between these two sets of colour indices that are valid for stars earlier than type M:

$$(V-R)_C = 0.715(V-R)_J - 0.02; \qquad (V-I)_C = 0.77(V-I)_J + 0.01 \tag{5.15}$$

reflecting the fact that the effective wavelengths have been changed significantly, from 700 nm to 640 nm for R, and from 900 nm to 800 nm for I. Bessell (1983) has extended these transformations into the very red M dwarfs and M giants, finding that different linear relationships are required, namely, for the M dwarfs:

$$(V-R)_C = 0.6(V-R)_J + 0.12; \qquad (R-I)_C = 1.045(R-I)_J - 0.094 \tag{5.16}$$

Consensus now seems to have been reached over the definition of the Johnson-Cousins *UBVRI* photometric system, with specific recommendations from Bessell (1990, 1995) for the coloured-glass filters to be used with alternative photomultipliers or CCDs so that observers can convert their observations to the standard *UBVRI* system in a straightforward manner. The effective wavelengths, for an un-reddened A0V star, are now as follows: U, 366 nm; B, 436 nm; V, 545 nm; R, 641 nm; I, 798 nm (Bessell 1990).

In addition to these wavelength changes, observers using larger telescopes, or CCDs on 1-m telescopes, have found that the improved sensitivity of their detectors has meant that the original standards generally have been too bright, and it has proved necessary to establish further sets of fainter standards. Landolt (1983, 1992) provided such a set of standards around the celestial equator so that they would be usable by both Northern and Southern Hemisphere observers. The original E-region standards at intermediate declinations in the southern sky established by Cousins in South Africa over many years were extended to fainter levels by Cousins and his colleagues at the South African Astronomical Observatory (SAAO), as described by Menzies et al. (1989). There are small differences between these two sets of newer standards, as demonstrated by Menzies et al. (1991), and Bessell (1995) has shown how the set of equatorial standard stars from Landolt can be transformed to the set of Southern Hemisphere E-region standards from the SAAO to provide a closer approach to an all-sky network of precisely defined standard stars for the *UBVRI* system with GaAs photomultipliers or CCDs. It is to be hoped that equivalent work of extending the set of original bright standards to fainter levels more appropriate for larger telescopes and CCDs will be pursued in the Northern Hemisphere at the intermediate declinations (45°) equivalent to the E-region.

The reasons for the enduring use of the *UBVRI* system are simply that its colour indices provide measures of interstellar extinction and of stellar temperatures over a wide range of spectral types, from B to K, and it offers the ability to conduct such multi-colour photometry to faint levels at usefully high speeds with modest telescopes. For example, integration times of 100 seconds with a CCD on a 1-m telescope provide *UBVRI* photometry to an uncertainty of less than 1% on stars with $V \sim 15$ mag. There also are large amounts of such data for many stars over the whole sky.

The original *UBV* system provided two colour indices: $(B - V)$, which measures the slope of the Paschen continuum in the visible part of the spectrum and is sensitive to stellar temperature, and $(U - B)$, which measures the height of the Balmer discontinuity at the series limit and is sensitive to both temperature and luminosity. The extension of the system to include *RI* pass bands introduced the colour indices $(V - R)$ and $(V - I)$, which sample the far-red part of the stellar-energy distribution and provide better temperature discrimination for the cooler stars. Recall Wien's law for estimating the wavelength λ_{max} of the peak in the energy distribution with wavelength of radiation from a black body at temperature T. It is given by $\lambda_{max} T = 2.898 \times 10^6$ nm K. Thus for the Sun, at $T \sim 6000$ K, we have $\lambda_{max} \sim 480$ nm, whilst for K stars at $T \sim 3500$ K, we have $\lambda_{max} \sim 830$ nm in the far-red part of the spectrum, and for early B stars at $T \sim 25{,}000$ K, we have $\lambda_{max} \sim 116$ nm, well into the ultraviolet. Measuring flux densities in various regions of the visible part of the spectrum can therefore discriminate temperatures in the range 3500–25,000 K with reasonable accuracy, and the *UBVRI* photometric system is quite successful in this regard, provided that one makes adequate corrections for the effects of interstellar extinction and reddening.

The effects of interstellar reddening in the *UBV* system were established empirically, primarily by observations of O and B stars classified visually from photographically recorded spectra that were clearly reddened or un-reddened. The scattering of starlight by interstellar dust preferentially removes blue light, because the amount of scattered light is proportional to λ^{-1} for Mie scattering by solid grains. Thus the extinction caused by intervening dust clouds increases the B magnitude more than the V magnitude, for example, so that $(B - V)$ always increases with increasing extinction. The *colour excesses* are defined as $E(B-V) = (B-V)-(B-V)_0$ and $E(U - B) = (U - B) - (U - B)_0$, where $(B - V)$ and $(U - B)$ are the observed colour indices of a star, and $(B - V)_0$ and $(U - B)_0$ are the intrinsic colour indices for that star. It has

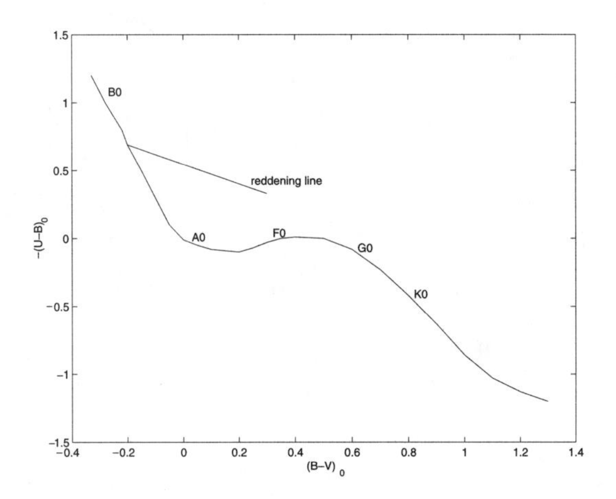

Fig. 5.1. The intrinsic colour line for zero-age main-sequence stars in the $(B-V)_0$–$(U-B)_0$ diagram for the *UBV* system, together with indications of the associated spectral types. The effect of interstellar reddening on the intrinsic colours is indicated by the *reddening line*, shown for $E(B-V) = 0.5$ mag on a star of spectral type B3, $(B-V)_0 = 0.20$ mag. Data from Schmidt-Kaler (1982).

been found that the ratio $X = E(U-B)/E(B-V) = 0.72 + 0.05E(B-V)$ for B stars, a value that is matched by the theoretical models of Bessell et al. (1998). Their paper also shows the need to include a colour dependence in the colour-excess ratios for stars of later spectral types, a consequence of the wide-band nature of the *UBVRI* system. The Q parameter was introduced to provide a means to separate interstellar reddening from observed colour indices and is defined as

$$Q = (U-B) - X(B-V) = (U-B)_0 - X(B-V)_0 \tag{5.17}$$

which is independent of interstellar reddening. Many years of observations have yielded mean colour indices and Q values as functions of spectral type, as tabulated, for example, by Schmidt-Kaler (1982). These intrinsic colour relationships for main-sequence stars are illustrated graphically in Figures 5.1 and 5.2, together with reddening lines. Thus, from observed *UBV* values, Q can be determined, and when it is combined with a spectral type to avoid possible ambiguities, the intrinsic colours and reddening can be determined for the observed star. The ratio of total interstellar extinction in the V band, A_V, to the colour excess, $E(B-V)$, has been found to be $R_{\rm ext} = A_V/E(B-V) \simeq 3.1$, from extensive studies, so that the observed V magnitude can be corrected for V-band extinction, A_V, before a distance is determined from

$$(V - A_V) - M_V = 5\log_{10} d - 5 \tag{5.18}$$

This value for $R_{\rm ext} \simeq 3.1$ is an average value that is valid for most directions across the sky. But $R_{\rm ext}$ does depend on the chemical and physical properties of the interstellar dust and is not precisely constant.

The foregoing considerations and the related figures show that the *UBV* system can separate out the effects of reddening from the observed colours of O and B stars because the reddening

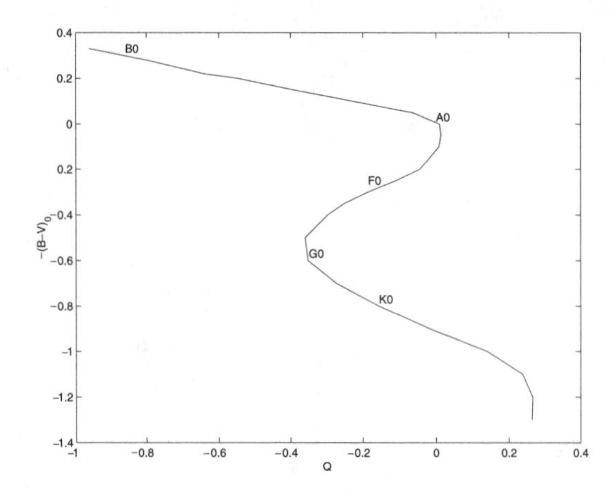

Fig. 5.2. The relationship between the reddening-free parameter Q and the intrinsic colour index $(B-V)_0$ for zero-age main-sequence stars, together with indications of the associated spectral types. The straight-line relationship for the B stars is given by $(B-V)_0 = 0.332Q$.

line lies at a substantial angle to the intrinsic colour line in the $(B-V)_0$-versus-$(U-B)_0$ plane. The $Q \sim (B-V)_0$ relationship is also single-valued for B stars. For intermediate spectral types, however, there is the potential for confusion, with a reddened late-B star, for example, appearing to have the same colour indices as an un-reddened late-A star. For the later spectral types, the angle between the intrinsic colour line and the reddening line is not large, and appeal should be made to the red colour indices, $(V-R)$ and $(V-I)$. In order to separate intrinsic colour and reddening lines for late-type stars, Pollacco and Ramsay (1992) have proposed the use of the loci in the $(V-I)$-versus-$(B-V)-(V-I)$ plane, together with the appropriate colour-excess ratios for $E(V-I)/E(B-V)$ that can be adopted from Bessell et al. (1998), for example.

Johnson (1966) provided a very useful tabulation of spectral types versus intrinsic colour indices for main-sequence stars and for late-type giants in the *UBVRIJKLM* system originally defined by Johnson and his colleagues. The R_J and I_J values do need to be transformed to the other *RI* systems mentioned earlier, by means of the Cousins (1980), Bessell (1983), or Taylor (1986) equations. The infrared colours have been superseded by more recent work, as discussed next.

5.3.2 Infrared *JHKLM* photometry

The *JHKLM* photometric system is a logical extension of the wide-band *UBVRI* system into the infrared region of the spectrum. The broad pass bands were selected to lie inside the windows of minimum water-vapour absorption in the Earth's atmosphere (i.e., at $J = 1.25\,\mu$m, $H = 1.65\,\mu$m, $K = 2.2\,\mu$m, $L = 3.4\,\mu$m, $M = 5.0\,\mu$m), and it follows that all serious infrared photometry must be conducted from very dry sites, usually at high altitude. As for *UBVRI* photometry, improving detector technologies have required revisions in the list of recommended standard stars, and some differences have developed between different telescope+detector combinations. The infrared *JHKL* standard stars published by Elias et al. (1982) have magnitudes and colour indices on the CIT system, which seems to be commonly used in the northern sky. Bessell and Brett (1988) have proposed that the various systems be made more compatible by application of various transformation equations, and they recommend the adoption of a system that is close to that defined by the work of Johnson and Glass and their associates; see the several references cited by Bessell and Brett (1988). The Johnson (1966) intrinsic-colour tabulation, noted earlier, has been revised in the light of more recent extensive studies, and Koorneef (1983) has provided intrinsic colours for main-sequence, giant, and supergiant stars as functions of spectral type for each of the infrared indices: $(V - K)$, $(J - K)$, $(H - K)$, $(K - L)$, and $(K - M)$. In addition, he has listed an adopted set of reddening ratios that are close to those calculated from theoretical models by Bessell et al. (1998). Leggett (1992) has discussed the published infrared photometry of low-mass red-dwarf stars with spectral types between M0 and M9, and she has provided a set of intrinsic infrared colours that, when combined with the empirical effective temperatures from Berriman and Reid (1987), will extend the calibration of the stellar-temperature range down to 2400 K. She has also added a substantial appendix listing the transformations used to convert various subsets of *RIJHK* photometry data to the CIT system.

It seems that wide-band infrared and optical photometric systems have now reached a level of comparability such that all-sky photometry findings can be compared at about $\pm 1\%$ uncertainty. With the detailed shapes of the pass bands defined in these homogeneous systems by the commonly adopted filters and detectors, the $S(\lambda)$ functions in equation (5.9), it is now possible to calculate the colour indices that are to be expected from model stellar-energy distributions, the $E(\lambda)$ functions in equation (5.9). There have been many attempts to perform such absolute calibrations of photometry by means of synthesizing colour indices from model-atmosphere flux densities calculated for a range of stellar temperatures, gravities, and metallicities. As both the observations and the models have been improved, the interagreements have become better. The intermediate spectral types or temperatures, roughly 5000–25,000 K, have enjoyed substantial agreement between observed and calculated wide-band colours for some years, but at both the higher and lower temperatures, substantial disagreements have persisted, often because of the difficulties of modelling the effects of spectral-line blanketing on the emergent stellar spectrum. The present-day understandings of this subject have been reviewed in IAU Symposium No. 189, *Fundamental Stellar Properties: The Interaction between Observation and Theory*, edited by Bedding et al. (1997). The latest calculations of wide-band colours, from U to L, and hence calibrations of stellar temperatures and of bolometric corrections, are given by Bessell et al. (1998). These can be used to assign an effective temperature to a star for which reliable intrinsic colour indices have been determined on a subset of the wide-band *UBVRIJHKL* system.

5.3.3 The Strömgren-Crawford $uvby\beta$ system

The $uvby\beta$ photometric system evolved out of numerous investigations conducted by Strömgren in the 1950s into the value of intermediate- and narrow-band photometry for providing stellar

Table 5.1. *Filter specifications for the uvbyβ system*

Filter name	Symbol	λ_c (Å)	$\Delta\lambda$ (Å)	Filter type
Ultraviolet	u	3500	380	glass
Violet	v	4100	200	interference
Blue	b	4700	200	interference
Yellow	y	5500	200	interference
Narrow β		4861	30	interference
Wide β		4861	100	interference

effective temperatures and surface gravities and for discriminating the effects of interstellar reddening. The system has been described by Strömgren (1963, 1966) and by Crawford (1966), who investigated the uses of narrow-band Hβ photometry (Crawford 1958; Crawford and Mander 1966) independently. The resultant *uvbyβ* system comprises four intermediate-band filters at different wavelengths and two filters centred on the Balmer Hβ line, one being narrow-band and the second intermediate-band. The specifications of those filters are given in Table 5.1.

Those filters were selected so that measured flux densities would represent the stellar spectral-energy distribution at regions that had been shown from earlier photometric and spectroscopic work to be sensitive discriminants of stellar temperatures, gravities, and metallicities. Thus, the u band measures flux densities entirely shortward of the Balmer discontinuity, but longward of the water-vapour cut-off from the Earth's atmosphere. For comparison, the wide-band U filter of the *UBV* system has a short-wavelength atmospheric cut-off for lower-altitude sites and includes some of the Balmer discontinuity on the longer-wavelength side. The v band measures the flux density received from that part of the stellar continuum that is most crowded with absorption lines of Fe I and II but is longward of the convergence of the H lines towards the Balmer-series limit. So the v band provides evidence of the *metallicity* of a star. The b band lies at λ4700, longward of B, in order to avoid the blanketing effects of the metal lines, and measures the continuum flux density. Likewise, the y band measures the continuum in the yellow region of the spectrum and can be expected to yield a simple transformation to the well-established V magnitude of the *UBV* system. The two β filters measure the flux densities centred on the Hβ line, the wide one, f_w, giving essentially the local continuum value, and the narrow one, f_n, the strength of the Hβ line. The β index is then defined by the relationship $\beta = -2.5 \log_{10}[f_n/f_w]$. In the original definition of the β system, Crawford (1958) used a narrow filter of only 15 Å at full-width half-maximum (FWHM), but later changed it to 30 Å in order to record more photons. The original definition of the system was retained, however, and so with $f_w \sim 10 f_n$ for stars with weak lines of Hβ, the resultant β index has a value of about 2.5, and it increases to about 2.9 for stars with the strongest H lines, around A0–A3.

Note that all the pass bands are defined almost in their entirety by the specifications of the filters alone, because the values of $\Delta\lambda$ in Table 5.1 are sufficiently narrow to ensure that the detector response will not change significantly within the filter pass band. By contrast, the *UBVRI* system has much wider pass bands, and so the detector makes a major contribution to the overall spectral-response curves, $S_j(\lambda)$, of the bands. As a direct consequence of the complete specification of the pass bands, there is only one standard version of the *uvbyβ* photometric system, and it is possible for theoretical flux densities and colour indices to be calculated from model-atmosphere flux densities with greater accuracy. The colour indices formed from these bands are as follows: (1) $(b - y)$, which measures the slope of the continuum in the blue–yellow region and is sensitive to stellar temperatures over the spectral-type range OB–G; (2) the *metallicity index*, $m_1 = (v - b) - (b - y)$, which indicates by how much the continuum in the v band is suppressed (because of line blanketing) relative to that expected from the b and y bands; (3) $c_1 = (u - v) - (v - b)$, which measures the height of the Balmer discontinuity and removes the influence of the metal-line blanketing as well; c_1 is temperature-sensitive for B stars and luminosity-sensitive for A and F stars; the colour index $(u - b)$ is also used for the hotter stars; (4) the β index, defined earlier, which measures the relative strength of the Hβ line and provides a temperature discriminant for stars between A2 and G, and more of a luminosity or surface-gravity discriminant for B and O stars. These latter dependences are direct results of the linear and quadratic Stark effects acting on the H atoms in a stellar atmosphere, otherwise called *pressure broadening*.

The original system of all-sky standard stars is that given by Crawford and Barnes (1970), and, like the *UBVRI* system, it contains stars of mostly naked-eye brightness. There is an additional, more extensive list of bright standards compiled by Perry et al. (1987). There are also the very extensive surveys of homogeneous photometry of A, F and G stars compiled by Olsen (1983, 1993) that may provide secondary standards over the sky. In order to satisfy more recent demands for fainter, secondary standard stars usable for CCD photometry on telescopes of 1 m and larger, a number of papers have provided observations to define secondary standards in selected areas of the sky. Thus, Twarog (1984) has presented *uvby* photometry for 36 stars near the South Galactic Pole, Kilkenny and Laing (1992) have reported *uvby* photometry for 201 stars that are already *UBVRI* standards in the southern E-regions, and Clausen et al. (1997) have presented *uvbyβ* photometry for 73 stars, many of which are E-region standards. As noted by those last authors, there is a pressing need for a complete network of equatorial *uvbyβ* standards similar to those from Landolt that would be suitable for the standardization of CCD photometry.

The colour-excess ratios for *uvbyβ* photometry have been determined by Crawford (1975a) from observations of large numbers of O stars. Because these stars have nearly identical intrinsic colours in this system and are observable over a large range of distances because of their high luminosities, they display, as a group, a wide range of interstellar reddening that is simply linear in each of the two-colour plots used to form the colour-excess ratios. Thus, he found $E(c_1) = 0.20E(b - y)$, $E(m_1) = -0.32E(b - y)$, $E(u - b) = 1.5E(b - y)$, $E(b - y) = 0.74E(B - V)$, and the total extinction in the V band, $A_V = 4.3E(b - y)$. In a subsequent paper, Crawford and Mandwewala (1976) calculated the colour-excess ratios in several photometric systems by combining stellar-energy distributions with the response curves of the filter-detector combinations. For *uvbyβ* photometry, they found very good agreement between those predictions and the earlier observations of reddened O stars. They provided these average colour-excess ratios for main-sequence stars of all types: $E(c_1) = 0.19E(b - y)$, $E(m_1) = -0.33E(b - y)$, $E(u - b) = 1.53E(b - y)$, and $A_V = 4.28E(b - y)$, only slightly

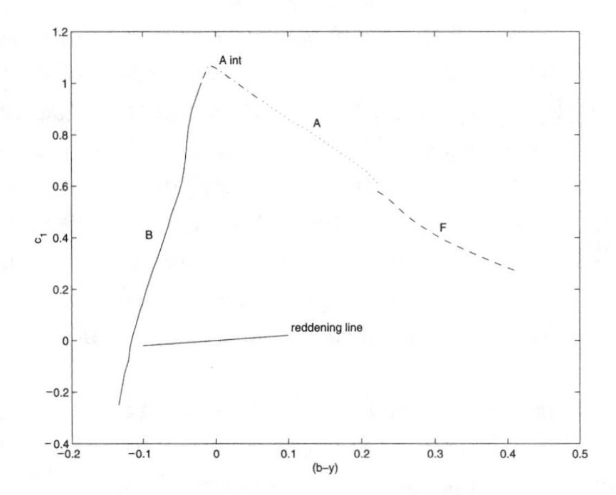

Fig. 5.3. The relationship between the intrinsic colour indices $(b-y)$ and c_1 of the *uvbyβ* system for B, A, and F main-sequence stars (Perry et al. 1987) and that for the A intermediate (A0–A2) stars (Hilditch et al. 1983). The reddening line indicates how interstellar reddening affects these colour indices, according to $E(c_1) = 0.19E(b-y)$ for $E(b-y) = 0.2$, from $[(b-y), c_1] = (-0.1, 0.0)$.

modified from the O-star values. Of particular interest is the fact that the effect of reddening on the c_1 index is only 20% of that on $(b-y)$, which means that it is easy to separate the reddening from the observed colours and determine intrinsic colours.

The procedure of determining the standard relationships between the various intrinsic-colour indices, along with absolute visual magnitudes, spectral classes, and so forth, was carried out in an entirely empirical manner, starting with F main-sequence stars in the solar neighbourhood for which reliable trigonometric parallaxes (distances) were known and where there was zero interstellar extinction. From that basis, well-observed open clusters were used to extend the relationships to A and B stars. Full accounts of that calibration work have been provided by Crawford (1975b) for F stars, Crawford (1979) for A stars, Crawford (1978) for B stars, and Crawford (1975a) for O stars. Because the β index is formed from a ratio of flux densities all observed at a single wavelength, it is independent of extinction within the atmosphere of the Earth, and it is also independent of interstellar extinction and reddening. Accordingly, it is used as the reference index for these empirical standard relationships between the various colour indices of the *uvbyβ* system. A summary list of these relationships is given for O, B, A, and F stars by Perry et al. (1987), and they are plotted in two-colour diagrams in Figures 5.3–5.5.

The observed *uvbyβ* colours of stars need to be processed through an algorithm to yield interstellar-reddening values and the intrinsic colours and absolute visual magnitudes of those stars. Full details of the procedures are found in papers by Crawford (1975b, 1978, 1979) for F, B, and A stars, respectively. For the A intermediate stars, the better procedure is to use the Strömgren (1966) $[a, r]$ method, with the corrections noted by Grøsbol (1978). The method suggested by Hilditch, Hill, and Barnes (1983) works for stars with normal metal abundances or with metal deficiencies, but fails for metal-rich stars. In outline, these procedures are as follows.

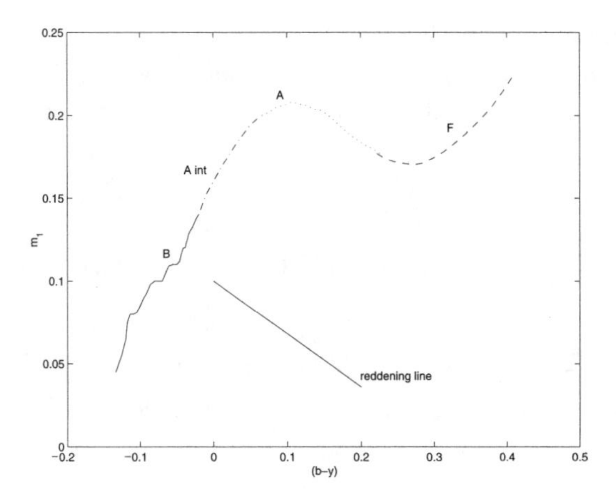

Fig. 5.4. The relationship between the intrinsic colour indices $(b - y)$ and m_1 of the *uvbyβ* system for B, A, and F main-sequence stars (Perry et al. 1987) and that for the A intermediate (A0–A2) stars (Hilditch et al. 1983). The reddening line indicates how interstellar reddening affects these colour indices, according to $E(c_1) = 0.19E(b - y)$ for $E(b - y) = 0.2$, from $[(b - y), m_1] = (0.0, 0.1)$.

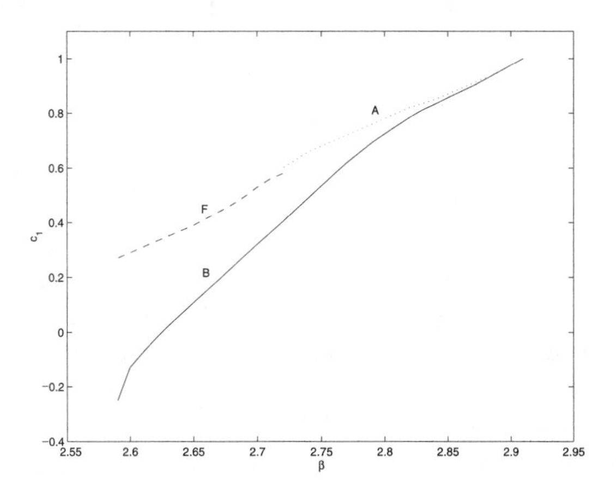

Fig. 5.5. The relationship between the intrinsic colour indices β and c_1 of the *uvbyβ* system for B, A, and F main-sequence stars (Perry et al. 1987). For A and F stars, β is temperature-sensitive, whilst c_1 is luminosity-sensitive. For B stars, c_1 is temperature-sensitive, whilst β is luminosity-sensitive.

Label the observed values for a star as $(b - y)$, m_1, c_1, and β, and the de-reddened quantities as $(b - y)_0$, m_0, and c_0. Calculate the reddening-free quantities

$$[m_1] = m_1 + 0.33(b - y) \tag{5.19}$$

$$[c_1] = c_1 - 0.19(b - y) \tag{5.20}$$

$$[u - b] = (u - b) - 1.53(b - y) \tag{5.21}$$

which are known as *bracket* m_1, *bracket* c_1, and *bracket* $u - b$, respectively.

F stars are characterized by having $\beta < 2.720$ and $[m_1] > 0.15$. For the observed value of the reddening-free parameter β, read the values of m_1 and c_1 from the intrinsic-colour relationships given by Crawford, labelled $m_1(\beta)$ and $c_1(\beta)$ in the following discussion. Form the quantities

$$\delta m_1 = m_1(\beta) - m_1; \qquad \delta c_1 = c_1 - c_1(\beta); \qquad \Delta\beta_F = 2.720 - \beta \tag{5.22}$$

The colour excesses $E(b-y)$, and so forth, and the de-reddened indices $(b-y)_0$, and so forth, are determined from

$$(b-y)_0 = 0.222 + 1.11\Delta\beta_F + 2.72\Delta\beta_F^2 - 0.05\delta c_1 - (0.1 + 3.6\Delta\beta_F)\delta m_1$$
$$E(b-y) = (b-y) - (b-y)_0; \qquad E(c_1) = 0.2E(b-y);$$
$$E(m_1) = -0.32E(b-y) \qquad c_0 = c_1 - E(c_1); \qquad m_0 = m_1 - E(m_1)$$

Recalculate δm_1, δc_1, and hence $(b-y)_0$, c_0, and m_0 to convergence.

The absolute visual magnitude is calculated from

$$M_V = M_V(\beta) - (9 + 20\Delta\beta_F)\delta c_1 \tag{5.23}$$

where $M_V(\beta)$ is the intrinsic relationship for unevolved stars, as given by Crawford.

A stars are identified as having $2.720 < \beta < 2.880$ and $[m_1] > 0.15$. Calculate δm_1 and δc_1, as defined earlier, and hence

$$(b-y)_0 = 2.946 - \beta - 0.10\delta c_1 - f\delta m_1 \tag{5.24}$$

where $f = 0$ if $\delta m_1 > 0$, and $f = 0.25$ if $\delta m_1 < 0$. Hence we have $E(b-y)$, $E(c_1)$, $E(m_1)$, and the intrinsic colours c_0 and m_0. Then the absolute visual magnitude follows from $M_V = M_V(\beta) - 9\delta c_1$.

B stars are selected as having $[m_1] < 0.15$ or $(b-y) < -0.01$. Then, adopting $c_0 = c_1$ initially, it is possible to use an approximate relationship due to Crawford to determine the de-reddened colour index $(b-y)_0 = -0.116 + 0.097c_0$, and after two or three iterations, convergence is reached. However, the intrinsic relationship between $(b-y)_0$ and c_0 is not quite linear, and it is slightly more accurate to interpolate in the empirical intrinsic relationship to find $(b-y)_0$ from an initial c_0. Then calculate $E(b-y)$ and $E(c_1)$ to find a revised value of c_0, and repeat the procedure. Convergence is achieved within three iterations. Hence, we also obtain m_0. When calculating the absolute visual magnitude, we have to make allowance for the possibility of the star having evolved away from the intrinsic-colour line, as was done for the F and A stars. For F and A stars, the c_0 index acted as the luminosity discriminant, and we used the parameter δc_1 to reflect that evolution or displacement from the intrinsic relationship. For B stars, the roles of c_0 and β are reversed, with c_0/β as the temperature/luminosity discriminants. Thus for a given value of c_0, we form the parameter $\Delta\beta_B = \beta(\beta) - \beta$, where $\beta(\beta)$ is the value of β on the intrinsic relationship for the given c_0. The empirical relationship for absolute magnitudes for B stars is found to be $M_V = M_V(\beta) - 10\Delta\beta_B$.

Identification of A intermediate stars usually follows from a star's failure to meet the criteria for B or A stars; if $[m_1] < 0.221$ with $[c_1] > 0.917$, then a star should be an A intermediate type, because it fails to meet the criteria for B stars and for A stars. Comparison of $[m_1]$ with that calculated from the intrinsic relationships in Table 1 of Hilditch et al. (1983) provides a first estimate of $(b-y)_0$, and hence $E(b-y)$, $E(m_1)$, and a new value of m_0. Adopt a new value of $(b-y)_0$ from m_0 and redetermine $E(b-y)$, and so forth, until convergence is achieved. Absolute visual magnitudes follow from $M_V = M_V(\beta) - 10\Delta\beta_B$. This procedure works for stars that are not metal-rich, that is, those with $\delta m_1 \geq 0$. For all A intermediate stars, Strömgren's $[a, r]$ method works well, provided that the corrections noted by Grøsbol (1978) are included. The quantities a and r are defined as

$$a = 1.36(b-y) + 0.36m_1 + 0.18c_1 - 0.2448 = (b-y) + 0.18(u-b) - 1.36 \quad (5.25)$$

$$r = 0.35[c_1] + 2.565 - \beta \quad (5.26)$$

and the colour excess $E(a) = 1.288E(b-y)$. The quantity r is evidently reddening-independent. Strömgren (1966) proposed an empirical relationship for calculating a reddening-free value for a, say a_0, and noted that more observations of stars within 100 pc of the Sun would improve that relationship. Grøsbol (1978) has revised that relationship to be

$$a_0 = 1.54[m_1] + 0.74r - 0.27 \quad (5.27)$$

Hence, from observed colours, calculate $[m_1]$ and $[c_1]$, and therefore a_0 and r. Then we obtain $E(a) = a - a_0 = 1.288E(b-y)$ and values for $E(b-y)$ and $(b-y)_0$. The de-reddened colours of the star are found, and the absolute visual magnitude follows from the Strömgren relationship of $M_V = 1.5 + 6.0a - 17.0r$.

Calibrations of the $uvby\beta$ colour indices in terms of effective temperatures, surface gravities, and bolometric corrections for stars have been made numerous times. They all testify to the effectiveness of Strömgren's original plan for developing a photometric system that would provide stellar effective temperatures and surface gravities directly. Several sets of calculations of $uvby\beta$ indices have been made from model-atmosphere flux densities, one of the more recent being that of Lester et al. (1986), which presents tabulations over the range $5500\ \mathrm{K} \leq T_{\mathrm{eff}} \leq 20{,}000\ \mathrm{K}$ for a range of heavy-element abundances that cover the Milky Way galaxy and the Magellanic Clouds. The lines of constant temperature and surface gravity in the various two-colour planes, $(b-y)_0 \sim c_0$, $(b-y)_0 \sim m_0$, and $m_0 \sim c_0$, are nearly orthogonal over substantial areas of the diagrams, thereby yielding good discrimination of these stellar parameters. However, it seems that these calibrations via model atmospheres may need to be modified in the light of recent major advances in representing the effects of line blanketing, as discussed at IAU Symposium No. 189 (Bedding et al. 1997).

Several attempts have been made to calibrate the $uvby\beta$ indices by means of empirically determined stellar effective temperatures and surface gravities, using interferometric observations for angular diameters, spectrophotometry over wide wavelength ranges to determine stellar-energy distributions, and spectroscopy to study H-line profiles. A rigorous review of published work on this topic, together with a revised temperature calibration of the $uvby\beta$ indices based on empirical measures, is given by Napiwotski et al. (1993).

For stars in the range $9500 < T_{\mathrm{eff}} < 30{,}000$ K, the $[u-b]$ index, equation (5.21), is preferred because it has a somewhat larger range than c_0, and Napiwotski and associates give the nearly linear relationship

$$\Theta \equiv \frac{5040}{T_{\rm eff}} = 0.1692 + 0.2828[u-b] - 0.0195[u-b]^2 \quad (5.28)$$

yielding a scatter of observed points about this mean relationship of $\sigma = \pm 3\%$.

The relationship between $(b-y)_0$ and temperature is best represented for practical purposes by linear functions in two separate temperature ranges, connected by a small interpolation region. Thus, they give

$$\begin{aligned} (b-y)_0 &= 0.307\Theta - 0.175 \quad \text{for } 0.17 < \Theta < 0.5 \\ (b-y)_0 &= 0.709\Theta - 0.376 \quad \text{for } 0.5 < \Theta < 0.6 \\ (b-y)_0 &= 1.324\Theta - 0.745 \quad \text{for } 0.6 < \Theta < 0.9 \end{aligned} \quad (5.29)$$

In a comparison with previously published calibrations based on model-atmosphere fluxes, Napiwotski et al. (1993) found that their empirically based results agreed best with the calibration grid of Moon and Dworetsky (1985), with the need for some small corrections to the surface-gravity determinations in that grid. Those results have been used by Schönberner and Harmanec (1995) in a detailed comparison of the empirically determined masses, radii, and luminosities of stars in well-studied eclipsing binaries with the most recent stellar-structure models to show that they are in remarkable agreement and define a very precise relationship between effective temperatures and absolute visual magnitudes for unevolved stars extending over the range $6700 < T_{\rm eff} < 38{,}000$ K. This relationship agrees well with other relationships at the lower temperatures ($T_{\rm eff} < 10{,}000$ K), but departs systematically from other relationships at higher temperatures. This topic will be discussed further in Section 6.2.

The $uvby\beta$ photometric system has fulfilled its promise of providing determinations of stellar temperatures, surface gravities, absolute visual magnitudes, and metallicity indices, separated from the effects of interstellar reddening, all by means of cost-effective and time-effective multi-colour photometry. It is used extensively in studies of eclipsing binary stars in particular, as well as in studies of the distributions and kinematics of stars in the Milky Way galaxy.

5.3.4 The DDO system and other photometric systems

The DDO system is a narrow/intermediate-band photometric system for studies of G, K, and M stars, and in that sense it is complementary to the $uvby\beta$ system. It was developed at the David Dunlap Observatory by McClure and van den Bergh (1968) and is a useful tool for stellar temperatures, surface gravities, metallicities, and stellar kinematics in our galaxy, for studies of the integrated flux densities from other galaxies, and for late-type binaries. The specifications of the six filters have been given by McClure (1976), and their main features are listed in Table 5.2.

The 48 filter is used to define a magnitude, $m(48)$, whilst the three nearby filters, 45, 42, and 41, are used together with 48 to define a set of colour indices in the standard manner, following equation (5.8), written as $c(45-48)$, $c(42-45)$, $c(42-48)$, and $c(41-42)$. The 45 pass band acts as a nearly complete continuum measure close to the other filters, which each measure the

Table 5.2. *Filter specifications for the DDO system*

Filter name	λ_c (Å)	$\Delta\lambda$ (Å)	Filter type
48	4886	186	interference
45	4517	76	interference
42	4257	73	interference
41	4166	83	interference
38	3815	330	glass
35	3460	383	glass

strengths of spectral features. Thus the 48 pass band lies in a region of absorption due to the molecule magnesium hydride, and $c(45 - 48)$ has been found empirically to be sensitive to stellar surface gravity. The 42 pass band measures the break in the continuum shortward of the G band of the molecule CH, at 4300 Å, and hence $c(42 - 45)$ acts as a temperature parameter. The 41 pass band lies within the CN molecular band, and so $c(41 - 42)$ measures its strength, which is sensitive both to surface gravity and to the abundances of C and N. These four filters are all that are required for investigations of mid- and late-type Population I stars.

The remaining pass bands, 38 and 35, are used to provide further discriminations of heavy-element abundances in late-type stars, with $c(38 - 42)$ providing a metallic-line blanketing index, and $c(35 - 38)$ measuring the Balmer discontinuity and acting as a surface-gravity or luminosity index.

The set of standard stars published by McClure (1976) was carefully selected to ensure good coverage of the region of the sky within $\pm 10^\circ$ of the equator, and it was added to the original set of Northern Hemisphere standards. That equatorial set ensured that the photometric system could be extended into the Southern Hemisphere, and that has been reported by Cousins and Caldwell (1996) in their DDO photometry of many E-region standards. The set of standards includes many stars of magnitudes $V \sim 7$–8, and with the narrow pass bands it follows that DDO photometry can reasonably be pursued with 1–2-m telescopes with CCD cameras.

The standard colour-index diagrams are $c(42 - 45) \sim c(45 - 48)$, or temperature versus surface gravity, and $c(42 - 48) \sim c(41 - 42)$, which gives temperature versus the CN index, or a combination of abundance and surface gravity. Calibrations of these colour-index planes in terms of spectral types and luminosity classes are given by McClure (1973), and in terms of effective temperatures and gravities by Claria et al. (1994). The relevant reddening ratios are also given by McClure (1973). It seems unfortunate that this well-defined intermediate-band system for late-type stars has not been used more often for photometry of binary stars, specifically because it can discriminate so well between main-sequence and giant stars and can provide good temperature determinations.

There are many other photometric systems that have been developed for specific or general use, and it would not be appropriate to attempt a summary of all of them here. Golay (1974) has provided a major review, including a detailed discussion of the *Geneva 7-colour system*, and that text should be consulted for further details, together with the books by Henden and Kaitchuk (1982), Warner (1988), and Sterken and Manfroid (1992). But we shall briefly review the *Utrecht photometric system*, because it is used specifically for detailed, accurate photometry of eclipsing binaries with late-type components. Provoost (1980) has shown that this narrow-band ($\Delta\lambda \sim 50$–100 Å) system provides effectively monochromatic flux densities at various optical wavelengths between 470 nm and 870 nm. This wavelength range is well suited to the GaAs photomultiplier and to standard CCD detectors. Of the original set of filters, it seems that most of the work on eclipsing binaries has used the four narrow-band filters centred at 472, 672, 782, and 871 nm, each of which samples continuum regions of typical stars. Because the system provides nearly monochromatic magnitudes, the precision of the photometry is excellent, and the sensitivity to different temperatures amongst cool stars is considerable because of the wavelength range that is sampled. Examples of the work with this system have been given by de Landtsheer (1983) and van Gent (1989), demonstrating how effectively a monochromatic system can be modelled. Mention should also be made of the Walraven *VBLUW* system, developed at Leiden and used extensively via a five-channel photometer on a 0.9-m telescope in the Southern Hemisphere, in South Africa and later in Chile. The properties of the system have been discussed extensively by Lub and Pel (1977), whilst an example of the use of the system for binary stars is the study of the intermediate polar EX Hya by Siegel et al. (1989).

5.3.5 Optical and ultraviolet spectrophotometry

Before the advent of CCD detectors, spectrophotometry had to be performed with spectrometers that scanned the spectrum across an entrance slot to a photomultiplier. That sequential acquisition of a stellar spectrum demanded huge amounts of telescope time, as well as great care with the elimination of systematic errors from the data that resulted from, for example, changes in atmospheric transmission during each scan. That also meant that spectrophotometry of eclipsing binaries was not performed often because of their photometric variability over time scales that were similar to those needed to acquire the data. There is an enormous literature on the determination of standard stars for spectrophotometry that are distributed around the sky like photometric standards, though far fewer in number. With the introduction of CCDs for optical spectrophotometry, it has proved possible to follow the lead indicated by the ultraviolet spectrophotometry carried out with the International Ultraviolet Explorer (IUE) satellite and its spectrometers and establish new spectrophotometric standards that have their flux densities determined to uncertainties of about $\pm 1\%$ over a wide wavelength range. Such developments also mean that spectrophotometry is becoming an increasingly important observational tool for studies of close binaries, because entire spectra can be recorded in seconds rather than over many minutes or hours.

Absolute spectral-energy distributions were determined by Oke (1990) for 25 stars observed with the Hale 5-m telescope and its double spectrograph equipped with a pair of very stable CCD detectors. The entrance slit was 10 arcsec wide, and the wavelength range covered by those spectra was from 3150 Å to 9400 Å. Many of those stars have also been studied extensively by the IUE satellite, and it has proved possible to combine those data from the UV and optical regions, with the help of theoretical model-atmosphere fluxes for white dwarfs with pure-hydrogen atmospheres. That latter work has been completed by Bohlin (1996) to create spectrophotometric standards that cover the wavelength range from 1150 Å to 9400 Å, and to be extended to 3 μm. These standards should then be usable by the Hubble Space Telescope (HST) over its entire operating range from the ultraviolet to the near infrared. Entrance apertures on the HST Faint Object Spectrograph are in the range 0.4–3.7 arcsec, which allows spectrophotometric accuracy with diffraction-limited images of 0.1 arcsec, and wavelength resolution of about 6 Å. The standard 'pipeline' reduction procedures for HST data include absolute flux calibrations, so that the final spectra are provided in units of millijanskys as a function of wavelength. The high-speed mode allows spectra to be secured at intervals of about 5 seconds, which ensures that even the rapid changes in flux densities observed in cataclysmic variables can be monitored successfully, as shown, for example, by Knigge et al. (1998a,b) and Baptista et al. (1998a,b).

With ground-based 4-m-class and larger telescopes, and CCD spectrographs, the same sort of time-resolved spectroscopy is now being used, although it is typically conducted as high-speed *slit* spectroscopy, rather than any attempt at proper spectrophotometry. Nevertheless, such rapidly acquired spectra can be studied for changes in absorption-line and emission-line profiles as functions of time relative to the spectral continuum taken as unity.

5.3.6 Ultraviolet photometry

In contrast to the situation for optical and infrared photometry, there are no established photometric systems for the ultraviolet (UV) region. Photometric surveys have certainly been carried out, such as that on 53,000 stars by the S2/68 experiment on the TD1 satellite, which provided colour indices from various wide-band regions in the ultraviolet. Carnochan (1982) has discussed these data and provided various (UV $-$ V) colour indices as functions of spectral type and luminosity class over the entire spectral range. Similarly, the ultraviolet experiment on board the ANS satellite (van Duinen et al. 1975) provided five-colour photometry from a spectrometer with band passes at 155, 180, 220, 250, and 330 nm. Intrinsic ultraviolet colours for stars over the entire spectral-type range were determined by Wu et al. (1980). The use of these satellites, together with the Orbiting Astronomical Observatory 2 (OAO-2), to determine light curves for eclipsing or photometrically variable binary stars has been quite limited, perhaps in part because of the requirement for large amounts of observing time for such work. An example of good ANS photometry is that published by Eaton and Wu (1983), where they secured complete UV light curves for the 1.6-day binary δ Pic. Most of the effort during those times was devoted to spectroscopic studies of binaries, via the IUE satellite and its UV spectrometers, to determine radial velocities from the high-resolution data, as noted earlier, and spectrophotometry from the low-resolution data.

In the 1990s, two ultraviolet experiments took the field to new areas, in terms of both speed of acquisition and wavelength coverage. The launch of the Hubble Space Telescope initially allowed some high-speed UV photometry to be performed, but that particular instrument was removed at the time of the Costar upgrade. More importantly, however, the Faint Object Spectrograph (FOS) was confirmed to be a most versatile instrument, with the capability of recording entire UV spectra (1100–2500 Å, and resolutions of $\lambda/\Delta\lambda \sim 200$) at integration times of about 5 seconds and readout times of less than 1 second on stars with brightnesses typical of well-studied CVs. Not only could such data be used for investigations of overall line-profile variability, but also they were spectrophotometric, and hence photometry could be deduced from the data by careful selection of true continuum regions between potential contaminants like broad emission features or deep absorption lines and molecular bands. Examples of such work have been reported by Knigge et al. (1998a,b) and Baptista et al. (1998a,b). The Goddard High Resolution Spectrograph (GHRS) has been used in a similar fashion when the sources have been bright enough to permit such time-resolved spectrophotometry to be performed. Here the spectral resolution is higher, at $\lambda/\Delta\lambda \sim 10{,}000$, so that true narrow-band photometry has been derived from such data, as shown, for example, by Baptista et al. (1995).

The second UV experiment was the deployment of the Extreme Ultraviolet Explorer (EUVE) satellite in 1992. This spacecraft is equipped with a deep-survey (DS) instrument for photometry in the EUV energy band of 75–175 Å, together with three spectrometers providing low-resolution spectra in three bands: SW from 70 to 190 Å, MW from 140 to 380 Å, and LW from 280 to 760 Å, with resolutions in the range from 200 to 400. The photon energies corresponding to these wavelengths are sufficient to ionize hydrogen atoms, and so it might be expected that the distribution of neutral hydrogen in interstellar space would severely limit the available path lengths at EUV wavelengths. However, the local distribution of hydrogen is found to be quite clumpy, as shown by Warwick et al. (1993) from the first EUV all-sky survey carried out by the ROSAT wide-field camera (Pounds et al. 1993) that found 383 bright sources. Whilst the direction towards and above the galactic centre is limited to about 10 pc at EUV wavelengths, other directions are quite free of absorption out to at least 80 pc, and some white-dwarf stars have been detected out to 300 pc. The EUVE all-sky survey revealed 356 bright EUV sources (Malina et al. 1994), and the first EUVE source catalogue with 410 point sources was published by Bowyer et al. (1995). Those two surveys are complementary and do not use the same filter pass bands. The EUVE satellite is providing the first systematic exploration of the far-ultraviolet spectrum, and its instruments are capable of working at the high speeds necessary for successful investigations of the short-period binaries, where substantial amounts of EUV radiation are to be expected, namely, amongst the magnetically active stars in RS CVn and W-UMa-type contact systems, and amongst the intermediate polar and AM Her systems, where the accretion flow is magnetically channelled onto the surface of the white dwarf. Recent examples of such work have been reported by Brickhouse and Dupree (1998) and Rucinski (1998b) on W-UMa-type systems, by Szkody et al. (1999) on outbursting *CVs*, and by Vennes et al. (1995) and Warren et al. (1995) on AM Her systems. As a theoretical complement to these new EUV observations, Wade and Hubeny (1998) have calculated model EUV spectra for accretion discs in cataclysmic variables for direct comparison with the EUVE spectrometry, as well as the HST FOS and GHRS spectra.

5.3.7 X-ray photometry

Just like UV photometry, x-ray photometry has to be performed from outside the Earth's atmosphere via experiments on Earth-orbiting satellites. According to the tabulation by Charles and Seward (1995), a total of 20 x-ray-astronomy satellite missions were launched between

1969 and 1994, and several more have been launched since then, with two major launches (XMM and AXAF) successfully completed in 1999. Each satellite typically has carried several x-ray experiments, and it would not be appropriate to attempt to list even an outline of all the photometers, detectors, and spectrometers that have been employed. The text by Charles and Seward (1995) provides a good introduction to many of the different instruments deployed on the major astronomical satellites. The most common x-ray-photon energy ranges that have been investigated fall into two broad categories, the *soft-x-ray band* between about 1 and 3 keV and the *hard-x-ray band* from about 4 keV. These correspond to wavelengths of 12–4 Å and 3 Å, respectively, notably higher energies than the EUV ranges discussed earlier.

The precise energy bands used by all such experiments have depended on the latest advances in detector technology at the time of launch, and so direct comparisons between x-ray-flux densities for sources from different experiments conducted at different epochs are not simple. Received flux densities from x-ray sources typically are quoted in units of counts per second within a specified energy range for a given detector, in the same manner as optical photometry that has not been converted to any standardized system. In some cases, these x-ray-flux densities have been converted into millijanskys when a well-defined calibration of the instrument has been established in laboratory tests before launch and continually monitored after launch by observations of standard x-ray sources such as the Crab Nebula. The energy bands used in x-ray photometry are broad, typically kilo-electron-volts wide, so that the concerns expressed in Section 5.3, involving equation (5.14) in particular, about the justification for performing wide-band optical photometry apply even more strongly here. Furthermore, there is no guarantee, unlike the case for many ordinary stellar sources, that x-ray sources will be devoid of individual spectral peculiarities. Such individual differences, like broad emission bands that may display time-dependent variations, for example, render attempts at transformations to a standard system of questionable value. Appropriately, the results from x-ray photometry are given in terms of the instrumental system, defined solely in terms of the individual characteristics of the instrument and its detector, which are in turn specified as precisely as possible from the pre-launch calibrations and from continued monitoring, in orbit, of known x-ray sources.

The data-sampling rates in x-ray photometry have improved since the first experiments, so that the current Rossi X-ray Timing Explorer (RXTE) is capable of providing integration times from milliseconds upwards, and over a wide range of photon energies, from 2 to 60 keV. The range of discoveries from x-ray astronomy is remarkable and is well documented in many popular and advanced texts. The *Uhuru* satellite established the existence of pulsed x-radiation from x-ray stars with periods as short as 1.2 seconds, as well as occurrences of periodic eclipses of the x-rays, and hence recognition of the existence of x-ray binaries. The fact that the x-ray eclipses were so sudden showed that the x-ray sources had to be small relative to their orbiting companions, and their status as neutron stars or black holes accreting matter from their RLOF or stellar-wind companions was soon confirmed. We now have three sets of x-ray binaries, the high-mass (HMXB) systems with O–B giant/supergiant companions, the low-mass (LMXB) systems with later-type F–K components (including the soft-x-ray transients, SXTs), and the *supersoft* x-ray sources.

The x-ray pulse periods exhibited by known x-ray binaries extend from 0.07 second to 835 seconds, and the pulses from individual systems show enormous ranges of pulse shape and amplitude, even amongst consecutive pulses. There are also substantial variations in pulse shape and amplitude as functions of the energy band. A representative set of examples of such pulse profiles is given by White et al. (1995). It is the accurate timing of such pulses that provides us with evidence of periodic advances and delays in the arrival times of pulses, the light-travel-time effect across the barycentric orbit of the pulsed source in the binary system. The analysis of such data to determine the elements of the orbits has been discussed in Chapter 3.

Many other phenomena have been found amongst the x-ray binaries: x-ray on/off states, independent of stellar eclipses, that are considered to be due to eclipses by precessing, warped accretion discs; flickerings, bursts, and dips that are connected with the vagaries of the accretion processes. The phenomenon of *quasi-periodic oscillations* (QPOs) amongst x-ray binaries was discovered when the flickering behaviour of x-ray count rates was found to display broad peaks in any periodogram analysis of the data. These sources did not show oscillations at single well-defined periodicities, which would have been evident in periodograms as simple sharp peaks at given frequencies, but rather a wide range of oscillations that were aperiodic, thereby giving rise to broad distributions of power within periodograms. This QPO behaviour, which occurs at kilohertz rates – see, for example, Wijnands et al. (1997) on GX 17+2 – is considered to be part of the accretion phenomena occurring in the innermost orbits of the accretion disc, close to the accretor surface.

The x-ray colour-colour diagram is the equivalent of two-colour diagrams in optical photometry. Here we define x-ray *colour* or *hardness ratio* as the ratio of received flux densities in two energy bands. Hasinger and van der Klis (1989) used a *hard colour*, defined as the count-rate ratio of the 6–20-keV band to the 4.5–6-keV band, and a *soft colour*, defined as the ratio of the 3–4.5-keV band to the 1–3-keV band. In that seminal paper, they showed that the temporal behaviour of x-radiation from LMXB systems was strongly correlated with the x-ray spectral properties as delineated in the two-colour diagram, and they found that two sets of behaviours of LMXBs could be identified. Thus, in plots of hard colour versus soft colour, many LMXBs traced out a Z-shaped distribution of measured values, with the frequencies of the QPOs adopting different characteristics dependent on the location of the LMXB in the different parts of the Z distribution; these became known as *Z sources*. But other LMXBs were called *atoll sources*, because they traced out different distributions that resembled atolls of land in an ocean on Earth! Some of these atoll sources then yielded two-colour diagrams that sometimes looked like bananas, whilst at other times looked like separated small islands. So we read of atoll sources showing *banana* and *island* states in their spectral characteristics! These alternative states can occur for a single source at different times. Figure 5.6 illustrates schematically the type of two-colour diagram with the characteristic Z and atoll shapes, demonstrating that the terminology is at least very descriptive.

These variations in x-ray colour are linked to x-ray-flux densities, or x-ray luminosities, such that the Z sources are the LMXBs with higher luminosities ($L \sim 10^{31}$ W), whilst the atoll sources have $L < 10^{30}$ W. The Z sources show the QPO behaviour and a more complex x-ray spectrum, and it appears that the location of the source within the Z shape depends on the mass-transfer rate from the companion. Thus, at lower $\dot{m}$, Z sources have harder spectra and are found on the upper or horizontal branch (HB) of the Z; at intermediate or more typical $\dot{m}$, the sources are on the sloping middle part of the Z, the normal branch (NB); at higher $\dot{m}$, the sources have softer spectra and are located on the lower or flaring branch (FB) of the Z, because

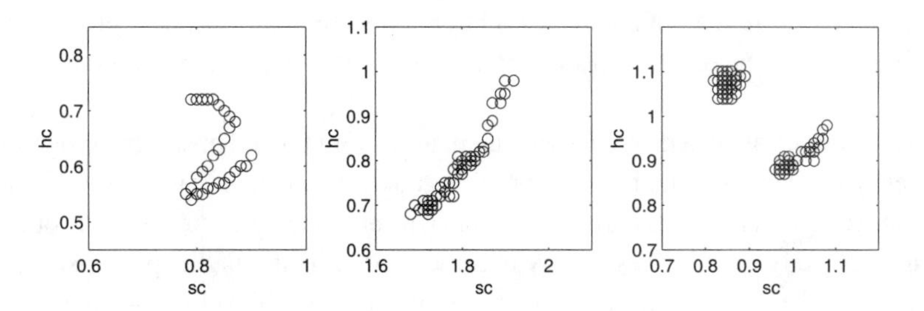

Fig. 5.6. Schematic x-ray colour-colour diagrams (soft colour versus hard colour) for Z sources and atoll sources amongst the low-mass x-ray binaries. From left to right: a Z source, an atoll source in the banana state, and an atoll source in the island state.

it was observed that some systems did show flaring activity at those stages. By contrast, the atoll sources have harder spectra than Z sources, readily modelled by a simple power law, and they display x-ray bursts rather than QPOs.

Recorded x-ray spectra are of low resolution and cover 1–30 keV. They are usually represented by simple power laws or by two-component models incorporating black-body radiation from an accretion disc with $T \propto R^{-3/4}$, and a second component representing the boundary layer between the accretion disc and the accretor. The corresponding black-body temperatures for this x-radiation are of the order of 10^7 K. Fluorescent iron lines have been discovered in the spectra of some systems, showing variations in flux levels that may be dependent on orbital phase. Further detailed discussion of the properties of x-ray binaries, from x-rays to radio wavelengths, is available in the research-level text *X-ray Binaries*, edited by Lewin et al. (1995).

5.4 Ground-based photometric observations and reductions

5.4.1 Photometric observations

Ground-based photometry of binary stars can be conducted with telescopes of modest aperture (0.5–2 m). Such telescopes are sometimes located at the best astronomical observing sites in the world, but there are many more of them located at less ideal observing sites, where the quality of the astronomical sky may be mediocre or even marginal. Photometry of high precision requires a photometric sky, namely, a sky that has no clouds, is free from specific layers of dust at different levels in the Earth's atmosphere, and is as free as possible of sources of light pollution. At the best astronomical sites, such stable photometric skies last all night on 70–75% of the nights of the year, and photometry conducted in such conditions should provide uncertainties of ±0.5–1% between sources located over large ranges in angular position. Photometry of this form is often called *all-sky photometry*, because many sources can be observed in many different parts of the sky during a night, and all of the observations of these programme sources will be interleaved with observations of the photometric standard stars referred to in the earlier subsections. At the observing sites more typically occupied by college observatories, the photometric conditions of clear skies may occur quite commonly for partial nights only, and for a number of complete nights, but only rarely for several consecutive nights. As a result, it may not be realistic to attempt to perform all-sky photometry, except on the best of nights.

The consequences of such different weather patterns for planned observing programmes are quite extreme. A relatively poor site may provide good data on binary systems that have orbital periods that are significantly less than 1 day, so that a complete light curve may be obtained in a single uninterrupted observing session. With some luck with the weather patterns, it is possible to extend this procurement of light curves as far as binaries with orbital periods of 1–2 days and still secure a complete light curve within a month or two. But for systems with $P \geq 3$ days, so that a typical observing night provides only 10% of the complete light curve, it may prove to be virtually impossible to secure a complete curve, because the probability of obtaining a clear night that covers a specific orbital-phase interval is small. Good observing sites do not present such problems, and the ability to use the good conditions to tie all observations to a selected standard system ought not be missed. A beautiful example of these factors is the set of *uvby* light curves for the eclipsing binary TZ For, with a 75.7-day period (Clausen et al. 1991). Standardization of the photometry was essential to permit the various parts of the light curve secured on all the nights to be tied together successfully.

Photometric observations of close binary stars usually are conducted in an intensive manner, *monitoring* one system continually during a night, and over several consecutive nights if the orbital period is greater than a few hours. For photoelectric photometers equipped with photomultiplier tubes, the recorded signal is that from a single source together with that from the surrounding sky contained within a circular aperture of diameter typically about 20 arcsec, located at the focal plane of the telescope. Because the recording equipment is never perfectly stable, and the atmospheric transparency is never perfectly constant, it is necessary to alternate observations of the binary star with observations of one or more *comparison stars*, usually within about 1° on the sky of the variable. It is to be hoped that these comparison stars will prove to be constant in apparent brightness, or will already have been established to be so from previous work. It is also necessary to record the sky background level, at regular intervals between observations of the variable and the comparison stars, by selecting a piece of blank sky close to the observed stars. The photons from each source are recorded for a pre-set integration time in order to achieve a desired value of signal to noise (S/N). Typical integration times are in the range 10–100 seconds. The values of S/N attained will depend on the received flux density from the source via the particular telescope and photometer used. Claudius and Florentin-Nielsen (1981) have presented a full discussion of the best methods for estimating integration times to achieve the photometry of the required accuracy. Such *sequential differential photometry* of variable and comparison sources, and the local sky, requires great care to achieve good results and can be performed reliably only in photometric conditions.

Techniques have been developed to improve the accuracy of differential photoelectric photometry by allowing for simultaneous observations of the variable and comparison sources, together with the surrounding sky defined by the photometer aperture. The Twin Photometric Telescope (Reddish 1966), originally located at the Royal Observatory, Edinburgh, and later moved to St. Andrews, achieved excellent results in limited photometric conditions. The light curves for eclipsing binaries that were obtained showed root-mean-square (RMS) scatter of the data points about a mean curve of only ±0.0045 mag (Bell and Hilditch 1984; Bell et al. 1987), a remarkable achievement for a sea-level site. The high-speed photometer developed by Nather and Warner (1971) included two photomultipliers, one to monitor the variable source continuously at rates as short as 1 Hz, and the second to follow any transparency variations revealed by a constant comparison star. That photometer confirmed earlier pioneering work on short-period oscillations seen in cataclysmic variables and dramatically improved the time resolution and quality of the data. Other two-star or multiple-star photometers have been developed to pursue this useful technique, which not only allows photometry of higher precision but also permits continuous monitoring of rapidly varying sources. A substantial discussion of the various techniques employed can be found in Warner's *High Speed Astronomical Photometry* (1988).

A major improvement in photometry came with the development of *direct imaging* systems that possessed the superb linearity of response of photomultipliers and additionally had greater dynamic range, namely, the charge-coupled devices (CCDs) installed into simple cameras. Photometry conducted with a CCD camera records, simultaneously, the variable source together with all the other stars in the field of view of the telescope and CCD chip, as well as the background sky. This simultaneity ensures inherently greater accuracy. In good conditions, sequential photometry will provide data with uncertainties of ±1%, whilst in the same conditions, CCD imaging will provide uncertainties of ±0.2–0.5%. The term *ensemble photometry* is emerging as a descriptor for this mode of astronomical photometry. The only significant disadvantage of CCD cameras lies in their limited response times. Whilst images can be recorded with integration times as low as a few seconds if small windows of an entire CCD are utilized, they cannot yet match the millisecond rates of high-speed photometry conducted with photomultipliers.

A consequence of the simultaneity of CCD photometry is that *differential photometry* can be performed in marginal conditions, when the sky is certainly non-photometric, such as during the presence of unstructured thin mist or cloud. Thus the viability of typical college observatories for worthwhile astronomical research has been extended with the introduction of these precise imaging systems.

This differential photometry, whether simultaneous or sequential, provides accurate measures of the variations in received flux densities from the binary star relative to one or more constant companion stars. Once an orbital period has been established, or is known from earlier work, it is possible to form a light curve for that binary from the total amount of recorded data. If the observations are broad-band and purely differential and no attempt has been made to tie these data to a standard photometric system by means of observations of standard stars in all-sky photometry, then the data set is somewhat compromised. My reason for saying this is that the data remain on the 'natural system' of the particular telescope and photometer combination, which cannot be transformed to a standard system if no standard stars have been measured during the same observing interval. Thus, the effective or constant-energy wavelengths of the system used are only approximately known in the case of broad-band filters, and any comparison between such data and earlier or subsequent work will be compromised by this lack of information. Therefore, it is good practice to ensure that the necessary empirically determined equations for transforming natural-system data to the standard system be found by many observations of the standard stars of that system.

For narrow-band filters ($\Delta\lambda \leq 100$ Å), for which there is not likely to be a standard photometric system anyway, the natural photometric system is well defined by the filter response curves alone, and hence the constant-energy wavelengths are well defined. Since CCDs have been introduced, it has proved possible to conduct narrow-band photometry with good accuracy on moderately faint sources (10–15 mag) with 1-m telescopes, and it would seem to be worthwhile to pursue such quasi-monochromatic photometry more extensively than seems to have been the case thus far, if only to ensure good quality photometry at specific wavelengths that can be selected to omit known broad emission or absorption lines or bands and can be easily matched with model-atmosphere fluxes. College observatory sites usually are not sufficiently reliable that these standard-star observations can be made regularly. So a strong suggestion might be made that such facilities should be used for narrow-band photometry of sources in order to minimize the effects of atmospheric variations and the background sky illumination from nearby street and security lights. At the better astronomical sites, it is clear that narrow-band photometry can be performed with extra precision, and intermediate and wide-band photometry can be tied to the selected photometric system via the standard star observations and used to study much fainter objects. Some might argue that sources with unusual spectra, such as CVs, with their broad emission lines, cannot have their multi-colour photometry transformed to a standard system in any event. So why bother? But the reply should be that the local comparison stars probably are normal stars, and their data can be so transformed, thereby making the data sets more useful for intercomparisons with earlier and subsequent data. The searches for longer-term variability and correlations with other information, such as cyclic magnetic activity, are the stronger for such attention to observation of standard stars. Henden and Honeycutt (1997) have provided secondary standards for CCD fields around 76 variables, mostly of the CV type.

The flux density from a source recorded by a photoelectric detector will be stored as an arbitrary number of 'counts' that normally will be linearly dependent on the received signal. It is, of course, possible to saturate some pixels of a CCD when a source is too bright for the selected integration time and the CCD response becomes somewhat non-linear before the saturation limit is reached. Accordingly, care must be exercised in selecting integration times and avoiding fields with excessively bright stars in them. Equally, a photomultiplier will react non-linearly to a very high photon flux density, and the observer must be alert to these possibilities so that they can be avoided.

A typical night of CCD photometric observations of a binary star might take the following form: At dusk, in the civil-twilight sky, clear-sky exposures with all required filters are obtained to provide the so-called *flat fields* for CCD photometry that are used to quantify the relative responses of all the pixels to a uniform illumination. This procedure sounds very easy, but in practice it can be quite difficult, because the sky brightness will be decreasing rapidly, and there may be, only a 10-minute window when the sky will be neither too bright nor too faint for the necessary observations to be made. Because most CCDs are red-sensitive, it is best to start with the most blue filters and work toward the reds at dusk, and the reverse at dawn. The zero level or bias level of the CCD must also be determined by means of several zero-second exposures with all camera shutters closed. Once the sky approaches the end of nautical twilight, it is possible to begin observations of selected standard stars to effect transformations from the natural system of the photometer to the standard system. It is sensible to aim for observations of about 10 standard stars that cover a reasonable range of colour indices, as well as a range of air masses X to determine the extinction coefficients. Once these observations are completed, work on the field of the selected binary star can start and can be continued throughout the night, with occasional excursions to record a nearby standard-star field so as to monitor the stability of the photometer. The end of the night is the reverse of the beginning, with more standards observed in late-astronomical/early-nautical twilight, bringing the total number of observations of standard stars to about 25, followed by more flat fields on the dawn sky, particularly if the dusk-sky observations were incomplete, and more bias frames. For observations using a photometer equipped with a photomultiplier, the procedures are the same, except that the flat-field exposures are replaced by sequential observations of the civil twilight sky through two focal-plane apertures with different diameters, in order to determine the so-called *dead-time correction* for that particular pulse-counting photoelectric photometer. Such a correction is required because at high pulse-arrival rates from brighter sources, sequential pulses or events can effectively merge, as a second event arrives in the 'dead time' whilst the first event is being processed by the electronics of the photometer. A helpful description of dead-time corrections and other aspects of pulse-counting photometers is given in Chapter 4 of Henden and Kaitchuk's *Astronomical Photometry*. It is crucial for all photometry that the time of each observation be recorded, either in local sidereal time or in universal time (UT).

When using single- and multi-channel photometers, the observing procedure for the standard stars is the same, except that the multi-channel system is considerably more efficient – simultaneous, rather than sequential, observations are made in n pass bands. The resultant colour indices are also more accurate. These advantages have been exploited well, for example, in the use of the Strömgren $uvby\beta$ 6-channel photometer of Copenhagen University at the ESO site at La Silla, as well as the Steining 5-channel instrument at MacDonald Observatory. These multi-channel instruments use a coarse grating to provide the initial dispersion of the incoming radiation from a single source, followed by filters to define the precise pass bands, and then n photomultipliers that can be operated over a wide range of integration times, from high-speed photometry to standard photometry. Clearly, this type of instrument cannot be used as an *imaging* system for CCD frames, so its gain in terms of simultaneous wavelength coverage is somewhat offset by its loss of simultaneity in recording the apparent brightnesses of all the sources in a field seen by a CCD camera as a single-channel instrument. Some sets of standard stars – for example, within the Landolt *UBVRI* equatorial standards – are to be found within small regions of the sky, so that a single CCD frame in a given pass band will provide simultaneous measures of several standard stars.

For observations of binary stars, the monitoring rate will be dictated by the required time resolution for the light curve, as well as by the expectation of any major difference in the temperatures of the components of the binary. Multi-channel instruments at good photometric sites are ideal for short-period binaries with variability due to a number of sources within the binary, as well as for short-period binaries with components of very different temperatures. As an example, a short-period ($P \sim 3$ hours) eclipsing binary comprising an O-type subdwarf

($T \sim 40{,}000$ K) and an M-type dwarf ($T \sim 3000$ K), with oscillations on the subdwarf having a period of 15 minutes, is best observed with a multi-channel instrument at a good, stable photometric site. But the rapid oscillations ($P \sim 60$ seconds) seen in some CVs and related objects require excellent time resolution, and it may be necessary to use a single-channel photomultiplier instrument to gain as much signal as possible from the source. For less desirable sites, the advantage of CCD imaging in providing simultaneous differential photometry of high accuracy as compared with the poorer time resolution in wavelength coverage from such a single-channel instrument, is to be preferred. As a further example, detailed eclipse mapping of the surface features in a W-UMa-type eclipsing binary with components of nearly equal temperatures is best achieved by careful CCD photometry, with which variations of 0.2–2% are quite readily detected. As always, an observing programme that is well planned before observations begin will yield much better scientific returns than one that is approached haphazardly.

5.4.2 Photometric reductions

The immediate reduction of data in photoelectric photometry conducted with photomultipliers is simple. The counts on all sources for all integration times should be converted to a single scale, usually as counts per second, and corrected for the effects of dead time in the photometer system. The observations of sky background taken close to each of the standard stars, and in all the utilized pass bands, are subtracted from the star + sky counts recorded for each source.

For CCD photometry, the extraction of useful data from the individual CCD frames is more complex. There are standard CCD reduction packages available to facilitate these procedures, such as DAOPHOT (Stetson 1987) and DOPHOT (Schechter et al. 1993) and those incorporated into the Image Reduction and Analysis Facility (IRAF) from the U.S. National Optical Astronomy Observatories, or the GAIA package in the U.K. Starlink software. The first operations are the subtraction of the background bias level and the flat-fielding of all the frames. Then these CCD photometry packages must be used to measure the total flux density recorded for each image above the sky background level. Various modes of photometry are available. Straightforward aperture photometry can be used, effectively simulating the traditional simple photometer, and it is entirely adequate for star fields that are uncrowded and for programme-star fluxes well above the sky background level. Otherwise, some form of three-dimensional profile fitting must be adopted to match the distribution of flux density in each star image (the *point-spread function*) and the background sky level in that region of the CCD frame. Either way, the net result is the same as for photomultiplier photometry, with a total number of counts per second above the local sky background for each required image in each CCD frame.

When the light from a star passes through the Earth's atmosphere, it suffers some attenuation, which is dependent on wavelength. The actual physical processes are complex, involving Rayleigh scattering from atoms and molecules and scattering due to dust and other aerosols. Despite this complexity, the net result in practice is that total extinction of the starlight can be approximated by a single extinction coefficient that is dependent on wavelength multiplied by the path length through the atmosphere. The observed flux density f of radiation is given by

$$f = f_0 \exp(-\tau X) \tag{5.30}$$

where f_0 is the flux density just outside the atmosphere, τ is the wavelength-dependent optical depth per unit air mass, and X is the *air mass* giving the path length through the atmosphere in terms of a unit column of air at the zenith. Then in magnitude terms we obtain

$$m - m_0 = -2.5 \log_{10}(f/f_0) = 2.5\tau X + \text{constant}; \qquad m = m_0 + kX + \text{constant} \tag{5.31}$$

where k is the extinction coefficient in units of magnitudes per unit air mass. The air mass is given to a high degree of accuracy by the secant of the zenith distance z. At high zenith distances, $z > 60°$, it is necessary to introduce correction terms that allow for the curvature of the Earth's atmosphere. Such a formula has been derived by Bemporad; see Hardie (1962). The expression for sec z is derived from standard spherical astronomy to be

$$\sec z = (\sin\phi \sin\delta + \cos\phi \cos\delta \cos h)^{-1} \tag{5.32}$$

where ϕ is the latitude of the observatory, and h and δ are the hour angle and declination, respectively, of the star being observed. The standard formula for the air mass X is then

$$X = \sec z - 0.0018167(\sec z - 1) - 0.002875(\sec z - 1)^2 - 0.0008083(\sec z - 1)^3 \tag{5.33}$$

The difference between X and sec z is only 0.005 at $z = 60°$, but it is 0.02 by $z = 70°$. Note that if the variable and the comparison star are within 1° of each other on the sky, then the difference in air mass is $\Delta X \sim 0.001$, and a differential extinction term is negligible. Equally, for the typical size of a CCD frame of only a few arcminutes on the sky, the differential extinction across the frame is negligible.

5.4.2.1 *Standard stars*

To convert observations made with a particular telescope + filter + detector combination to the standard system it is necessary to include zero-point terms in the transformation equations, as well as scale factors or transformation coefficients for the colour indices. Then the working equations for a magnitude m and a colour index c are, respectively,

$$m = -2.5 \log_{10}(f); \qquad c = -2.5 \log_{10}(f_1/f_2) \tag{5.34}$$

$$m_0 = m - k_m X + z_m; \qquad c_0 = s_c(c - k_c X) + z_c \tag{5.35}$$

where the measured flux densities f, f_1, and f_2 are in arbitrary units of counts per second and are converted to an instrumental magnitude m and a colour index c in the usual manner. The zero subscripts signify values corrected for atmospheric extinction. The extinction coefficients are k_m and k_c with air mass X. The zero points z_m and z_c are arbitrary factors that convert the arbitrary flux-density units utilized to the standard system of magnitudes. The scale factor s_c for the colour index c should be a value close to unity for carefully selected filters and detectors, and these transformation equations should be linear and single-valued.

For wide-band systems, the wavelength dependence of the extinction may demand a second-order term in k that is a function of the colour index of the star being observed; thus,

$k \simeq k' + k''c$. In the *UBV* system, $k'' \simeq +0.03$ for $(B - V)$, and -0.01 for $(U - B)$. For intermediate- and narrow-band systems, the extinction coefficients are independent of the colours of the stars, a factor that helps to ensure good-quality transformations.

The set of n standard-star observations for a single night, $n \sim 25$, provides n equations of the foregoing type for the magnitude and the colour indices, each in p unknowns, with $p \leq 3$, namely, $s_{m,c}$, $(s_{m,c}k_{m,c})$, and $z_{m,c}$. For a magnitude, s_m should be unity. A straightforward least-squares solution can then be carried out to obtain the best estimates for the p unknowns from the n equations for each colour index or magnitude of the standard system. It is clear that good solutions for the three unknowns will be obtained if the distribution of colour indices of the observed standards covers most of the range of each index and if the distribution of air masses covers the range $X = 1$–2. This latter requirement does mean that standards must be observed at $z \sim 50$–$60°$, and one should not rely on one or two very red or very blue standards at such large zenith distances.

Values of scale factors, extinction cofficients, and zero points are then defined for each night of observation. Typically, a particular programme will have been carried out over a number of (hopefully) consecutive clear nights, and it is to be expected that the derived coefficients will show good repeatability from night to night. In particular, the scale factors will be well defined and consistent from night to night; and at good photometric sites, the extinction coefficients should also display good consistency. Most photometric systems are defined by the filter + detector combination (i.e., independent of the telescope and the site), and hence for an observing session it is sensible to adopt mean values for the scale factors that will have errors of only 1–2%. Once the scale factors are so defined, the entire least-squares procedure should be re-run with only the extinction coefficients and zero points as the unknowns. The precision with which these values are defined on each night should improve markedly. Whether or not one should then adopt mean extinction coefficients for the session is a rather fine point. For example, if most observations of programme stars were obtained at low air masses, say $X \sim 1.2$, then a small change in extinction is equivalent to a small change in zero point. Hence, adopting a mean value for k will simply cause a compensating change in z for any night on which the extinction was in fact different from the standard value, and the net effect on the programme-star magnitudes and colours will be zero. For observations made over larger ranges of X, then clearly each night's value of k should be retained, provided it is well defined by standard-star observations over a good range of air masses. At sites of good photometric quality, the extinction coefficients are remarkably constant over long periods, and mean values may sensibly be adopted. On part nights, where there has been insufficient time to define the k values adequately, the use of mean extinction coefficients is a realistic procedure.

Having established final values for s, k, and z for each magnitude and colour index, it is then necessary to search for second-order effects that may become apparent only at this stage. Firstly, from all the nights of observation, mean $(O - C)$ residuals must be determined for magnitudes and colour indices for all the standard stars observed. Here O is the observed value for a star, and C is that calculated from the derived transformation equations noted earlier. These residuals are plotted against a suitable colour index [e.g., $(b - y)$ or $(B - V)$] to check for possible colour-dependent terms. In some cases, such terms are significant, and second-order terms of about $0.02(B - V)$ are common in V magnitude transformations. It is only by reducing the data on standard stars in the foregoing manner that such small-scale and systematic effects are detectable. A second check is to look for zero-point drift in the magnitude scale occurring during individual nights. Such drifts are usually of small amplitude (∼0.03) over many hours and are likely to be negligible for photometers with temperature-controlled detectors, as is standard practice at present.

In summary, the procedure for determining the transformation coefficients should be as follows:

1. Define individual values for s, k, and z for each night.
2. Adopt mean values for s, and re-run solutions to redetermine k and z.
3. At good sites, adopt mean values for k, and redetermine z for each night.
4. Calculate mean $(O - C)$ residuals for each colour index and magnitude for each star, and check for second-order colour terms and zero-point drifts.
5. If such second-order terms are found to be significant, then incorporate fixed values for these terms into the original equations and repeat steps 1–5. One or two iterations will suffice to establish the complete set of transformation coefficients.

5.4.2.2 *Programme stars*

With the final transformation equations established by observations of the standard stars, the reductions for the programme stars are easily carried out. Although dependent principally on the adopted equipment and observing procedure (e.g., simultaneous or sequential multi-colour observations), it seems appropriate to transform the colour measurements of the variable directly to the standard system, rather than forming differential colour indices relative to the comparison star. For example, such colour observations may be crucial in deciding whether or not an observed eclipse in a binary is an occultation or a transit.

For high-quality light curves, defined with an uncertainty of ±0.005 mag or better, it is important to use the closely spaced magnitude observations of variable and comparison stars from sequential photometry in a differential mode, because small-scale transparency and background sky variations should affect both stars equally and nearly simultaneously. A convenient procedure is to utilize a least-squares cubic-spline fitting routine to form differential magnitudes. Firstly, a spline is fitted to the set of local sky flux densities obtained throughout the night, and values for the background sky are interpolated to the times of observation of both stars before subtraction from the star + sky fluxes. Secondly, a least-squares spline is fitted to the comparison-star flux densities or magnitudes and is again used to interpolate such flux densities or magnitudes to the times of observation of the variable, in order to form differential flux densities or magnitudes. Such a procedure helps to improve the accuracy with which a light curve can be defined, and additionally it provides a realistic estimate of the accuracy of the observations from the RMS error of the comparison-star observations about the least-squares cubic spline. Finally, it should be established that the comparison star is constant in brightness both during a night and from night to night. When such constancy relative to the standard stars is confirmed, then any changes in the shapes of the light curves can be unequivocally established. From CCD imaging data, where the differential photometry is simultaneous, this part of the data reduction is more simple, but the need to check the constancy of the comparison stars remains as important. The number of well-established cases of changes in light-curve shapes from night to night is substantial and covers many different types of binaries as well. Careful photometry establishes the best light curves.

The photometry of the comparison stars can also be used as an independent check on the extinction coefficients determined from the observations of standard stars. If the field of the binary has been followed through a large range of air masses, from $X \sim 1$ to $X \sim 1.8$–2.0, then a plot of comparison-star magnitudes against X should yield a straight line of slope equal to the extinction coefficient for the wavelength observed. The graph will not be a straight line if the comparison star is a variable(!), or if the zero point of the photometer is not stable, or if there have been changes in extinction values during the night.

It is a tradition that light curves are formed as plots of differential *magnitudes* against time or orbital phase. The differential data can simply be left as flux-density ratios between the binary and the comparison star, because that is the form of data required by light-curve analysis codes, discussed in the next section. For binaries with light curves that have amplitudes of several magnitudes, it is a practical convenience in graph plotting to use the logarithmic scale of magnitudes. For small-amplitude light curves, plots of flux-density ratios are sensible. In some recent publications, light curves have been plotted with a flux-density scale given in millijanskys, without any justification stated as to how the transformation from arbitrary flux-density units of counts per second on a particular instrument was made to the absolute flux-density scale discussed briefly in Section 5.2. Analyses of light curves are conducted on a purely relative flux-density scale in any event, with the reference level usually being set at the flux-density ratio observed at the quadrature phases (0.25 and 0.75 for a binary system with a circular orbit).

5.5 **Light curves of eclipsing and non-eclipsing binaries**

5.5.1 Orbital periods from photometric data

The light curves for most eclipsing binary stars exhibit rather sudden eclipses, separated by substantial time intervals during which there is little change in apparent brightness. Thus, typical light curves cannot readily be represented by sine/cosine series, and hence the Fourier techniques of period finding described in Chapter 3 will not give the best results. The most simple procedure is to make use of the well-defined eclipses to establish precise times of eclipse minima. Then the problem of period finding reduces to a more straightforward issue of solving a linear equation for the best orbital period from a set of the times of minima, where the number of orbital cycles elapsing between two successive observed times of minima has already been estimated by means of a preliminary value for the orbital period. That is, some well-guided guesswork is involved in the initial stages of analysis! Such guides may be, for example, the durations of the eclipses, evidence for non-eclipse variations during a night, and variations in radial velocities. Monitoring an eclipsing system for an entire night will reveal or exclude periods of a few hours. For longer periods, and for non-eclipsing but photometrically variable systems in which that variation is a direct function of the orbital period, it may prove beneficial to make use of the non-parametric techniques involving phase-dispersion minimization (PDM) that were referenced in Chapter 3. These procedures rely on finding a best period amongst a range of possible periods by searching for a minimum dispersion in phase of the observed data points. Thus the question 'How long is a piece of string?' takes on true significance when that string joins together all the observed data points. The minimum length corresponds to the best period for that data set. Because these latter methods are presented in detail by Stellingwerf (1978) and Dworetsky (1983) and are incorporated into the code PERIOD (Dhillon and Privett 1995), we shall not discuss them further, but outline the simple method of period finding from sets of times of eclipse minima.

An eclipse exhibited by a binary star can be recorded with careful photometry at a sampling rate that will ensure an accurate representation of the true eclipse curve. For short orbital periods with small stars, an eclipse may last only a few minutes, in which case one may need a high-speed photometer to monitor the brightness changes quickly enough. For longer orbital periods of days or weeks, an eclipse may last for hours, or more than an entire night for longer-period systems with large stars. The sampling rate obviously can be more leisurely, but it is necessary to ensure that all the differential photometry is very consistent. A time of minimum light in mid-eclipse, generally referred to as a *time of minimum*, is most readily determined by a simple graphical procedure of plotting the differential magnitude, or the ratio of flux densities, between the binary and the comparison star against the time of observation (e.g., in UT or in heliocentric Julian dates, HJD) to yield a partial light curve. The curve will be only approximately V-shaped or parabolic, but by joining points of equal brightness on either side of the minimum, to define a set of horizontal chords on the graph, it is then easy to determine a vertical line that bisects all of those chords and defines a time of minimum by its intersection with the time axis. Provided that the partial light curve is symmetric about the minimum, then the bisector will indeed be vertical on the graph. But if the curve is asymmetric, a common occurrence in CVs, then this procedure will fail, and it will be necessary to accept a less precise determination of a time of minimum. The derived time of minimum should then be expressed in terms of HJD. Figure 5.7 illustrates this simple procedure, which makes use of all the data that define the shape of the eclipse curve.

A sequence of such measurements of times of minima will allow a determination of the orbital period from a simple linear ephemeris equation, as used in Section 4.5. It is essential, of course, that the eclipses be correctly identified as being primary or secondary eclipses, usually on the basis of their depths relative to the light-curve maxima at the quadrature phases. In a few systems, the eclipses are of equal depths, to within 0.01 mag, and so care must be exercised in interpreting the times of minima to determine an orbital period. It is necessary to make an initial estimate about an orbital period in order to calculate the cycle numbers ϵ in a linear ephemeris like equation (4.75). Then a least-squares solution for such an equation with observed times of minima at estimated cycle numbers will provide an improved determination of the orbital period. For binaries with orbital periods of a few hours, and hence complete light curves observable within a single night, orbital periods are unambiguously determined immediately. For any orbital period of longer duration, there is always the possibility of finding spurious values for the period that will fit the observed times of minima. In such cases, ephemerides with the different possible periods should be used to predict times of eclipses on different nights so that observations can discriminate between the true and alias values. A final test

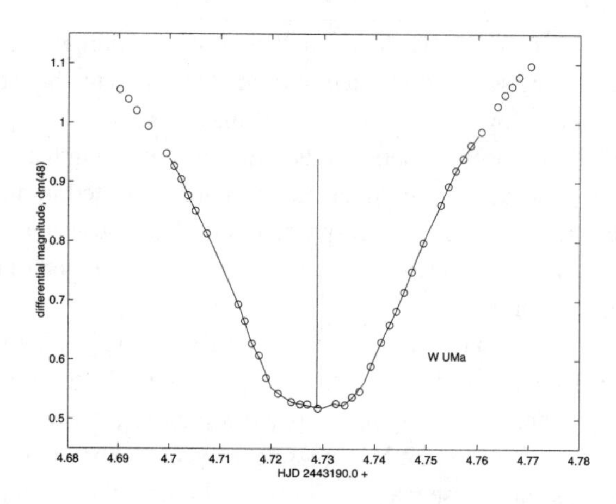

Fig. 5.7. Differential photometry of the contact binary W UMa in the '48' filter of the DDO photometric system obtained on HJD2443194, showing observations secured through a primary eclipse. A cubic spline has been fitted to the data, and a bisecting chord has been determined from the interpolated values of time and differential magnitude to yield a time of minimum brightness from the symmetry of the eclipse curve, as shown by the vertical line at HJD2443194.7288. (Data obtained from Hilditch, 1981.)

lies in collating all the data from different nights to form a complete light curve. If sections of data from different nights overlap in phase and have very different shapes, then the most likely explanation is that the selected orbital period is incorrect. An alternative, but less likely, explanation would demand additional variability in the binary system, which is nevertheless observed in many systems.

5.5.2 Some basic definitions and example light curves

There are several items of nomenclature that are used in descriptions of light curves and eclipse phenomena that need to be understood in order to be able to appreciate the physical picture that is being presented. The first set of these terms involves the mutual eclipses of a star and its companion. Figures 5.8–5.10 provide schematic illustrations of mutual eclipses in a binary system. For a binary system with its orbital plane close to or in our line of sight, that is, with $i \sim 90°$, and circular orbits, the alternating eclipses will be annular (the smaller star transiting across the face of the larger star) and total (the smaller star completely occulted by its larger companion). Thus we have an *annular transit* alternating with a *total occultation*; Figure 5.8 shows the annular transit, and its corresponding primary eclipse curve, as well as the secondary eclipse curve corresponding to the total occultation. For lower orbital inclinations, the eclipses become partial, but are still *partial transits* or *partial occultations*, as shown in Figures 5.9 and 5.10.

For spherical stars of radii R_1 and R_2 and separation a, the condition for eclipses to be seen at an orbital inclination i is simply

$$\sin(90° - i) \leq (R_1 + R_2)/a \tag{5.36}$$

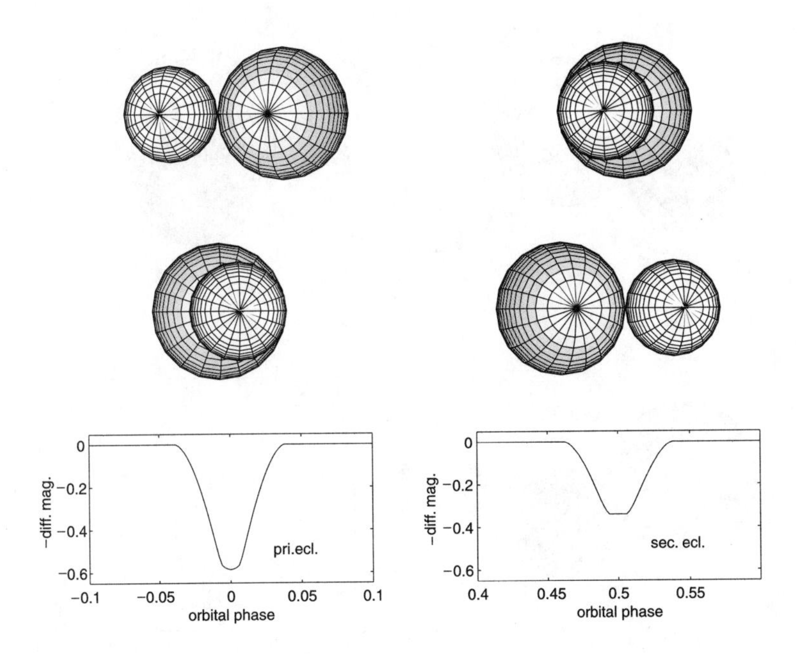

Fig. 5.8. Schematic representation of a detached binary-star system, with the stars having radii of 0.100 and 0.138 in units of the semimajor axis of the relative orbit and viewed at an orbital inclination of $i = 90°$, showing four orbital phases during an *annular transit* of a smaller, cooler star across the face of a larger, hotter star. The binary is seen at the beginning of the eclipse, *first contact*, at $\phi = -0.038$ (upper left), at *second contact*, $\phi = -0.006$ (upper right), at *third contact*, $\phi = 0.006$ (middle left), and at *fourth contact*, $\phi = 0.038$ (middle right). The corresponding *primary eclipse* in the light curve is shown at lower left, whilst the alternating *secondary eclipse* is shown at lower right. Note that the primary, annular-transit eclipse shows a slightly curved minimum due to the limb darkening on the eclipsed larger star and to the quite small difference in the sizes of the two stars. The secondary, total-occultation eclipse shows a flat minimum because the smaller star is completely hidden for a phase interval $\Delta\phi = 0.012$.

Thus, for the binary system illustrated in Figure 5.8, we have

$$\sin(90° - i) \leq 0.238; \qquad i \geq 76.2° \tag{5.37}$$

For a random orientation of binary orbits to our line of sight, eclipses are less likely to be seen in systems where the stars are small relative to their separation, and more likely for relatively larger stars.

Spherical stars, that is, those with relative radii ($r = R/a$) of about 0.1 or less, will show projected discs that are immaculate save for standard limb darkening resulting from a normal stellar atmosphere. (The complication of a star having true maculations, or starspots, will be discussed later.) Such a distribution of surface brightness is rotationally symmetric, so that for spherical stars in a circular-orbit binary, the light curve will show two symmetric eclipses of the same total duration, and intervals of constant brightness between the eclipses.

We can make simple estimates of the radii of the two stars, relative to their separation, from the times of first to fourth contacts illustrated in Figure 5.8, at least for systems where $i \sim 90°$ and the orbit is circular. It is clear from studying Figure 5.8 that the time interval from the first

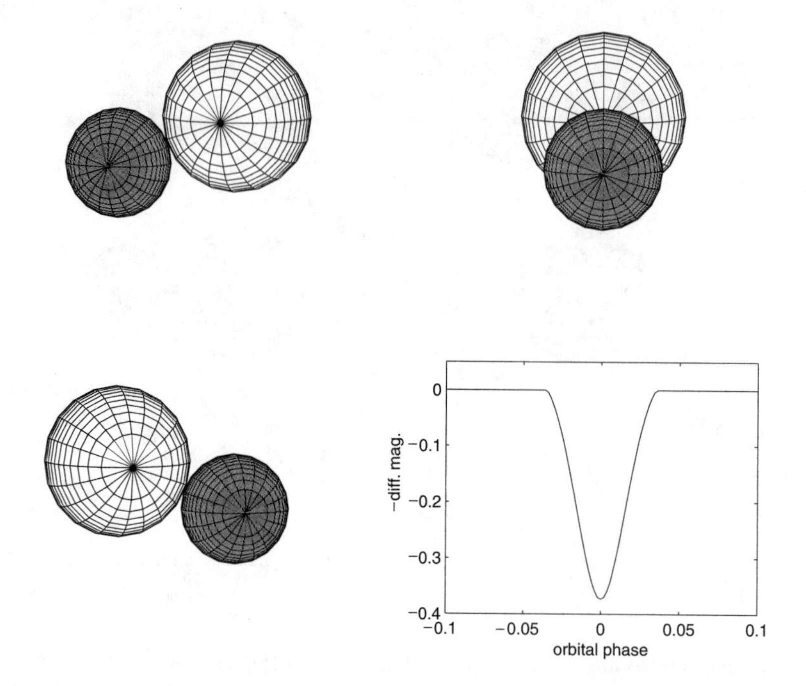

Fig. 5.9. Schematic representation of the same binary-star system as in Figure 5.8, but viewed at an orbital inclination of $i = 85°$, showing three phases during a *partial transit* of the cooler star across the face of the hotter star. The binary is seen at the beginning of the primary eclipse, *first contact*, at $\phi = -0.035$ (upper left), at *mid-eclipse*, $\phi = 0.0$ (upper right), and at the end of the eclipse, *fourth contact*, $\phi = 0.035$ (lower left). The corresponding primary eclipse in the light curve is shown at lower right. Note that the minimum is V-shaped, with no phase interval of nearly constant brightness.

to the second contact and that from the third to the fourth contact should be the same and will be a measure of the radius of the smaller, transiting star. Thus, we can write directly that

$$(\phi_2 - \phi_1) = (\phi_4 - \phi_3) = 2R_2/(2\pi a) \tag{5.38}$$

where $\phi_{1,2,3,4}$ are the four times of contact, expressed as fractional orbital phases, and a is the separation, or the radius of the circular relative orbit. Equally, the phase difference between the first and third contacts and that between the second and fourth contacts should be the same and will provide a measure of the radius of the occulted, larger star. Thus,

$$(\phi_3 - \phi_1) = (\phi_4 - \phi_2) = 2R_1/(2\pi\, a) \tag{5.39}$$

Applying these equations to our model binary in Figure 5.8 yields $r_2 = 0.100$, $r_1 = 0.138$, the values used to generate the light curves. Clearly, at lower inclinations and partial eclipses, this simple estimation procedure cannot be used.

If the orbit is significantly eccentric, then the orbital speeds of the two stars are dependent on orbital phase, and hence the durations of the two eclipses will, in general, be different. In addition, the location in orbital phase of the secondary eclipse relative to the primary eclipse

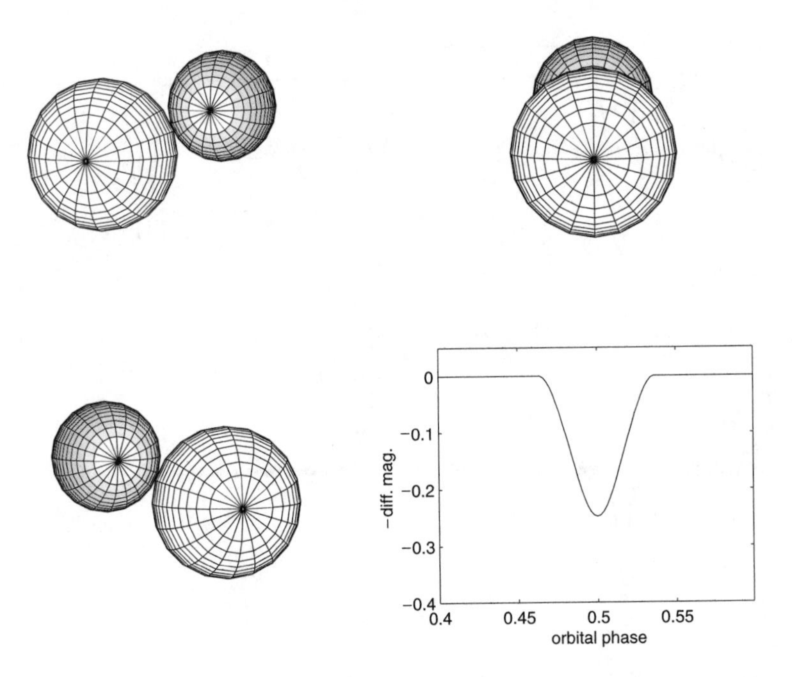

Fig. 5.10. Schematic representation of the same binary-star system as in Figure 5.8, but viewed at an orbital inclination of $i = 85°$, showing three phases during a *partial occultation* of the cooler star by the hotter star. The binary is seen at at the beginning of the secondary eclipse, *first contact*, at $\phi = 0.465$ (upper left), at *mid-eclipse*, $\phi = 0.500$ (upper right), and at the end of the eclipse, *fourth contact*, $\phi = 0.535$ (lower left). The corresponding secondary eclipse in the light curve is shown at lower right. Note that the minimum is V-shaped, with no phase interval of constant brightness.

will not be at phase $\phi = 0.5000$, but shifted by an amount that will depend on the shape and orientation of the relative orbit. An example light curve for such an eccentric-orbit system with spherical stars is GG Lup, as shown in Figure 5.11. The mathematical details of eclipses in eccentric-orbit systems are considered in Section 5.5.3.

Stars that are non-spherical and have their shapes described by the Roche model discussed in Chapter 4 will project a non-circular shape onto the sky and will exhibit limb darkening as usual, together with an effect known as *gravity darkening*, caused by the fact that the local surface gravity on a non-spherical star will vary, and the emergent flux from the star will be proportional to that local gravity (see Section 5.5.6). Thus a rotating non-spherical star will exhibit continual variations in flux density, in addition to any eclipses. The major contribution to that variation will be the changing projected area of the star onto the sky as it rotates, with the gravity darkening being a lesser contributor. The eclipses will normally be symmetric, because non-spherical stars, as described by the Roche model, are symmetric about the line of centres of the binary system. Because the projected area will be largest for both stars at the quadrature phases, and least at the conjunction phases, the total variation with orbital phase will show a double-peaked curve. Eclipses will be added at the conjunction phases if the orbital inclination is sufficiently high. The double-peaked variations outside the eclipse phases are usually called

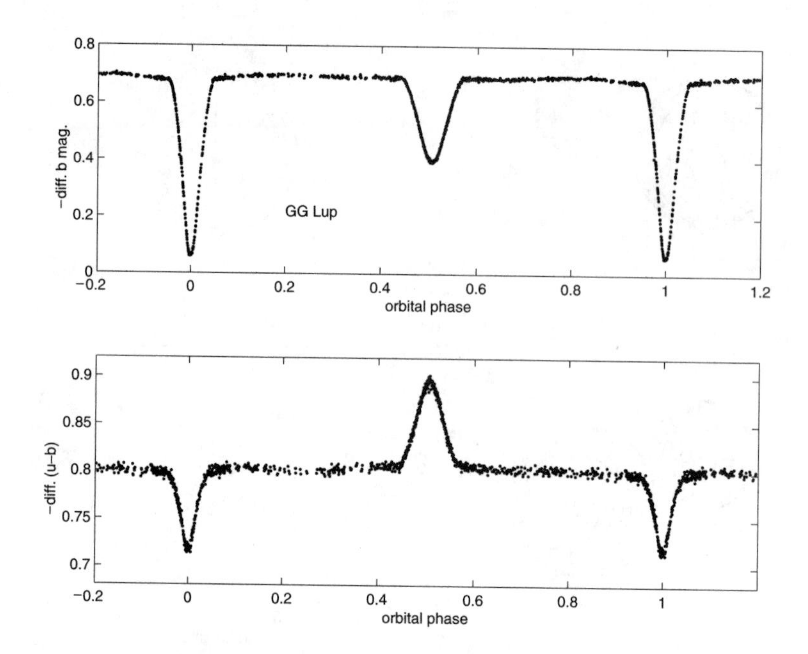

Fig. 5.11. Top: Differential Strömgren b photometry of the main-sequence binary GG Lup, with RMS uncertainties of ± 0.004 mag. The light curve displays nearly constant brightness outside the eclipse phases, indicating relatively well-separated spherical stars, and there is a short interval ($\sim$15 minutes) of constant brightness in mid-secondary eclipse, demonstrating a total eclipse of the smaller, secondary star by the larger primary. The two eclipses obviously have different durations, and the secondary eclipse was at phase 0.5064 when these observations were secured in 1985; these effects are due to the eccentric orbit ($e = 0.150$) with the major axis close to the line of sight in 1985 ($\omega = 86^\circ.4$). Bottom: Differential Strömgren $(u-b)$ colour index for GG Lup with a different vertical scale, showing the clear discrimination between the temperatures of the two stars afforded by the $(u-b)$ index and the alternating annular/total eclipses. The more massive, larger primary has $T = 14{,}750$ K, whilst the less massive, smaller secondary has $T = 11{,}000$ K. (Data obtained from Clausen et al., 1993.)

ellipsoidal variations, because the first attempts to model them made use of ellipsoidal shapes for the two stars. Such a shape does provide a good representation of the observed light curves, but the Roche model is more physically realistic. Thus, ellipsoidal variations in the light curve for a binary indicate that at least one of the stars in the system is relatively large, $r \geq 0.2$. Figure 5.12 illustrates the ellipsoidal variations and eclipses seen in the massive binary system SX Aur, which has both stars filling most of their Roche lobes. When stars overfill their Roche lobes, as in the contact binaries of W-UMa-type, we see continual variations in brightness of the system, and it is not possible to discern the start and end of either eclipse, as shown for W UMa itself in Figure 5.13.

If the temperature difference between the two stars in a binary is considerable (e.g., $\Delta T \geq$ 5000 K), and the two stars are tidally locked into synchronous rotation, then the radiation from the hotter star will raise the local temperature of the facing hemisphere of the cooler star, thereby creating a region of enhanced brightness centred on the substellar point of the cooler

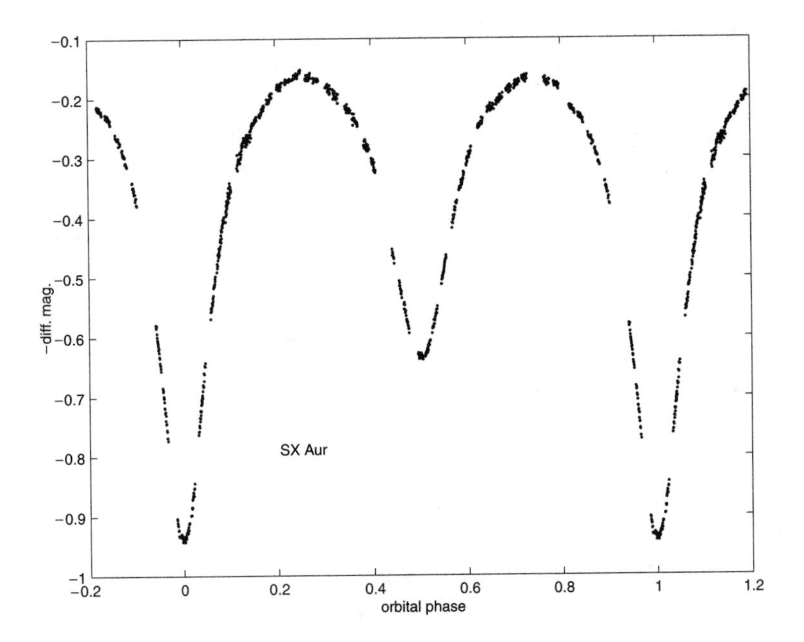

Fig. 5.12. Differential B photometry of the massive binary SX Aur with RMS uncertainties of ±0.005 mag. The light curve displays partial eclipses and a large *ellipticity effect* caused by the stars being non-spherical and filling most of their available Roche volumes. (Data obtained from Bell et al., 1987.)

star. The substellar point is on the line of centres joining the two stars, and if one were at that location, the hotter star would be seen to be permanently directly overhead. The effect on the light curve is to create continual changes in brightness outside of primary eclipse, reaching a maximum at phase 0.5, when the hemisphere of the cooler star that faces the hotter star is projected maximally into the observer's line of sight. This out-of-eclipse variation is called the *reflection effect*, a misnomer for most binary systems, because the effect is often one of surface heating, rather than reflected light, but the term is long-established. A most extreme example of the amplitude that the reflection effect can achieve is shown by the non-eclipsing, short-period binary KV Vel, composed of a sdO star (with a temperature of about 75,000 K) and a K dwarf star with an unperturbed, averted hemisphere temperature of about 3500 K, as illustrated in Figure 5.14. The details of this reflection-effect mechanism will be discussed in Section 5.5.6. A second example is the remarkable eclipsing binary PG1336-018, where the reflection effect is strong, and in addition there are pulsations on the sdB star with periods of 184 and 141 seconds(!), as demonstrated in Figure 5.15. There are many examples of pulsating stars in eclipsing binaries, as noted in Chapter 3.

In addition to pulsating stars in binaries, there are the stars that display solar-type magnetic activity, with associated cool starspots and chromospheric and coronal emission, often at levels that are much larger than observed on the Sun. The consequence is that light curves due to the binary alone become distorted by the maculated surface of one or both components, as illustrated very clearly in Figure 5.16 and Figure 5.17 for the RS CVn binary XY UMa. Such

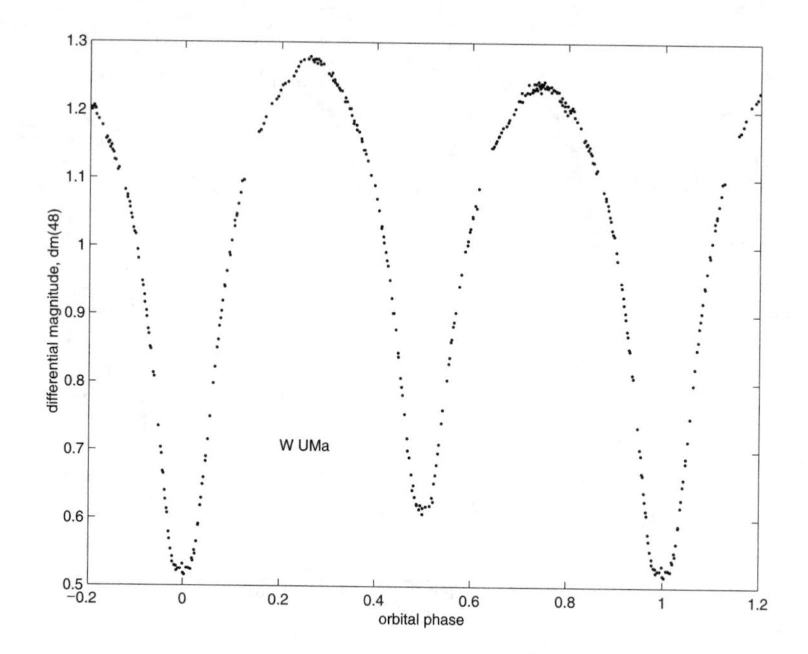

Fig. 5.13. Differential photometry of the contact binary W UMa in the '48' filter of the DDO photometric system, with RMS uncertainties of ±0.003 mag. This light curve is typical of the W-UMa-type contact binaries with spectral types of late F to K: eclipses of nearly equal depths, signifying that the stars have similar surface brightnesses (temperatures); continuous variation in brightness, demonstrating non-spherical stars filling their Roche lobes; unequal maxima, likely indicating cool starspots on these magnetically active, rapidly rotating stars with convective envelopes. (Data obtained from Hilditch, 1981.)

distortions render it difficult to determine the true sizes and shapes of the stars in such systems, and we shall discuss later in this chapter and in Chapter 7 how this problem can be approached.

The effects on binary light curves of the transfer of mass between the components of a binary system can vary from the barely detectable to the extreme. The first types of binaries in which such mass transfer was determined were the classical Algol systems, most frequently by the detection of emission lines or distorted absorption-line profiles in their spectra. The celebrated observations by Joy (1942) of red-/blue-shifted H emission lines during the ingress/egress phases of the primary eclipse of the Algol system RW Tau were correctly ascribed to the presence of an accretion disc around the primary component, and that led subsequently to major investigations of mass transfer in Algol binaries. Yet the effects of such gas streams and discs on the observed shapes of the light curves for Algol systems are quite small, with occasionally recorded dips in the light level before primary eclipses, and sometimes irregular or inconsistent light levels outside eclipses from data secured on different nights but at the same orbital phases. These differences typically are in the range of $\Delta m \sim 0.01$–0.05 mag. The recognition of the dwarf novae as binary systems, leading to major continuing efforts to understand the accretion processes in cataclysmic variables and in x-ray binaries, has changed that earlier perception

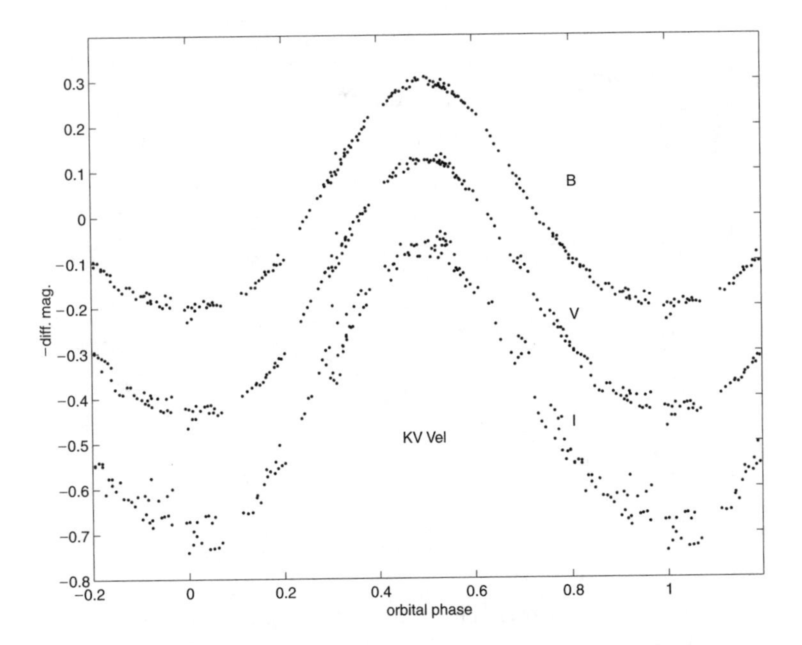

Fig. 5.14. Differential *BVI* photometry of the non-eclipsing sdO+K binary KV Vel. The V and I curves have been displaced by +0.2 and +0.4 mag, respectively, to avoid confusion. The reflection effect in this system is very large because of the extreme temperature difference, $\Delta T \sim 70{,}000$ K, between the two stars. (Data obtained from Kilkenny et al., 1988.)

dramatically. As examples of the effects of accretion processes, the following figures illustrate light curves in the optical, UV, and x-ray regions for binary systems in which the geometrical effects of eclipses and non-spherical shapes are progressively superseded by the processes of accretion.

The first example is the unusual binary system V361 Lyr, discussed earlier in Section 4.5 regarding observational evidence for orbital-period changes due to mass transfer. New observations, secured by Lister et al. (2000b), have confirmed the detailed shapes of the asymmetric optical light curves determined earlier by Kaluzny (1990, 1991) and by Hilditch et al. (1997), and these are shown in Figure 5.18. The analysis of such distorted light curves cannot proceed in a simple manner, but requires an initial geometrical model on which to determine an asymmetric distribution of surface brightness over one or both stars. In this case, despite the obvious asymmetries around first quadrature and through secondary eclipse, the primary eclipse has a remarkably symmetric shape, and its duration and depth ensure that a reasonable initial set of values (r_1, r_2, i, T_2) can be determined via the procedures described in this chapter. The asymmetric surface-brightness distributions are then determined from the techniques described in Chapter 7, and an iterative procedure is used to refine the model.

A second example is the cataclysmic variable EX Dra = HS1804 + 6753, for which simultaneous ground-based optical observations and HST UV data were secured in 1996 (Figures 5.19 and 5.20). Whilst the optical R-band data demonstrate a fairly typical and uneventful/light curve

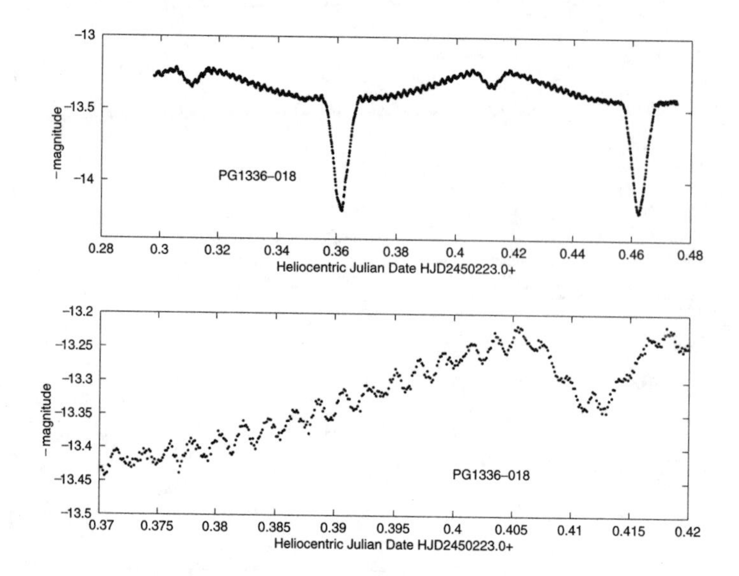

Fig. 5.15. Top: The remarkable discovery observations of a pulsating sdB star in an eclipsing binary, PG1336-018, made by Kilkenny et al. (1998). The reflection effect in this system is large because of the substantial temperature difference between the two stars. Bottom: Details of the phase range 0.37–0.42 illustrating that the pulsations continue through secondary eclipse and therefore are located on the primary sdB star. (Data obtained from Kilkenny et al., 1998.)

for a CV, with an obvious asymmetric eclipse of the white dwarf and an accretion hot spot on the outer part of the accretion disc by the Roche-lobe-filling companion, the UV data describe what is happening in the much hotter inner parts of the accretion disc around the accreting white dwarf. Here there is much evidence for rapid changes in the emitted flux density from this region, presumably linked to the process of accretion onto the white-dwarf surface. The scanning process inherent to the geometrical eclipses exhibited by binary stars can help to determine the locations of these sites of rapid variations in brightness.

The optical light curves for high-mass x-ray binaries and the infrared (IR) light curves of low-mass x-ray binaries reveal mainly ellipsoidal variations caused by the non-spherical shapes of the (usually) Roche-lobe-filling optical companions to the x-ray sources. For an HMXB the companion is an O or B supergiant, or a Be star, radiating most of its energy in the UV and optical regions, whilst for an LMXB the companion is a late-type dwarf star, seen most effectively in the near IR. Superposed on these ellipsoidal variations (Figure 5.21) there may be evidence of fluctuations in brightness from the inner parts of the accretion disc that may be dominant in the UV and influence the optical region, with fluctuations at levels of about 0.05 mag. Yet further examination of the averaged ellipsoidal variations will show that the two minima are not equal in most cases and often are reversed relative to that expected for a Roche-lobe shape. The x-radiation and UV radiation from the x-ray source and the inner accretion disc illuminate the facing hemisphere of the companion to provide a significant reflection-effect component to the total flux density. There is evidence also, for example, in LMC X-4 (Heemskerk and

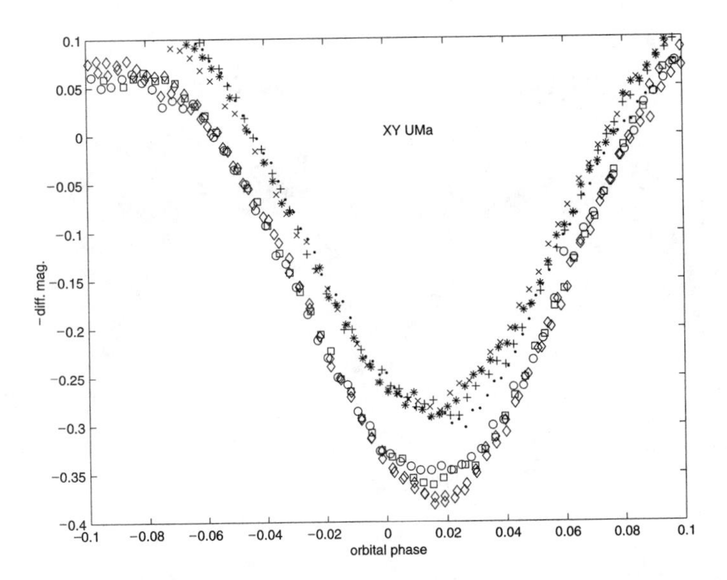

Fig. 5.16. Differential CCD photometry of the magnetically active RS CVn binary XY UMa in a narrow-band red filter. The orbital-phase range around mid-primary eclipse is plotted, showing the differences in eclipse shape over two observing seasons, 1997 (filled symbols) and 1998 (open symbols), relative to the same constant comparison star. The times of mid-eclipses are displaced from the adopted ephemeris used in 1995. Variations in the eclipse light curve are due to the presence of cool spots on the surface of the eclipsed star. (Data obtained from Lister et al., 2000a).

van Paradijs 1989), for precession of a tilted accretion disc, on time scales of about 30 days, that occults and reveals the x-ray source to the companion and thereby decreases and increases the reflection-effect component. An x-ray binary of intermediate mass, HZ Her = Her X-1, provides a more dramatic example of a tilted, precessing accretion disc.

Among the eclipsing x-ray binaries, the x-ray eclipses can display a range of shapes in the flux-density-versus-orbital-phase light curves. The ingress and egress phases may be nearly vertical, lasting less than $0.01P$, with the x-ray flux density approaching zero in the flat-bottomed total eclipse. Alternatively, the eclipses may be quite irregular and asymmetric, indicating that diminution of the x-ray flux density is occurring through absorbing columns both before and after the geometrical eclipse by the Roche-lobe-filling companion. Some systems, such as SMC X-1, exhibit both phenomena at different epochs, as indicated in the original *Uhuru* observations and in the recent *Ginga* and ROSAT data (Woo et al. 1995). The fact that the phase intervals between first and second contacts and between third and fourth contacts are so short demonstrates that the x-ray source must be very small; see equation (5.38).

As final examples of light curves for binary stars, Figures 5.22–5.24 illustrate optical and x-ray light curves for eclipsing polars. In these systems, a cool, low-mass star fills its Roche lobe and transfers mass through L_1 towards a white dwarf that has a strong magnetic field ($B \sim 10^4$ T). The gas in the stream is controlled by the magnetic field and may arch out

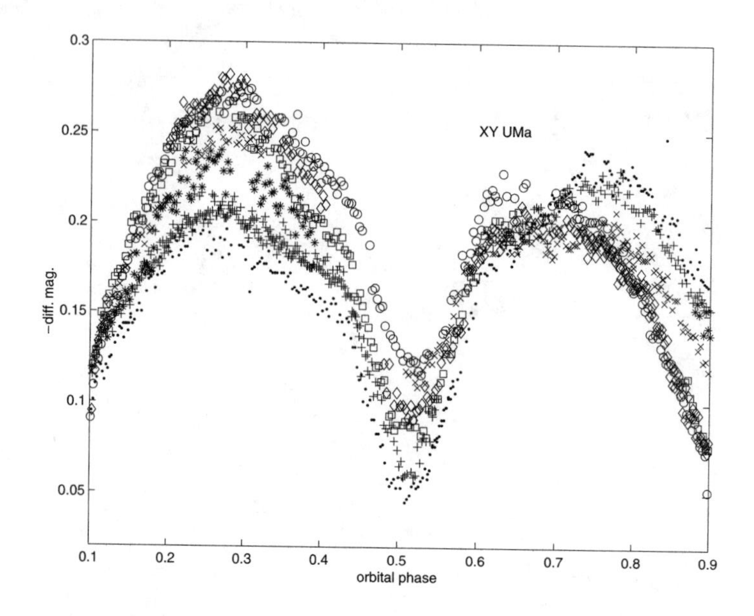

Fig. 5.17. Differential CCD photometry of the magnetically active RS CVn binary XY UMa in a narrow-band red filter. The orbital-phase range outside of the primary eclipse is plotted, showing the large changes in overall brightness of the system and in the shape of the light curve, caused by variable amounts of cool starspot activity on the primary star seen in 1997 (filled symbols) and in 1998 (open symbols). The same constant comparison star has been used throughout. (Data obtained from Lister et al., 2000a.)

of the orbital plane to impact close to the dominant magnetic pole on the white dwarf. The dominant source of x-rays and the major source of optical light is the accretion region at the end of the stream. There may be eclipses of the entire white dwarf by the companion, and of the accretion hot spot by the accretion stream. The system RE2107-05, HU Aqr, shows both phenomena in its orbital period of 0.086 day, whilst RXJ1802.1 + 1804 shows a stream eclipse in its 0.078-day period.

5.5.3 Eclipses of spherical stars: Basic analytical theory

This subsection provides a summary of the basic analytical theory that was developed in the first half of the twentieth century in order to derive the orbital inclination, the radii of the two stars in terms of the semimajor axis of the relative orbit, and the ratio of surface brightnesses. Because the theory is analytical, it serves as a stringent test of present-day numerical techniques when applied to spherical stars, and, in any event, some of the derived equations are useful for making preliminary estimates of parameters, as, for example, in the case of the x-ray eclipses in x-ray binaries.

In Figure 5.25, two stars are represented by their discs projected onto the tangent plane of the sky, with their centres separated by the distance a. The lengths $O_1N = y$ and $O_2N = x$ when the phase angle from mid-eclipse is $NO_2O_1 = \phi$. Thus, $x = a\cos\phi$, and $y = a\sin\phi$.

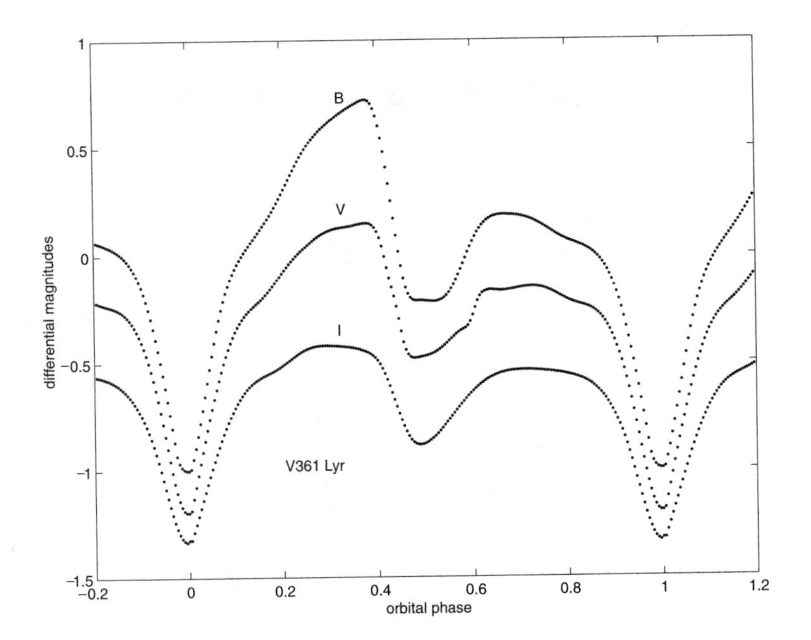

Fig. 5.18. Differential CCD photometry of the mass-exchanging binary V361 Lyr in B, V, and I pass bands. The I curve shows a typical ellipticity effect, caused in this system by the unusual situation of the main-sequence primary star filling its Roche lobe whilst the main-sequence secondary is detached. There is a small amount of distortion in the I curve around first quadrature, which is seen to become stronger in the V curve and very strong in the B curve. This distortion is interpreted, in conjunction with other spectroscopic data, as being due to a hot spot ($T \sim$ 9000–10,000 K) on the facing hemisphere of the secondary component caused by the impact of an accretion stream from the L_1 point. Note the effectiveness of the multi-colour photometry in establishing the temperature of that accretion region. (Data obtained from Lister et al., 2000b.)

If the orbit of the binary is observed to be at an inclination i, then we see that $x' = x \cos i$ and $y' = y$. The *projected separation* δ between the two stellar centres is then given by

$$\delta^2 = x'^2 + y'^2 = a^2(\cos^2 i \cos^2 \phi + \sin^2 \phi) \tag{5.40}$$

Defining the separation between the two stars as the unit of length, $a \equiv 1$, and defining $r_{1,2}$ in terms of a, so that they are called relative, or fractional, radii, and noting that $\cos^2 \phi = 1 - \sin^2 \phi$, then we obtain

$$\delta^2 = (\cos^2 i + \sin^2 i \sin^2 \phi) \tag{5.41}$$

An immediate application of this equation is to the special case of an x-ray binary in which there is a Roche-lobe-filling primary component and a much smaller x-ray source that is effectively a point-mass object. In such a binary, at the moment of first contact of the eclipse of the x-ray source by the companion, the projected separation δ equals the radius of the eclipsing star, and hence equation (5.41) provides a means of determining the orbital inclination if the

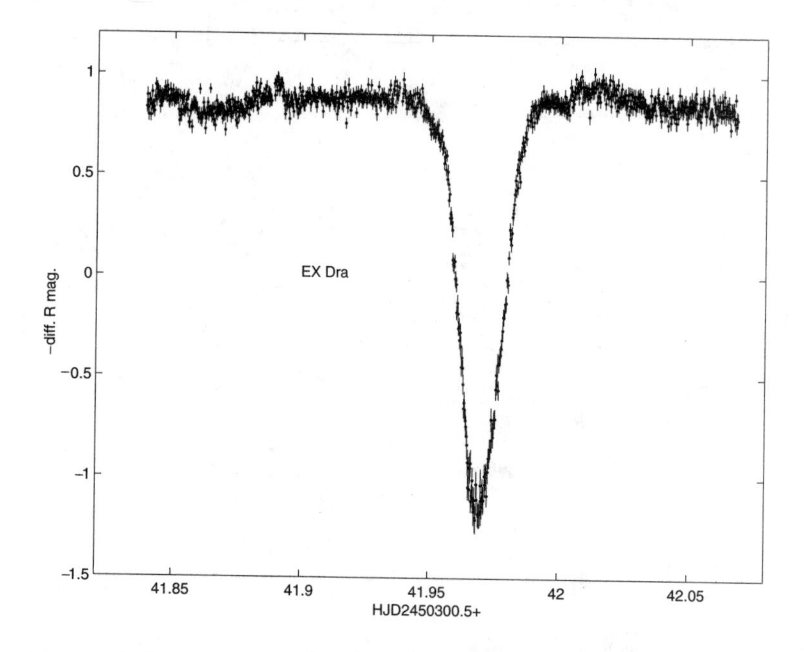

Fig. 5.19. Optical R-band CCD differential photometry of the eclipsing cataclysmic variable EX Dra = HS1804 + 6753, showing a typical asymmetric eclipse curve, and otherwise quite smooth variations with time. These data were obtained by D. Steeghs with the 0.9-m James Gregory Telescope at St. Andrews simultaneously with the HST UV observations that are shown in Figure 5.20. (From Steeghs et al., 1999, with permission.)

fractional radius of the companion is known. We have seen in Chapter 4 that the Roche model provides a value for the mean or volume radius of the Roche lobe in terms of the mass ratio q and the separation a of

$$r_L = \frac{0.49q^{2/3}}{0.69q^{2/3} + \ln(1 + q^{1/3})} \tag{5.42}$$

where the value of q is less than unity for the Roche lobe of the less massive star, and the inverse for the more massive star. Now the value of q is determined directly from the x-ray pulse-timing data and from the radial velocities of the companion star, as demonstrated in Chapter 3. Furthermore, the x-ray photometry provides accurate times or phases of the ingress and egress of the x-ray eclipse, thereby determining the value of ϕ at the first and fourth times of contact. Thus,

$$(\phi_4 - \phi_1)/2 = \phi_e \tag{5.43}$$

where ϕ_e is called the *eclipse half-angle*. Hence, we see that x-ray photometry determines i from

$$r_L = (\cos^2 i + \sin^2 i \sin^2 \phi_e)^{1/2} \tag{5.44}$$

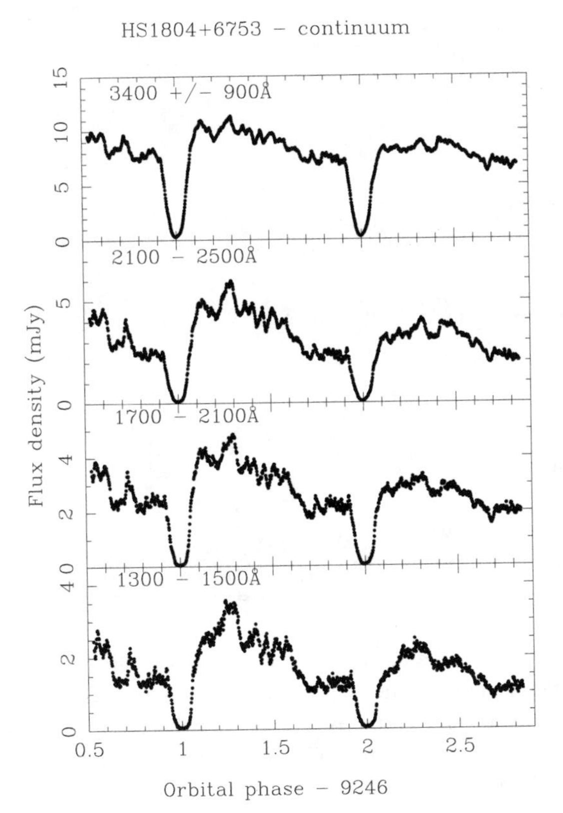

Fig. 5.20. HST UV observations in four bands of the CV, EX Dra, showing substantial short-time-scale variability in UV-flux densities that were not seen in the simultaneous optical data (Figure 5.19). (From Steeghs et al., 1999, with permission.)

Application of this technique by Primini et al. (1977) and Joss and Rappaport (1984) during x-ray photometry of SMC X-1, where ϕ_e was observed to lie in the range 26°.5–29°, yielded a value of $i = 68°.5–72°.5$ for $q = 0.06$. More recent investigations by Woo et al. (1995), which included modelling of the stellar wind in that system, have reduced these values for ϕ_e and i.

In general, however, both stars have substantial radii, and so we have to consider a wider prescription. We define a quantity called the *geometrical depth*, $p = (\delta - r_2)/r_1$, and the ratio of the radii, $k = r_1/r_2$, such that $k \leq 1$; then $\delta = r_2(1 + kp)$, and

$$\cos^2 i + \sin^2 i \sin^2 \phi = r_2^2(1 + kp)^2 \tag{5.45}$$

This is a fundamental equation relating the inclination of the orbit, i, and the two fractional radii $r_{1,2}$ to the geometrical-depth parameter p and the orbital phase ϕ. At the four times of contact, p equals $+1$, -1, -1, and $+1$, respectively. In a central eclipse, $\delta = 0$ at mid-eclipse, and $p = -1/k$.

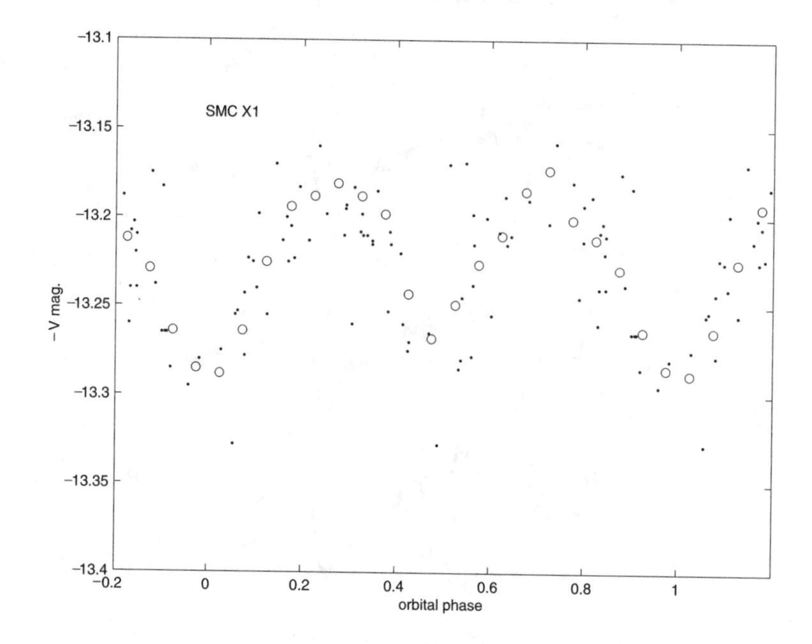

Fig. 5.21. The optical V light curve for the massive x-ray binary SMC X-1. The dots are individual values showing the substantial secular variability of the source superposed upon the mainly ellipsoidal variations, shown by the phase-binned averages (small circles) that are dependent upon the orbital phase. The RMS errors on the average values are typically 0.009 mag, about the size of the symbols used. (Data obtained from van Paradijs and Kuiper, 1984.)

The link between p and the actual amount of light lost at any phase within an eclipse requires a substantial mathematical investigation that has been well described in several monographs, such as those by Kopal (1946, 1950, 1959) and Russell and Merrill (1952). The reason for the complexity is that the projected discs of stars are not uniformly illuminated, but exhibit *limb darkening*, a decrease in the apparent surface brightness per unit area from the centre of the disc towards the limb, caused by the atmosphere of the star. A brief outline of the problem will enable the reader to appreciate why the mathematics becomes awkward, in addition to yielding some useful formulae.

The straightforward description of limb darkening that applies to normal stars is given in terms of the *linear limb-darkening law*, where the surface brightness at any point on the projected apparent disc of the star is

$$I = I_0(1 - u + u\cos\gamma) \tag{5.46}$$

where I_0 is the surface brightness per unit area at the centre of the disc, u is the limb-darkening coefficient, and γ is the angle between the surface normal at that point and the line of sight. The limb-darkening coefficient lies in the range $0 \leq u \leq 1$, with B stars having $u \sim 0.2$, and

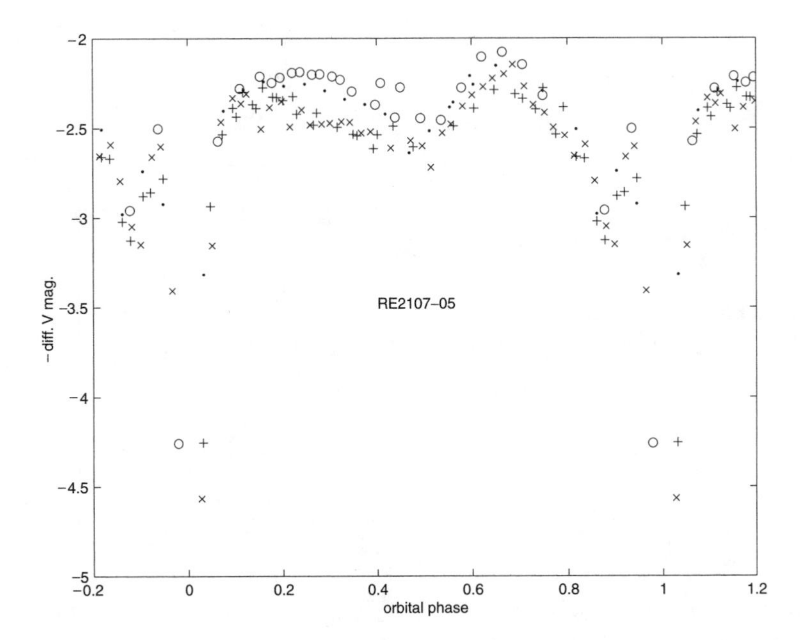

Fig. 5.22. Differential V CCD light curves for the eclipsing polar RE2107-05 = RXJ2107.90518 = HU Aqr, obtained by the author on 4 nights over a 10-night interval in 1993. The required integration time of 300 seconds limited the phase resolution, so that the deep eclipse of the white dwarf by the cool companion, lasting only 10 minutes, is not resolved. However, note that there is a second eclipse, repeated on all four nights, occurring at phase 0.875, which is attributed to eclipse of the accretion hot spot by the magnetically channeled accretion stream from the cool companion to the white dwarf. There are also substantial variations from night to night, and over seasons, with the overall brightness of the system varying by about 2 mag. Further details on this source, including phase-resolved photometry and spectroscopy, have been published by Hakala et al. (1993), Schwope et al. (1993, 1998), and Sohl et al. (1995).

G stars, $u \sim 0.6$. The total brightness of the entire apparent disc of radius r_2 is then

$$l_2 = \int_0^{r_2} (1 - u + u \cos\gamma) 2\pi r \, dr \tag{5.47}$$

$$= 2\pi I_0 (1-u) \frac{r_2^2}{2} + 2\pi I_0 u \int_0^{r_2} (\cos\gamma) r \, dr \tag{5.48}$$

At a position r on the projected disc surface, $r = r_2 \sin\gamma$, so that the second integral is easily found and gives the result that

$$l_2 = \pi r_2^2 I_0 [1 - (u/3)] \tag{5.49}$$

Outside of eclipses, the brightness of the binary system is constant for immaculate spherical stars, and so we can adopt that brightness level as the reference level for examining the eclipses.

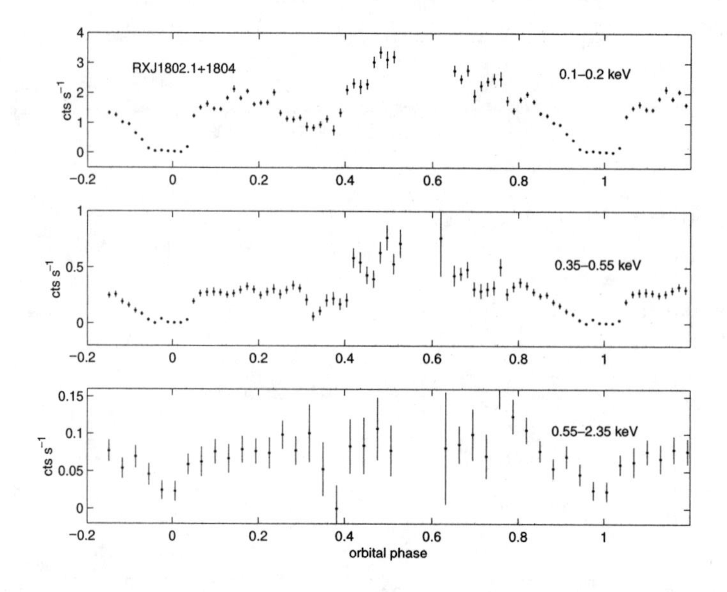

Fig. 5.23. The x-ray light curves for the x-ray-stream-eclipsing polar RXJ1802.1+1804 obtained by Greiner et al. (1998) in 1992–3 with the ROSAT PSPC in three energy bands. Note the clear but asymmetric eclipse, where the x-ray-flux density decreases to near zero in all energy bands. (Data obtained from Greiner et al., 1998.)

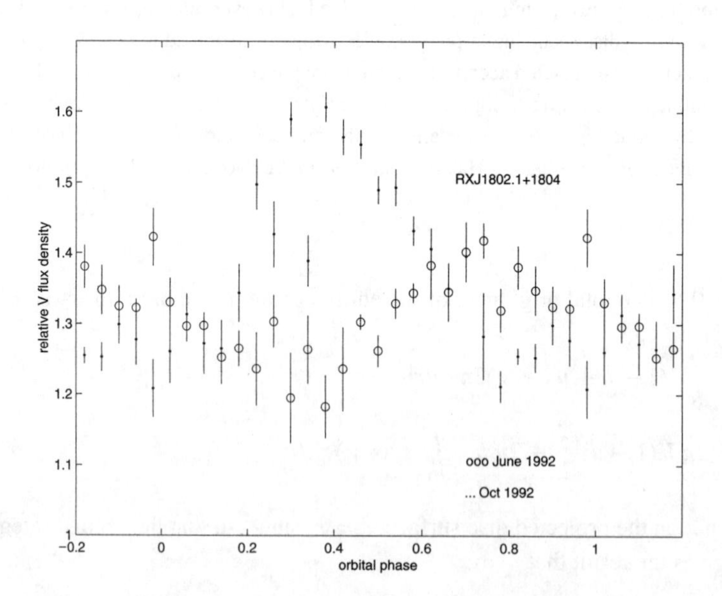

Fig. 5.24. The optical V-band light curves for the x-ray-stream-eclipsing polar RXJ1802.1+1804 obtained by Greiner et al. (1998) in June and October 1992, showing substantial changes in amplitude, shape, and phase of the light variations. (Data obtained from Greiner et al., 1998.)

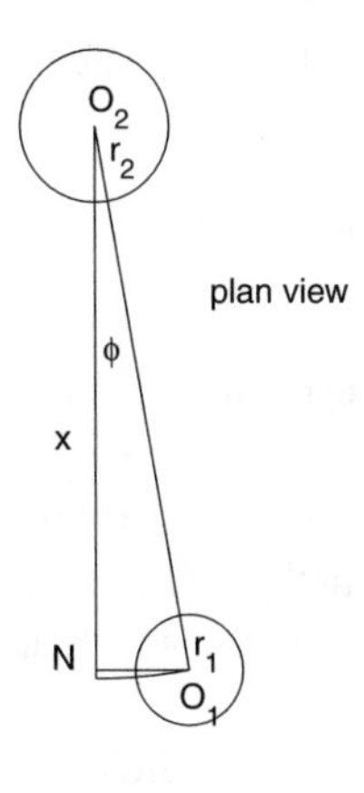

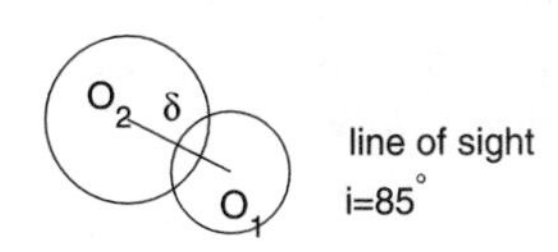

Fig. 5.25. Geometry of eclipses. In the *plan view*, the two stars, at O_1 and O_2, have radii r_1 and r_2, respectively. They are separated by a distance $a \equiv 1$ and are seen at a phase angle ϕ. This illustration is drawn for a transit of the smaller star across the face of the larger star; clearly, the two stars may be interchanged for an occultation. The *line-of-sight view*, here shown at $i = 85°$, demonstrates the two stars overlapping, and the projected separation between the centres of the two stars is δ at that phase angle.

The amount of light lost during any phase of an eclipse can be written as

$$\Delta l_2 = \int_S I \cos\gamma \, d\sigma \tag{5.50}$$

$$= I_0(1-u)\int_S \cos\gamma \, d\sigma + I_0 u \int_S \cos^2\gamma \, d\sigma \tag{5.51}$$

where the integral is taken over the *eclipsed area* of the star, and $d\sigma$ is an element of area on the eclipsed disc.

Now we define the *fractional loss of light* δf for star 2 when it is eclipsed by star 1 as

$$\delta f = \frac{\Delta l_2}{l_2} \tag{5.52}$$

and we can write

$$\pi r_2^2 \delta f = \frac{3(1-u)}{(3-u)}\int_S \cos\gamma \, d\sigma + \frac{3u}{(3-u)}\int_S \cos^2\gamma \, d\sigma \tag{5.53}$$

After some careful mathematics, it is found that this equation for the fractional loss of light by a limb-darkened star can be written as a function of two extreme cases, δf^U and δf^D, being the fractional losses of light by a uniformly illuminated star ($u = 0$) and by a completely limb-darkened star ($u = 1$), respectively, so that

$$\delta f = \frac{3(1-u)}{(3-u)}\delta f^U + \frac{2u}{(3-u)}\delta f^D \tag{5.54}$$

The expression for δf^U is established quite readily, and in the special case of an annular transit across a uniformly illuminated star, it is given by the simple expression $\delta f^U = (r_1/r_2)^2$, the ratio of the areas of the two discs. However, the determination of the expression for δf^D has been described by Kopal as requiring lengthy and troublesome algebra, and the final result is expressible only in terms of elliptic integrals, which are nevertheless analytic. Note that the mathematical complications arise solely because stars have limb-darkened projected discs. The final expressions for these fractional losses of light show that $\delta f = \delta f(r_1, r_2, \delta)$.

These δf functions all start at $\delta f = 0$ at the beginning of an eclipse, but thereafter they depend on the particular circumstances of an individual binary. To overcome this limitation, we can write $\delta f = \delta f_{2c}$ for the fractional loss of light at the moment of second contact and then introduce the *normalized fractional loss of light* α by

$$\alpha = \delta f / \delta f_{2c} \tag{5.55}$$

These α *functions* then can vary from zero to unity from first to second contacts, and they can be greater than unity only during an annular eclipse of a limb-darkened star. They are made to depend on the two independent parameters, $k = r_1/r_2$, with $r_1 < r_2$, and the geometrical depth p, where $-1 < p < +1$. Thus $\alpha \equiv \alpha(k, p)$, and they can be calculated for the two extreme cases, α^U and α^D, to find α for any value of u via

$$\alpha = \frac{3(1-u)}{(3-u)}\alpha^U + \frac{2u}{(3-u)}\alpha^D \tag{5.56}$$

Inversion of these tabulations then provides $p \equiv p(k, \alpha)$, which is put into the fundamental equation (5.45) relating (r_1, r_2, i) to the observed phase ϕ and the observed loss of light relative to the constant level outside of the eclipse.

Tabulations of the α functions were reported by Tsesevich (1939, 1940) and Merrill (1950, 1953) and were used, respectively, in the methods developed by Kopal (1950) and Russell and Merrill (1952) to evaluate (r_1, r_2, i) from the light curves of eclipsing binaries. [The Russell-Merrill method was developed from the original investigation by Russell and Shapley (1912a, b), and the references cited therein.] The two procedures adopted different approaches, with the Kopal method using least-squares solutions of n equations of the form in equation (5.45) for n observations within eclipses, and the Russell-Merrill method relying on graphical techniques via the nomograms published by Merrill (1953). Both required substantial effort to secure reliable estimates for these geometrical values.

Note that these methods deal only with the geometrical effects of eclipses by limb-darkened spherical stars. Thus, outside of eclipses, the brightness of the binary system is constant and is adopted as the reference level, so that the flux density from the binary is $f_{\mathrm{ref}} \equiv 1$. Within eclipses, the flux density f received from the system is smaller. If the observations were expressed in magnitude form, then we would write

$$m - m_{\text{ref}} = -2.5 \log_{10}(f/f_{\text{ref}}) \tag{5.57}$$

and this equation would be used to convert the magnitude differences $m - m_{\text{ref}}$ at each observed phase ϕ into relative flux densities $f/f_{\text{ref}} = f$ for comparison with the theoretical light losses due to eclipses. Clearly, if the original data were left in the form of relative flux densities, perhaps relative to a local constant comparison star, then it would be necessary only to form the flux-density ratio f/f_{ref}.

In the pass band used for the observations, at some effective or constant-energy wavelength, the eclipses tell us about the relative brightnesses of the two stars. Outside of eclipses, the two stars provide

$$l_1 + l_2 = 1 \tag{5.58}$$

whilst at some arbitrary phase within an eclipse of star 1,

$$f = 1 - \delta f l_1 \tag{5.59}$$

For a total eclipse, with an interval of constant brightness at the minimum, we see only the flux density f_{min} from the *eclipsing* star at that time. In these units, with $f_{\text{ref}} = 1$, a total eclipse gives $f_{\text{min}} = l_2$ if star 2 is the larger star, and hence $l_1 = 1 - f_{\text{min}}$. It then follows that the *ratio of the mean surface brightnesses of the two stars* is given by

$$\frac{J_1}{J_2} = \frac{1 - f_{\text{min}}}{k^2 f_{\text{min}}} \tag{5.60}$$

For partial eclipses, the maximum fractional loss of light at mid-eclipse becomes an additional unknown, so that the solution may be more difficult.

Once preliminary values for the geometrical elements $(r_{1,2}, i)$ have been determined, it is then possible to make use of the *differential-correction* procedure introduced in Chapter 3 for analyses of radial-velocity curves. Here one has n equations of the form in equation (5.59) for the n observations in an eclipse. Hence, in differential form,

$$\Delta f = -\delta f \, \Delta l_1 - l_1 \Delta(\delta f) \tag{5.61}$$

and

$$\Delta(\delta f) = \frac{\partial(\delta f)}{\partial r_1}\Delta r_1 + \frac{\partial(\delta f)}{\partial r_2}\Delta r_2 + \frac{\partial(\delta f)}{\partial i}\Delta i + \frac{\partial(\delta f)}{\partial u}\Delta u \tag{5.62}$$

All of the partial derivatives in equation (5.62) are specified analytically, so that this technique can be applied to the photometric observations in the same manner as for radial velocities.

These analytical techniques for determining the geometrical elements of an eclipsing binary (r_1, r_2, i) and the ratio of surface brightnesses (J_1/J_2) were used in preliminary graphical form in the early part of the twentieth century and had been developed into accurate techniques for spherical stars by about 1950. Subsequently they were used extensively as the techniques of photoelectric photometry and the quantum efficiency of photomultipliers were improved to permit light curves to be defined to uncertainties of about $\pm 1\%$, which demanded rigorous solutions of those light curves.

Attempts were made to broaden the application of this analytical theory to include stars that were non-spherical, so that sensible analyses could be made of systems displaying ellipsoidal variations. By replacing the spherical prescription with one based on triaxial ellipsoids, it was possible to calculate the out-of-eclipse variations caused by the continually changing projected areas of such bodies. As an analogy, consider turning a rugby football (or an American football) about its shortest axis. When the long axis is at a right angle to your line of sight, the projected area is at a maximum, and when it is in your line of sight, the projected area is at a minimum. Thus, ellipsoidal variations provide two maxima and two minima per orbital cycle. At the eclipse phases, when the long axes are close to the line of sight, the shape projected onto the sky is close to being circular, and hence the analytical theory for spherical stars *may* be applicable. The procedure adopted was to fit a least-squares sine/cosine series to the variations in brightness outside the eclipses and then divide the observed flux-density values by the sine/cosine-series values at each phase point in order to make the out-of-eclipse flux densities all close to unity. This procedure was called *rectification*. The sine/cosine series was extended across the eclipse phases as well, so that the eclipse shapes were altered to some extent, depending, obviously, on the amplitude of the ellipticity effect. With the observed light curve rectified, the data could then be analysed to yield (r_1, r_2, i) as before. These derived radii would be approximately the radii across the axis at right angles to the line of centres and in the orbital plane of the binary. These results were then combined with the analytical equations for ellipsoids to determine the shapes of the stars.

The procedure worked, at least for stars that were only moderately non-spherical. More distorted stars are not well approximated by ellipsoids, but are defined by the geometry of Roche surfaces, and application of the rectification procedure to those stars led to systematic errors in their derived sizes and shapes. In any event, there was an uneasy feeling associated with this rectification procedure, an aversion to forcing the observational data to fit the theoretical model that was being used to analyse them. It was therefore inevitable that new methods were sought to analyse the light curves of binaries with non-spherical stars, where the variations in brightness outside of eclipses were regarded as being as much a part of the description of the system as the eclipse shapes themselves. These new techniques required the use of digital computers because of the large number of calculations that would be required per observed phase point, and so their emergence was rapid during the late 1960s, when most universities and research establishments had acquired computers. Before considering these developments in the next subsections, we must digress briefly to take account of eclipses in binaries with eccentric orbits.

5.5.4 Eclipses in systems with eccentric orbits

We noted in Chapter 4, in the discussion of observations of apsidal motion (Section 4.2.1), that primary and secondary eclipses occur in an eccentric-orbit system when $\theta + \omega - \pi/2 = 0$

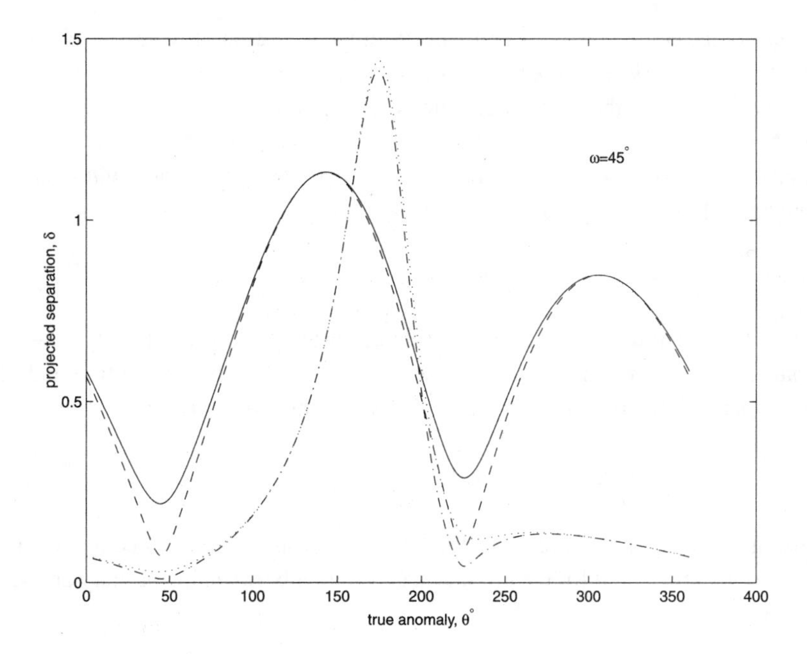

Fig. 5.26. The projected separation δ as a function of the true anomaly θ for binaries with inclination $i = 75°$ and eccentricities $e = 0.2$ (solid line) and $e = 0.9$ (dotted line); as well as inclination $i = 85°$ and eccentricities $e = 0.2$ (dashed line) and $e = 0.9$ (dash-dot line). Note that the minima in δ are close to the conjunction values $\theta = 45°$ and $225°$ in each case.

and $\theta + \omega - 3\pi/2 = 0$, respectively. We also quoted an equation for the projected separation between the centres of the two stars, δ, equation (4.9), that is somewhat different from equation (5.41), derived earlier. The two are reconciled simply by noting that the separation a in a circular orbit, in equation (5.41), has to be replaced by the instantaneous value of the separation, r', by means of the standard formula given in Chapter 2, and the phase angle ϕ for the circular orbit is now the sum $\theta + \omega$. Thus, the general expression for δ for an orbit of eccentricity e and longitude of periastron ω is

$$\delta = \frac{1 - e^2}{1 + e\cos\theta}[1 - \sin^2 i \sin^2(\theta + \omega)]^{1/2} \tag{5.63}$$

The form of this expression for $\omega = 45°$, inclination $i = 75°$ and $85°$, and $e = 0.2$ and 0.9 is shown in Figure 5.26. It is clear from the figure that the value of the true anomaly at each minimum is somewhat different, by about 1–2°, from the conjunction values of $\theta = 45°$ and $225°$ in this example.

It is easy to confirm, from equation (5.63), that when $i = 90°$ and $\delta = 0$, $\theta + \omega = 90°$ and $270°$. Whilst these effects are rather subtle, the position in terms of orbital phase $(t - T_{\rm ref})/P$ of the secondary eclipse in a standard light curve relative to the primary eclipse in an eccentric-orbit binary can be displaced substantially from the circular-orbit value of $\phi = 0.5$. The displacement depends upon the orbital eccentricity and the longitude of periastron. In addition, the *durations* of the eclipses will generally be different, as demonstrated by the observations in Figure 5.11 of the binary GG Lup, noted earlier. In order to calculate the sizes of such effects, it is worthwhile deriving the relevant equations for the simplest case of the orbit plane in the line of sight ($i = 90°$).

The link between orbital phase, or the mean anomaly η, and position within an eccentric orbit is given by Kepler's equation

$$\eta = \frac{2\pi(t - T_{\rm ref})}{P} = E - e\sin E \tag{5.64}$$

and we need to calculate the difference in η between primary eclipse, when $\theta + \omega - \pi/2 = 0$, and secondary eclipse, when $\theta + \omega - 3\pi/2 = 0$. The link between the true and eccentric anomalies, θ and E, can be made via the equations for the radius vector, namely,

$$r' = \frac{a(1 - e^2)}{(1 + e\cos\theta)} = a(1 - e\cos E) \tag{5.65}$$

and by using the preceding expressions for θ at primary and secondary eclipses. After some algebra, we obtain the result for the orbital-phase interval between the primary and secondary eclipses (Kopal 1959):

$$\frac{2\pi(t_{\rm sec} - t_{\rm pri})}{P} = \pi + 2\tan^{-1}\frac{e\cos\omega}{(1 - e^2)^{1/2}} + \frac{2e\cos\omega(1 - e^2)^{1/2}}{(1 - e^2\sin^2\omega)} \tag{5.66}$$

which is applicable only for values of i close to 90°. The left-hand side is determined directly from observations. The right-hand side can be rewritten in terms of the quantity

$$X = \pi + 2\tan^{-1}\frac{e\cos\omega}{(1 - e^2)^{1/2}} \tag{5.67}$$

so that

$$\frac{2\pi(t_{\rm sec} - t_{\rm pri})}{P} = X - \sin X \tag{5.68}$$

and the equation can be solved numerically for X via a Newton-Raphson iteration, just as for Kepler's equation. Thus the displacement of secondary eclipse from orbital phase 0.5 provides a determination of the quantity $e\cos\omega/(1 - e^2)^{1/2}$ that is independent of other determinations of e and ω from, for example, radial-velocity data. It was also shown by Kopal (1959) that the durations $d_{\rm pri}$ and $d_{\rm sec}$ of the two eclipses (that is, the time intervals between first and fourth contacts) provide a determination of the quantity $e\sin\omega$ from

$$e\sin\omega = \frac{(d_{\rm sec} - d_{\rm pri})}{(d_{\rm sec} + d_{\rm pri})} \tag{5.69}$$

again for the case when $i \sim 90°$. Thus (e, ω) are determinable from photometric observations alone, and these simple formulae can be used to establish preliminary values for entry to a more thorough solution for the entire light curve via present-day techniques. Clearly, equations (5.68) and (5.69) show that the displacement of secondary eclipse from orbital phase 0.5 is zero whenever $\omega = 90°$ or 270°, whilst the difference in durations of the two eclipses is maximal at that orientation. The binary system GG Lup illustrates this point very well, with secondary

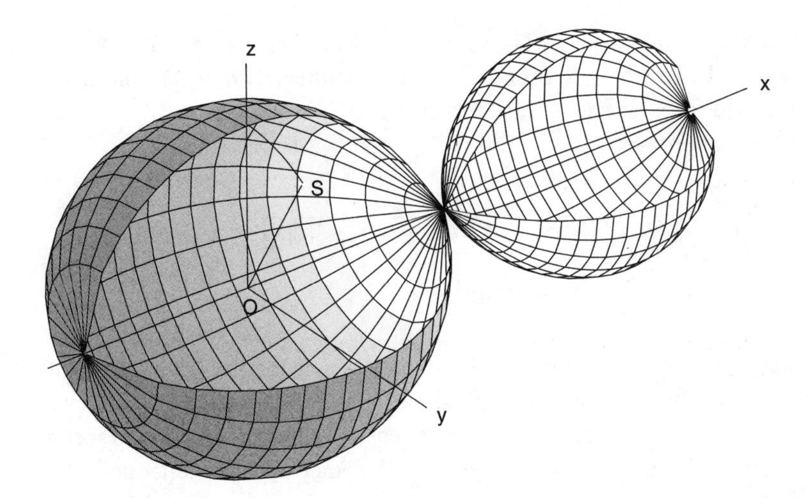

Fig. 5.27. A cutaway model showing the surfaces of stars just touching their Roche lobes for a mass ratio $q = 0.4$. The origin O of the coordinate system is at the centre of mass of the more massive star, with the x axis along the line of centres (positive towards the secondary star), the y axis in the orbit plane perpendicular to the x axis, and the z axis perpendicular to the orbit plane. At a point S on the surface of a star, $OS = \rho$, and $zOS = \theta$, so that $y = \rho \sin\theta$ and $z = \rho\cos\theta$.

eclipse at phase 0.5064, and the eclipse durations differing by about 0.03 in phase. The final values determined by Andersen et al. (1993) are $e\cos\omega = 0.0098$ and $e\sin\omega = 0.15$, so that $e = 0.150$ and $\omega = 86^\circ.4$.

Binary systems can have their orbits oriented to the line of sight such that they display only one eclipse, if the orbit has a sufficient eccentricity. Such is the case for the massive young binary NY Cep (Scarfe 1979; Holmgren et al. 1990b), where $e = 0.48$ and $\omega = 58^\circ$.

5.5.5 Modelling of non-spherical stars via the Roche model

Several prescriptions can be found for making numerical models of binary stars based on the Roche-model geometry that was introduced in Chapter 4. Here we describe the geometry presented by Mochnacki and Doughty (1972a), which can readily be used for all stars, whether in detached, semidetached, or contact states.

In Figure 5.27, the surfaces of stars that just fill their respective Roche lobes are shown for a mass ratio of $q = 0.4$. The coordinate system is centred on the centre of mass of the more massive star at O, and the x axis is along the line of centres, positive towards the less massive star, with the y axis at a right angle to the x axis and in the orbital plane, and with the z axis perpendicular to the orbital plane.

The expression derived in Chapter 4 for the normalized potential Φ_n that describes equipotential surfaces in the Roche model is

$$\Phi_n = \frac{2}{(1+q)r_1} + \frac{2q}{(1+q)r_2} + \left(x - \frac{q}{(1+q)}\right)^2 + y^2 \tag{5.70}$$

for a position $S(x, y, z)$, where $r_1 = (x^2 + y^2 + z^2)^{1/2}$, and $r_2 = [(x-1)^2 + y^2 + z^2]^{1/2}$. This equation can now be written in terms of cylindrical coordinates (ρ, θ, x), where $\rho^2 = y^2 + z^2$, and θ is measured from the z axis, such that $zOS = \theta$. Then $y = \rho \sin\theta$, $z = \rho\cos\theta$, and the expression for Φ_n becomes

$$\Phi_n = \frac{2}{(1+q)} \frac{1}{(x^2+\rho^2)^{1/2}} + \frac{2q}{(1+q)} \frac{1}{[(x-1)^2+\rho^2]^{1/2}} + \left(x - \frac{q}{(1+q)}\right)^2 + \rho^2 \sin^2\theta \tag{5.71}$$

To calculate the shapes of the two stars in a binary system, it is necessary to specify the mass ratio q and a radial distance from the origin of the coordinate system to a point on each stellar surface. In this cylindrical coordinate system, a sensible choice of location to specify Φ_n is on the line of centres, the x axis, at $\rho = 0$ and $\theta = 0$, namely, the substellar point for each star. A chosen value of x then permits Φ_n to be determined from equation (5.71), and that value of Φ_n specifies the equipotential surface describing the shape of the entire star. The selected value for x at the substellar point is the limiting value for $x = x_s$ for the stellar surface in that hemisphere. The limiting value for x on the averted hemisphere will occur at $x = x_b$, again on the line of centres. The value of x_b for the primary star is determined by finding the negative value of x that satisfies

$$\Phi(\rho, \theta, x) = \frac{2}{(1+q)} \frac{1}{(x^2+\rho^2)^{1/2}} + \frac{2q}{(1+q)} \frac{1}{[(x-1)^2+\rho^2]^{1/2}} + \left(x - \frac{q}{(1+q)}\right)^2 + \rho^2 \sin^2\theta - \Phi_n = 0 \tag{5.72}$$

with the value of $\Phi_n = \Phi_{n,p}$ already determined from the substellar point, where $x > 0$. Thus, we have $x_{\max} = x_s$ and $x_{\min} = x_b$ defining the extent of the primary star in the x direction. The surface of the star is then specified by cycling around $\theta = 0 \to 2\pi$, whilst stepping through $x = x_s \to x_b$, and determining each value of ρ by means of a Newton-Raphson iterative solution of equation (5.72), namely,

$$\rho_{j+1} = \rho_j - \frac{\Phi(\rho_j, \theta, x)}{\Phi_\rho(\rho_j, \theta, x)} \tag{5.73}$$

where

$$\Phi_\rho = \partial\Phi/\partial\rho = -\frac{2}{(1+q)} \frac{\rho}{r_1^3} - \frac{2q}{(1+q)} \frac{\rho}{r_2^3} + 2\rho \sin^2\theta \tag{5.74}$$

Each point on the stellar surface is then specified by $x = x$, $y = \rho\sin\theta$, and $z = \rho\cos\theta$.

For the secondary star, the starting point is again the substellar point on the x axis, where in this case $x_s = x_{\min}$. Again, Φ_n for the secondary is determined by use of equation (5.71) for the substellar point, and then the corresponding maximum value $x_{\max} = x_b$ is that value of x greater than 1 that satisfies the specified value of $\Phi_n = \Phi_{n,s}$. The same procedure is then followed to determine the values of (x, ρ, θ) and hence (x, y, z).

There is an upper limit to the size of each star, namely, the Roche limit that is specified on the line of centres by the location of the inner Lagrangian point L_1. The value of $x = x_{L_1}$ for a given mass ratio is determined from

$$x_{L_1,j+1} = x_{L_1,j} - \frac{\Phi_x\left(0, 0, x_{L_1,j}\right)}{\Phi_{xx}\left(0, 0, x_{L_1,j}\right)} \tag{5.75}$$

where

$$\Phi_x = \frac{\partial \Phi}{\partial x} = -\frac{2}{(1+q)}\frac{x}{r_1^3} - \frac{2q}{(1+q)}\frac{(x-1)}{r_2^3} + 2\left(x - \frac{q}{(1+q)}\right) \tag{5.76}$$

$$\Phi_{xx} = \frac{\partial^2 \Phi}{\partial x^2} = \frac{2}{(1+q)}\frac{3x^2}{r_1^5} + \frac{2q}{(1+q)}\frac{3(x-1)^2}{r_2^5} - \frac{2}{(1+q)}\frac{1}{r_1^3} - \frac{2q}{(1+q)}\frac{1}{r_2^3} + 2 \tag{5.77}$$

For completeness, we note two other partial derivatives that are required, in addition to the foregoing, for determining values of the local surface gravity in later subsections, namely,

$$\begin{aligned}\Phi_y &= \frac{\partial \Phi}{\partial y} \\ &= \rho \sin\theta \left(-\frac{2}{(1+q)}\frac{1}{r_1^3} - \frac{2q}{(1+q)}\frac{1}{r_2^3} + 2\right)\end{aligned} \tag{5.78}$$

$$\begin{aligned}\Phi_z &= \frac{\partial \Phi}{\partial z} \\ &= \rho \cos\theta \left(-\frac{2}{(1+q)}\frac{1}{r_1^3} - \frac{2q}{(1+q)}\frac{1}{r_2^3}\right)\end{aligned} \tag{5.79}$$

Other partial derivatives of Φ can be found from Mochnacki and Doughty (1972a).

To determine the shapes of contact binaries, where the common envelope lies outside the Roche-lobe surfaces, it is necessary to find only one common value of the normalized equipotential Φ_n. This task is achieved by specifying an additional parameter, given the illustrative name of *fill-out factor*. Unfortunately, there are two definitions of this term in common use. That used by Mochnacki and Doughty is the quantity

$$F = \frac{(\Phi_{n,1} - \Phi_n)}{(\Phi_{n,1} - \Phi_{n,2})} + 1 \tag{5.80}$$

where the subscripts 1 and 2 refer to the values of Φ_n at the inner contact surface passing through L_1 and the outer contact surface passing through L_2, respectively. Thus, a contact binary with the two stars just filling their Roche lobes has $F = 1$, whilst an overcontact system that fills the entire outer contact surface has $F = 2$. The alternative definition is due to Rucinski (1969), where

$$f = \frac{(\Phi_n - \Phi_{n,2})}{(\Phi_{n,1} - \Phi_{n,2})} \tag{5.81}$$

so that $1 \geq f \geq 0$, from inner to outer surfaces, and $F = 2 - f$. Typical contact binaries have approximate values $F \sim 1.1$–1.2, or $f \sim 0.9$–0.8, respectively.

Thus, when Φ_n is determined according to an adopted value of the fill-out factor, the maximum and minimum values of x for the common envelope can be found as before from equation (5.72), and they will lie on the averted hemispheres of the two components, at $x_{\max} > 1$ and $x_{\min} < 0$.

This parameterization of the Roche model has been used to generate the figures of binary-star shapes that are found in this text.

5.5.6 Limb and gravity darkening and the reflection effect

We noted earlier that a linear law of limb darkening can be adopted to represent the distribution of brightness across a stellar disc. As theoretical models of stellar atmospheres have been improved in overall accuracy, more detailed studies have been made of that distribution, including examinations of different limb-darkening laws. Whilst many investigations of eclipsing binaries have adopted linear laws as standard, with appropriate values of the limb-darkening coefficient, there has remained a question of whether or not that representation is sufficiently accurate. In his light-curve code LIGHT, Hill (1979) incorporated a polynomial interpolation of limb-darkening tables calculated from model atmospheres, so that values of $I(\mu)$ would be recovered for each surface element for any set of (T_{eff}, $\log g$, λ), where λ was the wavelength considered. The most comprehensive and up-to-date discussion of limb-darkening laws would seem to be that by Van Hamme (1993), which is based on the Kurucz (1991) ATLAS-model stellar atmospheres. He has found that a better representation of the brightness distribution is given by a logarithmic law of the form

$$I(\mu) = I_0(1 - u_a + u_a\mu - u_b\mu \ln \mu) \tag{5.82}$$

which follows closely the linear law from the centre out to angles of about 70° to the local surface normal, but deviates systematically beyond that value, and provides an excellent representation of the emergent intensity. Present-day photometric observations have reached the level of accuracy required to distinguish between these two limb-darkening laws, particularly when data of very high S/N are being used to establish surface-brightness maps, as discussed in Chapter 7. Van Hamme has provided values of the coefficients $u_{a,b}$ for the effective-temperature range $3500 \leq T_{\text{eff}} \leq 50{,}000$ K and the surface-gravity range $0 \leq \log g \leq 5.0$ from models with solar abundances only. The coefficients for linear and logarithmic laws are given for bolometric calculations, as well as for monochromatic and pass-band-specific calculations. There is, perhaps, some caution to be taken over whether or not some of the broad pass bands adopted are close to those recommended by Bessell (1995). For other chemical compositions, such as those in the LMC ($Z = 0.004$) and the SMC ($Z = 0.008$), it would be necessary to consider using the linear or quadratic values calculated by Wade and Rucinski (1985) from the Kurucz (1979) atmospheres with heavy-element abundances at 0.1 and 0.01 of the solar values.

For stars that are spherical and are unperturbed by the presence of magnetic fields or other surface anomalies, the limb-darkening law provides a complete description of the distribution of emergent intensity $I(\mu)$ in any direction μ. Then the emergent flux is an integral over the solid angle Ω, namely,

$$F = \int_{\Omega} I(\mu)\mu \, d\Omega \tag{5.83}$$

For stars that are non-spherical, the emergent intensity and flux F at each point are no longer independent of position on the stellar surface. It was shown by von Zeipel (1924) that the emergent flux at any point from a stellar atmosphere in radiative and hydrostatic equilibrium was directly proportional to the local gravity g, that is, $F \propto g$, and because the bolometric flux $F \propto T^4$, then the local temperature $T \propto g^{0.25}$. Thus, in general terms, it has become standard practice to write either that $F \propto g^{\beta}$ or that $T \propto g^{\beta}$, with $\beta = 1.0$ or 0.25, respectively, where β is referred to as the *gravity darkening exponent*. It is also referred to as the *gravity brightening exponent*, if one prefers to say that the emergent flux increases as the gravity increases, rather than that it decreases as the gravity decreases. Recall that the shape of a star is defined in the Roche model by an equipotential surface, so that $\vec{g} = \nabla\Phi_n$, and is given by the equations in the next subsection. In a helpful discussion by Rucinski (1989), it is commented that this value of β for radiative atmospheres may well be an upper limit for non-spherical stars, and it certainly has been the case that a number of researchers have attempted to determine the value of β from analyses of well-defined light curves for early-type binaries that have high-quality complementary radial-velocity curves. In general, it seems that $\beta = 0.25$ (when expressed in terms of temperature) is a realistic value to adopt for radiative atmospheres. For convective atmospheres, however, the dependence of emergent flux on local gravity was calculated by Lucy (1967) to be much less significant, with $F \propto g^{0.32}$, or $T \propto g^{0.08}$. For 20 years or so after that, various observational tests were used to examine that proposed value of β for convective atmospheres, with particular attention being paid to the contact binaries of the W UMa type, which have the most distorted shapes amongst all binaries. Rafert and Twigg (1980) concluded from light-curve analyses of several systems that the gravity darkening exponent for convective envelopes lay in the range $0.04 \leq \beta \leq 0.12$, approximately confirming the theoretical value. Eaton et al. (1980) derived the value $\beta = 0.03 \pm 0.01$ from UV photometry of the prototype system W UMa, whilst Hilditch (1981) used DDO photometry to study four systems and found that $\beta \sim 0$. However, in those latter studies it was clear that the non-symmetric shapes of several of the light curves, evidently caused by the presence of starspot activity on those stars, were hampering the solutions to the light curves and rendering the derivations of β values somewhat questionable (Rucinski 1989). A further theoretical investigation of the problem was undertaken by Sarna (1989), who concluded that Lucy's original study had determined the correct value of $\beta = 0.08$ for convective atmospheres. Most recently, Pantazis and Niarchos (1998) have reinvestigated the observational determination of β for radiative and convective atmospheres by studies of 36 systems with well-defined symmetric light curves and have concluded that the theoretical values of $\beta = 0.25$ or 0.08, respectively, are consistent with the available data. In a new development, Claret (1998) has calculated new limb-darkening coefficients and new gravity darkening exponents from his stellar models. He has shown that the foregoing canonical values of β are sensible, but that, for example, during the evolution of a 2-$M_{\odot}$ star across the main sequence and into the giant region, the value of β will change smoothly between limits of about $0.25 > \beta > 0.05$ because of substantial changes in its effective temperature, whilst a 1-$M_\odot$ star will maintain $\beta \sim 0.09$ across the main sequence.

In conclusion, gravity darkening is an effect that is additive photometrically to the variations in projected area of a non-spherical star during an orbital cycle. For stars with convective atmospheres, with $\beta = 0.08$, that addition is small, but is nevertheless significant at the 2–3% level. For stars with radiative atmospheres, like the Roche-lobe-filling companions to the x-ray sources in HMXBs, that addition is substantial, and because the local gravity is lower near the substellar point than on the averted hemisphere, that addition leads to a substantial difference in the minima of the ellipsoidal variations exhibited by those objects in the *UV* and optical regions.

The remaining effect to be discussed in this subsection is the reflection effect, about which much has been written over several decades. The term is used to describe the mutual irradiation of the facing hemispheres of the two stars in a binary system, even though, in general, the physical process is not one of true reflection of incoming radiation. For stars of intermediate temperatures, $T \sim 10^4$ K, incoming radiation is understood from theoretical studies to be completely absorbed by the atoms and ions in the irradiated atmosphere, thereby altering the temperature structure within the atmosphere and resulting in increased emergent intensity and flux, such that the bolometric *albedo* $\alpha = 1$. For higher temperatures, the density of free electrons in the atmosphere increases, so that the proportion of the irradiating flux that undergoes wavelength-independent Thomson scattering increases, reaching about 50% of the total by $T \sim 30{,}000$ K (Hutchings and Hill 1971b); the remainder is absorbed and re-emitted, so that $\alpha \sim 0.5$ for that temperature. For the cooler stars with convective atmospheres, $T \leq 6500$ K, theory suggests that $\alpha \sim 0.5$, whilst the remainder of the incoming radiative energy is converted to bulk motions of the atmosphere gases (Rucinski 1969). There have been several observational attempts to determine α for convective atmospheres, and all have concluded that a value of about 0.5 is a sensible estimate (Rucinski 1989). So the term *reflection effect* is really applicable only to that subset of hot binaries where electron scattering plays a dominant role in the total opacity of the irradiated atmosphere.

The most comprehensive review of the historical development of the subject is probably that presented by Vaz (1985). In two subsequent papers, Vaz and Nordlund (1985) and Nordlund and Vaz (1990) have discussed theoretical calculations of the response of an atmosphere to irradiation, concluding that (1) the bolometric albedo for a radiative atmosphere at intermediate temperatures is unity, whilst that for a convective atmosphere is very close to 0.5, and (2) the usual decrease in temperature with decreasing optical depth through a photosphere that is not irradiated can be changed by irradiation to where the temperature will *increase* for decreasing optical depths less than unity, that is, a temperature inversion. The same effect was found by Brett and Smith (1993) in their investigation of irradiation of a dwarf-M-star atmosphere, such as those found in CVs. Those temperature inversions occurred at small optical depths of $\tau \sim 0.01$–0.001 in those calculations, where the irradiating star was taken to have a quite low temperature of $T \sim 10,000$ K. It would be very interesting to know what would be calculated for the cases of sdO stars (with $T \sim 70,000$ K) irradiating dwarf-M-star atmospheres, as seen, for example, in KV Vel (Hilditch et al. 1996), where the M dwarf displays emission lines of C IV (see Figure 3.14 in Chapter 3).

Practical calculations of the reflection effect are necessarily fundamentally different from those involving limb and gravity darkening. Whilst these latter effects can be included for calculating the flux seen by a distant observer independently of the other star, apart from the effects of eclipses, the calculation of the reflection effect intimately involves the visibility of the irradiating star from *every* surface element on the facing hemisphere of the irradiated star, and vice versa. For the surface element at the substellar point, the irradiating star is seen directly overhead at all times as a limb-darkened disc, and the total received flux density is equal to that from a point source obeying the inverse-square law. As one moves away from the substellar point, the irradiating star sinks lower in the local sky, and more and more is eclipsed by the local horizon. By the time one reaches azimuth values around $A \sim \pm 90°$, that is, the azimuth circle through both poles at right angles to the line of centres, most or all of the irradiating star will be below the local horizon. Only if the irradiating star is larger than the irradiated star will there still be irradiated surface elements beyond $A \sim 90°$, and these will see only the extreme and darkened limb of the irradiator.

Clearly, the heating effect of the radiation on the irradiated photosphere is maximal at the substellar point and decreases to zero for all points around the aforementioned azimuth circle. So the effects of the irradiation are just the same as the Sun's effects on the Earth's surface at the equinoxes – scorchingly hot on the equator at midday, and seriously cold at the poles. This common observation does seem to have been forgotten in some of the representations of the reflection effect, where the irradiated star has been treated as an illuminated flat disc, with the polar regions receiving as much radiation as the substellar point! Thus the amplitude of the calculated reflection effect, not surprisingly, has been grossly over-estimated, by factors of 2–3.

So, let us be realistic. The reflection effect can be calculated numerically by means of various prescriptions, all of which stem from the equation (Kopal 1959) that the irradiation of a surface element is

$$F_{\rm irr} = \int\int \frac{I(\cos\gamma_1)\cos\gamma_1\cos\gamma_2}{d^2} dS \tag{5.84}$$

where the double integration is taken over all surface elements dS on the irradiator that are visible from the element on the irradiated star. The specific intensity $I(\cos\gamma_1)$ is that sent in the direction γ_1 to the local surface normal and received at the element at an angle γ_2 to the local surface normal at that element; the distance between the two surface elements is d. Figure 5.28 illustrates the geometry of the reflection effect and clearly shows that surface elements on the two stars will be mutually irradiating if and only if $\cos\gamma_1 > 0$ and $\cos\gamma_2 > 0$.

The original numerical calculations of the reflection effect that were incorporated into the first light-curve synthesis codes faced the problem of making summations over $n > 1000$ elements on the irradiating star for *every element* on the irradiated star. With perhaps $n > 1000$ irradiated elements, a straightforward addition of all the contributions then depended on n^2, whereas the calculation of the light curve depended only on n. Because of the limited computing power of that time (1968–1970), approximations were adopted to speed up those calculations, and substantial discussion of those approximations appears in the relevant papers, together with tests showing that such approximations provided results that were accurate to about $\pm 5\%$ or so. Certainly the effects of irradiation for a binary system in a circular orbit need to be calculated only

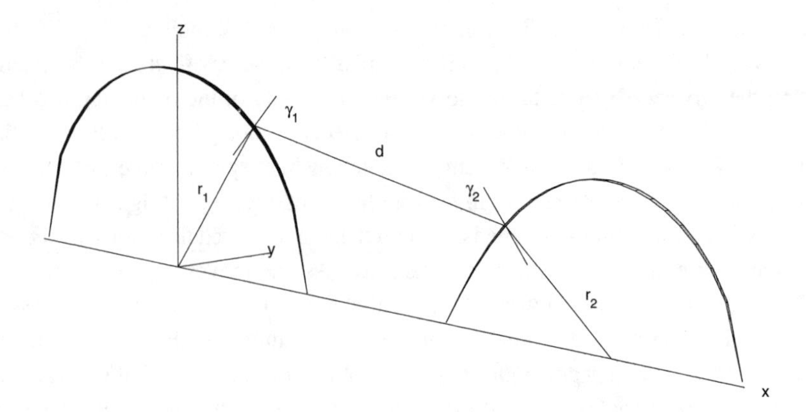

Fig. 5.28. A slice through a semidetached binary with $q = 0.3$ along the x–z plane, with $z > 0$, in order to illustrate the calculation of the reflection effect. A surface element on star 1 is at a radial position r_1 and at a distance d from a surface element on star 2 at a radial position r_2, both located in the x–z plane in order to simplify the diagram. The angles between the local surface normals and the directions to the elements are $\gamma_{1,2}$, respectively.

once, because the mutual separation does not change. Additionally, the problem is symmetric about all four quadrants, although care is needed to assign the correct signs to various angles. For eccentric-orbit binaries, the problem is lengthened substantially because of the continuously changing separation. Wilson (1990) has rediscussed the techniques to achieve improved accuracy and efficiency in numerical computation of the reflection effect for light curves. That paper has set a new standard for accuracy, including the need to iterate to convergence the multiple reflections inherent in the mutual irradiation of the two components. Cranmer (1993) has also discussed the need to iterate the effects of mutual heating of components.

Thus, Wilson uses

$$F_{\mathrm{irr}} = \alpha_2 \sum\sum \frac{\Re_1 \Gamma_2 I(\cos\gamma_1)\cos\gamma_1\cos\gamma_2}{d^2}\Delta S \tag{5.85}$$

where α_2 is the bolometric albedo of star 2, and $\Re_1$ is the reflection factor for star 1, being the ratio of the bolometric radiated flux (including any that results from irradiation by the companion) to that emitted when there is no irradiation from the companion; that is, $\Re = 1$ when there is no irradiation. The quantity Γ_2 is the gravity darkening factor, the ratio of the local bolometric flux at the particular surface element to that at some reference element, commonly taken to be the rotational pole. The irradiation of star 1 by star 2 has the same equation with the subscripts interchanged. The values for the bolometric albedo used in the foregoing equation range from unity for radiative atmospheres to 0.5 for convective atmospheres. Wilson (1990) did not discuss whether or not the unit albedo for radiative atmospheres should be modified in the presence of substantial electron scattering, whereas Hill (1979) adopted the prescription from Sobieski (1965) that allows specifically for both electron scattering and absorption and thereby a reduced value for the bolometric or heat albedo.

As an illustration of the effects of irradiation on the facing hemisphere of one component in a binary system, Figure 5.29 shows the variation, as a function of orbital phase, of the

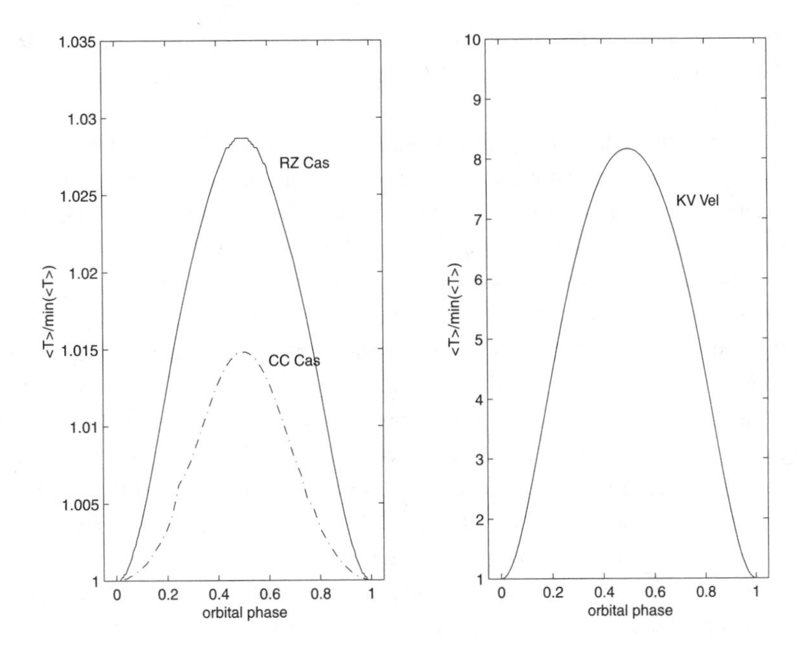

Fig. 5.29. Plots of the ratio of the mean (flux-weighted) temperature $\langle T \rangle$ to the minimum value $\min(\langle T \rangle)$, integrated over the visible hemisphere seen at different orbital phases of the cooler components in three binary systems: Left: CC Cas, a typical O+B-type detached binary; RZ Cas, a typical A+K-type Algol system. Right: The extreme-reflection-effect system KV Vel. Note the difference in the vertical scales between the two panels.

mean, flux-weighted temperature of the cooler component in each of three binary systems, as calculated by the light-curve synthesis code LIGHT of Hill (1979). For most binaries, the reflection effect adds only a few percent to the total brightness of the system. But for some systems with very extreme temperature differences between the sources, the reflection effect can be dominant, as for KV Vel. The same types of variations are seen in the x-ray binaries, where the x-irradiated hemisphere of the companion is considerably hotter than the averted hemisphere. For example, in the x-ray binary, HZ Her, the irradiated hemisphere has a spectral type of early B, together with emission lines of C III and N III, whilst the averted hemisphere displays a late-A-type spectrum (e.g., Reynolds et al. 1997). Amongst x-ray binaries, there can also be further complications, where the tilted and precessing accretion disc may intercept the x-radiation from the compact source, resulting in varying amounts of x-ray heating of the facing hemisphere of the Roche-lobe-filling companion (e.g., Heemskerk and van Paradijs 1989).

5.5.7 The synthesis of symmetric light curves

Whilst the Roche model was invoked to provide the basis for the Kopal (1955) classification scheme (detached, semidetached, contact) for binary stars and became central to the development of the theories of mass transfer and of binary-star evolution in the 1960s, it was not until Lucy's papers (1968a,b) that the Roche model was used as the basis for calculating a theoretical light curve for a binary star. Lucy's model was developed to explain the light curves for contact binaries like W UMa by introducing the concept of a common convective envelope surrounding both stellar cores. Thus, whilst the two stars had very different masses, the observed binary showed very little change in surface temperature as the binary revolved in its orbit. The shape of the common convective envelope was described by an equipotential surface lying somewhat outside the individual Roche lobes for the two stars.

The term *light-curve synthesis* is used to describe the procedure whereby a light curve for a binary system is calculated via a theoretical model, with values of various parameters for the binary being specified as input values for that model. The theoretical model may include all the known physical processes that define the temperatures at all the surface elements on each star, as well as the geometrical effects of orientation to the observer's line of sight, the orbital phase angle, and the visibility of each surface element at each phase angle. A light curve is regarded as symmetric if the light variations occurring between phases 0.5 and 1.0 are the mirror images of those between 0.0 and 0.5; the light curve can be folded about phase 0.5, and both halves of the curve should overlap completely.

Lucy's model used Roche geometry to define the shapes of the stars and grey atmospheres to define the temperature as a function of optical depth τ by the Milne-Eddington solution

$$T^4 = \frac{1}{2} T_{\text{eff}}^4 \left(1 + \frac{3}{4} \tau \right) \tag{5.86}$$

and hence the emergent intensity at each angle according to the Eddington-Barbier relationship

$$I_\lambda(\mu) = B_\lambda[T(\tau = \mu)] \tag{5.87}$$

where B_λ is the Planck function at wavelength λ and temperature T. Gravity darkening was included via $T/T_{\text{mean}} = (g/g_{\text{mean}})^{0.08}$, whilst the reflection effect for such stars with nearly equal values of T_{eff} was ignored.

Lucy's work was followed very quickly by other light-curve synthesis models that became progressively more generalized. Thus, Hill and Hutchings (1970), Hutchings and Hill (1971a,b), and Wilson and Devinney (1971) presented general numerical codes for the synthesis of light curves for eclipsing binary stars of all types that were based on the Roche geometry and featured complete accounts of gravity darkening, evaluations of the reflection effect that were shown to be accurate to better than $\pm 5\%$, and the options of using emergent fluxes, specific intensities, and limb darkening calculated from Planck functions with grey atmospheres or from model-atmosphere grids. Mochnacki and Doughty (1972a,b) adopted the cylindrical coordinate system already introduced for the Roche geometry and considered the light curves for W-UMa-type contact systems again, with the addition of mutual irradiation and calculations of the variations in absorption-line profiles due to the variations in local surface values of temperature and gravity.

To calculate a light curve, it is necessary first to specify the line of sight of the observer in terms of the orbital phase angle ϕ and the orbital inclination i. It is usual to follow Kopal and

Kitamura (1968) to write the direction cosines

$$\begin{aligned} l_0 &= \cos\phi \sin i \\ m_0 &= \sin\phi \sin i \\ n_0 &= \cos i \end{aligned} \tag{5.88}$$

so that the line of sight is defined by the unit vector $\hat{l}_s = (l_0, m_0, n_0)$ (Mochnacki and Doughty 1972a,b), which, of course, changes continuously as the binary rotates, or, if you prefer, as you walk around the binary. Then for a given surface element, the angle between the local surface normal and the line of sight is given by

$$\cos\gamma = -\hat{l}_s \vec{g}/g = -\frac{(l_0\Phi_x + m_0\Phi_y + n_0\Phi_z)}{\left(\Phi_x^2 + \Phi_y^2 + \Phi_z^2\right)^{1/2}} \tag{5.89}$$

where $\vec{g} = \nabla\Phi_n$ is the surface-gravity vector of magnitude g, and $\Phi_{x,y,z}$ are the partial derivatives introduced in equations (5.77)–(5.79). Thus $g = (\Phi_x^2 + \Phi_y^2 + \Phi_z^2)^{1/2}$. It is usual practice to express g in terms of its value at the rotational pole of the star concerned. The area of each surface element at the position (ρ, θ, x) is then

$$\Delta S = -\frac{g\rho\Delta\theta\Delta x}{(\Phi_y \sin\theta + \Phi_z \cos\theta)} \tag{5.90}$$

Then the total emitted flux in the direction of the line of sight is the summation over all visible surface elements of

$$F = \sum_j I_j(\cos\gamma)\cos\gamma\Delta S_j \tag{5.91}$$

where the specific intensities emitted in the direction γ will include all contributions discussed earlier from the effects of limb darkening, gravity darkening, and the reflection effect that alter the local temperature observed. Adopting the mean effective temperature of the star, $T_{\rm eff} = (L/\sigma S)^{1/4}$, for a star of total luminosity L and total surface area S, then the polar temperature can be specified by

$$T_{\rm pole} = T_{\rm eff}\left(\frac{S}{\sum_j \vec{g_j}^{\beta}\Delta S_j}\right)^{0.25} \tag{5.92}$$

so that the local temperature due to the effects of gravity darkening, for example, is given by $T/T_{\rm pole} = (g/g_{\rm pole})^{\beta}$.

Various numerical techniques are used to determine whether or not a particular surface element on a star is visible at a particular phase angle. At a given orbital phase, the star that is nearer the observer, the eclipsing star, is considered first. Those surface elements with negative values of $\cos\gamma$ lie on the averted face of the star and need not be considered further. Those with positive $\cos\gamma$ do contribute to the total flux observed. Wilson and Devinney used the set of surface elements just beyond the horizon of the eclipsing star to define the shape of the eclipser projected onto the sky at that phase angle. For the star that is farther away from the observer, the eclipsed star, the same procedures apply, except that it is necessary to determine also whether or not any of the surface elements with $\cos\gamma > 0$ have a line of sight to the observer that passes through the eclipsing star. The projected distance onto the sky of the surface element on the eclipsed star, using the centre of the eclipsing star as the origin of the coordinate system, is compared with the shape of the eclipser at that phase. If the projected distance is less than the radial size of the eclipser, then the surface element is not visible at that phase angle and does not contribute to the total flux observed. Mochnacki and Doughty used a somewhat different algorithm. The same tests were made on the values of $\cos\gamma$ for each surface element, and those on the eclipsed star that were facing the observer were examined in the following way. Along the line of sight, the value of the normalized potential Φ_n is sampled at discrete points. If any of these values are greater than the value of Φ_n that defines the surface of the *eclipsing* star, then clearly the line of sight is passing through the eclipsing star, and hence that surface element on the eclipsed star is not visible. Linnell (1984) introduced his own program for light-curve synthesis in a paper that details, extensively, the mathematical formalism that he used for the adopted spherical polar geometry and the stellar physics. In that paper, he adopted a modification of the Mochnacki and Doughty method for establishing eclipsed surface elements in order to avoid certain possible errors. For all of those numerical codes, tests were made of the synthesized light curves, with the exact analytical results based on the α-functions for spherical stars. Agreements were secured to uncertainties of ± 0.001 mag or better. In addition, several independent checks were made concerning the interagreement between those various codes, and all proved very satisfactory.

The input parameters that define a light curve are as discussed earlier in this chapter: the polar or mean radii of the two stars, $r_{1,2}$, expressed in terms of the semimajor axis of the relative orbit; the orbital inclination i; the mass ratio q; the polar or mean temperatures $T_{1,2}$; the albedos $\alpha_{1,2}$; the gravity darkening exponents $\beta_{1,2}$; and the limb-darkening coefficients $u_{1,2}$ (and extra terms for eccentric orbits). Thus, the variations in flux density received by the observer that are due to the binary orbital motion and are expressed relative to an adopted level are given by

$$\Delta f = \Delta f(r_{1,2}, i, q, T_{1,2}, \alpha_{1,2}, \beta_{1,2}, u_{1,2}) \tag{5.93}$$

a total of $m = 12$ parameters. The dominant, or controlling, parameters for the shape of the light curve, $\Delta f \sim \phi$, are evidently those that define the sizes and shapes of the stars ($r_{1,2}, q$), the orbital inclination i, and the ratio of surface brightnesses, which is related to the two temperatures $T_{1,2}$. The effects of varying $\alpha_{1,2}$, $\beta_{1,2}$, and $u_{1,2}$ are more like second-order effects.

Observed light curves are defined by sets of n observations of (ϕ, f), where f is a flux-density ratio between the binary star and one or more local comparison stars. A reference value of $f = f_{\mathrm{ref}}$ is adopted, usually at one of the quadrature phases, $\phi = 0.25$ or 0.75, but the reference value may be anywhere within the light curve as long as it is used consistently for both the observations and the theoretical light curve that is calculated from an initial set of the m parameters. Thus we can form $\Delta f = f - f_{\mathrm{ref}}$ and make a direct comparison between the observed light curve and the calculated light curve. Provided that the initial set of m parameters is chosen carefully, the differences $(O - C)$ between the two curves will be small, and so a solution of the observed light curve can be sought by means of the procedure of *differential corrections* that was introduced in Chapter 3 for the solution of radial-velocity curves. In that latter case, the partial derivatives of a radial-velocity curve with respect to each of the orbital parameters were specified by analytical equations. In this case of light curves for non-spherical stars, such differentials cannot be specified analytically, but they can be calculated numerically by means of difference equations such that for the dependence of Δf on the parameter $p_i, i = 1, \ldots, m$, we have

$$\frac{\partial \Delta f}{\partial p_i} = \frac{\Delta f(p_i + \frac{1}{2}\delta p) - \Delta f(p_i - \frac{1}{2}\delta p)}{\delta p_i} \tag{5.94}$$

To form these difference terms, it is necessary to calculate a set of light curves with each parameter p_i varied in turn by $\pm\frac{1}{2}\delta p_i$. These simple differentials are linear and can represent only the local region of the $\chi^2 = \sum(O - C)^2/(n - m)$ m-dimensional hypersurface for m light-curve parameters around that solution.

Such differential-correction procedures were used in the earliest light-curve synthesis codes and were continued within the Wilson-Devinney (WD) code and the Rucinski (1973, 1974, 1976a,b) WUMA3 code for contact binaries. To make them work effectively, it is necessary to make good estimates of the initial parameters by using all the observational evidence available, not just the shapes of the light curves at different wavelengths. Thus, in the next two subsections, we discuss the need for careful evaluations of colour indices and spectra that can define the effective temperatures of the stars, as well as the need for good-quality radial-velocity curves for both components that can define the mass ratio more precisely than can be achieved through light-curve synthesis, although care must be exercised in removing possible systematic errors, generally called *non-Keplerian corrections*. These general comments apply to all techniques for the solution of light curves, and time spent carefully assessing all the observational data before embarking on a light-curve solution is, in my experience, time spent very productively, and certainly advantageously.

The χ^2 hypersurface will generally contain a number of 'local' minima at different combinations of the light-curve parameters. Several parameters have correlated behaviours, the most obvious and problematic being those between the orbital inclination i and the mass ratio q, and between r and u. Recall that the *shapes* of the stars of given sizes are defined by q in the Roche model. Thus a light curve defined by a binary with a given q and i may be very similar to that for a binary with slightly more distorted stars but viewed at a slightly lower inclination, or less distorted stars at a higher inclination. A spectroscopic mass ratio can remove this ambiguity, if it is determined carefully. If a differential-correction procedure is used far from the correct solution, then solutions may converge very slowly, converge to the incorrect minimum, or possibly diverge. *Grid-search* techniques have been adopted whereby large numbers of light curves have been calculated for wide ranges of several parameters, each calculated curve being compared with the observed light curve to form a χ^2 value and hence a χ^2 hypersurface. If the search area is large enough, then the global minimum will be found, but at substantial cost in computing time.

Improvements in procedures to secure optimum solutions for light curves have continued to be implemented. Hill (1979), in his revised LIGHT code, introduced the CURFIT routine (Bevington 1969), which implements the Marquardt (1963) method that finds χ^2 minima more effectively.

It still relies on the calculation of differentials, which was time-consuming on the slower machines of the 1980s, though with present-day computers this problem has essentially been removed. This code has proved to be very stable, permitting correct solutions to be found from quite crude starting values. At lower inclinations, $i \sim 65^\circ$, convergence may be slow, due, in part at least, to the correlations between the various parameters. Kallrath and Linnell (1987) introduced a new search procedure, based on the simplex method, that does not require the calculation of differentials. A simplex is a geometrical figure with $m + 1$ vertices whose form alters and shrinks as it finds its way towards the global minimum. It was applied by them to the WD code and later (Linnell and Kallrath 1987) to Linnell's code, with excellent results. Kallrath (1993) has further discussed the merits of the simplex method, with its speed and accuracy, and has recommended switching to a differential-correction procedure like the Levenberg-Marquardt algorithm when the minimum zone has been found. Most recently, Metcalfe (1999) has developed an optimization scheme based on a genetic algorithm, coupled to the WD code, that permits a more nearly global search to be conducted.

Numerous other programs have been written over the past few decades, each with its specific advantages for particular tasks, and many are referenced in the conference proceedings *Light Curve Modelling of Eclipsing Binary Stars*, edited by Milone (1993). Whilst the WD code has been the most frequently used procedure for light-curve synthesis over the past 30 years or so, other codes, such as EBOP (Etzel 1980, 1993), for well-detached systems with spherical components, and LIGHT2 (Hill and Rucinski 1993), have been used extensively. Stagg and Milone (1993) have also detailed improvements and extensions to the WD code, such as including more model-atmosphere fluxes. *Binary Maker*, by Bradstreet (1993), is very helpful as a self-teaching guide for the solution of light curves. Most recently, a new text, *Eclipsing Binary Stars: Modelling and Analysis*, by Kallrath and Milone (1999), has been published.

The overall success of the various light-curve synthesis codes has been remarkable, with the field benefiting from very accurate representations of symmetric light curves of all types. Thus the properties of several evolved and interacting binaries are known as well as we know those of detached and unevolved stars. Example light curves with their best-fit theoretical-model curves are illustrated in Figures 5.30–5.32. The typical uncertainties attained, from light curves of high quality with RMS errors of 0.003–0.01 mag, are better than ± 0.001 in the relative radii and $\pm 1^\circ$ in orbital inclination when the mass ratio is known to an accuracy of ± 0.01.

An issue that has repeatedly cropped up during the past 25 years is the effect of radiation pressure, amongst the hotter B and O stars, on the shapes of the equipotential surfaces in such binary systems. A discussion of the importance, or otherwise, of this effect, including many references to earlier literature, has been presented by Howarth (1997). He has concluded that the shapes and the limiting volumes for stars that obey the von Zeipel law (*i.e.*, those with radiative envelopes) are unchanged by their own radiation pressure. If the companion has a very significant radiation field, then perhaps that radiation will affect the shape of the irradiated star, in addition to the reflection effect discussed earlier, but these issues are yet to be settled.

In summary of this subsection, it is worth noting that the analysis of light curves should not be considered as an isolated exercise in fitting a model to a set of data. If that approach were adopted, so that all of the parameters theoretically could have values throughout their possible ranges, then the sophistication of the codes required for finding the global minimum in χ^2

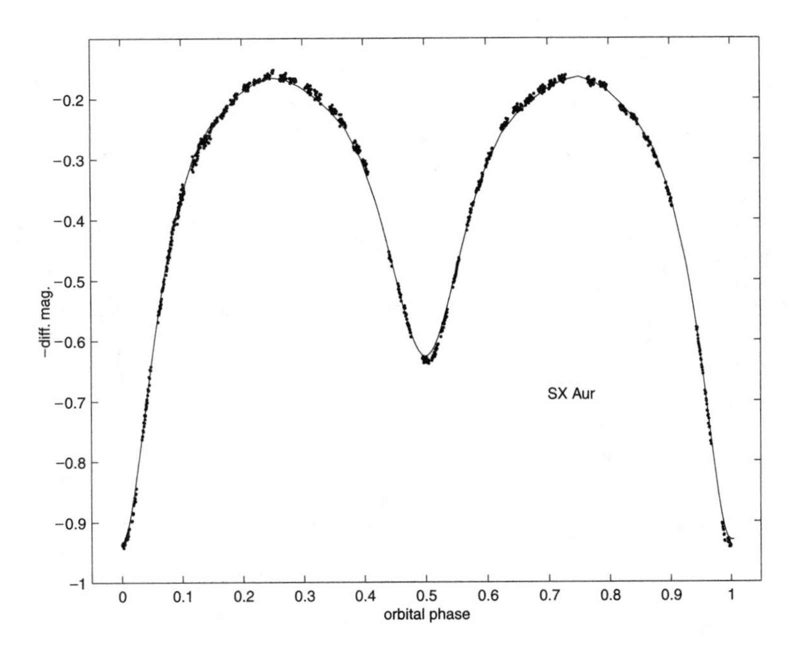

Fig. 5.30. The same data as in Figure 5.12, together with the final LIGHT2 synthesized light curve, shown as a continuous line, illustrating the very good fit of the model to the data for this near-contact early-type binary.

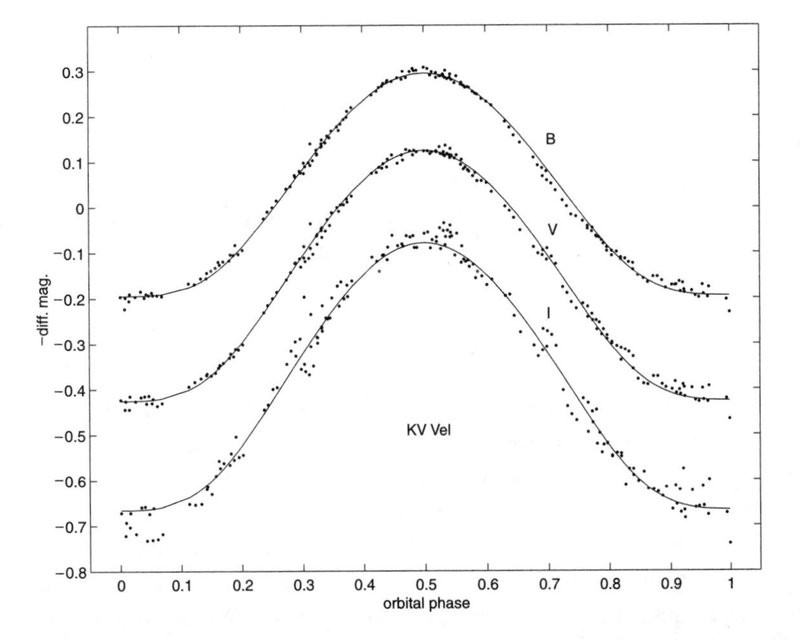

Fig. 5.31. The same data as in Figure 5.14, together with the final LIGHT2 synthesized light curves, shown as continuous lines, illustrating the very good fit of the model to the data for this detached sdO+K-star binary with a very large reflection effect.

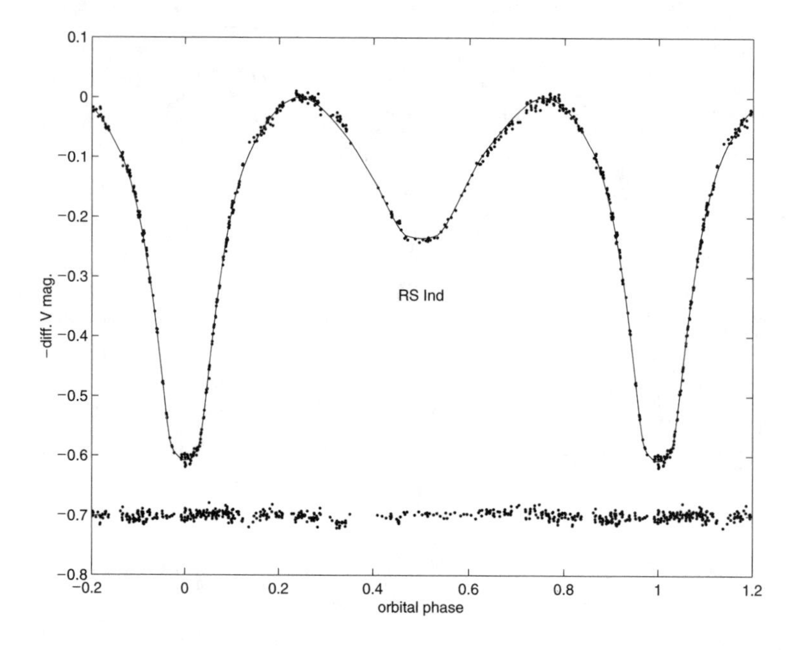

Fig. 5.32. The V light curve for the near-contact binary RS Ind, together with the LIGHT2 synthesized curve, shown as a continuous line. The lower set of symbols at $\Delta V \sim -0.7$ are the $(O - C)$ values scattered about $\overline{(O - C)} = 0.000 \pm 0.007$ mag. Primary eclipse is annular, and secondary eclipse is total. (Data taken from Hilditch and King, 1988.)

hyperspace would have to be very substantial. However, in practice, solutions for light curves should not be sought in isolation, but should be guided by knowledge about a particular binary gained from other data, such as the radial-velocity curves that can provide a mass ratio accurate to a few percent, and the spectra and colour indices that can constrain the possible ranges of temperatures for the stars. Multi-colour light curves provide the best constraints, particularly when they are coupled with accurate radial-velocity data.

5.5.8 Temperatures of stellar components

In many of the foregoing subsections it was evident that knowledge of the surface brightnesses or temperatures of the two components is an essential part of the problem. Yet it is also clear from the discussion of the geometry of eclipses for spherical stars that light curves cannot provide values for surface brightness or temperature directly. From a light curve at a single effective wavelength, one determines only the ratio of surface brightnesses as seen at that wavelength. Such a ratio does not define a ratio of effective temperatures, because the light curve will not be bolometric. If a set of light curves can be determined at different wavelengths, then the dependence of surface-brightness ratios on the temperatures of the two stars will constrain the temperature ratio, but still may not define *separate* temperatures, except in the circumstances of total/annular eclipses. So we need to find a source from which to determine the effective temperature of at least one of the components, namely, an additional observationally determined quantity. Then the light-curve solutions will be able to provide surface-brightness ratios, which in turn will allow a determination of the temperature of the companion.

The obvious observational source for determining the effective temperature of one of the components is the spectral type of the more luminous star, or a set of colour indices determined over a range of orbital phases in one of the established photometric systems discussed in Section 5.3. Spectral types for stars in the range early B to early K can be defined to accuracies of ±1 subtype from good data, and empirical calibrations of spectral types with temperatures allow $T_{\rm eff}$ to be determined to accuracies that may be as good as ±200 K amongst the cooler stars, and more like ±500–1000 K for the hotter stars (Böhm-Vitense 1981). For O stars, the empirical calibrations become progressively more tenuous from B0 to O5, with uncertainties in the range ±1000 K to ±2000 K (Böhm-Vitense 1981; Harries and Hilditch 1997a).

Colour indices can also be used to establish stellar temperatures, provided that care is taken to ensure that the effects of interstellar reddening are removed first. The colour indices are, of course, composite values that depend on the temperatures of the two stars and their relative flux densities at the wavelengths used. For eclipsing binaries where the eclipses are alternately total and annular, then the observed colour indices at the middle of total eclipse will provide direct measures of the effective temperature averaged over the hemisphere of the star that is still visible, namely, the hemisphere that is averted from its companion. For spherical stars, this mean value is the value to be adopted for the complete star. For distorted stars, the temperature of the averted hemisphere will be lower than the polar value, and perhaps somewhat higher than the average value in the absence of heating by the companion; but it may be closer to its undisturbed value if there is a strong reflection effect in the system.

The observed colour indices reflect the spectral-energy distributions of both components added together in proportion to their received flux densities. At any given orbital phase, an intrinsic colour index $(x - y)$ will be formed from the addition of the flux densities of the two components in each of the x and y pass bands, namely,

$$(x - y) = -2.5 \log_{10} (f_{1,x} + f_{2,x})/(f_{1,y} + f_{2,y}) \tag{5.95}$$

A measured colour index must first be corrected for the effects of interstellar extinction. For many binaries, the standard single-star methods described in Section 5.3 will suffice, either because the two components have similar colours or because one component is dominant over the wavelength ranges of the adopted photometric system. If that is not the case, then it may be difficult to separate the interstellar extinction from the observed colours.

How do we separate the two contributions in order to determine the effective temperatures of both components? If one eclipse is total, then clearly the observed colour indices at mid-eclipse will represent the flux-weighted mean colour of the hemisphere of the eclipsing star, plus any effects of circumstellar material observed at that phase, after correcting for the effects of interstellar extinction. At any other phase, both stars contribute. Estimates of the spectral types for both components will provide an initial set of intrinsic colour indices from the empirical calibrations noted earlier (Section 5.3). An initial light-curve solution in one colour, say y, will provide an estimate of the flux-density ratio at y between the two stars, commonly referred to as the *light ratio*, $l_y = f_{1,y}/f_{2,y}$. Because any colour index is a ratio of flux densities, we can adopt, say, $f_{1,y} \equiv 1.0$, and hence $f_{2,y} = 1/l_y$, whilst the standard definition of a colour index, equation (5.8), plus the initial estimates of the individual colour indices for each star will provide $f_{1,x}$ and $f_{2,x}$. Thus the foregoing equation can be used to calculate an expected value of $(x - y)$ at that orbital phase. Sets of observed values for $(x - y)$ at various orbital phases can be compared with calculated values from different intrinsic colour indices and light ratios until overall agreement is reached. This empirical, iterative procedure can be quite definitive in many cases, and somewhat ambiguous in others. Success depends on whether or not the photometric system used can provide the necessary discrimination for that particular combination of stars in that binary. As a service towards improving the determination of temperatures for stars in binaries, Hilditch and Hill (1975) presented *uvby* photometry on 308 Northern Hemisphere binaries, each observed at several orbital phases, and Hill et al. (1975) provided the complementary spectral classifications for 186 of those systems from classification spectra, secured again at several orbital phases. Later, Wolf and Kern (1983) determined *uvby*β indices at quadrature and minimum phases for 288 binaries in the Southern Hemisphere. Those observing programmes have proved to be very helpful in the analysis of many binary systems by many investigators.

Maxted (1994) and Maxted et al. (1994a) have used *pairs* of colour-index observations at two orbital phases, a and b, where the light ratios differ substantially, to determine the separated colour indices for both components directly from de-reddened observations. For de-reddened colour indices of the binary $(x - y)_a$ and $(x - y)_b$, and corresponding light ratios in the y pass band of l_a and l_b, the colour indices of the individual stars $(x - y)_1$ and $(x - y)_2$ are

$$(x - y)_1 = -2.5 \log_{10} \left[\frac{(1 + l_b)10^{-0.4(x-y)_b} - (1 + l_a)10^{-0.4(x-y)_a}}{(l_b - l_a)} \right] \tag{5.96}$$

and

$$(x - y)_2 = -2.5 \log_{10} \left[\frac{l_b(1 + l_a)10^{-0.4(x-y)_a} - l_a(1 + l_b)10^{-0.4(x-y)_b}}{(l_b - l_a)} \right] \tag{5.97}$$

With the difference $l_b - l_a$ occurring in the denominator, it is obvious that these pairs of observations must be selected carefully so that the difference in light ratios will be substantial – a factor of at least 2 being recommended by Maxted.

An additional source of evidence used to help determine stellar temperatures comes from the relative strengths of stars' spectral lines. The *equivalent width* of a spectral line provides a measure of the total absorption due to the line relative to the local continuum. Let the measured values for a line from the two stars be $\mathrm{EW}_{m,1}$ and $\mathrm{EW}_{m,2}$. These measures are taken relative to the combined continuum, and it is straightforward to show that the true values are given by

$$\mathrm{EW}_{t,1} = \mathrm{EW}_{m,1}(1 + l)/l; \qquad \mathrm{EW}_{t,2} = \mathrm{EW}_{m,2}(1 + l) \tag{5.98}$$

where l is the continuum light ratio at, or close to, the line wavelength.

The ratio of true values is equal to the ratio of measured values only when the light ratio is unity. For binary systems composed of stars of similar spectral types, such as two B stars, there will be lines of H, He, Mg, and Si present in both spectra. Hill (unpublished data) compiled a data base of EW measures on O5–B9 stars to establish the dependences of EW on spectral type and on the reddening-free Q index. Then intercomparisons between such values and those secured for many lines in a double-lined B-type binary allowed the intrinsic colours (via Q) and the spectral types for the two stars to be determined, as well as the mean light ratio over the typically 400 Å of spectrum that was used. Examples of such results have been reported by Hill and Khalesseh (1993) and Hill and Holmgren (1995).

For contact binaries, where the temperatures of the two stars are closely similar, the light ratio can be estimated from the cross-correlation functions (CCFs) of the various spectra. Hilditch and King (1986) showed, by co-adding spectra at different velocities and relative flux densities, that the ratio of the areas under the two CCFs from the two components gave a good estimate of the light ratio for the case of two equal spectra; this simple procedure was confirmed by Hill (1993).

Howarth et al. (1997), in a major study of ultraviolet spectra for O- and B-type binary systems obtained from the IUE, showed that the CCF area also could be calibrated against spectral type for systems with dissimilar components. They also confirmed the use of CCFs for determining the projected rotational velocities of stars (see Chapter 3 and Section 5.5.9).

The techniques of tomographic separation of spectra introduced in Chapter 3 should, at least in principle, allow one to determine separate and accurate spectral types for both components, provided the spectral data set has sufficient quality; see the study by Hill et al. (1994), where tomography clearly revealed that the binary CC Cas was composed of O8.5+B0 stars. But this method does require foreknowledge of the light ratio to work effectively and to avoid complications. However, tomographically separated spectra should also allow more quantitative measurements to be made of absorption-line profile shapes and strengths for direct comparison with other empirically determined relationships between line-strength ratios and temperatures, or even for matching against synthetic spectra calculated from model stellar atmospheres. The model-atmosphere codes of I. Hubeny, TLUSTY and SYNSPEC (Hubeny 1988; Hubeny and Lanz 1992, 1995; Hubeny et al. 1994), are being used increasingly to determine effective temperatures and surface gravities for stars from synthesized spectra matched to high-S/N spectra. These techniques use the detailed shapes of the absorption lines relative to the continuum in order to determine $T_{\rm eff}$ and $\log g$. A good set of spectra determined for the purposes of radial velocities means that a combined S/N $\gg$ 100 should be attained in the tomographically separated spectra, as evidenced, for example, in the work of Bagnuolo et al. (1992a, 1994), Thaller et al. (1995), Penny et al. (1997), and Liu et al. (1997). Such applications are in their early stages of development and bode well for continued improvement in determinations of stellar parameters from binary stars.

Relatively, very little absolute spectrophotometry is performed on binary stars, though that may be changing with the improving technologies. In such circumstances, it is possible to use the shape of the spectral continuum, as well as the lines, in order to determine temperatures. Such procedures have been used when one component of the binary is strongly dominant over a specific wavelength range, such as that found for white-dwarf + red-dwarf binaries. A recent example is the work of Orosz et al. (1999), where spectra in the blue region define the properties of the white dwarf, which contributes 97% of the blue light from the system.

In the x-ray bands, the dominant contributions to the total x-ray flux from x-ray binaries come from the accretion disc, which displays a range of temperatures according to $T \propto R^{-3/4}$, all modelled as black-body fluxes via the standard Planck formula

$$B_\nu(T) = \frac{2h\nu^3}{c^2}\frac{1}{\exp(h\nu/kT) - 1} \quad \mathrm{W\,m^{-2}\,Hz^{-1}\,sterad^{-1}} \tag{5.99}$$

and from a second, hotter, isothermal black-body source representing the boundary layer between the disc and the neutron star. Such two-component models, the disc and the boundary layer, have been used routinely in analysing the x-ray spectral-energy distributions from x-ray binaries. But as x-ray spectroscopy has improved in resolution, the requirements for more sophisticated models have grown to the levels illustrated by the work of Zycki et al. (1999) on the x-ray transient source GS2023+338 = V404 Cyg, where disc spectra have to be combined with a harder x-ray component plus Compton scattering and fluorescent Fe K-line emission. These considerations are noted more appropriately in Chapter 7.

5.5.9 Rotational velocities and non-Keplerian corrections

Rotational velocities of stars were discussed in Chapter 3 in the context of determining radial velocities from cross-correlation techniques. It was shown that sensible correlations existed between the FWHM values of the CCFs and the projected rotational velocities of the stars, and once the calibration had been determined, such an empirical procedure was very effective for stars that were rotating sufficiently rapidly. These measures provide $V_{\mathrm{rot}} \sin i$ values that can be useful in placing additional constraints on the radii of the stars via

$$V_{\mathrm{rot}} \sin i = \frac{2\pi R \sin i}{P} \tag{5.100}$$

for synchronous axial rotation and orbital revolution. Although such constraints are weak in comparison with light curves for eclipsing systems, they may prove very helpful in non-eclipsing systems where the stellar radii are not so constrained. An example is the work of Gies and Bolton (1986), where they used the spectra of the optical companion to the black-hole candidate in the x-ray binary Cyg X-1 to establish the velocity semiamplitude and the projected rotational velocity of the companion. Those data then provided a determination of the mass ratio of the system, and hence minimum masses for both components if the companion fills its Roche lobe and is rotating near synchronism.

The standard rotational profile from Gray (1992) was introduced and used in Chapter 3. A full account of the historical development of the study of rotational velocities of stars has been given by Gray (1992, ch. 17). Suffice it to note here that Gray discusses the Fourier techniques that can be applied to determining rotational velocities from spectral-line profiles when stars are rotating slowly, $V_{\mathrm{rot}} \sin i < 20$ km s^{-1}. In such cases, the contributions to the overall line profile from photospheric effects such as macroscopic turbulence are significant and need to be separated from the effects of stellar rotation. But for close binary stars, the rotational velocities usually are synchronous and therefore large, so that the rotational profile dominates. Slettebak et al. (1975) have reviewed the somewhat subjective methods for estimating V_{rot} that make use of the ability of the human eye to compare patterns and judge photographic spectra against those of rotational-velocity standards, in the same manner as spectral-classification techniques. The bright standard stars have well-determined values of $V_{\rm rot}$ from detailed comparisons of spectral-line profiles from high-resolution spectra with line profiles calculated from stellar-atmosphere theory. Present-day technology allows researchers to secure digital spectra at notably higher S/N values and at higher spectral resolution, so that direct comparisons between observed profiles and theoretical profiles can be made on fainter stars. For spherical stars in solid-body rotation, the rotational profile is derived as described in Chapter 3 (Gray 1992). For non-spherical stars in binaries, and with spectra of high resolution, it is necessary to make use of a light-curve synthesis model of the binary to calculate the expected intensity-weighted line profile as a function of orbital phase, so that the variations in surface gravity and temperature across the face of the visible stellar disc at each phase are included.

Such an approach has been noted by Mochnacki and Doughty (1972a) and proposed more substantially by Van Hamme and Wilson (1985). It requires some set of preliminary values for the parameters of the binary, the radii, mass ratio, orbital inclination, and so forth, to calculate the light curve and the intensity-weighted line profiles for each star. A first determination and analysis of the velocity curves, by means of the techniques described in Chapter 3, will provide an initial mass ratio. A light-curve synthesis code can be used to find a first solution to the observed light curve(s), and hence the remaining parameters. In the WD and LIGHT2 codes, the line profiles are calculated by assuming that the same Gaussian profile is observed from all parts of the stellar disc, which then contribute to the intensity-weighted mean profile as a function of orbital phase. This approximation works well in practice when the spectra being analysed for radial velocities have moderate resolution (~ 0.5 Å px^{-1}), which has been typical of most work on binary stars to date. The contribution to rotational velocity from each surface element is found by calculating the projected distance of each element from the rotation axis, at each orbital phase, and multiplying it by $\omega \sin i$. The resultant mean profile, called a *synthetic profile* by Hill (1993), can be used for three purposes.

Firstly, the synthetic profile takes account of the likely non-symmetric distribution of surface brightness over the star and therefore can provide the average correction to a measured radial velocity that is required by that non-symmetric distribution of surface brightness. Such corrections are called *non-Keplerian corrections* because they can be used to remove the obvious departures from Keplerian two-body motion seen in normally measured velocity curves for distorted stars. Examples of such corrections are illustrated in Figures 5.33 and 5.34. (Note that the mass ratio is best defined by accurate radial-velocity curves for both components, rather than being left as a free parameter in the light-curve solution. Only in the case of complete eclipses does a light curve provide sufficient discrimination between the effects of varying q and varying i.) These corrections are systematic and can be as large as a few percent of their respective semiamplitudes, $K_{1,2}$. Because those values of K enter the equations for the stellar masses raised to powers of 2 or 3, it is essential that they be fully corrected for non-Keplerian effects. It should be noted that there are essentially two factors combined into the term 'non-Keplerian corrections'. The first factor, well illustrated through the primary-eclipse phases (-0.15 to 0.15) in Figure 5.33, is the *Rossiter effect*, where the observed radial velocity of a partially eclipsed star (symmetric surface brightness distribution or otherwise) is

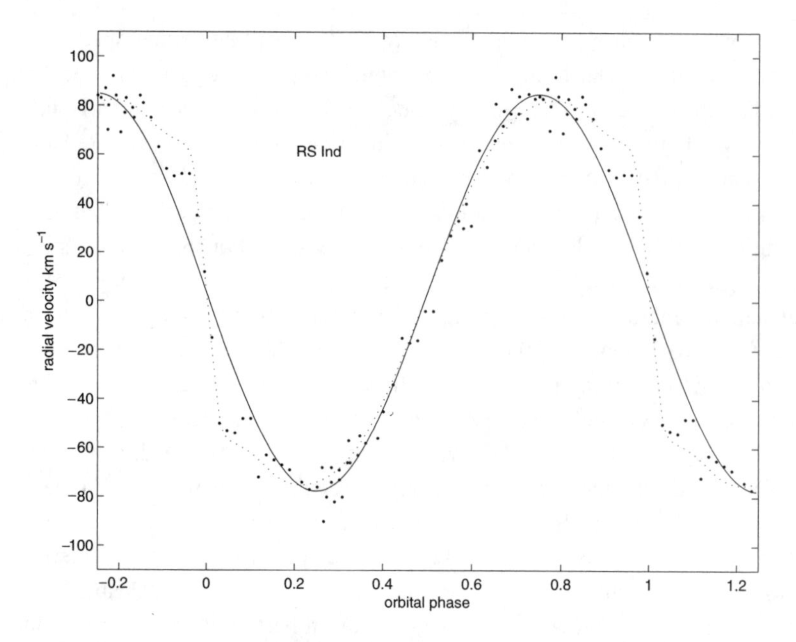

Fig. 5.33. The radial velocities of the primary component of the near-contact binary RS Ind plotted against photometric orbital phase. The velocities were derived from cross-correlation analyses of intermediate dispersion spectra recorded with a diode-array detector. The individual velocities have RMS uncertainties of about 7 km s^{-1}, but still illustrate the rotational distortion of the radial-velocity curve through the phases of primary eclipse (0.9–0.1). The solid line is a simple circular-orbit solution from only the data around both quadratures. The dotted line shows the *non-Keplerian corrections* added to that curve and calculated from the LIGHT2 solution to the V light curve shown in Figure 5.32. (Data obtained from Hilditch and King, 1988.)

systematically biased towards receding/approaching velocities as its approaching/receding hemisphere is eclipsed. The second factor is the lower-amplitude but nevertheless systematic effect discussed earlier for stars with asymmetric brightness distributions that influences the determination of the semiamplitudes principally from the quadrature regions of the velocity curves.

Secondly, the synthetic profile can be used within the cross-correlation analysis described in Chapter 3 to determine an improved quality of fit to the observed CCFs. For very distorted stars, such as those in contact binaries, the observed CCFs are found to be asymmetric when sufficient spectral resolution is employed. Thus the symmetric profiles employed in Chapter 3 do not provide an excellent fit to the data. The work of Hill (1989) and Hill et al. (1989a,b), as summarized by Hill (1993), has shown that application of these asymmetric synthetic profiles to measurements of CCFs from contact binaries results in radial velocities that are immediately Keplerian, because the non-Keplerian distortions have been accounted for in the fitting process. Hill and associates found that only two iterations were necessary to yield final values for all parameters from the radial-velocity and light-curve data. A similar approach was adopted by Rucinski (1992), Lu and Rucinski (1992), and Rucinski et al. (1993) in their reinvestigation of the spectral-line *broadening functions* for contact binaries, first studied by Anderson and

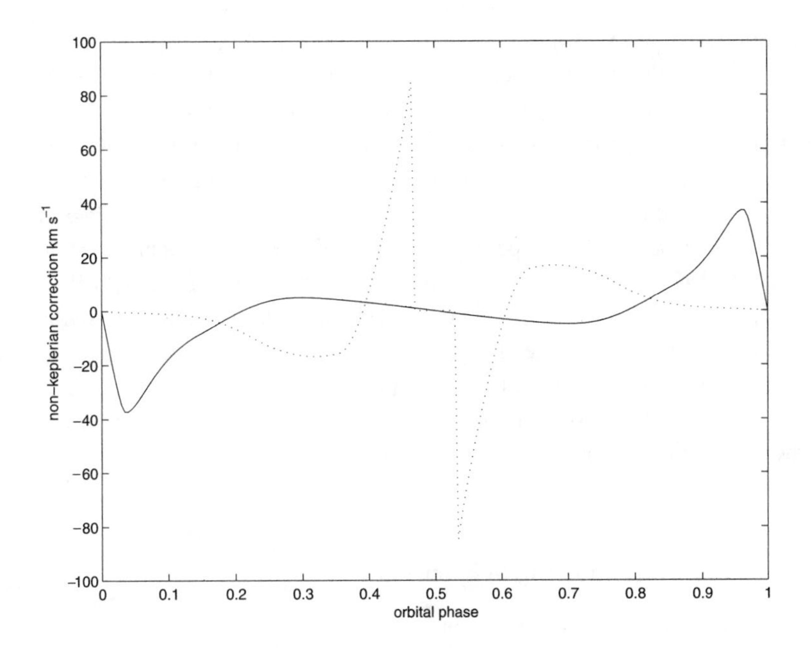

Fig. 5.34. The phase-dependent *non-Keplerian corrections* calculated by LIGHT2 for the eclipsing binary RS Ind: primary component (solid line), secondary component (dotted line). The rotational distortions (Rossiter effect) through the eclipses are obvious and large, but the much smaller corrections around the quadrature phases are very important because they affect, systematically, the final values of the semiamplitudes $K_{1,2}$ and therefore the minimum masses for the two stars.

Shu (1979) and Anderson et al. (1980, 1983). The broadening function is an alternative name for the Doppler profile, the intensity distribution versus radial velocity, which may not be the same as that versus position on the surface of the star if there is significant differential rotation. Full descriptions of the procedures used have been given in the papers cited, and a further helpful description has been given by Rucinski (1999). These same issues must apply to strongly distorted stars in other types of binaries, such as the Roche-lobe-filling secondary components in Algol systems, cataclysmic variables, and low-mass x-ray binaries, as well as the primary components in high-mass x-ray binaries. High-resolution spectra at high S/N are required to study them properly.

Thirdly, the synthetic profile will provide a direct measure of the rotational velocity of each star if the observed spectra have sufficient spectral resolution. This latter purpose has been discussed extensively by Muckerjee et al. (1996) in their study of Algol-type binaries, in which the effects of mass transfer are expected to have produced some systems in which rotational velocities will be substantially asynchronous.

But, of course, we have not reached the end of such developments in studies of spectral-line profiles. The great improvements in detector technology now mean that we can secure échelle spectroscopy at resolutions of approximately 0.1 Å over about 3000 Å in time scales that are less than 1% of the short orbital periods seen amongst W-UMa-type contact binaries, the short-period RS CVn systems, of typically $V \sim 10$ mag. Late-type stars have about 2000 absorption lines over such a wavelength range, mainly due to Fe, and all of them contain information about the distribution of surface brightness across the stellar surfaces. Extraction of a mean intensity-weighted line profile from all these contributing lines is now possible, thanks to the increasing power of computers and the use of newer procedures in numerical analysis, such as *least-squares deconvolution*. This technique is fully described in *Numerical Recipes* (Press et al. 1992), and its power for studies of late-type stars has been demonstrated by Donati et al. (1997) in their spectropolarimetric investigations of magnetically active stars. Such developments belong with the topics discussed in Chapter 7, the imaging of the surfaces of the stars, those points of light in the night-time sky. Suffice to state here that a mean line profile determined by means of least-squares deconvolution from a single échellogram can provide the radial velocity, the rotational velocity, and information on the distribution of surface brightness seen at the orbital phase sampled by that échellogram.

5.5.10 Modelling of asymmetric light curves

5.5.10.1 *RS-CVn- and W-UMa-type systems*

During the late 1970s and 1980s, attention turned from solving the light curves for symmetric systems to finding methods that could address the asymmetric light curves so often exhibited by the RS-CVn-type binaries and the W-UMa-type contact systems resulting from magnetic activity. In such systems, the existence of cool regions, or starspots, was inferred from light curves, colour-index curves, and spectral signatures such as emission lines of Ca II and Hα, indicative of active chromospheres (Hall 1972; Guinan and Giménez 1993). Supporting evidence for very active chromospheres and coronae came from UV and x-ray observations, demonstrating that these stars are much more magnetically active than the Sun.

The first complete mathematical description of the photometric effects of starspots was due to Budding (1977), who derived analytical expressions for light variations from rotating spherical stars with circular/elliptical spots at any longitude and latitude. Such a procedure was also followed by Vogt (1981), but was applied simultaneously to V and R light curves in order to derive spot temperatures. Eaton and Hall (1979) used a numerical light-curve synthesis code and incorporated circular spots on the surfaces of stars, with the spot size and location (longitude, latitude) being solved for in the fitting process. Multi-colour photometry (e.g., *UBVRI*) could provide the necessary discriminating observations to determine the spot temperatures. Successful modelling of many light curves for the detached RS CVn eclipsing binaries was achieved because it was shown that all of the magnetic activity was confined to one star, usually the G- or K-type giant.

The effects of such cool regions on the light curves for eclipsing binaries are to introduce smooth, quasi-sinusoidal distortion waves, with the amplitudes and phasing dependent on the spot locations on the rotating surface. An example is shown in Figures 5.16 and 5.17, where the basic symmetric-eclipse light curve for XY UMa is distorted by an additional smooth variation, with time-dependent amplitude and phase. For well-detached RS CVn stars, with all the magnetic activity confined to one component, it is *relatively* easy to separate the spot component from the (otherwise) symmetric light variations due to the particular geometry of the binary.

The time interval over which there has been consistent photometric monitoring of these RS CVn binaries has now reached about 35 years, aided particularly in the past 15 years by the development of *automatic photometric telescopes* (APTs). Over such intervals, the evolutions of the spot distributions in several systems have been studied, particularly via *UBVRI* photometry, because that is a good straightforward diagnostic for spot activity. The availability of such data has enabled Strassmeier and Bopp (1992), Strassmeier et al. (1994), and Oláh et al. (1997) to investigate the lifetimes of spots or spot groups and the evidence for spot migrations, spot cycles, and differential rotation.

The Strassmeier (1988) light-curve synthesis code has been used extensively for modelling the variable light curves of RS CVn systems. It is based on triaxial ellipsoids, rather than Roche geometry, and is similar to the Wood (1971) code WINK and the Napier (1981) code. Spots are postulated on only one component, usually the primary, and differential rotation of the stellar surface is accounted for via the same type of relationship that is used for the Sun, namely,

$$\omega = \omega_0 - \omega_2 \sin^2 \theta \tag{5.101}$$

where $\omega_0 = 2\pi/P_{\mathrm{eq}}$, with the equatorial period $P_{\mathrm{eq}} = P_{\mathrm{orb}}$, the orbital period, and θ is the latitude on the stellar surface. The quantity $\omega_2 = \omega_0 k/(1-k)$, where $k = \omega_{\mathrm{pole}}/(\omega_{\mathrm{eq}} + \omega_{\mathrm{pole}})$, is the differential-rotation coefficient.

Typical orbital periods for RS CVn systems are in the range of days to tens of days, and consequently many of the known systems in the catalogue of Strassmeier et al. (1993) are non-eclipsing systems displaying only these quasi-sinusoidal light variations with amplitudes of about 0.2–0.5 mag. It has been found possible to represent these variations with quite simple two-spot models, as reviewed by Henry et al. (1995), and to follow the evolution of the spot distributions from long series of observations, some extending over 15 consecutive years (Strassmeier and Bopp 1992; Strassmeier et al. 1994; Oláh et al. 1997). The spot migration rates have been found to be slower than those for the Sun (which has $k = 0.18$), at values of $k = 0.02$–0.05, whilst spot temperatures have been deduced to be about 1200 K below the photosphere level. The spot sizes seem to be substantial, in the range of 2–30% of the stellar surface area, with polar regions often included – similar to the results from Doppler imaging (Chapter 7). Eaton et al. (1996) have cautioned that these two-spot models, which describe the light variations very well, may just be simplified configurations that represent more complex distributions of spots. Appropriately, they have calculated a substantial number of simulations of circular spots, randomly distributed over a stellar surface, with a range of lifetimes τ and a constant probability of allowing spots to disappear of $1/\tau$. Their simulations show that a differentially rotating star with many (10–40) randomly placed spots with finite lifetimes could sensibly explain the observed light curves and their modulation with time. These results would seem to be consistent with the Doppler images of magnetically active stars, which suggest quite complex distributions of active regions over the stellar surface. In a study of the long-term photometric behaviour of the short-period RS CVn system XY UMa, Hilditch and Collier Cameron (1995) concluded that a sensible explanation for all the data lay in a substantial polar crown, with a second belt of spot activity at low latitudes, and suggestions of a spot cycle comparable to the solar magnetic-activity cycle. That particular survey is on-going, and the data in Figures 5.16 and 5.17 illustrate the continuing variability in light-curve shape and in the overall brightness of the system.

The short-period RS CVn systems ($P_{\rm orb} < 1$ day), typified by XY UMa, are usually eclipsing systems, where the spot distribution can seriously alter the appearance of the eclipse light curve. Difficulties are experienced in extracting the geometry of the system from the combined effects of eclipses and spot distributions. However, the primary star is usually the dominant source of brightness, $l_1/l_2 > 10$, so that solutions can be achieved unambiguously if there is a sufficient number of observed light curves, where, for example, one at least has a symmetric primary eclipse. Such eclipse data can be used to determine preliminary estimates for stellar radii and the orbital inclination. In combination with a spectroscopic mass ratio, the system geometry can be specified, and spot distributions can be found from each light curve. An iterative procedure usually will yield sensible results quickly, particularly when eclipse mapping techniques (Chapter 7) are utilized in preference to the spot models noted earlier.

For contact binaries composed of convective-envelope stars, the solutions for their commonly asymmetric light curves can become quite ambiguous, because the two stars will have very similar surface brightnesses. The example of W UMa itself was shown earlier, Figure 5.13, where the two maxima of the light curve differ by 0.037 mag. A more extreme illustration is shown in Figure 5.35, where the two maxima differ by 0.09 mag. This well-observed phenomenon is sometimes referred to as the *O'Connell effect*, after its first recognition (O'Connell

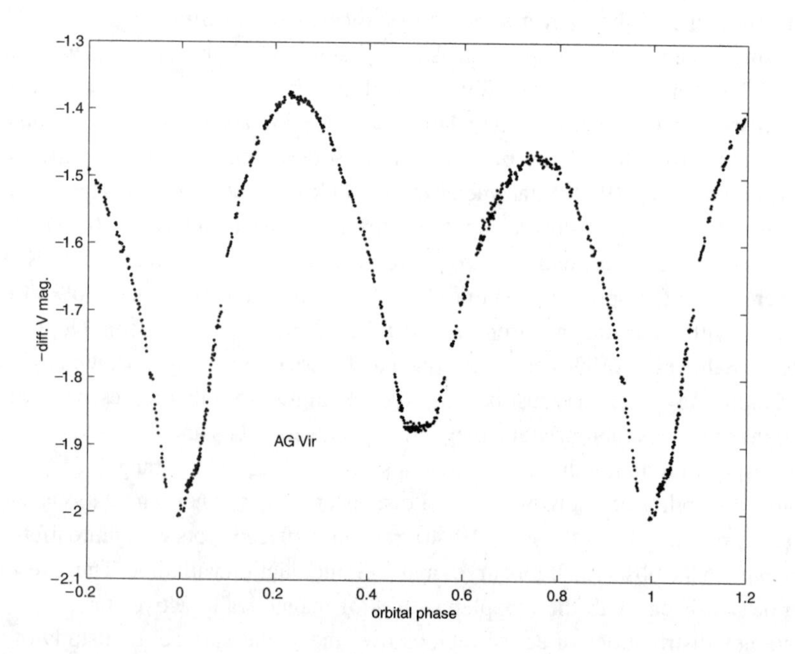

Fig. 5.35. The differential V light curve for the contact binary AG Vir showing differences in the brightness levels at the two quadratures of 0.09 mag. Secondary eclipse seems to be total/annular, whilst primary eclipse is substantially asymmetric around the minimum. (Data obtained from Bell et al., 1990.)

1951). The two stellar hemispheres presented to the observer at phase 0.25 evidently do not have the same surface brightness as those presented at phase 0.75. The question of whether the brightness difference between the two quadratures is caused by cool or hot spots on one or both stars can be answered by multi-colour photometry that is sensitive enough to the changes in mean temperature of the binary as the system rotates. The shape of the light curve in that half of the orbital cycle containing the higher maximum will provide a different determination of the geometry of the system than will that containing the lower maximum. For the W UMa light curve, the two halves of the light curve are quite similar, but for AG Vir the two halves are very different and yield very different shapes for the stars and for the orbital inclination. Bell et al. (1990) showed that solutions for the single-colour light curve of AG Vir could be obtained with different geometries and cool starspots on either star, an unacceptably ambiguous result. Hendry et al. (1992) and Maceroni and van't Veer (1992) showed, by means of many simulations, that analysing the asymmetric light curves of W-UMa-type contact binaries with cool starspots distributed over their surfaces was generally ambiguous, resolvable only via combinations of Doppler imaging and eclipse mapping. Progress in understanding these systems will require both accurate multi-colour photometry and high-resolution spectroscopy, together with the techniques of analysis described in Chapter 7.

5.5.10.2 *Systems with accretion discs*

The collective name 'cataclysmic variables' (CVs) encompasses a substantial variety of photometric behaviours. Some of the orbital light curves are amenable to analysis by means of the techniques already described in terms of Roche geometry, such as those of the *dwarf novae* when they are in quiescence, rather than in outburst, and the nova-like CVs. The codes used by Zhang et al. (1986) and Wood et al. (1989) contain a standard model for a CV with a Roche-lobe-filling secondary component, which is normally a red-dwarf star that contributes little to the overall brightness of the system in the optical region, but is important in the far red and infrared. The primary component (the more massive star) is a white dwarf that is surrounded by an accretion disc that is taken to be geometrically thin and confined to the orbital plane. The accretion stream from the inner Lagrangian point L_1 of the secondary component follows the trajectory calculated by Lubow and Shu (1975) and impacts the rim of the accretion disc in a *bright spot*. The location of the hot spot is determined by the deflection of the accretion stream by Coriolis effects and by the radial size of the accretion disc. The angle of deflection has been determined by Lubow and Shu to lie in a rather narrow range specified by the mass ratio of the binary system. In the optical region, the brightest component often is the bright spot, followed by the accretion disc, and then the white dwarf. Accordingly, the dominant effect on the light curve due to orbital motion is the upper half of a quasi-sinusoidal hump due to the rotation into the observer's view, and out again, of the hot spot on the accretion disc. If the orbital inclination is favourable, then there will be eclipses of the accretion disc, the hot spot, and the white dwarf by the red-dwarf star that will be obvious in the optical region. In the infrared, we can see the ellipsoidal variations of the red-dwarf star.

Because the orbital periods for CVs typically are a few hours and the stars are small, the durations of any eclipses are short, of order 20 minutes. High-speed photometry is essential for detailed studies of CVs. Figure 5.36 shows the mean light curves for the dwarf nova OY Car

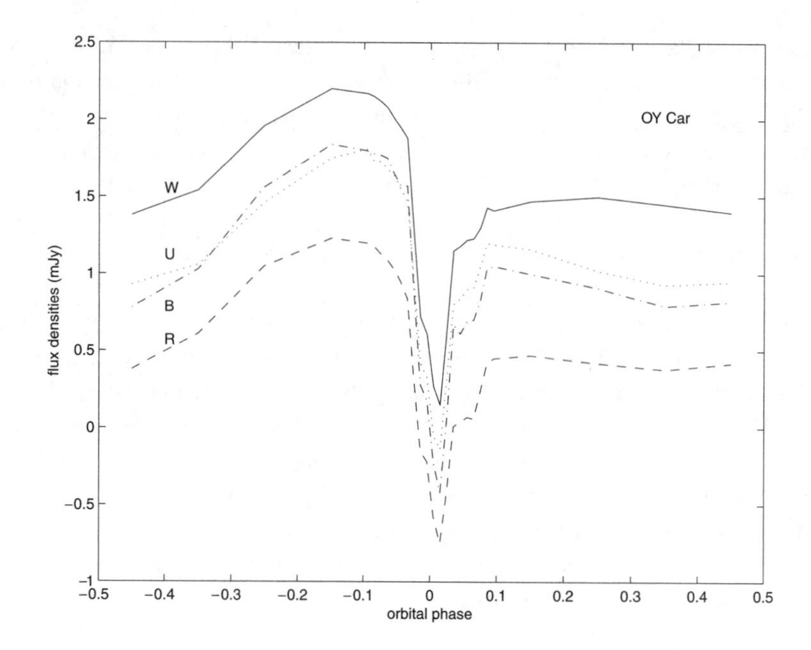

Fig. 5.36. Mean light curves in white light (W) and in *UBR* filters for the dwarf nova OY Car observed in quiescence. The *UBR* curves have been displaced downwards in the figure by amounts of 0.3, 0.6, and 0.9 mJy, respectively, to show the curves more clearly. The original high-speed photometry was obtained at 1-second time resolution and therefore contains many details not illustrated in these mean curves. But the hump due to the rotation into and out of the line of sight of the bright spot is obvious, lasting about $0.5P_{\rm orb}$, as is the rapid eclipse ingress/egress of the white dwarf at phases $\phi = \mp 0.025$, each lasting about 1 minute. The eclipse of the bright spot has ingress at $\phi = +0.001$, and egress at $\phi = +0.075$. (Data obtained from Wood et al., 1989.)

deduced from high-speed photometry obtained in white light and with *UBR* filters by Wood et al. (1989) when the binary was in quiescence. It shows very clearly the hot-spot hump and the rapid eclipses of the white dwarf, followed by the bright spot. By fitting a least-squares cubic spline to these data and then computing numerical differentials, it is possible to determine quite precisely the orbital phases of the eclipse contacts, as introduced earlier in this chapter. These timings yield values for the white-dwarf radius and the extent of the hot spot along the rim of the accretion disc, both in terms of the semimajor axis of the relative orbit and dependent on the orbital inclination. The simple equations noted in Sections 5.5.2 and 5.5.3 can be used to secure estimates of (i, q) from such times of eclipse ingress and egress. Careful examination of the light curve will also reveal the times of eclipse of the accretion disc by the secondary star, thereby giving a size estimate for the disc. This decomposition of a CV light curve into its various contributors is very well illustrated in the paper by Wood et al. (1989), and the system OY Car is favourably oriented for such techniques to be successful.

Spectroscopy of CVs, particularly in the far red (~8000 Å), shows absorption features corresponding to the cool secondary component that yield quite reliable radial-velocity curves for those stars and hence a mass function $f(m)$ for the binary. The white-dwarf spectrum has very shallow absorption features, due to the very high surface gravity, $\log g \sim 6.0$, and in any event is buried within the broad emission lines from its surrounding accretion disc. Attempts have been made to use the outermost parts of the emission lines, those parts corresponding to the rapidly rotating inner parts of the accretion disc close to the white dwarf, to determine the orbital motion of the white dwarf. But those attempts have proved to be somewhat unsatisfactory, and it has been quite usual to adopt a theoretical radius–mass relationship for a white dwarf to infer a mass for the white dwarf. The mass of the secondary star then follows from the mass function and the orbital inclination estimated from the light curve at the derived mass ratio.

If the photometry has been tied to some standard system such that flux densities can be expressed in millijanskys, for example, then estimates can be obtained for the accretion luminosity seen from the impact of the accretion stream onto the accretion disc. The formula that links the accretion luminosity to the mass-transfer rate $\dot{m}$ in the stream is simply

$$L_{\text{acc}} = \frac{G(m_1 + m_2)\dot{m}\Delta\Phi_n}{2a} \tag{5.102}$$

where $\Delta\Phi_n$ is the potential difference between the L_1 point and the edge of the accretion disc. Thus an estimate can be obtained for the mass-transfer rate in such CVs from accurate, standardized photometry.

But very few CVs are so favourably oriented to our line of sight that we can see the separated eclipses of both of the main sources of radiation in a binary system. Most are seen at lower inclinations, which allows more detail of the distribution of light across the accretion disc to be determined, but which, in general, does not permit the crisp separation of components seen for OY Car. The techniques of eclipse mapping and Doppler tomography or imaging (Chapter 7) can be employed here with great success.

The optical and infrared light curves for x-ray binaries may become distorted from the usual ellipsoidal variations (Figure 5.21) provided by the non-spherical mass loser when the accretion disc around the x-ray source alternately obscures and reveals the x-radiation with respect to the facing hemisphere of the mass loser. The consequent decreasing and increasing irradiation of the companion by the x-ray source leads to a variable contribution to the total light from the reflection effect, with that variation being of order 1–2% of the total light of the system. Examples have been found of cyclic behaviour in such variations that have been described in terms of precession of a tilted accretion disc, such as those in HZ Her and LMC X-4. With a mass ratio determined from radial velocities of the optical companion and data on the pulse arrival timing for the x-ray source in HMXBs, the ellipsoidal variations are readily modelled by the standard Roche model already described, and the orbital inclination can be determined if the companion fills its Roche lobe. Here is one example of the usefulness of $V_{\text{rot}} \sin i$ values determined from line profiles or from *CCF*s that can constrain the range of possible solutions for the optical or infrared light variations for (i, q, r).

5.5.10.3 *O+O-star, Wolf-Rayet, symbiotic, and high-mass x-ray binaries*

An established effect seen in some interacting O+O-star binaries is the orbital-phase-dependent variation in strength of the absorption lines from the secondary star. Such variation takes the form of line depths being deeper when the secondary is approaching the observer. It has been identified in a number of systems over many years, and Howarth et al. (1997) recorded that it was found in 8 of 15 O+O-star binaries studied in the UV. The system AO Cas is particularly well known in this regard (Stickland 1997; Howarth et al. 1997), and as noted in the introduction to Chapter 7, the phenomenon was discussed by both Struve and Sahade in the 1950s for several systems. Howarth et al. (1997) and Gies et al. (1997) agreed on the term *Struve-Sahade effect* in their later investigations of the phenomenon. Gies and associates have proposed that the cause of this line-strength variation lies with the collision of the two stellar winds between the two stars, with the position and shape of that collision zone being determined by the momentum balance between the two winds. When the wind from the primary is dominant, the collision zone will be bow-shaped close around the secondary star, and its maximum density will be shifted by the Coriolis effect to the leading hemisphere of the secondary. Because the typical wind speeds from O stars are about 1000 km s^{-1}, the collision zone will be shocked, and the photosphere of the secondary will be heated by the back-scattered photons from that shock zone. The calculations by Gies and associates showed encouraging agreement with the line-profile observations and appeared able to accommodate the observed asymmetries on the light curve for AO Cas. They recommended that further work be done to improve that proposed model.

Wolf-Rayet (WR) stars, with their dense stellar winds, are found in binary systems, often with O-type main-sequence-star companions and orbital periods of days to tens of days. Whilst some have orbital inclinations that permit eclipses, others demonstrate orbital light variations that are due to the O-star radiation passing through varying path lengths in the WR wind on the way to the observer. The *photospheric* eclipses of some systems are decidedly asymmetric, perhaps because of varying opacities through the winds of both the WR star and the O-star and the likelihood of collisional shocks between the two winds. The *wind* eclipses in wider-orbit systems are generally quite symmetric, though not necessarily constant, and there is only one eclipse per orbit, because the WR wind is the only source of substantial opacity for the O-star radiation.

Cherepaschuk (1973) has developed a method for analysing the light curves for systems that undergo *atmospheric eclipses*, in addition to occultations/transits of the two stellar photospheres. His method is based on a model of spherical stars in circular orbits and incorporates the essential parts of the analytical theory of spherical-star eclipses discussed earlier. With values of the stellar radii, and the orbital inclination derived from other sources of data, Cherepaschuk's method solves light curves in different colours to obtain the distribution of optical depths as seen along the various lines of sight through the extended atmosphere or stellar wind seen at each eclipse. The radiation from the occulted star is exponentially attenuated by its interactions with the extended atmosphere or wind particles, and that extinction is found to depend on the wavelength of light studied. Thus, for V444 Cyg, Cherepaschuk et al. (1984) have found substantial differences in wind opacity over the wavelength range from 0.25 to 3.5 μm. In the UV and blue regions of the spectrum the opacity is independent of wavelength, thereby confirming electron scattering as the dominant source of opacity. But at 0.64–3.5 μm, the opacity increases substantially and is due to free-free opacity of H and He.

The single V-shaped dips in the light curves of some WR binaries have been interpreted by Lamontagne et al. (1996) as atmospheric eclipses of the O star by the extended wind of the WR star. The standard model for a spherically symmetric stellar wind from a WR star is described by expressions for the radial density distribution $\rho(r)$ and the radial wind-velocity distribution $v(r)$. These expressions are

$$\rho(r) = \dot{m}/4\pi\, r^2 v(r); \qquad v(r) = v_\infty(1 - R_\star/r)^\beta \tag{5.103}$$

where $\dot{m}$ is the rate of wind mass loss from the WR star, v_∞ is the terminal velocity in the wind, $R_\star$ is the core radius of the star, and β is an index for the form of the velocity distribution, with a typical value of $\beta \sim 0.8$–1.0. (The term *core radius* rather than photospheric radius is used for WR stars because the spectral continuum forms at positions that depend on the wind-velocity law and on the wavelength observed.) The density ρ is related to the number of free electrons in the wind, which give rise to Thomson scattering of radiation, by

$$\rho(r) = \mu m_p n_e(r)/\alpha \tag{5.104}$$

where α is the ratio of the number of free electrons per nucleon (so that $\alpha = 0.5$ for pure He II, for example), μ is the mean atomic weight ($\mu \approx 4$ for WR stars), and m_p is the mass of a proton. Figure 5.37 illustrates the dependence of $v(r)$ and $n_e(r)$ on radial distance r for typical values of the other parameters. Note that wind velocity reaches 90% of its terminal value within 10 times the radius of the WR star, when $\beta = 1$, and correspondingly the density decreases

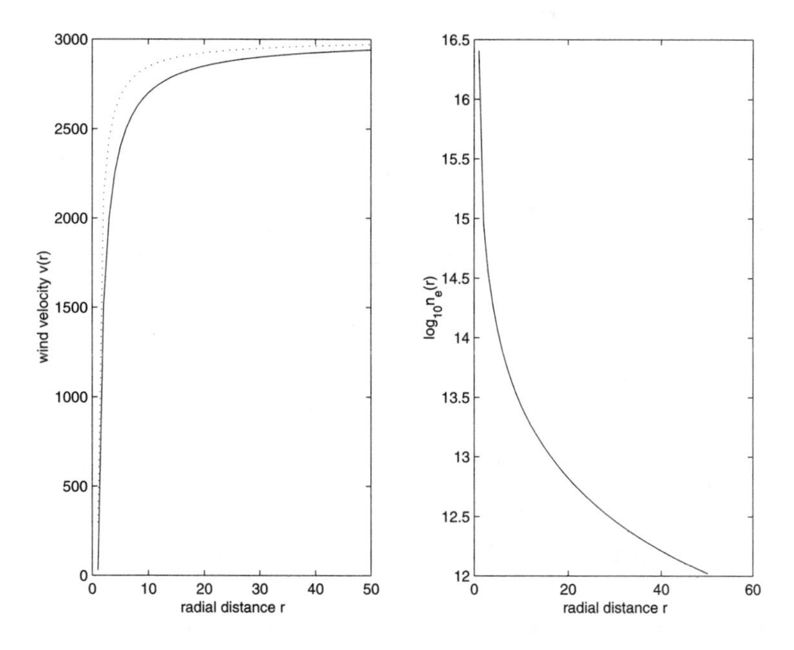

Fig. 5.37. The radial dependence of the wind velocity $v(r)$ and the electron density $\log_{10} n_e(r)$ m^{-3} for a typical WR wind for values of the exponent $\beta = 1.0$ (solid lines) and $\beta = 0.5$ (dash lines). In this illustration, $\alpha = 0.5$, $\dot{m} = 5 \times 10^{-5}$ M$_\odot$ yr^{-1}, $v_\infty = 3000$ km s^{-1}. The radial distance r is expressed in units of the WR-star radius.

extremely rapidly. Most of the scattering of radiation takes place close to the photosphere of the WR star, where the free electrons are provided from doubly ionized helium.

Then the incremental optical depth along the line of sight (x) due to electron scattering with free-electron cross section $\sigma_e = 6.648 \times 10^{-29}$ m^2 is

$$d\tau = \sigma_e n_e(r)\,dx \tag{5.105}$$

In these wind-eclipsing WR+O binaries we see only one eclipse, that of the O-star radiation by the WR wind. (The O-star wind is about 2 orders of magnitude less dense than the WR-star wind.) Thus any decrease in the observed magnitude of the binary due to the scattering by the wind is described by exponential attenuation of the O-star radiation as seen through an optical depth τ and is given by

$$\Delta m = -2.5 \log_{10}[f_{\mathrm{WR}} + f_{\mathrm{O}} \exp(-\tau)] \tag{5.106}$$

If the wind is optically thin ($\tau \ll 1$), then equation (5.106) can be modified, by using a polynomial expansion for the log term to first order only, to yield

$$\Delta m = +2.5(\log_{10} e)\frac{f_{\mathrm{O}}}{f_{\mathrm{WR}} + f_{\mathrm{O}}}\tau \tag{5.107}$$

Thus we need to evaluate τ for each phase of the binary orbit. Lamontagne et al. (1996) have shown that the expression for

$$\tau = \int_{x=x_0}^{\infty} d\tau \tag{5.108}$$

can be evaluated analytically for integer values of β, so that for $\beta = 1$ the expected decrease in magnitude is

$$\Delta m = A\left[\frac{2}{[(1-c^2)(1-b^2)]^{1/2}}\left\{\arctan\left(\frac{1+b}{1-b}\right)^{1/2} + \arctan\left[\left(\frac{1+b}{1-b}\right)^{1/2}\tan\left(\frac{\arcsin c}{2}\right)\right]\right\}\right] \tag{5.109}$$

where

$$A = \frac{2.5(\log_{10} e)\alpha\sigma_e\dot{m}}{4\pi\, m_p v_\infty a(1 + f_{\mathrm{WR}}/f_{\mathrm{O}})} \tag{5.110}$$

where a is the radius of the (adopted) circular orbit, $c = \sin i \cos(2\pi\phi)$, and $b = (R_\star/a)/(1-c^2)^{1/2}$.

The resultant light curves are illustrated in Figure 5.38 for a range of values of the orbital inclination at a fixed mass-loss rate. Lamontagne et al. (1996) have shown that the shape of the eclipse curve is essentially insensitive to the value of β between 0 and 1, a fortunate circumstance, because the value of β is not well established. Clearly, the depth of each eclipse is determined by the maximum optical depth through which the O star is observed, which in turn is defined by the mass-loss rate $\dot{m}$ incorporated into the expression for the quantity A. These

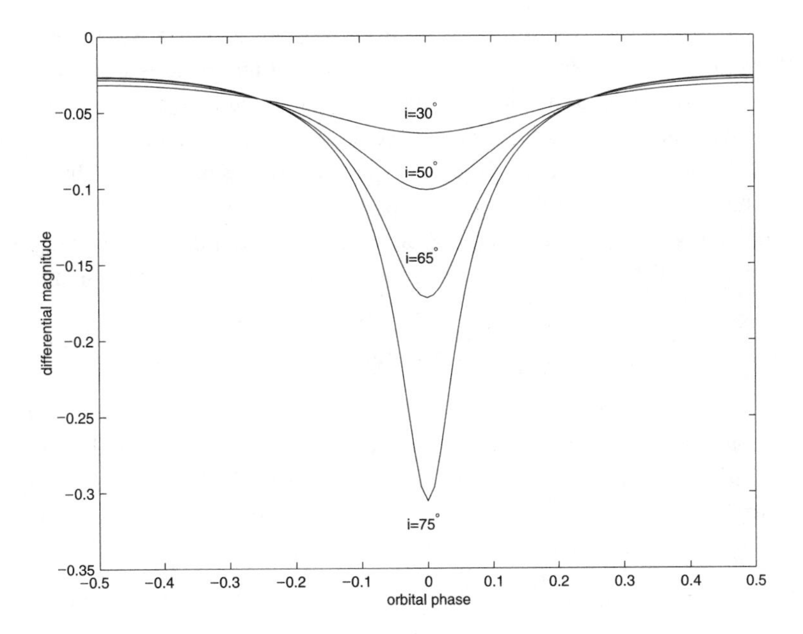

Fig. 5.38. Theoretical light curves for a WR+O binary in which eclipses occur because of the changing optical-path length of the WR wind in front of the O star, as the observer rotates around the binary. A single eclipse per orbit is observed, and its shape determines the orbital inclination, as illustrated. The actual depth of an eclipse depends on the maximum optical depth traversed by the O-star radiation, which in turn depends on the mass-loss rate from the WR star. In this illustration, $\alpha = 0.5$, $\dot{m} = 5 \times 10^{-5}$ $M_\odot$ yr^{-1}, $v_\infty = 3000$ km s^{-1}, $R_\star/a = 0.02$, $f_{WR}/f_O = 1$, $\beta = 1$. Changing the mass-loss rate by, say, a factor of 2 will change the amplitude of the eclipse curve by the same factor.

wind eclipses can therefore be used to establish the mass-loss rate and the orbital inclination for such WR + O binaries, and hence the masses for the stars if both velocity curves have been determined.

The symbiotic binaries are composed of cool, M-type giant primary stars with hot companions that typically are evolved subdwarfs. The composite spectrum for such a system is dominated in the optical region by the M-giant spectrum with its molecular bands. But in the UV region, the hot companion provides a blue continuum, and superposed upon that continuum is a set of nebular emission lines arising from the interaction of the energetic photons with the outflowing wind of the giant star. It is well established that the source for the scattering of the UV photons in the wind is Rayleigh scattering by neutral H atoms, given by the standard expression

$$\sigma_R = \sigma_e(\lambda_0/\lambda)^4 \tag{5.111}$$

where σ_e is the electron-scattering cross section, as before, and $\lambda_0 = 1026$ Å for H atoms. Such a strongly wavelength-dependent cross section means that wind eclipses may be obvious in the UV, where $\sigma_R \sim \sigma_e$, though negligible in the optical region. Vogel (1991) first demonstrated that the velocity law for the outflowing wind from the cool giant star could be determined empirically by means of analysing the changes in the UV continuum fluxes as the hot companion was observed to move behind the giant star and its wind. Assuming, as before, that the wind is spherically symmetric and that the orbit is circular, the rate of wind mass loss is as given by equation (5.103), but an expression for the wind-velocity law has not been established. To find an empirical representation, Vogel (1991) and Pereira et al. (1999) considered the varying path length through the giant's wind that the UV photons travel from the hot companion on their way to the observer. The photons encounter a column density, due to neutral hydrogen, of

$$N_{\mathrm{H}}(\delta) = \frac{2\dot{m}}{4\pi\mu m_{\mathrm{H}} R_{\star} v_{\infty}} \int_{\delta}^{\infty} \frac{dr}{(r^2 - \delta^2)^{1/2} r v(r)} \tag{5.112}$$

where $\delta^2 = x^2(\cos^2 i + \sin^2 i \sin^2 \phi)$ is the projected separation between the components at orbital phase ϕ, and x is the semimajor axis of the relative orbit expressed in terms of the radius $R_\star$ of the giant star; m_{H} is the mass of the H atom, μ is the molecular weight in the wind, and $v(r)$ is the radial wind-velocity law. These column densities can be determined from observations by fitting continuum UV spectra of the form

$$F_\lambda(\phi) = F_\lambda^0 \exp[-N_{\mathrm{H}}(\phi)\sigma_{\mathrm{R}}] \tag{5.113}$$

at each observed orbital phase. These fits have been found to be particularly successful when proper account is taken of the dependence of σ_{R} on λ close to the Lyman resonance lines (Vogel 1991).

Inverting the foregoing integral equation to find $v(r)$ from a set of such derived column densities is not a simple problem. Suffice to state here that Pereira et al. (1999) have shown that a convenient analytical expression that well represents the empirical results is given by

$$v(r) = v_\infty\{\Gamma_{\mathrm{inc}}[B(r-1), A] + A\exp(-Br)\Gamma_{\mathrm{inc}}(Br, A)\} \tag{5.114}$$

with the incomplete gamma function Γ_{inc} defined by

$$\Gamma_{\mathrm{inc}} = \frac{1}{\Gamma(a)} \int_0^r t^{a-1} \exp(-t)\, dt; \qquad \Gamma(a) = \int_0^\infty t^{a-1} \exp(-t)\, dt \tag{5.115}$$

The parameters A and B fix the scales of the expression to fit the empirical velocity law. Figure 5.39 illustrates the foregoing analytic velocity law, which is seen to be very different from that for WR stars, involving a very slow change in velocity close to the surface of the giant star, followed by a rapid rise around $r \sim 5R_\star$ towards the terminal velocity v_∞. The column density corresponding to that velocity law is also shown.

Mass-loss rates of $\dot{m} \sim 10^{-9} M_\odot\ \mathrm{yr}^{-1}$ have been derived from these studies of attenuation of the UV continuum light. These wind effects are separate from the true photospheric eclipses that are observed to occur in a few of the symbiotic binaries. Because the orbital periods for symbiotic binaries typically are measured in years, photometric measurements of eclipses are usually sparse. Just as in WR + O binaries, there is evidence for variability in the properties of the winds from these cool giants.

Finally, in the high-mass x-ray binaries, where the x-ray pulsar is orbiting an O-type supergiant or a Be star, it has been found that the x-ray source provides an excellent probe of

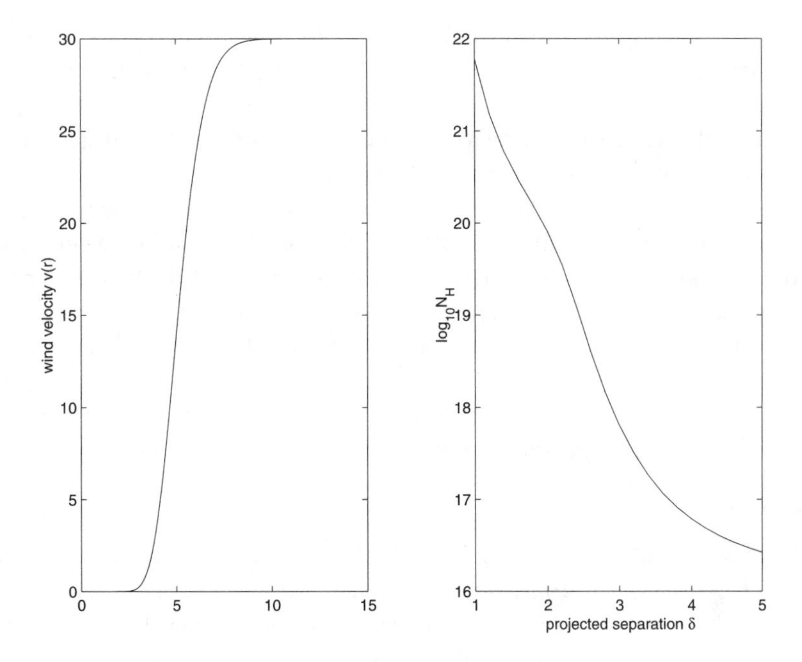

Fig. 5.39. Left: A repesentation of the empirical wind-velocity law as a function of radial distance r from the red giant, determined for the cool giant stars in symbiotic binaries (Vogel 1991; Pereira et al. 1999), with $v_\infty = 30$ km s^{-1}. Note that it is very different from that adopted for WR and O stars. Right: The corresponding column density of H, per square metre, as a function of the projected separation δ. Both r and δ are expressed in terms of the radius $R_\star$ of the red giant for the system SY Mus. Here $A = 14.20$, $B = 3.383$, $R_\star = 86\mathrm{R}_\odot$.

the H-column density distribution in such systems. Several investigations have studied the changes in spectral-energy distribution in the x-ray bands as the neutron star has passed behind the outflowing wind from the O star. An example is the work of Clark et al. (1994) on the HMXB 4U1538-52 = QV Nor, using data collected with the *Ginga* x-ray satellite. They have found the column-density enhancements (by factors of about 30) detected in other HMXBs close to the ingress/egress phases of the x-ray eclipse that are caused by the passage of the neutron star through the O-star wind, creating a bow shock. But during the egress phases, beyond the rapid increase/decrease in the measured column density of H just before and just after the occultation of the pulsar, the x-rays from the pulsar do probe an undisturbed part of the outflowing wind and permit a direct determination of the radial density distribution in the wind. They have found it necessary to invoke a hybrid law involving an exponential term close to the stellar surface and a power law farther out:

$$N_\mathrm{H}(r) = \frac{\dot{m}}{4\pi\mu r^2 v_\infty}\left[1 + \left(\frac{r}{r_1}\right)^2\left(1 - \frac{r_\mathrm{o}}{r}\right)^\beta \exp\left(-\frac{r - r_1}{h}\right)\right]\left(1 - \frac{r_\mathrm{o}}{r}\right)^{-\beta} \qquad (5.116)$$

where r_o is the radius of the O star, and h, β, and r_1 are adjustable parameters to fit the function to the data. When $r \leq r_1$, the exponential function is positive, and values of $r_1 = 1.2 r_\mathrm{o}$, with

$h = 0.05r_o$, are found to be necessary. For $r \geq r_1$, the exponential becomes negative, and either the normal power law takes over, with $\beta \geq 0$, or, with $\beta = 0$, we obtain a $1/r^2$ law of a constant-velocity wind.

Thus, close to the stellar surface, the results from x-ray binaries suggest that the theory of a steady-state radiatively driven wind requires some improvements in the subsonic region, whilst at greater distances the data from QV Nor are consistent with a constant-velocity wind. Clark et al. (1994) note that their density function should be considered 'an approximate representation of the average density in an inherently turbulent flow'. Amongst these x-ray binaries, the compact companion itself affects the properties of the wind, both because of its gravitational influence and because its x-ray emission alters the ionization balance within the wind.

5.6 **Polarimetry and spectropolarimetry**

5.6.1 Observations and standard stars

The topic of astronomical polarimetry has been discussed extensively in a recent text, *An Introduction to Astronomical Polarimetry* (Tinbergen 1996), and so only a few remarks will be made here. The polarized nature of an incoming beam is determined by means of various combinations of retarders and analysers that are placed in front of a standard photometer (for photopolarimetry) or a standard spectrograph (for spectropolarimetry). With photopolarimeters, it is usual to switch rapidly between two orthogonally polarized beams in order to remove the disturbing effects of the Earth's atmosphere and to continue an observation until an uncertainty of $\pm 0.01\%$ is achieved for the degree of polarization p, and $\pm 0.1°$ for the position angle. Retarders composed of $\lambda/2$ plates are used for measuring linearly polarized beams, and $\lambda/4$ plates for circularly polarized beams. Descriptions of efficient photopolarimeters used for studies of binary stars are given, for example, by Bailey and Hough (1982), Cropper (1985), and Manset and Bastien (1995). These have been used to make observations, with very short time resolutions, of the received flux densities, the percentages of linearly and circularly polarized light, and the position angles, particularly for the short-period *intermediate-polar/polar* systems, which display variations in all of these quantities on time scales of minutes/seconds. They have also been used to study the rather more leisurely changes in polarization properties as a function of orbital phase exhibited by the WR binaries.

Spectropolarimetry is becoming an increasingly used mode of observation because of the introduction of large-format CCD cameras that can yield photometric precision in sensible integration times of minutes. An excellent description of the practicalities of spectropolarimetric observations is given by Tinbergen and Rutten (1992). Here the typical uncertainty achieved for p is $\pm 0.1\%$. The advantage of spectropolarimetry is that the levels of continuum polarization can be decoupled from those traced across individual lines in a spectrum, a particularly strong diagnostic feature when applied to the complex emission- and absorption-line spectra exhibited by WR+O binaries.

For standard stars, a list of those showing zero polarization of their radiation has been compiled by Tinbergen (1979). For stars with different levels of linear polarization, Hsu and Breger (1982) and Bastien et al. (1988) have monitored a substantial number of stars well distributed over the sky and have shown that some of them can display variations at levels up to $\pm 0.3\%$ in p and $\pm 1^\circ$ in position angle over intervals of days to weeks. Thus, polarimetric standard stars are more difficult to define than photometric standards. Turnshek et al. (1990) have presented a compilation of calibration sources, including those for polarimetry, whilst Whittet et al. (1992) have observed 105 reddened stars to determine the wavelength dependence, from U to K, of interstellar linear polarization.

Polars and intermediate polars exhibit high levels of linear and circular polarization (10–30%) that are dependent on the binary orbital phase and vary on time scales of minutes/seconds. These high levels of variable polarization are interpreted in terms of cyclotron radiation from accretion columns of gas transferred from a Roche-lobe-filling star to its white-dwarf companion that has a very strong magnetic field. Such discussion of the three-dimensional distribution of accretion structures around binary stars is left more appropriately until Chapter 7. In keeping with the theme of this current chapter, discussing the means by which the sizes and shapes of stars in binaries are determined, we consider the use of linear polarization, and the Stokes Q and U parameters specifically, in providing estimates for stellar sizes in the WR+O binaries, where the usual procedures of light-curve analysis are found to be inadequate. The reasons for such failings are simply the optical-depth variations as functions of orbital phase caused by the varying projections of the stars against the outflowing WR wind, and perhaps the collision of that more dense wind with the much less dense wind from the companion O star.

5.6.2 Linear polarization in WR binaries

We discussed in Chapter 3 the use of the variations in linear polarization with orbital phase to determine the orbital parameters i and Ω. That polarimetric method relied on the double-sine-wave oscillation of p observed during an orbital cycle and caused by the scattering of the O-star photons off the free electrons in the dominant, outflowing WR wind. In that method, based on the prescription given by Brown et al. (1978), it is assumed that the orbital inclination is sufficiently low that eclipses do not take place. However, for eclipsing systems, we see two effects in linear polarization, the first being the standard double-wave variation discussed fully in Chapter 3, and the second being an additional variation occurring only during the eclipse of the core of the WR star by the O star. Chandrasekhar (1946) showed that a significant degree of polarization should be observed when the O star occults a sufficient number of scatterers in the WR wind, such that the assumed spherical symmetry and hence the net zero polarization of the WR wind should be removed by the occultation. Robert et al. (1990) reported the first detection of such eclipse effects in the eclipsing WR+O system V444 Cyg and presented a method to interpret the polarimetric data. That method was improved by St. Louis et al. (1993), and we shall discuss a summary of that procedure here.

St. Louis et al. (1993) have shown that an observable polarization is caused by the occultation of electrons in the WR wind by the shadow cylinder of the O star as seen by the observer at different orbital phases. The only orbital phases where this occultation is significant and sufficient to destroy the otherwise spherical symmetry is during the eclipse of the core region of the WR star, where the electron-column density in the wind is greater than elsewhere. For high electron densities, such that the optical depth is substantial, it is argued that multiple scattering will take place and thereby cancel any net local polarization. But Figure 5.37 shows that n_e decreases very rapidly with increasing distance from the WR star, such that optically thin conditions prevail only a short distance from the WR core. Thus the electrons in a shadow cylinder that passes near to the WR core will be more effectively prevented from contributing to the total amount of scattered, polarized light than will electrons farther out in the wind. Appropriately, the photons from the WR star that are scattered off the electrons in its nearby wind and are occulted by the shadow cylinder are more obviously removed from the total signal than are the O-star photons scattered off those same electrons, because the O star is much farther away. For the system V444 Cyg, St. Louis et al. (1993) showed that the contribution of the O star to the polarization signal was less than 0.5% of the WR-star contribution. In summary, the resultant values for Q and U that would be expected for a WR+O binary with a spherically symmetric wind from only the WR star would be those due to the WR-star contribution alone, as given by

$$Q_{\mathrm{WR}}(\phi) = \left(\frac{-3\sigma_e}{16\pi}\right) f_{\mathrm{WR}} \int_{-\infty}^{-R_{\mathrm{O}}} \int_0^{R_{\mathrm{O}}} \int_0^{2\pi} \frac{n_e(r_{\mathrm{WR}})}{r_{\mathrm{WR}}^4} \left(1 - \frac{R_{\mathrm{WR}}^2}{r_{\mathrm{WR}}^2}\right)^{1/2} \times [(\rho \sin\theta + \Delta\, y)^2 - (\rho\cos\theta + \Delta\, z)^2] \rho\, dx\, d\rho\, d\theta \quad (5.117)$$

$$U_{\mathrm{WR}}(\phi) = \left(\frac{-3\sigma_e}{16\pi}\right) f_{\mathrm{WR}} \int_{-\infty}^{-R_{\mathrm{O}}} \int_0^{R_{\mathrm{O}}} \int_0^{2\pi} \frac{2n_e(r_{\mathrm{WR}})}{r_{\mathrm{WR}}^4} \left(1 - \frac{R_{\mathrm{WR}}^2}{r_{\mathrm{WR}}^2}\right)^{1/2} \times [(\rho \sin\theta + \Delta\, y)(\rho\cos\theta + \Delta\, z)] \rho\, dx\, d\rho\, d\theta \quad (5.118)$$

where the notation is described as follows.

The derivation of these equations from the original theory of Brown et al. (1978) is clearly indicated by St. Louis et al. (1993) and need not be repeated here. The quantity σ_e is the electron-scattering cross section, and f_{WR} is the flux from the WR star at any orbital phase ϕ relative to the total flux density at that phase from the whole binary star in the pass band being used for the polarimetric observations. The integrand contains the electron density n_e in the WR wind at the radial position r_{WR} measured from the centre of the WR star, which has a radius R_{WR}. The triple integral is taken over the volume V, with volume element $dV = \rho\, dx d\rho\, d\theta$, of the shadow cylinder cast by the O star. This volume is defined in terms of the cylindrical coordinate system (x, ρ, θ), where x is positive in the direction to the observer at orbital phase ϕ measured from the centre of the O star, ρ is perpendicular to x at an angle θ to the z axis, which is perpendicular to the observer's line of sight. Then the integration limits extend over the full circle at each (x, ρ), with ρ taking values between 0, the x axis, and the radius of the O star, R_{O}, and the shadow cylinder extending from the far side of the O star, at $x = -R_{\mathrm{O}}$, to the effective outer limit of the WR wind. The position of the O star relative to the WR star is given by $(\Delta\, x, \Delta\, y, \Delta\, z)$, which depend on the orbital phase and are given by the following expressions in the case of a circular relative orbit of radius a:

$$\Delta\, x = a \sin i \cos\eta$$
$$\Delta\, y = a \sin\eta \quad (5.119)$$
$$\Delta\, z = a \cos i \cos\eta$$

where $\eta = 2\pi(0.5 - \phi)$. A standard expression is adopted for the electron density in the wind,

$$v(r_{WR}) = v_\infty(1 - R_c/r_{WR})^\beta \tag{5.120}$$

where the radius R_c is a cut-off radius, adopted so that the wind velocity law and hence the density will not include the dense region of the wind close to the WR star. The constant terms outside the integrals are combined with other multiplying factors from the expression for the wind density to yield a quantity

$$K = (-3\sigma_e/16\pi)(\dot{m}/8\pi\, m_p v_\infty) \tag{5.121}$$

which has units of metres unless one takes the convenient step of dividing it by the units used for r_{WR} inside the (then) dimensionless integrals. Thus K is a scaling constant that defines the amplitudes of the variations of Q and U with orbital phase, and it is evidently connected to the properties of the wind itself. For a mass-loss rate of $\dot{m} = 0.75 \times 10^{-5}\ \mathrm{M_\odot\ yr^{-1}}$ and a terminal velocity in the wind of $v_\infty = 1785\ \mathrm{km\ s^{-1}}$ determined for the binary system V444 Cyg from various different observations, the value of K is approximately 0.036.

Figure 5.40 illustrates the expected variations in the values for Q and U, calculated from equations (5.117) and (5.118) as functions of orbital phase ϕ, through the eclipse of the WR star by the O star. The forms of the Q and U variations are only moderately sensitive to the adopted values of the two stellar radii, as illustrated by St. Louis et al. (1993), and the amplitudes

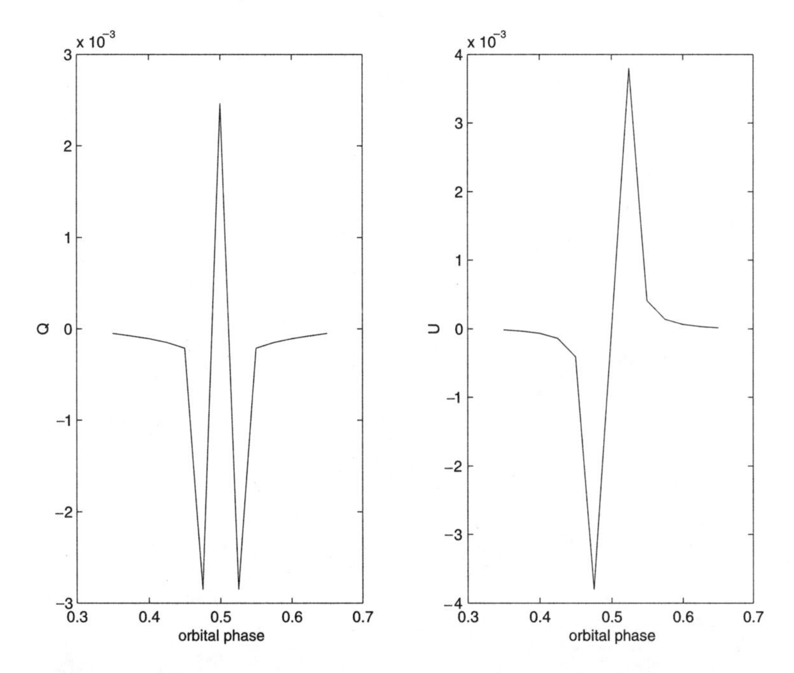

Fig. 5.40. Expected variations in the Stokes parameters Q and U as functions of orbital phase through secondary eclipse for a WR+O binary, calculated from equations (5.117) and (5.118). The values of the parameters used are close to those adopted by St. Louis et al. (1993) for the system V444 Cyg.

of both curves depend on the ratio $\dot{m}/v_\infty$ that describes the properties of the WR wind. Thus polarimetric data of high accuracy for such eclipsing WR+O binaries can allow determinations of, or at least substantial constraints on, the values for both stellar radii, as well as constraints on the orbital inclination because of the additional double-wave variation, and such data can provide a test of the wind parameters determined from the profiles of lines formed in the wind and from radio observations. The recent text by Lamers and Cassinelli (1999), *An Introduction to Stellar Winds*, provides much discussion of the properties of outflows from stars and should be consulted for further details on these topics.

All of the foregoing discussion is based, of course, on the assumption of a spherically symmetric WR wind that does not contain any density enhancements, or *blobs*. However, there is plentiful observational evidence to suggest that such density contrasts exist as transient features passing through the wind envelope over intervals of hours/days. Consequently, polarimetry of high accuracy may be obtainable for such binaries, but may not be repeatable because of the intrinsic variability of the source of polarization. Because such massive binaries have orbital periods of days to tens of days, it may not readily be possible to secure a set of polarimetric data that will define the stellar radii and the orbital inclination with high accuracy. However, in combination with the photometric data and with the spectroscopic and spectropolarimetric data now forthcoming for such difficult systems, it is possible that more secure values for these fundamental parameters will be obtained for some WR+O binaries in the near future.

6 Masses and absolute dimensions for stars in binaries

6.1 Introduction

For many observational astronomers who study the properties of binary stars, the ultimate goal of their work is to make direct determinations of the masses, radii, shapes, temperatures, and luminosities of the component stars, often referred to as the *astrophysical parameters*. The term *absolute dimensions* has been used to indicate that analyses of the radial-velocity curves and light curves for binaries really do provide descriptions of the stars in SI units, regardless of the distances of the binaries from us. As noted in Chapter 1, because the luminosities of the stars in binaries are determined directly, they act, potentially, as excellent *standard candles* for distance determinations amongst nearby galaxies. Much effort has been devoted to finding ways of ensuring that such data are free from systematic errors and have the smallest possible random errors, so that direct comparisons can be made between these empirical results and the predictions from stellar-structure and stellar-evolution theories applied to binary stars. The main theme underlying Chapters 3–5 in this text has been to demonstrate the ways in which systematic errors can be overcome, and random errors minimized, by making use of spectroscopy and photometry at the best spectral and temporal resolution consistent with the observational task at hand. This chapter will summarize the progress that has been achieved in these directions amongst the different subclasses of binary stars.

As a helpful summary, the formulae used to calculate the masses, radii, and so forth, from the various observed quantities discussed in Chapters 3–5 are collected here, together with the formulae required for calculating the associated uncertainties.

From radial-velocity curves, with both components visible, the minimum masses are given by

$$m_{1,2}\sin^3 i = (1.0361\times10^{-7})(1-e^2)^{3/2}(K_1+K_2)^2 K_{2,1} P\,\mathrm{M}_\odot \tag{6.1}$$

whilst the two semimajor axes of the barycentric orbits are

$$a_{1,2}\sin i = (1.9758\times10^{-2})(1-e^2)^{1/2} K_{1,2} P\,\mathrm{R}_\odot \tag{6.2}$$

and the semimajor axis of the relative orbit is $a = a_1 + a_2$.

For single-lined binaries, the mass function is

$$\begin{aligned} f(m) &= (1.0361\times10^{-7})(1-e^2)^{3/2} K_1^3 P\,\mathrm{M}_\odot \\ &= m_2^3\sin^3 i/(m_1+m_2)^2 \\ &= m_1 q^3 \sin^3 i/(1+q)^2 \end{aligned} \tag{6.3}$$

Systems containing sources of pulsed radiation permit the determination of the projected semimajor axis $a_p \sin i$ of the pulsed source, and when combined with radial-velocity observations of the companion giving $f(m)$, they yield the mass ratio and the minimum masses $m_{p,c} \sin^3 i$. Resolved binaries permit direct determinations of the parameters of the relative orbit (a, e, i, ω, Ω), and hence the total mass, from Kepler's third law, if the distance (parallax) has also been determined. When those are combined with radial-velocity data, the individual masses are determined.

The determinations of the uncertainties on these derived minimum masses and projected semimajor axes follow directly from the standard procedures of propagation of uncertainties. The fractional uncertainty in minimum mass is given by

$$\left(\frac{\sigma_{m_{1,2}\sin^3 i}}{m_{1,2}\sin^3 i}\right)^2 = \frac{9e^2}{(1-e^2)}\left(\frac{\sigma_e}{e}\right)^2 + \left(\frac{\sigma_{K_{2,1}}}{K_{2,1}}\right)^2 + 4\left(\frac{\sigma_{K_T}}{K_T}\right)^2 + \left(\frac{\sigma_P}{P}\right)^2 \tag{6.4}$$

where $K_T = K_1 + K_2$, and the fractional uncertainty for a projected semimajor axis is

$$\left(\frac{\sigma_{a_{1,2}\sin i}}{a_{1,2}\sin i}\right)^2 = \frac{e^2}{(1-e^2)}\left(\frac{\sigma_e}{e}\right)^2 + \left(\frac{\sigma_{K_{1,2}}}{K_{1,2}}\right)^2 + \left(\frac{\sigma_P}{P}\right)^2 \tag{6.5}$$

From solutions of light curves we obtain (i, $r_{1,2}$, $T_{1,2}$), as well as the mass ratio q in favourable situations, thus facilitating tests of atmospheric parameters such as limb darkening and gravity darkening. The absolute radii or dimensions of the stars are given by $R_{1,2} = ar_{1,2}$, and the propagation of uncertainties from a, i, and r is straightforward. It is the case that masses and radii are the most accurately determined fundamental parameters for stars in binaries. The relative radii determined from solutions of good light curves are accurate to $\pm 1\%$ or better, and the major source of uncertainty about masses and radii arises from the accuracies of the velocity semiamplitudes.

One of the temperatures, usually T_1, will have been adopted from other data, such as (1) de-reddened colour indices at a total eclipse, (2) using several values of de-reddened colour indices at different orbital phases, as described earlier, (3) spectral types, or (4) the fitting of stellar-atmosphere models to tomographically separated spectra. The other temperature, T_2, may have been derived from the foregoing techniques and confirmed by the light-curve solutions, or determined from the light-curve solutions directly. For the purposes of establishing the total luminosity of each star, the mean radius and the mean temperature are required to be entered into

$$L_{1,2} = 4\pi R_{1,2}^2 \sigma T_{1,2}^4 \tag{6.6}$$

or, more usually,

$$L_{1,2}/\mathrm{L}_\odot = (R_{1,2}/\mathrm{R}_\odot)^2 (T_{1,2}/\mathrm{T}_\odot)^4 \tag{6.7}$$

where $\mathrm{T}_\odot = 5780$ K. The mean radius is usually taken to be the so-called *volume radius*, or the radius of the sphere of the same volume as the (commonly) non-spherical star. The mean temperature is an intensity-weighted average over the star. The uncertainties on the adopted/derived temperatures obviously have a strong influence on the accuracy of the determination of luminosity, because they enter at the fourth power. Because of the large range in stellar luminosities, it is usual to use a logarithmic scale and convert to bolometric magnitudes, so that

$$M_{\mathrm{bol}\,1,2} - \mathrm{M}_{\mathrm{bol}\,\odot} = -5\,\log_{10}(R_{1,2}/\mathrm{R}_{\odot}) - 10\,\log_{10}(T_{1,2}/\mathrm{T}_{\odot}) \tag{6.8}$$

or

$$M_{\mathrm{bol}\,1,2} = C - 5\,\log_{10}(R_{1,2}/\mathrm{R}_{\odot}) - 10\,\log_{10}T_{1,2} \tag{6.9}$$

where the constant $C = \mathrm{M}_{\mathrm{bol}\,\odot} + 10\,\log_{10}\mathrm{T}_{\odot}$. If one adopts the standard values for the Sun of $\mathrm{M}_{\mathrm{bol}\,\odot} = +4.75$ and $\mathrm{T}_{\odot} = 5780$ K, then $C = 42.369$. But Popper (1980) has argued that we should not place too much reliance on the Sun, and instead he used a set of stars with known angular diameters in addition to the Sun to determine $C = 42.255$, with $\mathrm{M}_{\mathrm{bol}\,\odot} = 4.69$ and $\mathrm{BC}_{\odot} = -0.14$. This latter value for C has been used, together with the Popper (1980) bolometric-correction (BC) tabulation, to yield absolute visual magnitudes M_V for many eclipsing binaries via $M_V = M_{\mathrm{bol}} - \mathrm{BC}$. Schönberner and Harmanec (1995) used the Code et al. (1976) BC tabulation, which differs somewhat from that of Popper, and $\mathrm{M}_{\mathrm{bol}\,\odot} = 4.75$, with $\mathrm{BC}_{\odot} = -0.07$, to determine M_V values that are essentially the same as those via Popper's values.

The uncertainties in these M_{bol} and M_V values are given by

$$\sigma^2_{M_{\mathrm{bol}}} = \left(\frac{5 \times 0.4343\sigma_R}{R}\right)^2 + \left(\frac{10 \times 0.4343\sigma_T}{T}\right)^2 \tag{6.10}$$

$$\sigma^2_{M_V} = \sigma^2_{M_{\mathrm{bol}}} + \sigma^2_{\mathrm{BC}} \tag{6.11}$$

The importance of minimizing the uncertainties in the temperatures is obvious. The distance d to a binary system is then usually determined by combining the two M_V values via, for example,

$$M_{V,\mathrm{total}} - M_{V,2} = -2.5\,\log_{10}\left(1 + 10^{-0.4(M_{V,1}-M_{V,2})}\right) \tag{6.12}$$

and using the standard distance-modulus equation

$$V - A_V - M_{V,\mathrm{total}} = 5\,\log_{10}d - 5 \tag{6.13}$$

with the apparent V magnitude taken to be that seen at maximum brightness of the system, commonly the quadrature value.

The mean surface gravity g of a star is given by $g = Gm/R^2$ and is traditionally written as a logarithmic quantity in centimetre-gram-second (*CGS*) units. Thus, for the Sun, $\log_{10}\mathrm{g}_{\odot} = 4.438$, and we can write generally that

$$\log_{10}g = 4.438 + \log_{10}(m/\mathrm{M}_{\odot}) - 2\,\log_{10}(R/\mathrm{R}_{\odot}) \tag{6.14}$$

This quantity is determinable directly for each component in double-lined eclipsing binaries and can be compared with values derived independently from colour indices that are sensitive to surface gravity and from detailed analyses of absorption-line profiles, particularly the H lines, where the Stark effect can be strong. Likewise, the projected rotational velocity of a star, $V_{\rm rot} \sin i = 2\pi R \sin i / P_{\rm orb}$, can be determined from line profiles, as noted earlier, and compared with the values derived from the stellar radius and orbital inclination to see whether or not a star is rotating synchronously. These opportunities for cross-checks between the final results from radial-velocity-curve and light-curve analyses, on the one hand, and the independent values from, for example, line profiles, on the other, can be very valuable for improving the overall quality of the final results on a particular binary system.

6.2 Detached binaries

For *detached* binary stars, the celebrated review by Andersen (1991) sets the modern theme in this area. From determinations for detached, double-lined, eclipsing binary stars we know the masses and radii for main-sequence stars in (currently) non-interacting binaries to accuracies of better than $\pm 2\%$ for all main-sequence spectral types between about B0 and G8. For O stars and stars cooler than about K0, many of the available data are not sufficiently precise or sufficiently numerous to pass the stringent quality-control tests that Andersen imposed in order to achieve those impressively low uncertainties. Those requirements were complete multi-colour light curves, with RMS errors of ± 0.01 mag or less, and radial-velocity curves for both components defined by about 25 observations, each with RMS errors of ± 1 km s^{-1}, that sample both quadratures very well. Such a simple statement hides the considerable demands placed on the observer/analyser: securing large amounts of telescope time, sometimes on the largest telescopes, ensuring that all observations are obtained in well-controlled conditions on many long nights, and then analysing all the data in a careful, systematic fashion that may need to be repeated to guarantee the achieved accuracy.

Since that compilation, several more of the detached eclipsing binaries have had their masses, radii, temperatures, and luminosities determined to equivalent accuracy, and the total number of well-determined systems is currently about 57, or 114 stars. Figures 6.1 and 6.2 show the values for the 114 stars plotted in the standard planes of mass–luminosity, temperature–luminosity (the HR diagram), mass–radius, and mass–surface gravity. It is from such data that we can make empirical estimates of relationships between these fundamental stellar parameters. In the $\log_{10}$(mass)–$\log_{10}$(luminosity) plane, the relationship is slightly curved, allowing it to be represented by two straight lines of the form $\log_{10}(L/{\rm L}_\odot) = 4.0 \log_{10}(m/{\rm M}_\odot)$ for stars with $m \leq 10$ M$_\odot$, and with the exponent reduced to about 3.6 for stars of higher mass. In the $\log_{10}$(mass)–$\log_{10}$(radius) plane the zero-age main-sequence (ZAMS) relationship is approximately $\log_{10}(R/{\rm R}_\odot) = 0.6 \log_{10}(m/{\rm M}_\odot)$, except for lower-mass stars, where the exponent is nearer 0.9. Such relationships are useful to gain approximate estimates of the likely sizes and luminosities of stars for a given mass provided that they are in the main sequence. They should not be used for any more precise work, for the simple reason that the evolution of a star across the main-sequence band itself leads to substantial changes in its radius and effective temperature, an issue that was admirably explained by Nordström (1989) in her discussion of relationships between mass and spectral type for luminosity class V stars. A mass estimate from a main-sequence spectral type will be uncertain by at least $\pm 15\%$.

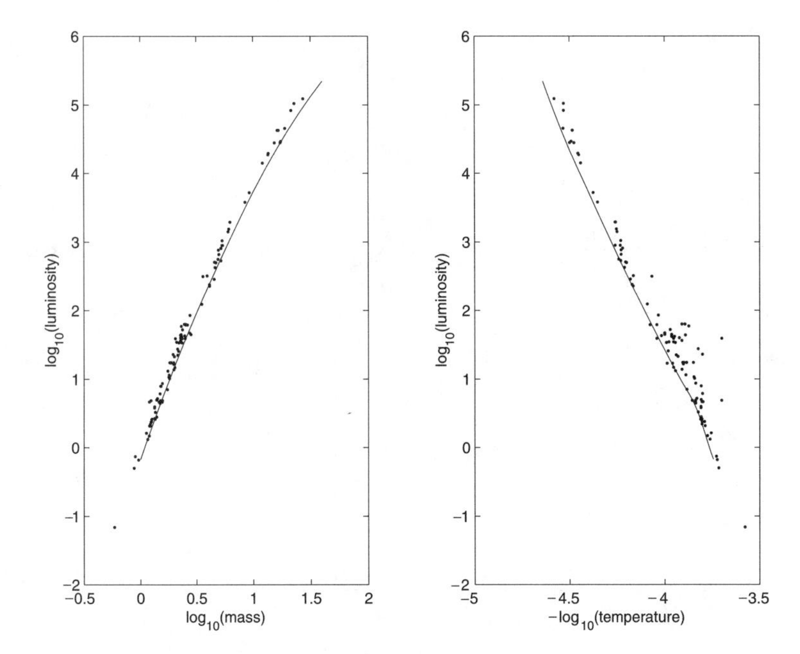

Fig. 6.1. Empirical masses ($M_\odot$), luminosities ($L_\odot$), and temperatures (K) for the stars in detached, double-lined eclipsing binaries, with average uncertainties of ± 0.006 dex in $\log_{10} m$, ± 0.055 dex in $\log_{10} L$, and ± 0.013 dex in $\log_{10} T$, from the review by Andersen (1991) and subsequent publications by Holmgren et al. (1990a, 1991), Clausen (1991, 1996), Andersen et al. (1993), Giménez and Clausen (1994), Nordström and Johansen (1994), Hill and Holmgren (1995), Lacy (1997), Lacy et al. (1997), Torres et al. (1997), Vaz et al. (1997), and Marschall et al. (1997). Also shown is the theoretical ZAMS line from the models with ($X = 0.70$, $Z = 0.02$) by Claret and Giménez (1992).

Also plotted on those planes are the ZAMS lines for chemical abundances, with $X = 0.70$ and $Z = 0.02$, from Claret and Giménez (1992). It is very clear that the agreements between the ZAMS lines and the lower envelopes of the distributions of observed data are excellent. Standard stellar evolution across the main-sequence band is towards increasing radius and hence increasing luminosity, though somewhat decreasing temperature, at constant mass. Consequently, we would expect to see a distribution of points away from a lower envelope, as observed. But on closer inspection, as emphasized by Andersen (1991, 1997), we find that the relationships like that of luminosity to mass are not simple monotonic ones with a little observational scatter. It would seem that chemical-composition differences between individual stars in the sample are sufficient to introduce additional scatter in these figures, at levels of up to 5%. Unfortunately, to date, very few stars in eclipsing binaries have had accurate determinations of their chemical abundances. But the developing techniques of tomography should lead the way in rectifying that paucity of accurate information.

So detached systems can provide excellent empirical data on masses, radii, and luminosities if the temperatures are known, and hence distances. With the recent publication of accurate parallaxes from the *Hipparcos* astrometric satellite, do the distances for these eclipsing binaries

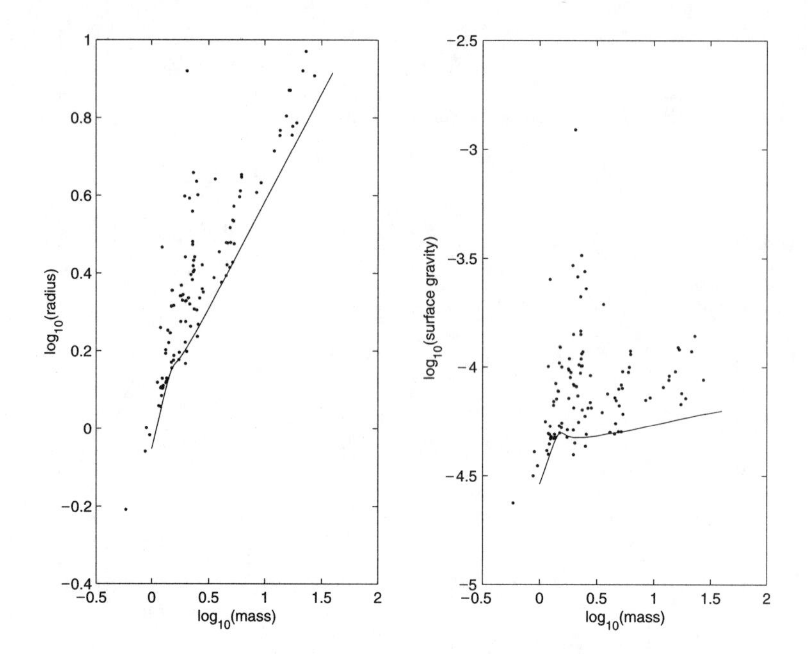

Fig. 6.2. Empirical masses ($M_\odot$), radii ($R_\odot$), and surface gravities [log g (CGS)], for the stars in detached, double-lined eclipsing binaries, with average uncertainties of ± 0.006 dex in $\log_{10} m$, ± 0.006 dex in $\log_{10} R$, and ± 0.015 dex in $\log_{10} g$, from the review by Andersen (1991) and subsequent publications by Holmgren et al. (1990a, 1991), Clausen (1991, 1996), Andersen et al. (1993), Giménez and Clausen (1994), Nordström and Johansen (1994), Hill and Holmgren (1995), Lacy (1997), Lacy et al. (1997), Torres et al. (1997), Vaz et al. (1997), and Marschall et al. (1997). Also shown is the theoretical ZAMS line from the models with ($X = 0.70$, $Z = 0.02$) by Claret and Giménez (1992).

agree between the two sources? Or, to invert the problem, given astrometric distances and therefore luminosities for stars with accurately determined radii, do the resultant temperatures agree with the calibrations via multi-colour photometry? Ribas et al. (1998) examined this issue by selecting 20 systems with *Hipparcos* parallaxes that were contained in the earlier review by Andersen (1991), plus AR Aur from Nordström and Johansen (1994). Their conclusions were that for the range 5000 K $\leq T \leq$ 10,000 K, the temperatures showed only a small systematic offset of 0.01 ± 0.01 dex, whilst for higher values up to 25,000 K, the offset was indeterminate at 0.015 ± 0.060 dex. Clearly, the data for the stars of higher temperatures (types B and O that are poorly represented amongst the solar-neighbourhood region of the Milky Way, and therefore are farther away, with less precise parallaxes) are not sufficient for such a stringent test.

Many of the previously mentioned detached systems have mass ratios near to unity, which made them easier to discover, because their spectra were double-lined, and the velocity amplitudes were large. These have tested the basic stellar-structure models and now confirm good agreement between observation and theory for unevolved stars, so that we can feel confident that the basis for understanding subsequent stellar evolution is well founded. However, the theory of differential evolution between two stars of the same age but different initial masses remains untested by such binaries. Some of the detached systems do contain stars of quite different masses, and these can be used for critical tests of differential evolution, because the stars should lie on the same isochrone line in the mass–radius or mass–log g plane and in the HR diagram. An example of excellent agreement is that between the results on AI Phe from Andersen et al. (1988) and the theoretical models of VandenBerg (1985), where the primary component is classified as a normal subgiant that has not (yet) filled its Roche lobe nor undergone any mass loss/exchange. A second and more evolved example is the system TZ For, where the primary is a normal giant (G8III) in a 75-day orbit, and the secondary is a subgiant (F7IV), both fitting the same theoretical isochrone, a major success. As emphasized by Andersen, these detached systems with substantial differential evolution serve as crucial tests for models of stellar structure and evolution, because all the parameters are specified except for the helium abundance and the mixing-length parameter, which are not directly observable.

Amongst the O-star binaries, which remain under-represented in the tabulations of very accurate data, we find systems of equal-mass stars and of somewhat dissimilar stars. Harries and Hilditch (1997a) considered the current status of such systems and collected results accurate to $\pm$2–5% in masses and radii. The mass–log g and HR diagrams are shown in Figure 6.3. The line connecting the two stars in each system typically is parallel to the isochrone line in each diagram, particularly for the system V3903 Sgr (Vaz et al. 1997) containing dissimilar stars. The agreements between empirical masses and radii and the corresponding theoretical values are very good, with no systematic differences. But the temperatures according to the models are systematically lower than those from the usually adopted calibrations of T versus spectral types or colour indices amongst O stars, by about 1500–2000 K. These comparisons seem to be in the same sense and of the same magnitude as those being suggested by the latest stellar-atmosphere models that incorporate more line blocking in the UV.

At the other end of the main sequence, the paucity of accurate data is again obvious, because of the fact that *eclipsing* systems containing such small stars will be inherently difficult to find. One remarkable recent success has been the very accurate determination of the masses of the two stars in CM Dra, secured by Metcalfe et al. (1996), which are given as $m_1 = 0.2307 \pm 0.0010\ \mathrm{M}_\odot$ and $m_2 = 0.2136 \pm 0.0010\ \mathrm{M}_\odot$. The system appears to be a member of Population II in the galaxy, and the accurate specification of all of its parameters will allow it be used as a test of competing stellar-structure models at that low mass, and as a test of the primordial helium abundance.

Mathieu (1994) reviewed our knowledge of the properties of pre-main-sequence binary stars, pointing out in this particular context that empirical determinations of masses for such stars were still very limited at that time, essentially to EK Cep (Popper 1987), which contains a main-sequence primary with a pre-main-sequence secondary. Claret et al. (1995) have compared the predictions of evolution models with the properties of EK Cep, including its abundance of lithium, and have shown that the two components share the isochrone line at 2×10^7 years, with the secondary still contracting to the main sequence, and the primary at the beginning of the H-core fusion phase. The observed apsidal-motion rate is also in agreement with the model. Since then, the system TY CrA (Casey et al. 1998) has provided additional high-quality empirical data. In that system, Casey and associates have shown that the secondary component is a 1.64-$\mathrm{M}_\odot$ pre-main-sequence star at the lowest point on the Hayashi track, whilst the 3.16-$\mathrm{M}_\odot$ primary is on the main sequence, with both stars sharing the age of 3 million years. To quote Casey and associates, the secondary of TY CrA 'is the least evolved star with a dynamically

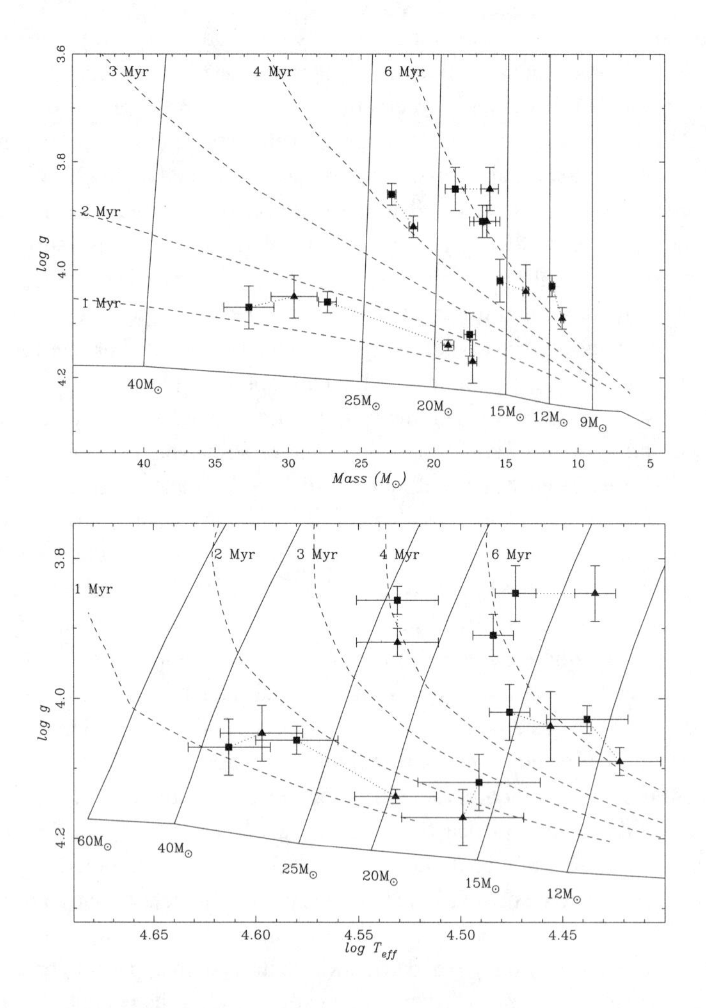

Fig. 6.3. The positions of the primary (squares) and secondary (triangles) components (joined by dotted lines) of the detached O-star binaries compiled by Harries and Hilditch (1997a) and plotted in the mass–log g plane (upper panel) and the log T–log g plane (lower panel). The solid labelled lines are lines of constant mass, whilst the lowest line is the ZAMS line, and the dash lines are isochrones of different ages, all taken from the models of Schaller et al. (1992).

measured mass'. Clearly, much work remains to be done in this field, with intriguing systems being discovered, such as DQ Tau (Mathieu et al. 1997), which is a classical T Tauri star that has been found to be a double-lined spectroscopic binary with a period of 15.8 days, an orbital eccentricity $e = 0.556$, and abundant evidence for circumstellar material.

The general term *chromospherically active binaries* (CABs) is now used to encompass all the detached systems that show levels of chromospheric activity far in excess of that seen on the Sun. Thus, RS CVn systems and BY Dra systems are included, all sharing the common characteristics of rapid axial rotation and deep convective envelopes, resulting in substantial magnetic activity. This activity is manifested as photospheric starspots, emission lines from

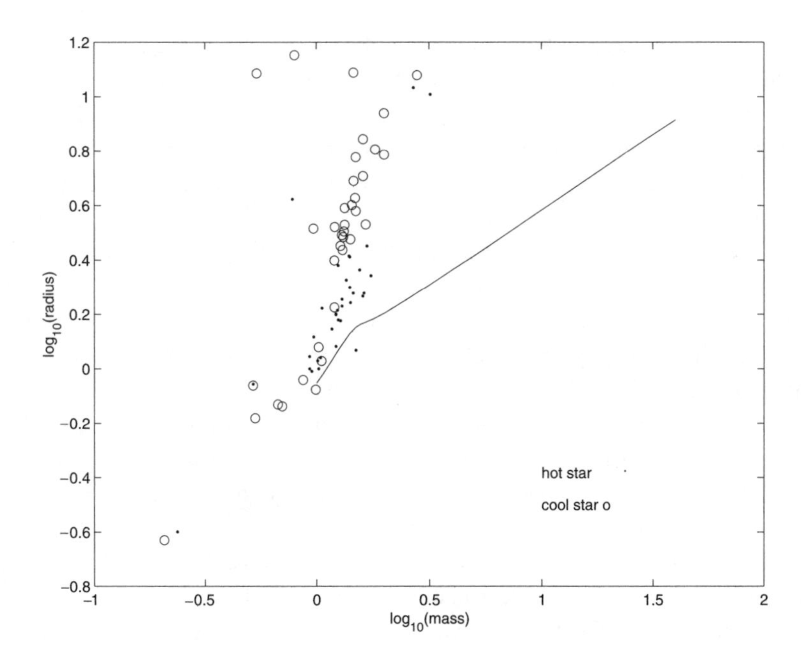

Fig. 6.4. The positions in the mass–radius plane of the components of chromospherically active binaries, as listed by Strassmeier et al. (1993). The hotter components are identified by dots, and the cooler components by open circles. The ZAMS line from Claret and Giménez (1992) is also shown.

the chromosphere and transition regions, and enhanced x-ray emission from active coronae. Strassmeier et al. (1993) published the second edition of their catalogue of chromospherically active binaries and provided summary data on the stellar masses and radii derived from those binaries that displayed eclipses. Those values are plotted in Figure 6.4, together with the same theoretical ZAMS line from Figure 6.2 in order to place these systems in some context. Although the accuracy of these data do not in general equal that attained for the uncomplicated detached systems noted earlier, these diagrams confirm that the *hotter* component in each system generally lies closer to the ZAMS, whilst the *cooler* components are the more evolved subgiants and giants. Note also the upper mass limit for such CABs around the transition from convective to radiative stellar envelopes. The catalogue contains both short-period systems ($P \leq 1$ day) and longer-period systems, and so we see unevolved main-sequence systems as well as systems whose orbital sizes permit significant evolution of the stars into the giant region before reaching their limiting Roche surfaces. The range of orbital periods for the systems plotted is from 0.5 to 39 days. Some of these longer-period systems could also be used as tests of differential evolution.

6.3 Semidetached binaries

If a binary system has one component filling its Roche lobe, that means that the system has passed through at least one stage of mass transfer between the two components, or one stage of mass loss from the system, or both. Consequently, it would be wrong to make comparisons between the empirical data on the masses and absolute dimensions for such systems and the theoretical stellar-evolution models for single stars alone. Rather, stellar-evolution codes must be modified to take account of the existence of an upper limit to the volume of a star in a binary and to allow for changes in orbital periods and sizes consequent upon the mass-exchange processes discussed in Chapter 4. The text by de Loore and Doom (1992), *Structure and Evolution of Single and Binary Stars*, describes how such theoretical modelling is conducted. The concern here is to ask how well we know the astrophysical parameters of such evolved systems, and whether or not they can serve as substantial or crucial tests for theories of binary-star evolution. Whilst we may argue about the accuracies of some of the empirical results, there is no doubt that evolved systems can serve as crucial tests of the theories.

The *classical Algol systems*, so named after the prototype Algol, or β Per, are known to each contain a main-sequence primary of intermediate mass that lies well within its Roche lobe and a Roche-lobe-filling secondary of substantially lower mass, so that $q \sim 0.3$, typically. The secondary is always classified as a subgiant and seemingly is more evolved than the more massive primary. This *Algol paradox* was resolved in the 1960s, following a suggestion from Crawford (1955), when it became clear that Roche-lobe overflow (RLOF) was an important mechanism in binary-star evolution and would lead to reversal of the mass ratio in a binary, such that the originally more massive star would become the secondary, whilst the originally less massive star would become the primary. (The possibilities for endless confusion over terminology in this subject abound! I shall follow the terminology of the Brussels group and use the terms *loser* and *gainer*.) The first models of binary-star evolution involved conservative mass transfer, sensibly so in order to limit the range of parameter space investigated. At first sight, it seemed that the Algol paradox could readily be understood as the consequence of differential evolution between the two stars of different masses, with the originally more massive star becoming the loser, and its companion the gainer, with all mass transferred conservatively. However, when it became possible to make quantitative tests (Refsdal et al. 1974), for one of the better-studied Algol systems, AS Eri (Popper 1973), it became clear that conservative mass exchange failed to explain the properties of this semidetached binary, and it became imperative to postulate non-conservative mass and angular-momentum transfer/loss.

Many grids for models of binary-star evolution have been calculated for the mass-transfer cases A, B, and C, as described in Chapter 1, and for varying amounts of mass and angular-momentum transfer/loss. Some examples, and many references, have been given by de Loore and Doom (1992). Specifically for Algol systems, De Greve (1989, 1993), De Greve and de Loore (1992), and de Loore and De Greve (1992) have considered a substantial range of evolution calculations, whilst Sarna (1993) has discussed in detail the evolutionary status of β Per itself. The overall conclusion is that conservative evolution cannot explain the properties of Algol systems and that non-conservative mass exchange in early case B is required. In an attempt to provide more accurate quantitative data for comparisons with models, Maxted (1994) and Maxted and Hilditch (1996) collected data on nine Algol systems for which astrophysical parameters had been derived from self-consistent solutions of the light curves and the radial-velocity curves of the two components, including taking account of non-Keplerian corrections.

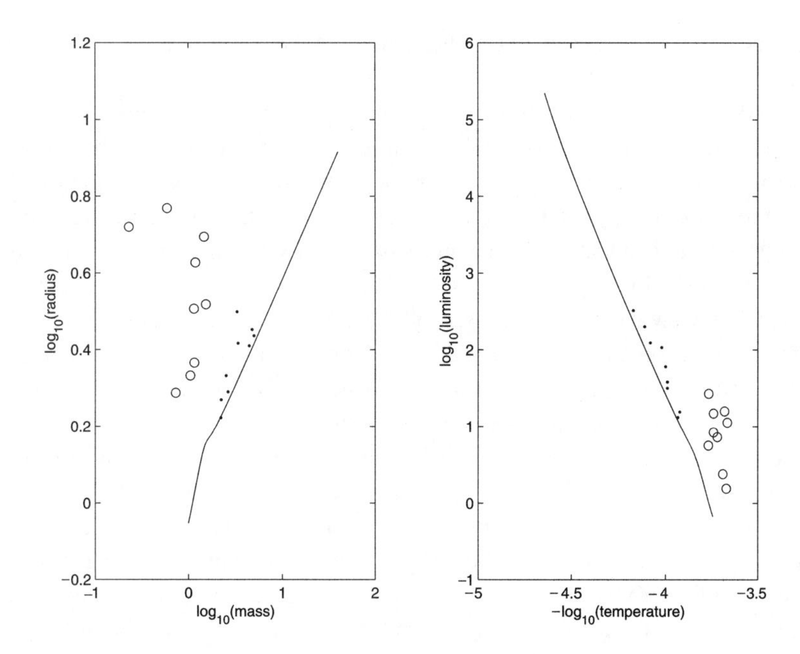

Fig. 6.5. The positions of the primary (dots) and secondary (open circles) components of nine classical Algol systems, selected to have very well-determined astrophysical parameters by Maxted and Hilditch (1996) and plotted in the mass–radius plane and the HR diagram, together with the same ZAMS lines as the other figures in this section. Mean uncertainties on the various quantities (pri/sec): masses, 0.033/0.017 dex; radii, 0.021/0.017 dex; temperatures, 0.014/0.017 dex; luminosities, 0.073/0.073 dex.

The resultant m, R, T, and L values are shown in Figure 6.5. They demonstrate very clearly the strong differences between these types of binaries and the detached systems discussed earlier. The current primary components, the gainers, appear as normal main-sequence stars, whilst the current secondary components, the losers, are all oversized and overluminous relative to main-sequence stars of the same mass, by amounts that range up to factors of 10. On the basis of detailed comparisons between these data and the available models, it would seem that more angular-momentum loss is required (perhaps by means of a magnetic stellar wind) than is currently employed within the models.

It is regrettable that only nine systems could be found for which the data approached the quality of that enjoyed in studying detached systems. Certainly it is very difficult to secure accurate measurements of the radial velocities of the low-luminosity secondary components, but the work of G. Hill and colleagues has shown how that can be achieved. With the application of tomography as well, it seems entirely feasible to extract well-defined spectra for the secondary components that could be used to determine the chemical abundances of these mass-losing stars. Sarna (1992, 1993) and Sarna and De Greve (1996) have argued that such determinations of the carbon abundance, in particular, for the losers in Algol systems can act as major constraints on the evolution models for these systems, provided that the data are sufficiently accurate. With the currently available data, Sarna and De Greve have shown that the agreement between the observed C deficiencies and the theoretical models is reasonable, but not definitive.

Budding (1984) presented a valuable catalogue of 414 binary systems whose properties placed them in the overall area of semidetached binaries similar to classical Algol systems. Some of them are well-known objects that have been studied for many years to investigate the processes of mass exchange, but it is regrettable that so few of them have precisely determined masses and other parameters. Further details about Algol systems in general have been published by Budding (1986), as well as in the conference proceedings *Algols*, edited by Batten (1989), and in a review paper by McCluskey (1993). Amongst the systems of earlier spectral type, there are several binaries containing O- or B-type primaries with somewhat later-type companions, giving mass ratios typically $q \sim 0.5$, whereas the classical Algols are nearer half that value. Comparisons between the properties of such stars and evolution models by Hilditch and Bell (1987) suggested that case-A models were required to explain them, rather than the early-case-B requirement to explain the low-mass losers in classical Algols. That seems also to be the case for the most massive semidetached systems, containing O-type primaries and late-O- or early-B-type secondaries, as noted by Harries and Hilditch (1997a). Figure 6.6 illustrates the data for five semidetached O+B binaries with well-determined parameters, together with isochrone lines from models for single stars. These binaries seem to have been formed via a case-A mass-exchange process, as determined by de Loore and Vanbeveren (1994), though there remains a need for specific modelling of individual systems, rather than reliance on calculations from model grids that are too coarsely spaced to address the few systems for which

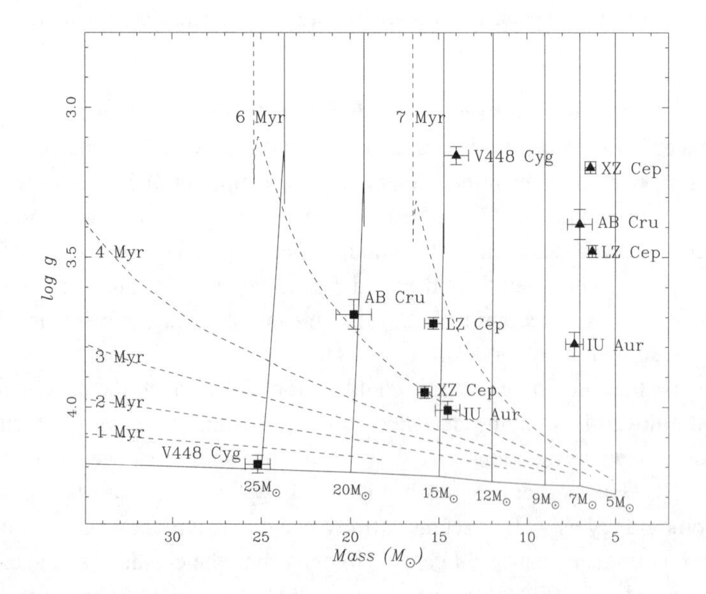

Fig. 6.6. The positions of the primary (squares) and secondary (triangles) components of the semidetached binaries with O-star primaries plotted in the mass–log *g* plane. The solid lines are the evolutionary tracks from Schaller et al. (1992) for single stars, and the dash lines represent isochrones for the same models.

there are good empirical data. The reviews by Figueiredo et al. (1994) and Vanbeveren et al. (1998) provide examples of comparisons between evolution calculations and the observed properties of O-star and B-star binaries.

The x-ray binaries and the cataclysmic variables fall into this category of semidetached binaries, but because their evolutionary state is more complex, involving more than one mass-loss episode, and one component is far removed from the main-sequence stage or giant stage of evolution, they are considered later, in their separate sections.

6.4 Contact binaries

The contact binaries compose the last main grouping of systems with both components clustered around the main sequence in the HR diagram. They range in spectral type from O to K and therefore encompass some quite different categories of binaries. By far the most common are the W-UMa-type contact binaries that are composed of stars with convective envelopes: spectral types late F to K. They display an obvious anomaly (mass ratio of about 0.5, and a surface-brightness ratio of about 1) that was first discussed by Kuiper (1941). Lucy (1968a) provided the first explanation in terms of the two stellar cores being surrounded by a common convective envelope at nearly constant temperature. Recent general reviews of the properties and evolution of W-UMa-type contact binaries are those by Rucinski (1993) and Eggleton (1996), in which many references can be found to earlier work.

The most recent survey of the parameters derived from analyses of light curves and radial-velocity curves by many researchers is that by Maceroni and van't Veer (1996), where 78 systems are listed. Of those 78, only 37 have radial-velocity curves as well, and Maceroni and van't Veer comment that perhaps 10 are based on spectra of good resolution and have been analysed by modern CCF-type techniques. In Figure 6.7, the masses, radii, temperatures, and luminosities are plotted for the components of 35 systems determined from the data of Maceroni and van't Veer (1996, tables 2 and 3). (Two systems were excluded because their mass ratios, derived spectroscopically and photometrically, were so discordant.) No assumptions have been made in determining these quantities; the equations summarized in the earlier part of this chapter were used directly. The uncertainties associated with these quantities are, in general, greater than those for detached systems, principally because of the shortage of very accurate radial velocities, but also because of the frequently asymmetric nature of the light curves, leading to different values for radii, and so forth, from different parts of the light curves, as noted in Chapter 5.

The distributions of the components about the ZAMS lines are very similar to those found by Hilditch et al. (1988) in an earlier survey covering contact and near-contact binaries and require some explanation. W-UMa-type binaries were subdivided into two groups by Binnendijk (1970), A-type and W-type systems, depending, respectively, on whether the more massive star or the less massive star was eclipsed at primary eclipse. It was noted that A-type systems seemed to have larger total masses, and to be rather more evolved than W-type systems, a factor well confirmed by Mochnacki (1981), as seen in the preceding figures. In the mass–radius plane, the primary components seem to act as normal main-sequence stars distributed through the main-sequence band. The secondary stars are oversized for their ZAMS masses by amounts

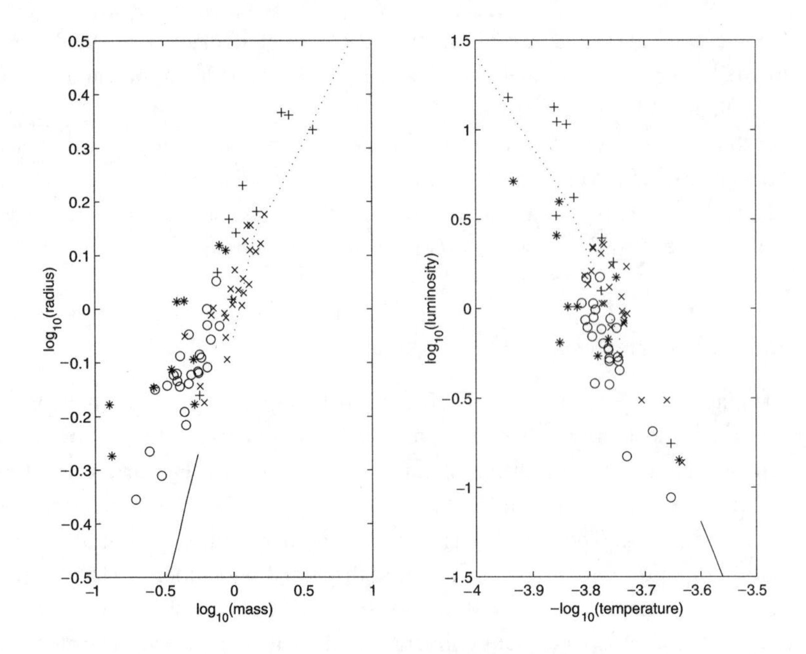

Fig. 6.7. The positions of the components of A-type and W-type W UMa contact binaries in the mass–radius plane and in the HR diagram, together with the ZAMS lines from Claret and Giménez (1992)(dotted line) and from Dorman et al. (1989)(solid line). Symbols: plus sign, A-type primary; asterisk, A-type secondary; cross, W-type primary; open circle, W-type secondary. Typical uncertainties: masses, ± 0.05 dex; radii, ± 0.025 dex; temperature, ± 0.02 dex; luminosities, ± 0.07 dex.

that range up to factors of 10. In the HR diagram, the secondary stars are, unusually, found to lie to the *left* of the main sequence, contrary to any expectations from standard stellar-evolution theory. The resolution of this anomaly is found by recognizing that the temperature assigned to the secondary from the light-curve solutions is that of the common envelope around its core, rather than the value that it would have if it were a separate star. Corrections for the expected *luminosity transfer* from the primary to the secondary amount to $\Delta \log T \sim -0.09$ and $\Delta \log L/L_\odot \sim -0.36$ for the secondary, and, correspondingly, $\Delta \log T \sim +0.05$ and $\Delta \log L/L_\odot \sim +0.20$ for the primary. These ensure that the locations of the secondaries are returned back to the right-hand side of the main sequence, and they change only slightly the positions of the primary components (Mochnacki 1981; Hilditch et al. 1988).

Whilst many accurate light curves for W-UMa-type binaries have been published, there remains a shortage of radial-velocity curves for both components obtained from spectra at resolutions of 10,000 and analysed via CCF or broadening-function analyses. Most recently, Lu and Rucinski (1999) and Rucinski and Lu (1999) have begun to rectify this issue by securing good radial-velocity data on (so far) 20 contact binaries. We can expect corresponding improvements in the derived properties for these systems.

The consensus seems to be that we understand W-UMa-type contact binaries to be formed from detached systems of low mass, with orbital periods of 1 day or less, that lose orbital angular momentum because of a magnetic stellar wind. As the orbit size shrinks, so do the Roche lobes, so that the limiting volume is compressed onto the larger, primary component first. This event causes standard RLOF to occur from the primary to the secondary on a relatively leisurely thermal time scale of about 10^7 years (Sarna and Fedorova 1989), rather than on the shorter time scale of 10^{4-5} years for the rapid phase of mass transfer seen in case-B RLOF. When only about 2% of the mass of the primary has been transferred to the secondary, the system forms a contact binary. It can remain in that state of quasi equilibrium, alternating about a state of quite marginal contact that might not be broken completely – the thermal-relaxation-oscillations (TRO) theory of Flannery (1976) and Lucy (1976) – for an uncertain interval of $n \times 10^8$ years, before coalescing into a single star.

The observational evidence to support such ideas is quite incomplete. Certainly there are many low-mass detached binaries in the appropriate period range of 1 day or less. The RS CVn systems, discussed earlier, contain a subset of short-period systems, for example, XY UMa, with $P = 0.6$ day, that display all the obvious signs of magnetic activity and hence angular-momentum loss caused by a magnetic stellar wind, even though we do not yet know the precise time scales for such mechanisms (e.g., Guinan and Giménez 1993). The transition to a contact state is expected to be quite fast and not easily observed, but perhaps one such system, V361 Lyr, is a prime example, as already discussed; see Hilditch et al. (1997) and the sources cited therein. W UMa systems display all the magnetic activity seen in RS CVn binaries – starspots, enhanced chromospheric and coronal emissions in the optical, ultraviolet, and x-ray regions, and occasional radio flares. As recent examples of such work, in a survey of near-contact and W-UMa-type binaries from the ROSAT all-sky survey, Shaw et al. (1996) showed that the x-ray-luminosity functions for the different types of systems (A-type and W-type W UMa systems, near-contact systems, RS CVn systems) were significantly different, thereby indicating substantially different coronal and magnetic activities; and Brickhouse and Dupree (1998) have used EUVE-satellite photometry to determine localized coronal structure in the contact binary 44i Boo. We have not yet succeeded in linking the seemingly semidetached/near-contact binaries that have orbital periods overlapping those of the contact systems (above $P \sim 0.45$ day) into a common evolutionary scheme that can encompass the theoretical expectations of alternating contact/near-contact states, as well as angular-momentum loss (AML) in contact states. Part of that solution will lie in securing very accurate data on the stellar surfaces by means of Doppler tomography so that we can study evidence for the transfer of mass and energy between the two components. There is already some evidence, in the form of enhanced brightness around the neck region of the common envelope, from several contact binaries, as discussed by Kaluzny (1985), Hilditch (1989), Hilditch et al. (1998), and Djurasević et al. (1999), but more detailed observational findings are required. Also contributing to solution of the problem of the evolution of contact binaries is statistical evidence in the form of relative numbers of contact and other systems found in unbiased surveys such as the OGLE data base, as discussed by Rucinski (1997a,b, 1998a,b). Perhaps, also, the evolution of contact binaries can explain at least some of the blue stragglers seen in old open clusters and even in globular clusters; see *The Origins, Evolution, and Destinies of Binary Stars in Clusters*, edited by Milone and Mermilliod (1996).

There are also contact binaries to be found at spectral types A, B, and O. Those that have been discovered amongst the A stars have been little studied, even though their properties must be quite different from those of W UMa systems, because they do not possess convective envelopes. Amongst the O- and B-type contact systems there are several with components of nearly equal masses and orbital periods, ranging from 0.66 to 5.5 days, that seem to fit with the theoretical models of Sybesma (1986), having ages of a few million years and evolving across the main sequence 'in parallel' (Hilditch and Bell 1987) . How they were formed remains an unanswered question. There are additional systems, like AO Cas and TU Mus, that seem to be marginal-contact systems with unequal components, typically $27 + 19\ M_\odot$, that have achieved that state by case-A mass-ratio reversal followed by alternating semidetached and marginal-contact configurations, with mass-transfer reversals between the components (Sybesma 1986; Hilditch and Bell 1987; Figueiredo et al. 1994). A recent addition to this small list of *OB*-type contact binaries, is the system, V606 Cen, (Lorenz et al. 1999). Vanbeveren et al. (1998) have provided a further review of the properties and evolution of massive stars in general.

6.5 Cataclysmic variables and related systems

The second edition of the *Catalogue and Atlas of Cataclysmic Variables* (Downes et al. 1997) lists 1020 CVs, together with accurate coordinates and valuable finding charts. The properties of CVs, with all their subtypes, have been reviewed in a monograph by Warner (1995), and so only a summary of the determinations of masses, and so forth, will be given here. The latest edition of the *Catalogue of Cataclysmic Variables, Low-Mass X-ray Binaries, and Related Objects* is the sixth edition by Ritter and Kolb (1998). Smith and Dhillon (1998) have conducted a critical analysis of the available determinations of the masses, radii, and spectral types for the secondary components in CVs and in LMXBs. Their final values for the masses and radii of these 14 secondaries (their table 3) are plotted in Figure 6.8, together with theoretical main sequences from two computations, as well as the masses and radii for the components of the detached eclipsing binaries YY Gem and CM Dra. It is clear that these stars, at least for systems with orbital periods less than 7–8 hours, have the same properties as normal main-sequence stars of the same mass, despite having passed through the evolutionary sequence required to produce a CV. That sequence involves evolution of the original primary (in a binary with a long orbital period) beyond the red-giant stage to case-C mass exchange, the formation and subsequent ejection of a common envelope from the system (the planetary-nebula phase, with its associated drastic reduction in orbit size), and the subsequent evolution of the red-giant core through the subdwarf O-star phase to the white-dwarf stage. Then orbital angular-momentum loss due to a magnetic stellar wind from the low-mass secondary brings the secondary into contact with its Roche lobe, and the consequent RLOF starts the CV phase of the evolution of the binary. The low-mass secondary, according to the current evidence, seems to be remarkably unaffected by these local turmoils. Regrettably, there is not yet enough data of sufficient quality to make a further test of the disrupted-magnetic-braking model that seeks to explain the existence of the gap in the distribution of orbital periods of CVs, between $P \sim 2$ hours and $P \sim 3$ hours. That test, as described by Smith and Dhillon, should determine whether or not the radii of the

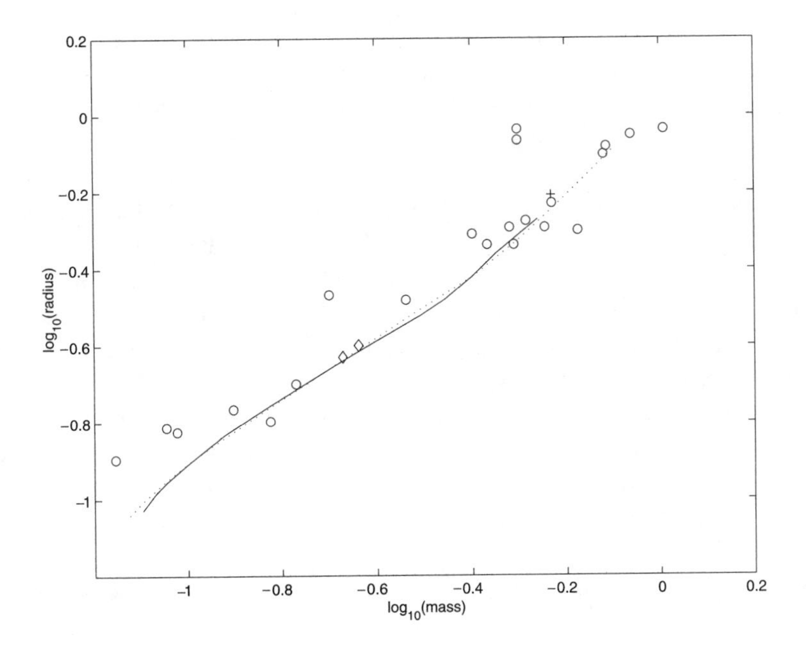

Fig. 6.8. The positions of the secondary components of cataclysmic variables in the mass–radius plane (open circles), from the compilation by Smith and Dhillon (1998), compared with theoretical models for low-mass stars from Dorman et al. (1989) (solid line) and from Chabrier and Baraffe (1997) (dotted line). Also plotted are the positions of the two components of the detached eclipsing binaries YY Gem (crosses) (Bopp 1974; Leung and Schneider 1978) and CM Dra (diamonds) (Lacy 1977; Metcalfe et al. 1996).

secondaries in systems just below the period gap are systematically smaller than the radii of those just above it. Smith and Dhillon have also demonstrated that it is not possible to establish a sensible relationship between the spectral type and the mass of a CV secondary. These findings are in agreement with those of Nordström (1989) for main-sequence stars in general.

The primary components of CVs are white dwarfs, which usually are well hidden in the optical and UV spectra by the bright accretion disc that encircles them. But during quiescence phases of a dwarf nova, for example, when the accretion disc is much reduced in size and brightness, the white dwarf can be studied, at least in the far-UV spectrum. A comprehensive review of the properties of white dwarfs in CVs has been published by Sion (1999), in which he notes that the distribution of determined masses shows a wide range between 0.4 and $1.2\,M_\odot$, with a broad peak around $0.8\,M_\odot$. Smith and Dhillon have reported mean values of $m_{wd} = 0.69 \pm 0.13\,M_\odot$ for those in systems below the period gap, and $m_{wd} = 0.80 \pm 0.22\,M_\odot$ for those above the period gap. Those findings should be compared with the determinations of isolated white-dwarf masses of $m_{wd} = 0.56 \pm 0.14\,M_\odot$ by Bergeron et al. (1995), who used $\log g$ values determined from H-line profiles together with a theoretical mass–radius relationship for white dwarfs and checked their results against a few determinations from gravitational redshifts.

If CVs are believed to have been formed as a result of a common-envelope stage of evolution, what evidence do we have that binary stars are to be found amongst the post-asymptotic-giant-branch (post-AGB) stars? Bond's review (1989) of the central stars of planetary nebulae (PNe) revealed that of the 1143 PNe known at that time, about 5 had central stars that were known to be binary systems. By the time of IAU symposium no. 180, *Planetary Nebulae*, edited by Habing and Lamers (1997), Livio (1997) reported that 15 PNe had close-binary central stars, of which 4 were known to be eclipsing binaries. These latter systems, of course, may provide the most comprehensive descriptions yet of the binary parameters with the minimum of assumptions, and it is interesting that UU Sge in Abell 63 (Pollacco and Bell 1993; Bell, Pollacco, and Hilditch 1994) and V477 Lyr in Abell 46 (Pollacco and Bell 1994) have both been found to contain sdO-star primary components, with low-mass K–M-type companions, in orbits of $P \sim 0.5$ day. These secondary stars are somewhat oversized for their ZAMS masses, perhaps indicating that they are not in thermal equilibrium as a consequence of the accretion of some material from the common envelope, and that might reveal evidence for different chemical abundances in the exposed inner regions of post-AGB stars. The orbital periods amongst these 15 close binaries in PNe cover a substantial range, 0.11–15.99 days, and three systems are known only on the basis of composite spectra. Much is yet to be learned about them.

There are also known binary systems composed of sdB or sdO primary components with low-mass main-sequence secondaries, such as HW Vir and KV Vel, respectively. These are generally regarded as *post-common-envelope binaries* (PCEBs), or indeed as pre-CVs, because they contain the accepted progenitors of white dwarfs as primary components, and the appropriately low-mass unevolved secondaries. This subject area in observational stellar-evolution theory is continuing to develop, with major surveys revealing substantial percentages of binary stars containing hot subdwarfs. For example, studies from the Edinburgh–Cape Blue Object Survey (Stobie et al. 1997; Kilkenny et al. 1997) have reported that about 53% of the objects are hot subdwarfs, and about 28% of them have composite optical spectra or colours. They note that because their follow-up techniques in the optical region normally will find only the companions of intermediate spectral types, and none of the later-type K and M stars, the binary frequency is likely to be substantially higher. Having said that, it is worth highlighting another one of several remarkable discoveries from the SAAO observers, the object EC13471-1258, found to be an eclipsing binary with a period of 3.6 hours, and composed of a DA white dwarf and a dMe main-sequence star. Multiple flares, thereby revealing the presence of the dMe star, were observed during the total eclipse of the white dwarf (Stobie et al. 1997). Jeffery and Simon (1997) have remarked that whilst sdO stars are understood to be the core remnants of AGB stars that have passed through the planetary-nebula phase and will evolve to become white dwarfs, the sdB stars are more difficult to understand. The binary frequency amongst these stars seems to be high, between 50% and 100%, according to several photometric and radial-velocity surveys cited by those authors, and perhaps that is a requirement to produce sdB stars. Their future evolution is directly to the white-dwarf stage from the extreme blue end of the horizontal branch.

As an example of the results achieved for these evolved objects, HW Vir has been shown by Wood and Saffer (1999) to be composed of a primary sdB star of mass $m_1 = 0.48 \pm 0.09\,\mathrm{M}_\odot$

and a secondary star of mass $m_2 = 0.14 \pm 0.02\,M_\odot$ with a radius $R_2 = 0.180 \pm 0.011\,R_\odot$, as expected for an unevolved main-sequence star of that mass. The sdO+K binary KV Vel has also had radial-velocity curves determined for both components (Hilditch et al. 1996), thus providing, in combination with analysis of the extreme reflection effect, the masses and radii for both stars: $m_1 = 0.63 \pm 0.03\,M_\odot$, $m_2 = 0.23 \pm 0.01\,M_\odot$, $R_1 = 0.157 \pm 0.03\,R_\odot$, $R_2 = 0.402 \pm 0.05\,R_\odot$. These values, together with estimated temperatures and hence luminosities, have been compared with values from the evolution models of Iben and Tutukov (1993) and Iben and Livio (1993) and have enjoyed good agreement. In conclusion, the theoretical models for the late stages of evolution of stars in binaries leading to the CV stage are in reasonable agreement with the very limited set of complete observations of both components in such binaries. The need to understand the strong irradiation of the dwarf M stars in these systems by the primaries (ranging from white dwarfs to sdO stars), whose temperatures extend from 20,000 K to perhaps 100,000 K, has already been emphasized in Chapter 5. Wood and Saffer (1999) have reported further observational data on the irradiation of these cool-star atmospheres, involving phenomena that are seen also in standard CVs.

6.6 X-ray binaries, black-hole binaries, and pulsar binaries

The x-ray binaries are divided into two categories according to whether the mass-losing companion star to the accretor has a mass that is greater than about $10\,M_\odot$ (a high-mass x-ray binary, HMXB) or has a mass that is less than about $1\,M_\odot$ (a low-mass x-ray binary, LMXB). Just to muddy the water, however, the system HZ Her, Her X-1, is of intermediate mass, with the optical companion at about $2\,M_\odot$. Here we shall review the determinations of such masses and other parameters for the stars in these systems, noting that a major recent review of the subject, *X-ray Binaries*, edited by Lewin, van Paradijs, and van den Heuvel (1995), discusses many other properties of these systems as well.

It has been known for more than 20 years that the masses for neutron stars lie in the region of the upper limit for white dwarfs, the Chandrasekhar mass of $1.44\,M_\odot$. The masses for O and B supergiants are around $20\,M_\odot$. The orbital speeds for neutron stars in HMXBs have been observed from pulse-timing data to be about 300 km s^{-1}, and so we would expect the semiamplitudes of the velocity curves for the optical companions to be about 20 km s^{-1}, quite low by typical binary-star standards. Because of the rotationally broadened nature of the spectral lines for the optical companions ($V_{\mathrm{rot}} \sin i \sim$ 150–200 km s^{-1}), it is difficult to determine individual velocities accurate to better than ± 1 km s^{-1}. In addition, there is abundant evidence from many studies of single and binary O and early-B stars of transient asymmetric line profiles and systematic shifts of line centroids by amounts of the order of several kilometres per second. The clearest example of this effect in HMXBs is demonstrated by the work of van Kerkwijk et al. (1995a) on the HMXB HD77581, Vela X-1, where they show that correlated excursions from the mean orbital-velocity curve occur on individual nights, amounting to ± 10 km s^{-1} over time scales that appear to be greater than the length of a single observing night. An explanation for such intrinsic variability in velocity would seem to lie with non-radial oscillations of the stellar photosphere, with time scales that range from a few hours to a few days; see *Rapid Variability of OB Stars: Nature and Diagnostic Value*, edited by Baade (1991), and IAU Symposium no. 162, *Pulsation, Rotation, and Mass Loss in Early-Type Stars*, edited by Balona et al. (1994). The overall consequence is that until the stellar intrinsic variability can be determined and separated from the orbital radial-velocity variations, we are left with the uncomfortable situation of increased scatter in the observed radial-velocity curve due to processes beyond the observer's control. One approach is that adopted by van Kerkwijk et al. (1995a) of securing a very large number of high-resolution, high-S/N spectra sampling all orbital phases so that a mean velocity curve can be determined with maximum accuracy. Unfortunately, most radial-velocity curves for the O–B supergiants in HMXBs are significantly *under-sampled*, so that the addition of intrinsic variability renders the determination of velocity semi-amplitudes open to significant systematic error.

The most precisely determined orbits are those of the radio pulsars, because of the accuracy achievable from the pulse timing. The orbits of pulse-timed x-ray pulsars are also very accurate, and one would hope that the determinations of the masses of the neutron stars in HMXBs would be sufficiently precise to be used as serious tests of the theoretical models for the structure of these extremely compact objects. However, the limits imposed on the accuracy of the determination of masses by the uncertainties in determining the velocity semiamplitudes of the optical companions have so far restricted that expectation. Because a pulsar orbit can be so well defined, with an amplitude about 20 times that of its optical companion, the uncertainty entering into the determination of the neutron-star minimum mass, equation (6.4), is principally that of the semiamplitude of the velocity curve of the O–B star, typically 10–15%. In addition, the determination of the x-ray-eclipse half-angle is subject to errors caused by the sampling rate of the x-ray observations throughout the eclipse and by the presence of accretion streams, discs, and other structures that can alter the absorbing column between the x-ray pulsar and the observer. Consequently, determination of the orbital inclination is restricted to uncertainties typically of $\pm 10\%$. Also, the rotational velocity of an O–B star is required for determining its radius, another factor of $\pm 10\%$. The resultant masses for neutron stars are therefore found to have uncertainties of the order of $\pm 20\%$. Van Kerkwijk et al. (1995b) have surveyed the various determinations of masses for neutron stars in HMXBs and have found that the six systems, including Her X-1, that exhibit x-ray eclipses and pulsed x-radiation provide an average neutron-star mass of $1.29 \pm 0.33\,M_\odot$. If the newer determination for Her X-1 from Reynolds et al. (1997) is preferred, then this average value becomes $m_{ns} = 1.36 \pm 0.32\,M_\odot$. To quote van Kerkwijk and associates, 'clearly, more accurate determinations are necessary. For the radio pulsars, this refers to the quantity of determinations, while for the X-ray pulsars it is the quality that needs improvement'. Results for the radio-pulsar binaries will be discussed a little later.

The parameters determined for the optical companions in these six systems typically are accurate to $\pm$10–20% for masses and radii, which is sufficient only to make reasonable checks on whether or not the available models for the evolution of high-mass binaries through to HMXBs are pursuing the correct idea. As pointed out in a review of binaries in other galaxies (Hilditch 1996), the optical companions in SMC X-1 and LMC X-4 do seem to be in the appropriate region of the HR diagram to fit with the models of non-conservative evolution, at ages of 20–25 $\times 10^6$ years, of de Loore and Vanbeveren (1994), except that the orbital periods are shorter than the model predictions, implying a need for greater losses of mass and angular momentum. Portegies Zwart and Verbunt (1996) have discussed a population-synthesis approach to the production of supernovae and x-ray binaries.

The LMXBs have a structure similar to that for CVs, except that the white-dwarf accretor is replaced by a neutron star or a black hole. Amongst the LMXBs, Smith and Dhillon (1998) have found that only five systems thus far have yielded good determinations of masses and radii for the secondary stars. The data show that the four K-type secondaries are perhaps somewhat larger than those in CV systems, and perhaps a little more evolved, but the distinction is not clear. The fifth system, V1033 Sco, GROJ1655-40, has an F3–6IV secondary that clearly has evolved from the ZAMS. But all five of those systems have primary components whose masses have been found to lie in the range 4–12 $M_{\odot}$, that is, higher than those expected of neutron stars, and therefore they should be stellar-mass black holes. All of them belong to the subclass of LMXBs known as *x-ray transients*, for their episodes of strong x-ray emission. Bailyn et al. (1998) have reviewed the distribution of masses determined for black holes in these systems and have concluded that they cluster around 7 $M_{\odot}$, significantly different from the masses for neutron stars. An additional soft-x-ray transient, 4U1543-47, has since been added to the picture, with determinations by Orosz et al. (1998) of its orbital parameters, and a mass estimate for its compact object in the range 2.7–7.5 $M_{\odot}$. But this is not to say that all black holes are to be found amongst the x-ray transients in LMXBs. The first black-hole candidate system, Cyg X-1, is an HMXB and remains one of the main candidate systems, with a black-hole mass of at least 7 $M_{\odot}$ and possibly 16 $M_{\odot}$ (Gies and Bolton 1986).

Away from these complications of accreting matter around the compact objects in x-ray binaries there are the *pulsar binaries*, each composed of a radio pulsar and a companion that can also be a neutron star – though found to be non-pulsing, or a white dwarf, or indeed a main-sequence star, in the two known cases thus far, PSRB1259-63 and PSRJ0045-7319, referred to earlier in this text. Thorsett and Chakrabarty (1999) have presented a timely review of the earlier studies of some 50 known radio pulsars in binary systems and have evaluated the accuracies of the mass determinations for those pulsed sources. They have identified 5 binaries that are each composed of two neutron stars and 40 binaries that are each composed of a pulsar (neutron star) and a white-dwarf companion. The remaining system is PSRB1257+12, with its planetary-mass companions. Most of these mass determinations cluster around $m_{\rm ns} = 1.35 \pm 0.04\,M_{\odot}$. Thus, we know the mass of a typical neutron star only for systems in which there seems to be no evidence that any significant accretion of mass has occurred (Thorsett and Chakrabarty 1999). Methods should be found to improve the observational data on the mass-transferring x-ray binaries to see if their neutron-star masses are significantly different.

6.7 Other evolved binaries

The remaining evolved binaries are systems that generally are enshrouded in gas that has been transferred from one component to the other in large quantities and/or ejected from the binary system. That is the case for the WR+O systems, with their radiatively driven stellar winds ejecting matter at rates of about 10^{-5} to $10^{-6}\,M_{\odot}\,\mathrm{yr}^{-1}$, and sometimes they show evidence of collisions between those winds. It is also true for systems like β Lyr, where a high mass-transfer rate is observed despite the locations of its component stars in the HR diagram being in agreement with models at the later stages of the slow mass-transfer phase of non-conservative case-B evolution. Many of these issues will be considered in the context of accretion processes in Chapter 7, but here we shall briefly survey the known masses amongst such objects.

Smith and Maeder (1989) have reviewed the determinations of the masses and absolute magnitudes for WR stars in binaries, concluding that those measurements could be uncertain by amounts in the range of 50–100%. Such lack of accuracy is not for want of trying, but is due to the sheer difficulty of securing reliable measurements of radial velocities in complex spectra. For the WR components, the radial velocity of the WR core, that is, well inside the stellar envelope, is accepted as being revealed by absorption lines such as N IV, at λ4058, which typically is well defined. But the WR wind supplies copious numbers of very broad emission features that render the composite spectrum of a WR+O binary very complex and tend to hide the normal O-star spectrum. Much effort has been devoted to searching for evidence of absorption lines attributable to the O-star companions, with unfortunately conflicting results. As an example, in the case of the WR+O binary CQ Cep, which has a remarkably short period, four separate determinations of velocity semiamplitudes yielded values for the WR star in the relatively small range $K_{\mathrm{WR}} = 285$–303 km s^{-1}, whilst those for the O star lay in the much wider range $K_{\mathrm{O}} = 170$–360 km s^{-1} (Harries and Hilditch 1997b). The consequences for the determinations of masses and radii obviously are enormous. In addition, but to a less dramatic extent, the difficulty of determining a proper representation for the observed light and polarimetry curves has already been noted (Chapter 5). Spectra that have very high S/N and resolution will be required to improve the situation.

There is a significant number of long-period binaries ($P \sim 1$–n years, with $n > 10$) that are each composed of a cool red-giant or supergiant star and a hotter and sometimes more evolved companion. These binaries include the following: ζ Aur and VV Cep types of systems, where the primary is a giant or supergiant and the companion is a main-sequence star; the *symbiotic binaries*, divisible into the S-type systems containing giants and the D-type systems containing Mira-type variables, and each with a companion that has the same properties as the central stars of planetary nebulae; and the barium (Ba), CH, and S stars, each containing a giant with abundance peculiarities thought to have been provided by mass transfer from a former AGB companion that is now a white dwarf. Reviews of their overall properties can be found in IAU symposium volume no. 151 (Kondo et al. 1992) in chapters by Carpenter, Kenyon, Smith, and Johnson, respectively. In nearly all cases, the binarity of such stars is now well established, and recent studies with coudé and échelle spectrographs have yielded accurate radial-velocity curves with semiamplitudes of only 3–15 km s^{-1}, and hence the mass functions $f(m)$ have been determined with accuracy (e.g., McClure and Woodsworth 1990; Schmutz et al. 1994; Schild et al. 1996; McClure 1997a; Schmid et al. 1998). The distributions of orbital eccentricities amongst the Ba and CH stars are different from those for normal G–K giants and demonstrate that these evolved binaries have undergone some orbital dissipation, probably due to mass and angular-momentum losses by means of a stellar wind from the former AGB star, or by some RLOF. But the fact that eccentricities lie in the range $0 < e < 0.2$–0.6 shows that a common-envelope event probably has not occurred in these systems. The values for $f(m)$, assuming a random distribution of orbital inclinations, show a narrow range, around 0.04 for Ba stars, and 0.095 for CH stars (McClure and Woodsworth 1990). With an estimated mass of $1.5\,M_\odot$ for Ba stars, their companions then have masses of $0.6\,M_\odot$, sensibly consistent with those of white dwarfs. Likewise for the Population II CH stars, with masses of about $0.8\,M_\odot$, their companions have masses also close to $0.6\,M_\odot$. Stars in the separate grouping of subgiant sgCH stars have also been studied by McClure (1997a), with the same conclusions being drawn.

The new method for polarimetric orbits, discussed in Chapter 3, provides good determinations of the orbital inclinations for some of the symbiotic systems. Members of the Zurich group have made careful analyses of unblended photospheric absorption lines to yield estimates of the rotational velocities for the giants in symbiotic stars, which then provide their radii. Using those data together with temperatures determined from spectrum-synthesis methods, the luminosities are derived, and comparisons of those results with evolution models for single stars in the HR diagram suggest very good agreement, so that masses can be inferred. Hence the companions' masses are determined from $[f(m), i]$ values, and at least amongst the detached symbiotics, those masses are typical of single evolved subdwarfs, at $0.55–0.60\,M_\odot$. Interestingly, the symbiotic novae, systems that show recurrent nova-type outbursts, signifying the accretion of matter onto the white-dwarf surface, followed by thermonuclear runaway events, seem to have masses more typically around $0.8–1.0\,M_\odot$ (Kenyon 1986; Shore et al. 1996). But not all of these stars with unusual photospheric abundances are found to be binary stars. McClure (1997b) has reported that in a survey of 22 R-type carbon stars there has been no evidence for binary motion during an observing programme lasting for 16 years. By contrast, normal red giants show a spectroscopic-binary frequency of about 20%, and Ba and CH stars are 100% binaries.

Our understanding of these evolved stars with peculiar abundances has been revolutionized over the past 20 years as a direct consequence of being able to conduct high-resolution spectroscopy on fainter objects, thereby yielding accurate radial velocities, and hence the recognition of orbits with long periods and small semiamplitudes. Additionally, the new method of determining orbital inclinations from spectropolarimetry of the Raman lines (Chapter 3) in symbiotic stars adds to the specifications of such binaries. Multi-wavelength spectroscopy, from the UV to the IR, yields temperatures for the two components, and recent investigations of eclipses have provided additional data on stellar radii, as well as the properties of the outflowing winds. Whilst our knowledge of these evolved systems falls far short of that for detached main-sequence systems, it is a revolutionary improvement over what was known 20 years ago.

7 The imaging of stellar surfaces and accretion structures

7.1 Introduction

Imaging, or mapping, is a natural part of our studies of binary stars, because of two factors: Stars spin on their rotation axes, usually in synchronism, and they orbit the centre of mass of the system, together with any other structures that belong to the binary. In effect, an observer 'walks around' the binary once per orbital period and is able to view the system from all of the orbital phase angles that are recorded via photometric, spectroscopic, and polarimetric means. In addition, because of the modern developments in astronomy, an observer has a more nearly bolometric view, and the radiation emitted from gases at very different temperatures, even within one binary system, can now be studied properly. Unfortunately, no close-binary system with interesting interactions between its components can yet be spatially resolved directly from Earth, because of the large distances to all the stars, so an observer sees only an integrated total amount of radiation at each observed wavelength from all the contributors in the binary – our points of light in the night-time sky again. But, as we have seen, the mutual eclipses of the components in suitably oriented binaries offer us a natural scanning mechanism that reveals the presence of those separate components, which can then be studied via all our observational tools. The Doppler effect, acting through the absorption/emission lines, provides velocities and hence inferred locations for the various contributors as well, and the polarization of the radiation reveals the distributions of scatterers and of magnetic fields. This chapter provides a description of the current mapping techniques used to determine the locations of the various radiation sources in binaries, with effective angular resolutions of about 10^{-6} arcsec! This chapter also summarizes our current understanding of stellar surface features and of the structures of accretion between the two stars, as well as the outflows from some systems. In this introduction, a brief history will illustrate how these subject areas have developed together and how they are important for determining more precise stellar masses and astrophysical parameters.

A review of early studies of the distribution of *circumstellar matter* within binary systems was presented by Sahade (1960), who discussed the observational evidence for gaseous rings, streams, concentrations near the Lagrange points, and expanding shells. The presence of gaseous rings within Algol binaries was first established by Joy's observations (1942, 1947) of red-shifted H emission lines just before the mid-eclipse of the primary star in RW Tau, followed by blue-shifted lines just after mid-eclipse. By the time of Sahade's review, 15 systems had been shown to have prograde rings around their primary components, all in Algol systems. The rings had low gas pressure, as shown by the absence of Stark broadening of their H lines, and they had ionization levels typical of stellar atmospheres. Streams of matter associated with the L_1 point and moving towards the trailing hemisphere of the primary at speeds of 200–400 km s^{-1} were discovered in some Algol binaries and were shown to explain, at least qualitatively, the observed *dips* in some light curves just before primary eclipse, as well as false eccentricities in measured velocity curves. But the mechanism of mass loss from the secondary through the L_1 point was decidedly unclear at that time. There was also evidence for the presence of shells of gas surrounding entire binary systems such as β Lyr, W Ser, and the WR stars and expanding at rates of 200–1300 km s^{-1}, and there were less dramatic but nevertheless persistent reports of stellar absorption-line strengths that varied with orbital phase, as in O-star binaries like AO Cas and some W UMa stars. Sahade commented that Struve 'had in mind an unsymmetrical common envelope with greater thickness in front of the leading hemisphere of each star of the system' to explain those observations. In Chapter 5 we noted that this phenomenon has been called the Struve-Sahade effect, which now receives a reasonable interpretation in terms of the colliding winds between the two stars (Gies et al. 1997). Some symbiotic stars had been suspected to be binaries, as Aller (1954) had suggested in interpreting the spectroscopic data for BF Cyg, and the first recognitions of DQ Her (Walker 1954, 1956) and AE Aqr (Joy 1954) as binary stars with circumstellar matter had been made. These latter systems are now regarded as *classical* examples of cataclysmic variables.

Ten years later, the observational data on gas flows in binaries were somewhat more quantitative, with estimates of gas densities at 10^{13} cm^{-3} in streams, and 10^{11} cm^{-3} in circumstellar rings, which by then were called discs (Batten 1970, 1973). In IAU symposium no. 51, *Extended Atmospheres and Circumstellar Matter in Spectroscopic Binary Systems*, edited by Batten (1973), the free-ranging discussions reported there revealed substantial changes in the quantitative analyses of some systems, such that binaries with circumstellar matter were no longer seen as pathological objects that hampered progress in determining the properties and evolution of binaries, but were being recognized as fundamental to that understanding. By the time of the next conference, IAU symposium no. 73, *Structure and Evolution of Close Binary Systems*, edited by Eggleton et al. (1976), the revolution had been effected, and reviews of extensive calculations with theoretical models for the evolution of binaries and for gas flows between the stars, including possible causes of nova outbursts, stood beside the major observational reports on x-ray binaries, dwarf novae, contact binaries, and the likely existence of black holes in binaries. The terms *accretion stream* and *accretion disc* had become standard, and the idea of a common-envelope stage of evolution was introduced to explain the origin of the cataclysmic variables.

During the past two decades, increasing sophistication in the methods of analysis for light curves and for spectra has permitted proper investigations of more complicated binaries, as discussed in earlier chapters. Consequently, research work on close binaries has become more dominated by studies of accretion processes in interacting systems, studies of the evidence for activity in the form of starspots on rapidly rotating cool stars in binaries, and studies of winds from hot stars and wind–wind interactions. The collection of review papers in

Accretion-Driven Stellar X-ray Sources, edited by Lewin and van den Heuvel (1983), established a new level of sophistication in interpreting data from all parts of the electromagnetic spectrum from these systems, whilst the conference proceedings *Interacting Binaries*, edited by Eggleton and Pringle (1985), and the book *Interacting Binary Stars*, edited by Pringle and Wade (1985), both illustrate the great advances made in the subject in a decade. Since 1976, annual workshops have been held under the rubric Cataclysmic Variables and Low-Mass X-Ray Binaries, and a biennial series entitled Cool Stars, Stellar Systems, and the Sun was begun in 1980. Whilst the latter workshops have devoted only part of their time to discussions of close binaries, they are mentioned because of their introduction into that subject area, during the 1980s, of the idea of reconstruction of images of non-immaculate stellar surfaces from sets of appropriate data. Similar ideas were being introduced to study the accretion structures in binaries, particularly the accretion discs in CVs that dominate the optical and UV light in those systems. Applications of those ideas have been pursued extensively in many papers, and they are strongly evident in *Active Close Binaries*, edited by Ibanoglu (1990), IAU symposium no. 151, *Evolutionary Processes in Interacting Binary Stars*, edited by Kondo et al. (1992), IAU symposium no. 165, *Compact Stars in Binaries*, edited by van Paradijs et al. (1996), the reviews edited by Lewin et al. (1995) in *X-ray Binaries*, and the monograph *Cataclysmic Variable Stars*, by Warner (1995). Most recently, conferences have been devoted to *Accretion Phenomena and Related Outflows*, edited by Wickramasinghe et al. (1996), and *Astrophysical Discs*, edited by Sellwood and Goodman (1999), and the aforementioned workshops on CVs and cool stars continue.

7.2 Image reconstruction

From statistical mechanics we learn that a thermodynamic system will be in its most probable state when the entropy of that system is at a maximum. *Entropy* can be defined by the expression

$$S = -k \sum_j p(j) \ln p(j) \tag{7.1}$$

where $p(j)$ is the probability that a system is in state j. In the context of image restoration, the entropy of an image will be at a maximum when the distribution of the image pixels is the most probable distribution. Maximizing the entropy of an image imposes a constraint on the multitude of possible images that could describe a particular set, f_{obs}, of observational data. It ensures that the most probable image is that which gives the smoothest distribution, or the least perturbed distribution. For example, if f_{obs} is a set of brightness measurements from a source, the application of a maximum-entropy regularization will render an image from f_{obs} that will be the least perturbed distribution of brightness across a source. Maximizing image entropy is attempting to minimize image structure, and in that sense the term *maximum entropy* can be understood as ensuring the most disorder.

A full description of the *maximum-entropy method* (MEM) for reconstruction of images, particularly in astronomical contexts, is given by Skilling and Bryan (1984), who cite many references to the wider subject of imaging from many forms of data. They comment that MEM seems to be the most successful for providing restorations of images from the sets of incomplete and noisy data that are typical of astronomical observations. The code MEMSYS (Skilling 1981; Burch, Gull, and Skilling 1983; Skilling and Gull 1985) provides the computational basis for these analyses.

7.2.1 Eclipse mapping in binaries

7.2.1.1 *Accretion discs*

The first application of imaging by MEM within binary systems was that by Horne (1983, 1985), who developed a code that included MEMSYS to provide maps of the surface-brightness distributions across the accretion discs seen in cataclysmic variables from analyses of their broad-band light curves. In these CVs, the disc is the brightest source of optical and UV radiation, and for favourably oriented systems it is eclipsed by the cool secondary star once per orbit. We observe the disc in projection against the sky as a two-dimensional image that is scanned by the projected leading edge of the secondary star, followed soon after by the projected trailing edge of the secondary. At each phase of the binary orbit, two curves can be drawn across the accretion disc, marking the boundaries between those parts of the disc just being eclipsed by the leading edge and those just being revealed by the trailing edge. The shape of the resultant photometric eclipse provides only a one-dimensional source of information about the distribution of surface brightness across the disc, and therefore it cannot provide a unique image, because many possible distributions of surface brightness, together with different shapes of the secondary star and the orbital inclination, could produce the same eclipse curve. To obviate this problem of indeterminacy, the image of the disc can be constrained to a particular model. In his first application, Horne (1985) limited the model of the disc to be a flat system in the plane of the binary orbit and then used MEM to constrain the image in a variety of ways.

The disc can be divided up into $N \times N$ pixels, with pixel j having an emergent intensity $I(j)$. Correspondingly, one can define a *default image*, with pixel intensities $D(j)$, and then define the image entropy by

$$S = \sum_{j=1}^{N^2} \{I(j) - D(j) - I(j) \ln[I(j)/D(j)]\} \tag{7.2}$$

so that

$$\begin{aligned} \partial S/\partial I(j) &= \sum_{j=1}^{N^2} \{1 - \ln[I(j)/D(j)] - 1\} \\ &= -\sum_{j=1}^{N^2} [\ln I(j) - \ln D(j)] \end{aligned} \tag{7.3}$$

which will equal zero when $I(j) = D(j)$. In an idealized case, the value of S will reach a maximum of zero when $I = D$ for all j. More generally, a measure of consistency between the observations f_{obs} and a theoretical model $D(j)$ that predicts f_{cal} is obtained in the usual way by ensuring that a χ^2 function of the form

$$\chi^2 = \frac{1}{N_{\text{obs}}} \sum_{k=1}^{N_{\text{obs}}} \left[\frac{f_{\text{obs}}(k) - f_{\text{cal}}(k)}{\sigma(k)} \right]^2 \tag{7.4}$$

where $\sigma(k)$ is the estimated uncertainty in each of the $k = 1, \ldots, N_{\text{obs}}$ observations f_{obs}, reaches a target value $C_{\text{aim}} = \chi^2$.

The specification of the default image evidently controls the value of entropy obtained. The default can be a uniform image, with all pixels having the same intensity, or the pixels can be weighted in some way that reflects a constraint determined by other observations or by theory. Thus,

$$D(k) = \sum_{j=1}^{N^2} w(k, j) I(j) \Bigg/ \sum_{j=1}^{N^2} w(k, j) \tag{7.5}$$

For steady-state viscous accretion discs (Frank, King, and Raine 1985, 1992), theory expects there to be a radial temperature gradient, with higher temperatures close to the accreting white dwarf in a CV, and temperatures decreasing outwards. Likewise, azimuthal structure should be suppressed, at least in the inner regions of the disc, because of the viscosity in the disc and because the orbital speeds of particles in the disc should be about 1000 km s^{-1}. Horne (1985) found that the most useful weighting scheme for simple discs was

$$w(k, j) = \exp\{-[R(k) - R(j)]^2/2\Delta^2\} \tag{7.6}$$

where $R(k)$ is the radial distance of the kth pixel from the centre of the disc, and Δ is the half-width of a Gaussian function. By imposing such a weighting scheme, the entropy loses sensitivity to radial structure on scales larger than Δ and therefore allows that part of the image to be defined primarily by the data. The constraint imposed on the solution is solely that azimuthal structure is suppressed by maximizing the entropy, and the resultant MEM image is the most axisymmetric image allowed by the data. There is no imposition of a particular radial temperature gradient, nor surface-brightness gradient, and no imposition of a radial extent for the disc.

It was also shown by Horne that if the distances in the problem are specified in terms of the distance of the accreting star from the L_1 point, then the scale of the image is nearly independent of the mass ratio of the binary, an important consideration for CVs, where radial-velocity curves for both components are difficult to obtain. Both the orbital inclination and the mass ratio can be constrained by the observed width in orbital phase of the eclipse of the accretion disc by the Roche-lobe-filling secondary star.

The individual pixels in the image are independent, so that if multi-colour broad-band light curves are analysed, it is possible to assign a temperature to each pixel according to its surface brightness, colour index, or spectral-energy distribution. Thus, Horne was able to test one of the main predictions of early accretion-disc theory that the radial dependence of temperature was $T(R)^4 \propto m\dot{m}R^{-3}$, a prediction that was confirmed, at least for nova-like CVs and for systems in a high-activity state (Figure 7.3), transferring mass at a rate $\dot{m}$ to the white dwarf of mass m. In outburst states, dwarf-nova discs are optically thick, and with the sizes of the discs determined from the known geometry of the systems, they provide estimates of distances to these binaries. The mass-accretion rates are found to be about 10^{-9} M$_\odot$ yr^{-1}. By contrast, Wood et al. (1989) found that the disc observed in IP Peg during a quiescent phase exhibited a radial temperature profile that was much flatter than those in the steady-state accretion-disc models.

In the early 1980s, only broad-band photometry could be performed at the time resolutions required for accurate monitoring of eclipses in CVs and related objects, and so the derived temperature profiles were constrained by the limitations of broad-band photometry. But by the early 1990s, improved detectors and large telescopes allowed low-resolution spectra to be secured at equally high time resolutions, enabling Rutten et al. (1994) to report on *spectral-eclipse mapping* of the nova-like CV UX UMa. They divided their spectra, recorded over 3600–9800 Å, into narrow bands of continuum and selected spectral lines in order to determine eclipse maps for the accretion disc for each individual band. They were able to show that the inner disc had a continuum spectrum that peaked in the UV and had H lines in absorption, whereas the outer disc had a fainter, redder spectrum with H emission lines. The accretion bright spot, where the stream from L_1 impacts onto the accretion disc, had H absorption lines again, indicating optically thick gas at that location. This particular binary was studied again by means of HST UV and optical spectroscopy by Baptista et al. (1998a), with a time resolution of 5.3 seconds. They determined eclipse maps of the accretion disc for individual spectral lines and narrow continuum regions (Figure 7.1) and then derived spectra for integrated annular regions of the disc and the azimuthally smoothed radial temperature gradient, as shown in Figures 7.2 and 7.3. Such detailed data can serve as crucial tests of theoretical model atmospheres for accretion discs, such as those by Wade and Hubeny (1998).

In order to remove the limitation of a flat accretion disc, Rutten (1998) has recently developed a new code that allows for a three-dimensional accretion disc together with a standard Roche-lobe-filling companion star. Such a model permits the disc to increase in thickness with radial distance from the accreting star (a *flared disc*), to be smoothly truncated at the outer edge, and to be tilted relative to the orbital plane. All of these factors have been inferred from a wide variety of observations of accretion discs in CVs and LMXBs. This improved modelling ought to provide improved consistency of results. Rutten examined whether or not the rather flat radial temperature profiles found for some CVs could be spurious results of the eclipse mapping, with an assumed flat disc applied to a flared disc, as proposed by Smak (1994). He showed that the correct radial profiles could be recovered provided that the flaring did not occult the inner regions of the disc, but when inner regions were occulted by the outer disc, then the reconstruction of the image did show a flat temperature profile. Further applications of this improved model should help to resolve such issues.

7.2.1.2 *Stars*

The ideas of eclipse mapping can also be applied to standard eclipsing binary stars. Here the default image is taken to be that of the immaculate stellar surfaces, each at its own effective temperature, that would be determined from a standard light-curve synthesis code. If the observed eclipse curves are beautifully symmetric, as illustrated for many binaries in Chapter 5, then the distributions of surface brightness over the two stars should be immaculate, save for the effects already included in the light-curve synthesis code of limb darkening, gravity darkening, and the reflection effect. But if the eclipse curves show some irregularities or asymmetries, then an effective eclipse mapping code should be able to interpret them into stellar-surface

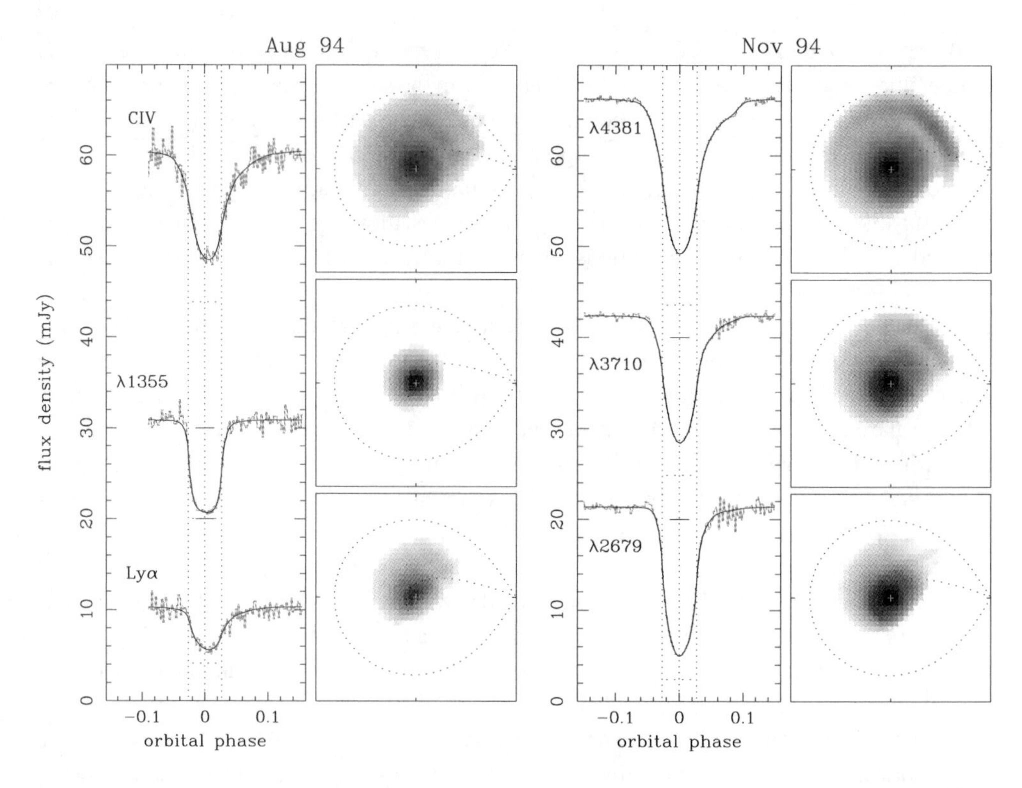

Fig. 7.1. The UV light curves through the eclipse of the cataclysmic variable UX UMa, obtained from HST spectra in August and November 1994 and the corresponding eclipse maps. Two light curves are from emission lines of C IV and Ly α, and the remainder are continuum regions. The photometric data are shown as grey histograms, and the fitted model light curves are solid black lines. The grey-scale maps have dark regions representing the brighter parts of the disc. The dotted curves on the maps represent the Roche lobe of the accreting star and the theoretical trajectory of the gas stream from the L_1 point to the accretion disc. (From Baptista et al., 1998a, with permission.)

maps that will reveal the presence of starspots or such like. There is not much to be gained from such detailed work unless the accuracy of the original photometry is particularly high, around ± 0.002 mag per observation. It should be fairly self-evident that if the values for stellar radii, temperatures, and so forth, determined from a synthesis code are not correct, then systematic artifacts will appear in the surface maps. For example, rings centred on the mid-eclipse longitude of the eclipsed star will be found in surface maps if one of the stellar radii is specified incorrectly. So, following the procedures discussed earlier in this text, the operation of a surface imaging code has to proceed in tandem with a synthesis code, iterating to find a best solution to the set of multi-colour light curves.

Hendry et al. (1992) applied MEM to the case of a contact binary where starspots were strongly evident from the distorted light curves. They showed, in agreement with earlier work referred to in Chapter 5, that it was impossible to determine on which of the two components a spot was located unless the spot-containing component was eclipsed by its companion. But for

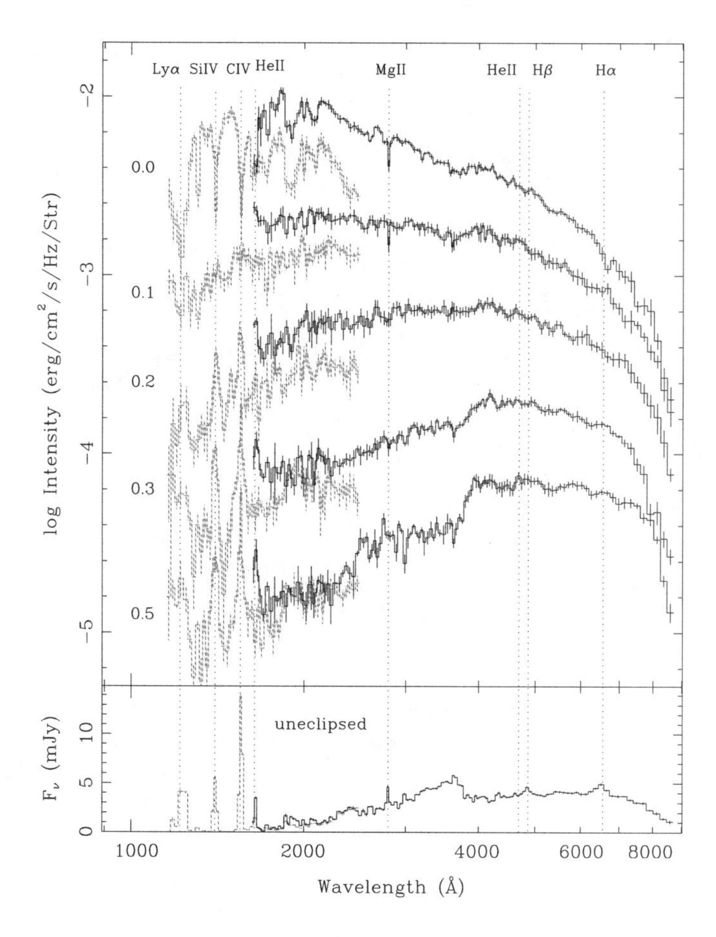

Fig. 7.2. Spectra for parts of the accretion disc of UX UMa (resolved spatially by means of eclipse mapping) located on the far side of the disc from the impact region with the accretion stream. The August 1994 data are shown in grey, and the November data in black. The distances between the accreting object and the annular rings used to determine the spectra are indicated on the left of each spectrum, in units of the Roche-lobe radius. Note that the flux densities become much lower and redder for the outer parts of the disc. (From Baptista et al. 1998a, with permission.)

the eclipsed regions, namely, the hemispheres of the two components that faced each other, the application of MEM in tandem with a light-curve synthesis code could recover test distributions of starspots with uncertainties of $\pm 5°$ in longitude and ± 2–$15°$ in latitude, depending on their locations in respect of eclipses. Collier Cameron (1997) has extended his Doppler-imaging code DOTS to include eclipse mapping by MEM of late-type eclipsing binaries where it is to be expected that magnetic activity will result in variable amounts of starspot coverage over either or both stars. Test reconstructions of an artificial image have shown that the eclipsed spots are recovered well, with a surface resolution of $\pm 10°$, whereas uneclipsed spots are less well determined, with high-latitude spots being substantially blurred and weakened, and lower-latitude regions being dispersed in latitude, although at the correct longitude. Figure 7.4

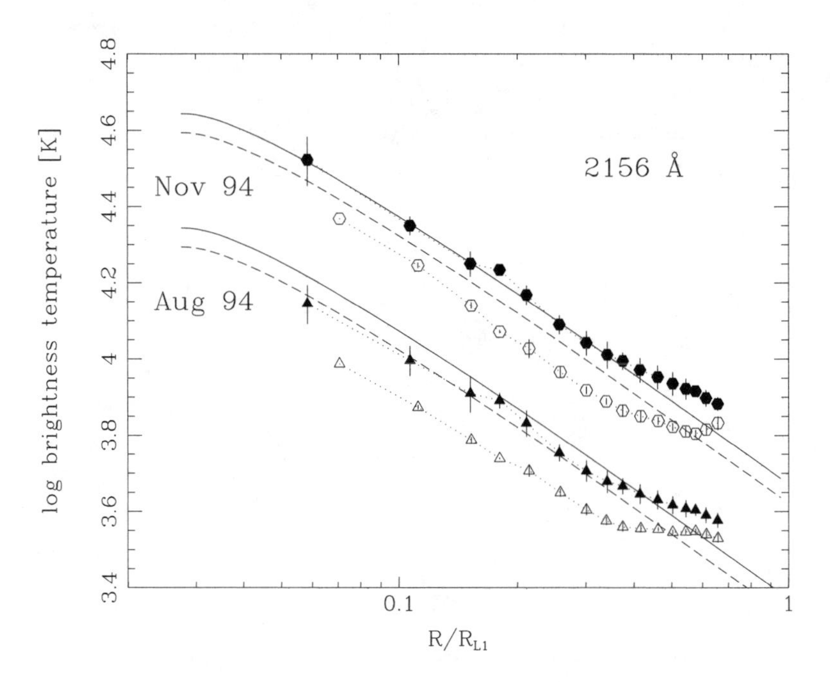

Fig. 7.3. Radial temperature profiles for the disc of UX UMa determined in August and November 1994, shown by various symbols. Expected temperature profiles from theoretical steady-state discs are shown as lines for two different mass accretion rates: $10^{-8.1}$ (solid) and $10^{-8.3}$ (dash) $M_\odot$ yr^{-1}. The upper curves have been displaced vertically upwards by 0.3 dex for clarity. The filled/open symbols represent data from the far/near side of the disc relative to the L_1 point. (From Baptista et al., 1998a with permission.)

illustrates the results on the RS CVn binary XY UMa for single-colour, narrow-band light curves from Lister et al. (2000a) and demonstrates that we are beginning to be able to map the evolution of spot activity from repeatedly securing light curves for such eclipsing binaries with RMS errors of ± 0.002 mag. This work is an extension of the monitoring programme started in 1992 on that star; see Collier Cameron and Hilditch (1997) and the references cited therein. These studies, together with earlier work, show that the global brightness level of that binary system varies by 0.5 mag and can be understood in terms of large-scale spot activity, including the likelihood of a substantial polar spot. Jeffries (1998) has analysed the x-ray fluxes determined from ROSAT over several orbital cycles of XY UMa and has shown that there are no x-ray eclipses occurring in this system. Either the coronal emission is extended over more than a solar radius or the corona is compact and is found at the high, uneclipsed latitudes where the presence of a substantial polar spot has been inferred from the long-term optical photometry.

A final example of eclipse mapping concerns the mass-exchanging binary V361 Lyr, for which recently acquired *BVI* light curves were shown in Chapter 5 (Figure 5.18). A simultaneous analysis of the three light curves was performed with the DOTS code to yield a map of the brightness distribution for the secondary star, which is gaining mass at about 10^{-7} $M_\odot$ yr^{-1} via a transfer stream directly from the L_1 point (Figure 7.5). That stream hits the surface of the secondary at a longitude of about 191° at an estimated (free-fall) speed of about 800 km s^{-1}.

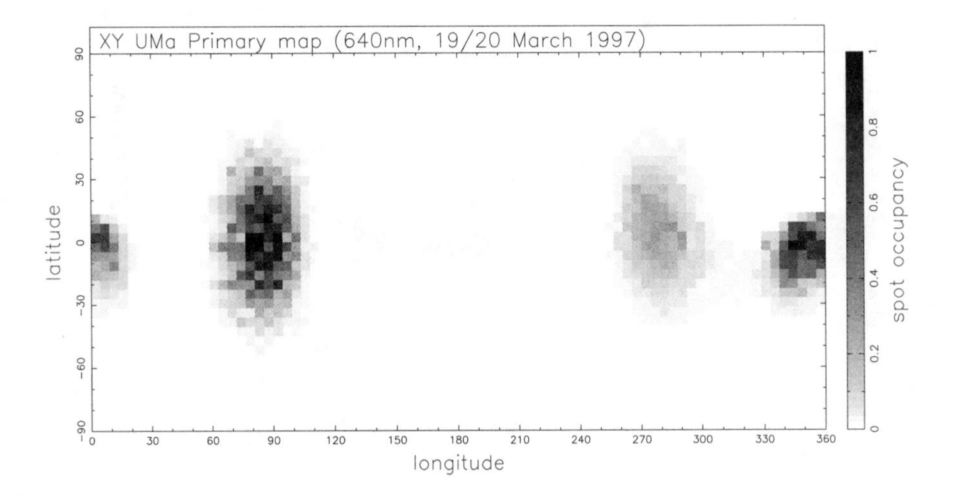

Fig. 7.4. An example eclipse map for the RS-CVn-type binary XY UMa determined from narrow-band optical light curves with RMS scatter of ±0.003 mag, secured in March 1997 by Lister et al. (2000a). The map is a projection of the surface of the primary, with black indicating full cool-spot occupancy of that pixel, and white indicating zero spot occupancy. The identified spot regions all occupy the equatorial zone, and nothing is found at high latitudes, because that part of the surface of the primary is not eclipsed by the secondary star. Longitude 0° is the substellar point on the surface of the primary component, and the longitude of the sub-Earth point decreases with increasing orbital phase. Thus longitude 270° is seen at maximum at photometric phase 0.25, or first quadrature, when the primary component is approaching the observer. (From Lister et al., 2000a, with permission.)

The observed luminosity of the hot region is approximately half of the accretion luminosity of about 0.5 $L_\odot$, suggesting that the rest of the energy is dissipated into the photosphere of the secondary star. The accretion region derived from the light curves illustrates the intuitively obvious shape for such an impact: a bright region at the impact site, with a well-defined leading edge at the left, and an extended region of lower brightness downstream from the impact site due to the fact that the accretion stream hit the photosphere at an angle of about 40° to the left of the surface normal. The location of the impact stream has not changed significantly over 10 years of observations.

The technique of eclipse mapping should not be regarded in isolation, because it is a complementary version of the other major application of surface imaging, namely, that of *Doppler imaging*, or *Doppler tomography*, from spectroscopic observations, as considered in the next subsection.

7.2.2 Doppler imaging in binaries

7.2.2.1 *Stars*

Doppler imaging, or Doppler tomography, makes use of the information contained within the profiles of spectral lines to infer a map of the distribution of brightness over the surface that contributes to those spectral lines. As noted in Chapter 3, the contributions to a Doppler-broadened

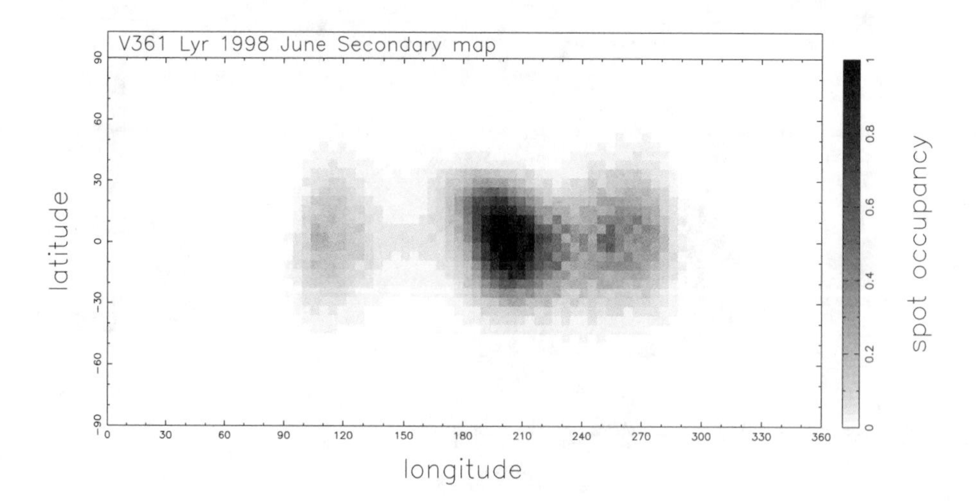

Fig. 7.5. An eclipse map of the secondary component of the mass-exchanging binary V361 Lyr obtained from the DOTS code applied to the *BVI* light curves from Lister et al. (1999b) and shown in Figure 5.18 in Chapter 5. The accretion hot spot has a temperature at maximum (black pixels) estimated at 10,000 K and is observed to be displaced from the line of centres (longitude 180°) to the trailing hemisphere, exactly in accord with the hydrodynamical models of Lubow and Shu (1975). Note also the cooler part of the region extending downstream from the impact zone. Longitude 180° is the substellar point on the surface of the secondary component, and the longitude of the sub-Earth point decreases with increasing orbital phase. Thus longitude 270° is seen at maximum at photometric phase 0.25, or first quadrature, when the secondary component is receding from the observer. (From Lister et al., 2000b, with permission.)

line profile for a rotating star are made by strips of the stellar surface parallel to the stellar-rotation axis, each displaced from the line centre by the Doppler shift corresponding to the star's projected rotational velocity for that strip. If the stellar surface is rotationally symmetric and immaculate, then the corresponding line profiles will also be symmetric. But if there are cool starspots at various locations, for example, then these will introduce small irregularities or bumps in the line profiles, whose positions will change as the spots rotate through the field of view of the observer. There is a strong correspondence between the longitudes of spots on the stellar surface and the bumps in the line profiles, so that a sequence of observations covering an entire rotation period will provide a two-dimensional set of data that will yield a two-dimensional image of the stellar surface. There is some ambiguity about whether spots occupy the northern or southern hemisphere of the star, being most problematic when $i = 90°$. But there is no need, as in the one-dimensional eclipse mapping of CV discs, to impose any restrictions on the surface map, though it is usual in Doppler imaging to adopt a smoothing function, like a Gaussian point-spread function, in order to limit the small-scale irregularities in the surface map.

Whether or not a starspot will produce an apparent emission bump or an absorption dip within a normal absorption-line profile will depend on the product of two factors, the continuum flux f from the region of the star where the spot is located, and the local equivalent width w of the spectral line considered, both for the spot photosphere and for the stellar photosphere, as noted by Vogt and Penrod (1983). If $(f \times w)_{spot} < (f \times w)_{star}$, then the spot will provide less absorption in the line at its appropriate Doppler shift, and an apparent emission bump will be seen. If $(f \times w)_{spot} > (f \times w)_{star}$, then the spot will provide more absorption in the line, and an absorption dip will be seen. These quantities will all depend on the apparent position of the spot on the stellar disc at the time of observation, so that spots may appear stronger close to the limb of the star than when viewed in the central parts of the disc.

For slowly rotating stars, say $V_{rot}\sin i < 10$ km s^{-1}, the Doppler broadening of the line is not sufficiently larger than other contributions to line broadening, like instrumental resolution, to provide a sensible number of resolved strips over the stellar surface. The Doppler broadening needs to be several times the overall Voigt profile for the procedure to work effectively. But for rapidly rotating stars, with $V_{rot} \sin i \sim 25$–100 km s^{-1} (Vogt and Penrod 1983), Doppler broadening is the dominant process in defining the shapes of spectral lines, and Doppler imaging can be used very effectively to determine the locations of surface features on stars. For the higher rotational velocities of normal stars, the central depths of spectral lines become very shallow, and it is more difficult to determine small irregularities in the line profiles.

The best imaging of stellar surfaces will be secured by means of high-resolution, coudé or échelle spectroscopy, giving resolutions of about 40,000, and with S/N values of several hundred. This observational demand would suggest that only the brighter stars can be investigated, but recent developments in applying the technique of *least-squares deconvolution* (LSD) to échellograms (Donati et al. 1997) allow Doppler tomography to be performed on stars down to the fourteenth magnitude via 4-m telescopes. This technique makes use of the fact that very large numbers of absorption lines, typically 1500–2000, are present in the blue–yellow part of the spectrum for a late-type star, mainly from Fe I and Fe II. All of the profiles of these lines will contain the signature of the distribution of surface maculations in the form of small irregularities. The set of all of these lines, appropriately broadened by the atmospheric properties of the (non-rotating) star and the instrumental broadening, can be convolved with the line-profile irregularities arising from a non-immaculate, rotating stellar surface to yield the observed line profiles as a function of phase within the rotational period of the star. The action of least-squares deconvolution is to extract a mean line profile from all of the 1500–2000 lines, so that its effective S/N is perhaps as much as 40 times that achieved from a single line, or S/N $\sim$ 2000 from an average spectrum! A substantial description of the technique is given by Donati et al. (1997) and Donati and Collier Cameron (1997). Again, this is an inverse problem in astronomy, where the observed mean profile has to be used to infer the map of the non-immaculate stellar surface. Implicit in the LSD technique is the assumption that all the lines used in the analysis respond in the same way to changes in surface properties, and, of course, the analyst should be wary.

As an example of the technique applied to a single star that is rotating quite quickly, Figure 7.6 shows the variations in the mean line profiles for the star PZ Tel (Barnes et al. 2000), determined from least-squares deconvolution, as a function of phase in the rotational period of about 1 day. The resultant surface map is shown in Figure 7.7, illustrating features that are commonly observed in magnetically active late-type stars: a substantial polar crown that is evidenced in

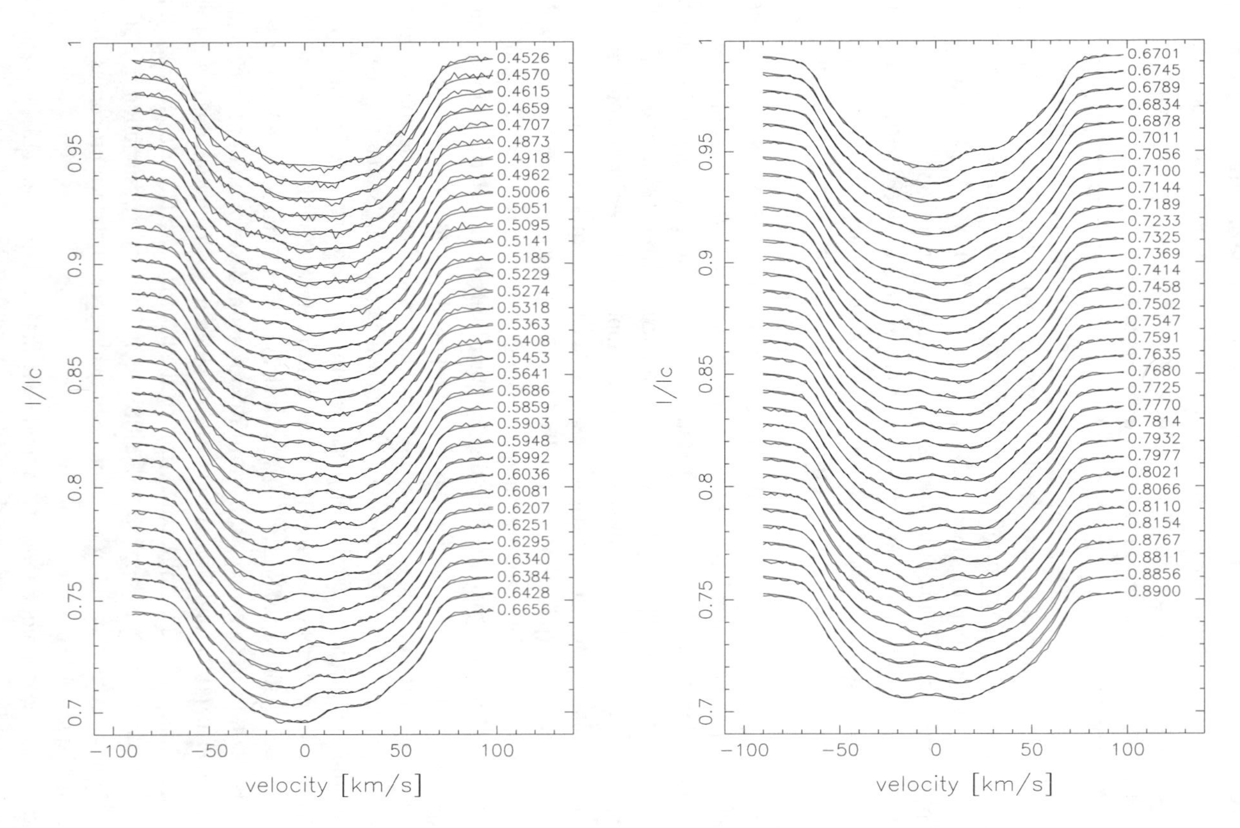

Fig. 7.6. Two sets of mean absorption-line profiles for the rapidly rotating G-type dwarf star PZ Tel determined via least-squares deconvolution from high-resolution ($\sim$6 km s^{-1}) spectroscopy obtained with the Anglo-Australian Telescope by Barnes et al. (2000). The data show clearly the broad, flat-bottomed profile, with small-scale irregularities or bumps travelling through the profile as the star rotates. (From Barnes et al., 2000, with permission.)

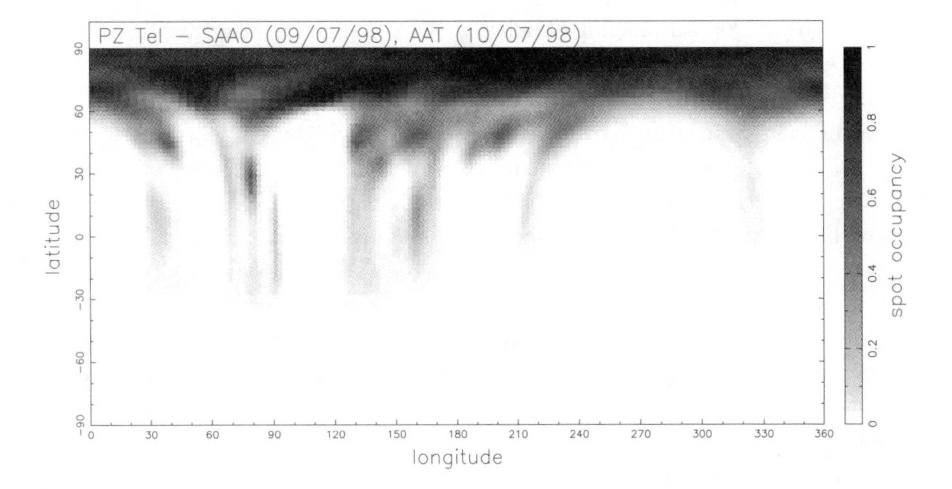

Fig. 7.7. The spot-occupancy map for the surface of the star PZ Tel derived via Doppler imaging from the set of mean line profiles shown in Figure 7.6, with black indicating full spot occupancy. The polar crown is very evident, and substantial numbers of lower-latitude features have been identified as a result of detecting the irregularities in the line profiles. (From Barnes et al., 2000, with permission.)

the mean line profiles by the presence of a substantial flat section in the deepest part of the lines, and discrete features at lower latitudes that provide the moving irregularities in the lines.

Piskunov (1991) has reviewed the historical development of the subject of surface imaging in astronomy as applied to the chemically peculiar Ap stars and magnetically active cool stars. Vogt and Penrod (1983) first applied the technique to RS-CVn-type binaries after the discovery by Fekel (1980) of distortions in the shapes of spectral lines in the binary system HR1099 (V711 Tau). In that first application, Vogt and Penrod adopted a trial-and-error approach to fitting their set of high-resolution spectra for HR1099 with various model distributions of spots around the star. Such a method works, though it may not provide the best solution to a given set of data, and of course it is very time-consuming. After discussions with K. Horne, Vogt, Penrod, and Hatzes (1987) published a substantial discussion of the ability of the maximum-entropy method to provide surface images of single stars from sets of high-resolution spectra, concluding that the images were very well constrained by the data alone and yielded excellent representations of test images under many conditions. An example of their procedures applied to RS CVn binaries is that by Vogt and Hatzes (1991), where Doppler images from three discrete sets of high-resolution spectra for the binary UX Ari demonstrated the existence of a large, stable polar spot and a quite complex arrangement of spots at many other positive and negative latitudes. With their three sets of data secured over a 5-month interval, Vogt and Hatzes were able to demonstrate also that the primary component of UX Ari was rotating differentially in a sense opposite to that of the Sun. Whilst the equatorial region is rotating in synchronism with the binary orbital period, the polar region is rotating more quickly, such that the ratio of equatorial to polar angular velocities is -0.020 ± 0.002, compared with $+0.2$ for the Sun.

The technique of Doppler imaging of binary systems containing magnetically active solar-type stars was discussed further by Vincent, Piskunov, and Tuominen (1993). Their model binary stars included all the parameters expected for a standard light-curve synthesis code, except that the stars were taken to be more or less spherical, as is usually the case for RS CVn binaries. Distributions of spots, or alphanumeric characters, were placed on both model stars, and the Doppler-imaging code was applied to recover the model images, with evidently good success on both stars. They found also that repeatable distortions of the maps were evident when some of the input parameters, such as stellar radii or orbital inclination, were set to incorrect values, thereby demonstrating that such geometrical parameters can be fine-tuned with eclipse mapping and Doppler mapping. Collier Cameron (1997) has also included the procedures of Doppler imaging into his surface-imaging code DOTS so that it can calculate images from photometric eclipse data, as noted earlier, or from sensibly complete sets of mean line profiles extracted via the LSD algorithm. The use of these complex codes requires care, because there are so many input parameters that can have varying effects on the derived images of the mapped surfaces. Whilst many words could be written about the effects on the derived images of varying individual parameters in turn, and several of these aspects are discussed in the aforementioned papers, it is very instructive to learn these matters from the first-hand experience of analysing a set of data with such a code.

Piskunov and Rice (1993) presented a review of surface imaging from spectral-line profiles for many different types of stars and noted that magnetic-field maps can be produced also via polarization data, a development that has seen success in the spectropolarimetric work of Donati et al. (1997), as applied to active single and binary stars, and including the LSD algorithm. That paper introduced the first direct determinations of magnetic-field configurations by means of *Zeeman-Doppler imaging* (ZDI), coupled to the LSD procedure, so that relatively faint stars could be studied. That successful use of the Zeeman splitting of absorption lines in the presence of a substantial magnetic field means that the strengths and the polarities of magnetically active regions can now be determined for stars other than the Sun. Donati (1999) has discussed extensive results from spectropolarimetry on HR1099.

There have been several detailed investigations of the changing distributions of maculations on active stellar surfaces with time that may extend our knowledge of the magnetic-activity cycles in stars other than the Sun and allow us to examine the influences that binarity and enforced tidal synchronism have on stars of different spectral types, and hence different depths of convection zones. The papers to be discussed next illustrate the success of such investigations.

Strassmeier (1994) studied the RS CVn binary HD106225, where Doppler images from lines formed within the photosphere and at different levels within the chromosphere were established, together with photometric maps from the rotational modulation. The spot temperatures were found to lie some 1000–1500 K below the photosphere values, and the spatial locations of active regions within the chromosphere relative to those of the photospheric spots were determined.

A comprehensive review of Doppler images, giving the distributions of brightness over the surface of the active K1 subgiant component of the RS CVn binary HR1099 (V711 Tau), has recently been published by Vogt et al. (1999). They have shown that over the interval 1981–1992 the polar spot has been persistent and remains largely unchanged, whilst complex distributions of spot activity have changed continually at lower latitudes, with few features remaining for more than a few months. They suggest that there is little or no evidence of spot migrations in this system, and they caution against interpreting migrating photometric waves as being due to longitudinal migrations of spots caused by differential rotation; rather, they propose that these variations are due to the formation and erosion of spots at different locations. They draw parallels between the structures and the stability of the spot distributions on HR1099 and the properties of the coronal holes on the Sun (little evidence of shear due to differential rotation, similarities in the structure of the polar crown and its extensions to lower latitudes) and propose that HR1099 and other fast rotators amongst the RS CVn binaries have axisymmetric, global dipole fields of several kilogauss. The slower rotators do not seem to possess these polar spots, and the conclusion is drawn that these high magnetic fields are induced by dynamo action, rather than remnant fossil fields. Donati (1999) has also studied HR1099 via spectropolarimetry to establish sequences of Zeeman-Doppler images for the K1 subgiant, giving not only brightness distributions similar to those described earlier but also the polarities and field strengths for the various regions. He shows that the magnetic field is predominantly azimuthal in at least one persistent, high-latitude feature, concluding that the dynamo processes operate throughout the whole convective envelope of the star, not just at the boundary between the radiative and convective regions as in the Sun. The average azimuthal field is about 6 kG, or 6 T, with the toroidal and poloidal components of the overall dynamo field having field strengths of a few hundred gauss. There are many references in that paper to successful detections of Zeeman signatures in other single and binary stars.

The 1.6-day binary system V824 Ara, which is thought to be composed of two pre-main-sequence stars, rather than the more evolved stars seen in RS CVn binaries, has been reported by Hatzes and Kürster (1999) to show magnetic activity in the form of starspots over both components, again with substantial polar spots, and with those distributions being mirror images of each other, perhaps indicating the influence of tidal forces in such a young system. A multi-wavelength study, from the UV to microwaves, on the RS CVn binary II Peg by Byrne et al. (1998) has established the presence of photospheric spots and associated chromospheric flares. Their observations, taken during a quiescent state of the binary, nevertheless demonstrate that the system is continually variable in chromospheric and coronal emissions. We can expect more multi-wavelength study campaigns for these objects, extending from x-rays to radio, particularly because of the high expectations for detailed x-ray spectra from AXAF (*Chandra*) and from XMM.

It is regrettable that these procedures have not yet been extended to investigations of the magnetically active contact binaries, even though all of the analysis and modelling codes are available. There is much to be learned in that field from the evidence for bright hotter regions around the adjoining neck between the two stars and the distributions and evolution of magnetically active regions on both of the rapidly rotating, cool stars in W UMa systems. The eclipse-mapping technique is being used on sequences of light curves, but Doppler images and Zeeman-Doppler images are needed too.

7.2.2.2 *Accretion streams and discs*

The existence of discs in dwarf novae and some nova-like CVs was recognized directly from the dominant broad, double-peaked emission lines of H and He in their spectra. In eclipsing systems, we see the blue peak eclipsed first during ingress, followed by the red peak, and

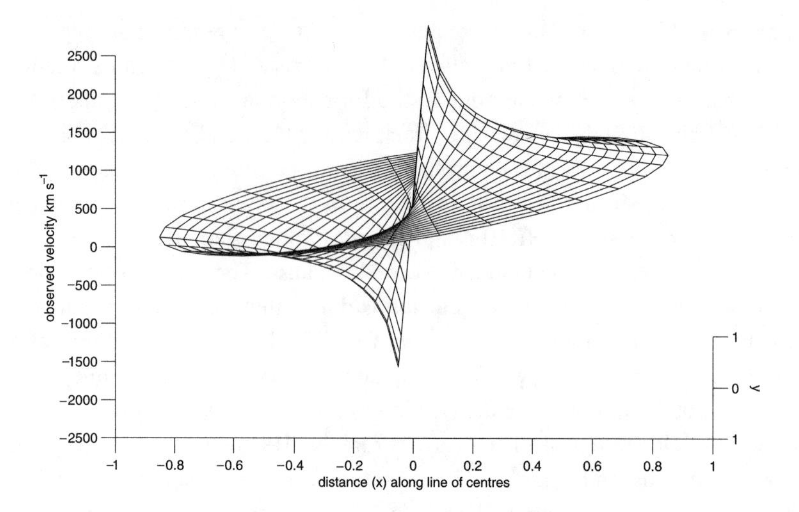

Fig. 7.8. A 3-D projection of the observed radial velocity as a function of (x, y) position in an idealized flat accretion disc extending from 0.05 to 0.85 in units of the Roche-lobe radius. The disc is assumed to be in Keplerian rotation about the accretor, and the velocities are representative of that expected in a CV with an orbital period of 0.2 day and both stars of 1 solar mass. The projection is viewed from an angle of 15° above the disc plane and perpendicular to the line of centres, in order to illustrate the dependence of the observed radial (line-of-sight) velocity V_y at orbital phase 0.25 on (x, y) position in the disc.

then the sequential reappearance of both peaks during the egress phases of the eclipse, thereby confirming that these discs surround the accretor, as in Algol systems. In the 1950s and early 1960s, Kraft, Crawford, and others showed that binary orbital motion could be measured in nearly all CVs, with the null results being due to low orbital inclination. The emission lines were due to gas orbiting a white dwarf in the orbital plane, with the gas being supplied from a Roche-lobe-filling cool star. Smak (1981) provided an overview of the earlier analytical work on predicting the shapes of emission lines from accretion discs, and he extended that work to illustrate the dependence of the line profiles on different representations of the distribution of emitted flux across axisymmetric discs. Figures 7.8–7.10 illustrate those expectations for a flat disc in Keplerian rotation about the accreting star and extending from 0.05 to 0.85 of the radius of the Roche lobe of the accretor. In Keplerian rotation [$V(R) \propto R^{-1/2}$], the innermost parts of the disc are rotating most rapidly, whilst the outer parts rotate much more slowly, as illustrated in the three-dimensional projection of the observed radial (i.e., line-of-sight) velocity V_y at orbital phase 0.25, plotted as a function of position (x, y) within the disc in the orbital plane (Figure 7.8). The angle of view has been selected at 15° in order to illustrate the dependence of V_y on (x, y), the rapid increase towards the innermost parts of the disc, and the large surface area of the more slowly rotating outer disc.

When the same disc is plotted in velocity coordinates (V_x, V_y) (Figure 7.9), where $V(R) = (V_x^2 + V_y^2)^{1/2}$, the disc is turned inside out, with the more slowly rotating outer parts contributing to the inner part of the figure, and the rapidly rotating inner parts spread around the outer part of the figure. Although the temperature of the disc decreases with increasing radial distance in

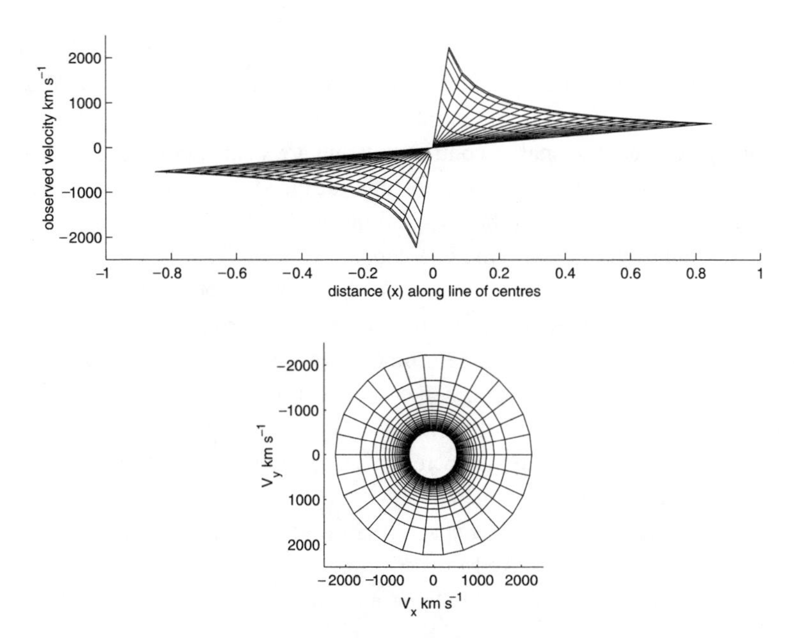

Fig. 7.9. Top: The same 3-D projection as in Figure 7.8 of the observed radial velocity as a function of (x, y) position in an idealized flat accretion disc, but seen in the disc plane. Bottom: Projection of the same disc onto the velocity plane (V_x, V_y) of the distribution of flux density from the pixels on the disc. The disc is turned inside out, with the outer disc at low velocities and having most of the flux density because of the larger surface area, and the innermost parts of the disc at high velocities but with a smaller surface area at higher temperatures.

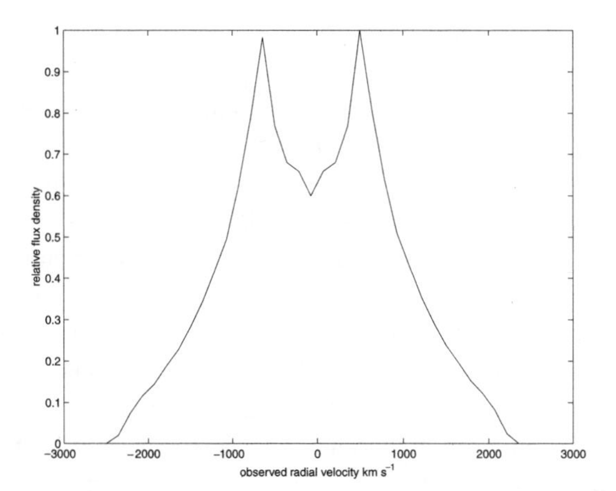

Fig. 7.10. An illustration of the expected double-peaked emission line from the Keplerian disc shown in the preceding two figures. From these figures it is clear that most of the emission comes from the outer disc, with its much larger surface area, whilst the hotter, rapidly rotating inner disc contributes only to the wings of the emission line.

the disc, the total surface area of the outer parts of the disc greatly exceeds that of the inner parts close to the accreting star. Accordingly, the outer pixels of the disc provide most of the received flux density, as indicated by the black central parts of the disc image in velocity coordinates, whilst the inner pixels make a smaller contribution and are shaded grey. The same results are seen in the expected emission-line profile for this idealized disc illustrated in Figure 7.10, where the double peaks result from the red- and blue-shifted emissions from the outer disc, and the inner disc contributes only to the line wings.

In addition to the double-peaked emission lines from axisymmetric discs, CVs often have an additional emission component that is observed to move from near one of the emission peaks to the other and back again in one orbital cycle. In a sequence of trailed spectra on photographic plates, this additional emission component traced out an S wave in one orbital cycle, frequently called the *S-wave distortion*, which was correctly attributed to emission from the bright region of impact between the accretion stream and the disc. Modern time-resolved digital spectroscopy allows us to do much more.

From the plot of disc flux in velocity space (the lower panel in Figure 7.9) it is clear that the observer will record a double-peaked emission line, a projection of the velocity-space image in the direction corresponding to the orbital phase at that time. For that reason, this technique is also referred to as *back-projection*. As the binary rotates, the entire emission line due to the disc will be Doppler-shifted according to the orbital motion of the accreting star. Such motion is likely to have a semiamplitude of 100–200 km s^{-1}, whilst the width of the emission line will be about 20 times that value. After subtraction of that bulk shift, the emission-line profile recorded at all orbital phases, in a properly phase-resolved manner so that image motion is essentially frozen, is equivalent to the nuclear-magnetic-resonance (NMR) scans in medical imaging. Application of the same mathematical techniques of inversion can provide images in velocity space of the source of the double-peaked emission-line profiles. Marsh and Horne (1988) first presented a detailed investigation and application of this Doppler tomography or Doppler imaging to the accretion discs in CVs. The use of velocity space for the resultant image is clearly a natural choice for the technique used, and the procedure assumes only that the observed binary system is in a steady state for the duration of the set of observations, namely, one orbital period ($\sim$0.2 day). The preceding figures should make it readily apparent that velocity-space images derived from a set of observations are simply inside-out images of real accretion discs, without the use of any form of transformation between velocity and position in that disc. In the foregoing simulations, that transformation was via a Keplerian velocity law, and theoretical models can be used to predict velocity-space images from a range of possible transformations for comparison with the images from real data.

Many Doppler images of CV discs are entirely compatible with the expectation of steady-state, thin, viscous accretion discs, including the locations of the mass-transfer stream from the L_1 point to the impact regions on the outer parts of the disc. The facing hemisphere of the mass loser, usually the late-type main-sequence star, is also often identified in these images from H emission lines, showing that the reflection effect is strong in these very close binaries. Good examples of such findings have been published by Marsh et al. (1990) for the dwarf nova U Gem and by Dhillon, Jones, and Marsh (1994) for the nova-like CV DW UMa, observed during a low-activity state. Marsh, Robinson, and Wood (1994) also presented Doppler images of the accretion disc for the black-hole binary A0620-00 as part of their derivation of accurate masses for both components in that system. The images were very similar to those seen in quiescent dwarf novae. Model spectra for steady-state accretion discs in CVs have been calculated by Wade and Hubeny (1998), covering the wavelength range of 850–2000 Å, as well as a substantial range of mass-accretion rate onto white dwarfs of different masses.

By contrast, the images secured by Steeghs, Harlaftis, and Horne (1997) for the accretion disc in the dwarf nova IP Peg during a rise to maximum light in an outburst are most remarkable. Steeghs et al. (1997) showed (Figure 7.11) a strong, two-armed spiral pattern in the disc images from Hα and He I λ6678 in velocity space, the first time that such a configuration had been detected. During a subsequent outburst maximum in the same object, Harlaftis et al. (1999)

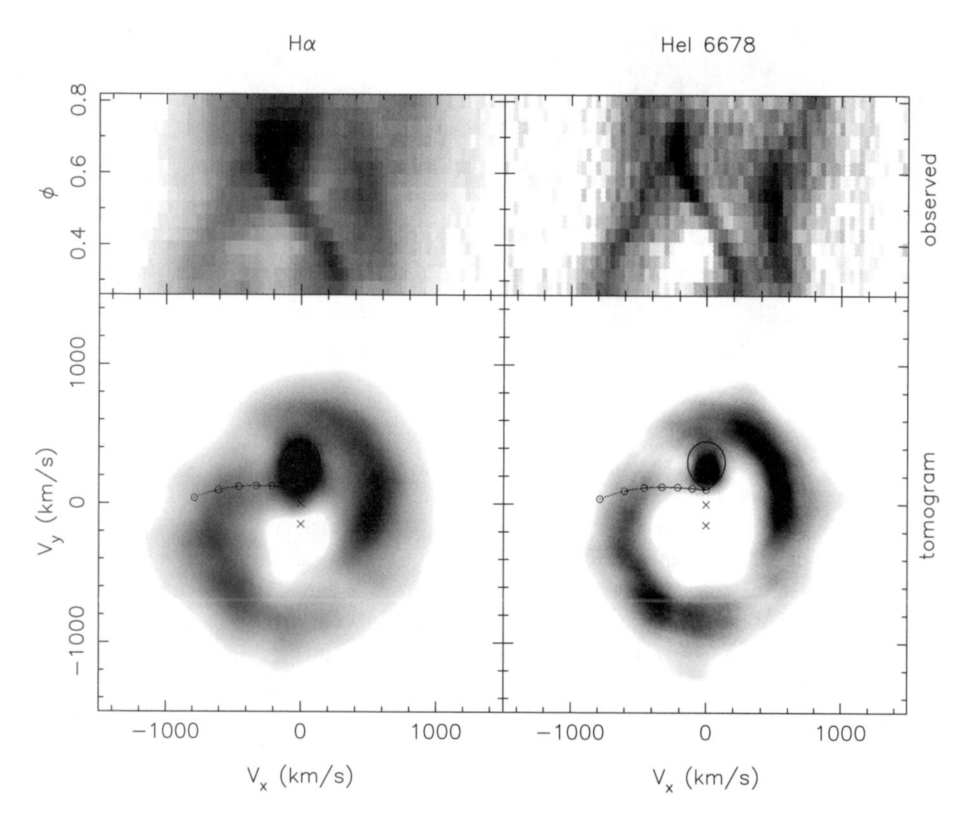

Fig. 7.11. Top: Montages of the phase-resolved spectra for IP Peg in the regions of the Hα and He I λ6678 emission lines showing multiple components with different dependences on orbital phase. Bottom: The corresponding Doppler maps for the two lines in velocity space (V_x, V_y) showing the obvious two-armed spiral pattern, together with the mass-losing secondary component. The Roche lobe of the secondary is indicated, and the location of a theoretical ballistic accretion stream from the L_1 point is shown as the curved near-horizontal line in the figure. The crosses indicate the locations of the centre of mass of the system and the centre of the white dwarf. (From Steeghs et al., 1997, with permission.)

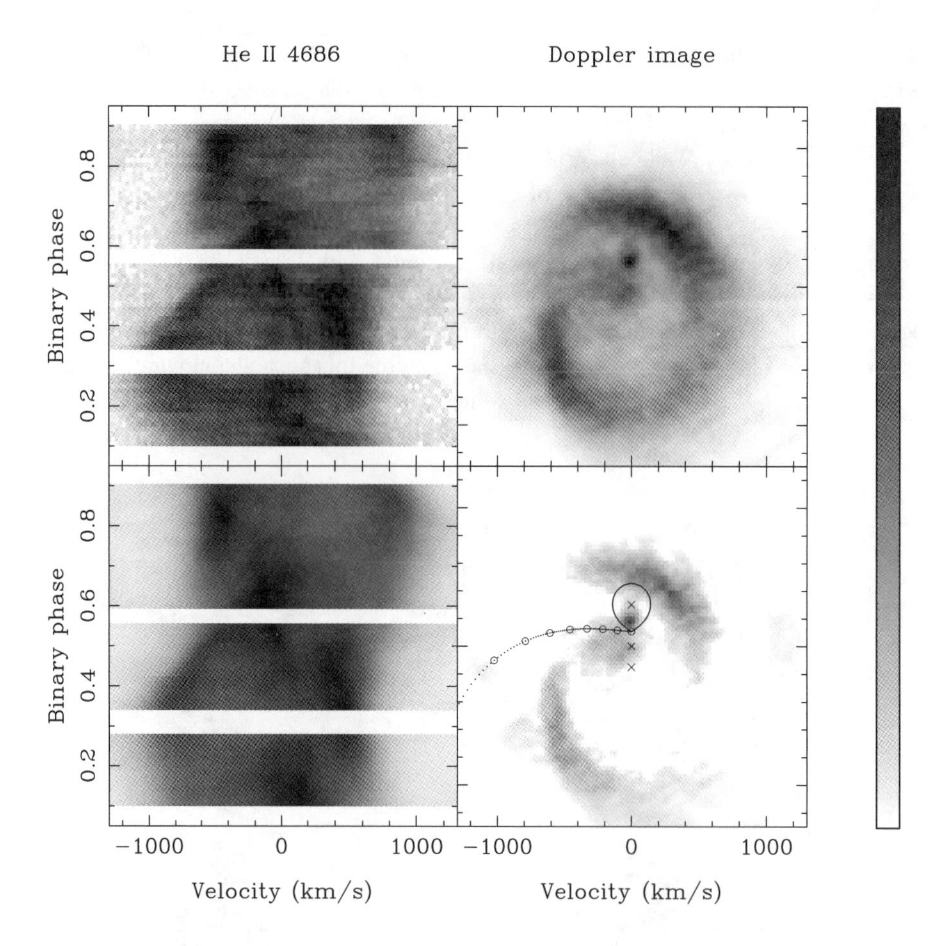

Fig. 7.12. Top left: Montage of phase-resolved spectra (a trailed spectrogram) for IP Peg during an outburst maximum, in the region of the He II λ4686 line, with two short phase intervals unobserved (0.28–0.34, 0.55–0.59). The complexity of the overall line profile is obvious. Top right: The corresponding Doppler map showing an axisymmetric image with a superposed asymmetric spiral pattern, a red-dwarf image, and an unresolved location with almost zero velocity. Bottom left: The projection of the Doppler image into an expected trailed spectrogram, for direct comparison with the original data. Bottom right: The asymmetric components of the Doppler image only, with the outline of the Roche lobe of the red dwarf, the location of a ballistic stream from L_1, and the centres of the two stars and the centre of mass marked with crosses. The grey-scale bar corresponds to a linear flux-density scale of 1–57 mJy. (From Harlaftis et al., 1999, with permission.)

observed the pattern again (Figure 7.12) and showed that the change in the He-II-line emission across the spiral, together with the velocity shifts of 200–300 km s^{-1}, demonstrated that those spiral patterns were caused by shock waves in the disc, where the sound speed was about 10 km s^{-1}. Steeghs and Stehle (1999) have constructed emission-line profiles for thin discs with tidal density waves that may reproduce the observations reported for IP Peg provided that the disc is large and hot.

There is also evidence for outflowing winds from accretion discs and perhaps diffuse sources of radiation that are not eclipsed by the red dwarf, giving rise to emission lines seen in mid-eclipse, some of which revert to absorption lines outside of eclipse, as shown, for example, by Knigge et al. (1998a) in their HST UV and optical observations of the nova-like CV UX UMa. There have been many recent reviews of the properties of accretion discs, particularly those in CVs and related objects, such as those by Horne (1998) and Marsh (1999), in which further details and many references can be found.

In the classical Algol binaries, where accretion discs were first discovered, Doppler tomography has again been successful in providing maps in velocity space of the distribution of disc emissions. Richards, Albright, and Bowles (1995) have made use of *difference profiles* for the Hα line in several short-period Algol systems to extract images of the accretion discs. Unlike the discs in CVs, where they often are the dominant sources of UV and optical radiation, the discs in Algol systems are difficult to detect. It was necessary for Richards and associates to subtract from the observed Hα profiles the expected photospheric spectra for the two stars in each system to yield a set of difference profiles sampling the entire orbital period. The disc velocities ($\sim$300 km s^{-1}) are also much lower than those in CVs because of the fact that the matter is being accreted onto a stellar surface that is much higher up the gravitational-potential well than is the surface of a white dwarf. Accordingly, such detailed work requires high-resolution spectra (resolutions of 10 km s^{-1}) in order to derive meaningful profiles for analysis. The resultant images show clear gas streams from the loser, in accordance with the theoretical hydrodynamical predictions, and the presence of weak asymmetric accretion discs around the gainer.

Some of the properties of discs in the x-ray binaries were mentioned in Chapter 5 in the discussion of analysis of their light curves. A helpful review is that by Verbunt (1999), from which some of the following summary is taken. For the HMXBs, the dominant source of UV and optical light is the O–B supergiant, and the contribution from the accretion disc around the companion is seen photometrically only at the 1–2% level. Precession of the accretion disc is understood to cause small changes in the amount of irradiation of the facing hemisphere by the x-ray source, leading to fluctuations in the overall light curve, as noted earlier for LMC X-4. The intermediate-mass system Her X-1 shows a more extreme 35-day variation, with an x-ray-on state for 10 days, and an x-ray-off state for 25 days, interpreted as resulting from a tilted and twisted accretion disc that precesses with a period of 35 days. Still et al. (1997) have discussed new spectrophotometry of this system, from which they have derived new Doppler maps for the accretion stream and disc around the neutron star and for the irradiated face of the companion via analyses of individual lines of H, He I and II, and C III.

Amongst the LMXBs, the systems in quiescence show emission lines like those of CVs, and their tomographic results are very similar, such as those found for A0620-00 by Marsh et al. (1994). From x-ray photometry, the presence of quite regular modulations of the orbital period have been detected and shown to result from variations in azimuth of the thickness of the accretion disc, causing changes in the column density seen by the observer. There are the sharp eclipses of short duration of the compact star by the Roche-lobe-filling companion, as noted in Chapter 5, and additional discrete intervals of x-ray absorption, or *dips*, that do not follow the orbital period exactly. These dips are taken to indicate time variations in the disc thickness, with the height of the outer disc being about 15% of the disc radius. An example is the object EXO0748-676 studied by Parmar et al. (1986). When these objects are in outburst, particularly those called the *soft-x-ray transients* (SXTs), also called *x-ray novae*, their spectra are very complex and are not well understood. An example of detailed studies of such systems is that by Hynes et al. (1998a), who reported on the 1996 outburst of the superluminal jet source GROJ1655-40 that lasted for the unusually long time of several months. With data from x-ray-to-optical wavelengths, they were able to investigate the relative contributions of the accretion structures, the irradiated secondary, and the compact x-ray source, as well as evidence for an outflowing wind from P Cygni profiles. In another paper, Hynes et al. (1998b) demonstrated that there is a time delay between the variable x-ray emission from this source and the UV and optical emissions that is interpretable in terms of reprocessing of the x-radiation from the compact source by its surrounding irradiated accretion disc. The time delays of about 14 seconds in this system provide an estimate for the size of the accretion disc, via the finite light-travel time to those reprocessing sites, that is consistent with the model for the system established by Orosz and Bailyn (1998) and van der Hooft et al. (1998). Those authors secured extensive multi-colour photometric and spectroscopic observations of GROJ1655-40 during quiescence and determined the masses and other parameters for the components from analyses of the ellipsoidal light variations, the spectroscopic orbit of the mass loser, and the spectra of the components. These showed that this SXT contains an accreting black hole of mass 7.0 $M_\odot$, with a mass-losing companion of 2.3 $M_\odot$ that is half-way between the main sequence and the red-giant branch. So we have an excellent description of the major components of this binary system, and the next task is to achieve an understanding of the causes of the outbursts and the subsequent phenomena. Such investigations are not yet determining surface images, but they are revealing the complexities of the outburst processes.

A serious anomaly amongst all the subclasses of CVs is the set of systems now called SW Sex stars, after the prototype (Thorstensen et al. 1991). These systems show only single-peaked emission lines, even in eclipsing systems where symmetry arguments would demand double peaks. The radial-velocity curves derived from these emission peaks lag the orbital motion of the white dwarf by 30–70°, and their corresponding location in the lower-left region in Doppler maps is not associated with either star nor with a standard disc. The remarkable and extreme CV AE Aqr seems, according to Horne (1998), to embody all of these anomalies and to suggest a probable solution to these issues. There have been many studies of AE Aqr; see Eracleous and Horne (1996), Wynn, King, and Horne (1997), Welsh, Horne, and Gomer (1998), and the references cited therein. From those studies it has been concluded that the white-dwarf component is strongly magnetic and is spinning rapidly at $P_{\rm spin} = 33$ seconds, so that $P_{\rm spin} \sim 1000 \times P_{\rm orb}$, and the action of its magnetosphere is to serve as a magnetic propeller that boosts the mass-transferring gas to escape velocities from the system. The Doppler maps of this system, secured from the Hα emission line by Welsh et al. (1998), recorded on 10,000 spectra with 6-second time resolution, show nightly changes and flux distributions that are quite unlike the pattern for any standard CV disc. The system is evidently linked to the subset of CVs discussed in the next subsection, the magnetic CVs that are called *polars*.

7.2.3 Imaging of polars and intermediate polars

One might wonder why this subclass of CVs is being given the preferential treatment of its own subsection. My reason for doing so is that there are three observational aspects to determining the structures of these systems that are somewhat different from those for most other binaries, namely, polarization, the different frequency components in the modulations of the x-ray, UV, and optical photometry, and Doppler imaging. My reasons for not discussing the polarimetric properties of these objects in Chapter 5 are the following: The main sources of x-ray, UV, and optical radiation from these systems seem to be the small, hot regions of accretion of matter on the magnetic white dwarf at the footpoints of the accreting field lines, near the magnetic poles. The polarimetric signatures contain information about those accretion hot spots, rather than the sizes and shapes of the stars themselves, which were the main aims of Chapter 5. The polarimetry does provide accurate determinations of orbital periods and also orbital inclinations, but that is not enough to describe an orbit from polarimetry, which was a topic in Chapter 3. So I decided to delay any discussion of these intriguing objects until this final chapter, because all of the information gleaned from the polarimetry, the frequency analysis, and the phase-resolved spectroscopy counts towards determining three-dimensional (3-D) images of these interacting binaries.

The subclasses of CVs now known as intermediate polars and polars were established as distinct groups as a direct consequence of the discovery by Tapia (1977) that the system AM Her exhibited rapidly changing degrees of circular and linear polarization, up to 10–30%. The circular polarization, in particular, suggested a strong magnetic field of order 10^3 T and hence spiralling electrons emitting cyclotron radiation. Soon other binaries were found to show similar behaviours of polarization variations dependent on orbital phase, and they became known as the AM Her stars, or *polars*. The canonical view of polars is that they are CV-type systems in which the white-dwarf accretor has a strong magnetic field that takes control of the flow of gas from the L_1 point, threads the gas onto its magnetic-field lines, and channels the flow along those 3-D arched field lines towards one or both poles of the presumed dipole magnetic field. A second group was also identified and called the intermediate polars, because their somewhat weaker fields, their greater orbital periods, and the asynchronous rotation of the white dwarf allowed a standard ballistic accretion stream to form and develop an outer accretion disc that was disrupted by the magnetic field of the white dwarf at some intermediate location between the outer disc and the white-dwarf surface. There have been many reviews of the properties of these fascinating binaries, amongst them being those by Patterson (1994), Warner (1995), Cropper (1998), and Hellier (1999a), with many references and details discussed therein. Note that both stars are intrinsically faint in these systems, as they are for CVs in general, but these polars do not have substantial and warm accretion discs to provide more luminosity, only small, but very hot, accretion spots. Accordingly, polars typically have apparent V magnitudes fainter than about the fourteenth magnitude, and the successes in studying them owe as much to ingenuity in observing techniques as to developments in technology.

In the polar systems, the axial-rotation or spin period of the white dwarf is locked to the orbital period of the binary, so that $P_{\text{spin}} = P_{\text{orb}}$. In intermediate polars, $P_{\text{spin}} \ll P_{\text{orb}}$, with $P_{\text{spin}} \geq 33$ seconds. Frank et al. (1992) and Warner (1995) have presented helpful reviews of many theory papers on magnetically controlled accretion that involved attempts to understand the properties of magnetic white dwarfs in CVs, as well as neutron stars in x-ray binaries. Because the devotees of this entire subject area seem to enjoy working in CGS units, modified to the scales found in CVs, we shall accept their practice in this brief sketch.

The term *magnetosphere* is used to describe the volume of space around a star with a magnetic field within which that field controls or significantly influences the motions of charged particles. The magnetic pressure at radial distance r from the centre of the star is given by $B^2(r)/8\pi$ for a field of strength $B(r)$. If the field is an overall dipole, then $B(r) = \mu/r^3$, where $\mu = B_\star R_\star^3$ is the *magnetic moment* of the star of radius $R_\star$ and surface field $B_\star$. If there is spherically symmetric accretion of matter onto the star at a rate $\dot{m}$, then we can write

$$\dot{m} = 4\pi\rho(r)v(r)r^2 \tag{7.7}$$

where ρ and v are the density and radial infall velocity at distance r, respectively. The radius of the magnetosphere, r_{m}, is then given by equating the magnetic pressure to the ram pressure of the infalling gas, so that

$$B^2(r)/8\pi = \rho(r)v^2(r) \tag{7.8}$$

and $v(r) \sim (2Gm/r)^{1/2}$, because the flow will be close to the free-fall velocity. Then the equilibrium radius, called the *Alfvén radius*, is given by

$$r_{\mathrm{m}}^7 = \mu^4/(8Gm_{\mathrm{wd}}\dot{m}^2) \tag{7.9}$$

or, in the units seemingly beloved of CV pundits,

$$r_{\mathrm{m}} = (9.9 \times 10^{10})\mu_{34}^{4/7} m_{\mathrm{wd}}^{-1/7} \dot{m}_{16}^{-2/7} \text{ cm} \tag{7.10}$$

That is, the magnetic moment is given in units of 10^{34} G cm^3, and the mass-accretion rate onto the white dwarf of mass m_{wd} is in units of 10^{16} g s^{-1}. For polars, $\mu_{34} \sim 1$, and $B_\star \sim$ $few \times 10^7$ G, or $few \times 10^3$ T. For a binary composed of two 1-$M_\odot$ stars in a 4-hour orbit, the volume radius of a Roche lobe is about 4×10^{10} cm, so the magnetosphere can occupy most of the accretor's Roche volume and therefore control the flow of gas from L_1 to the surface of the white dwarf. For cases in which the accretion is not spherical, which is the situation obtaining in polars, Ferrario et al. (1989) have provided a more complex formula for r_{m} that still has the same dependence on $\dot{m}$, and nearly the same on m_{wd}.

From the basic theory of accretion channelled onto the region around a magnetic pole (Frank et al. 1992) it is understood that the structure of the accretion column contains a shocked region that is situated somewhat above the surface of the accretor. Here the infalling gas is slowed dramatically before settling onto the white-dwarf surface, and the temperature within that shocked region is about 10^8 K. The radiative-cooling mechanisms for the shock include the following: bremsstrahlung or free-free emission by relativistic electrons, mostly in x-rays; cyclotron radiation from spiralling electrons moving semi-relativistically; and Compton scattering of photons by electrons. Some of the radiation from the shock region will be emitted into space after traversing various path lengths through the accretion column, and some will be absorbed/scattered into the accreting surface, where it will be thermalized and re-emitted, mostly as soft-x-ray,

UV, and optical radiation corresponding to a black-body temperature of $few \times 10^4$–10^5 K. Thus the overall spectrum for a polar can be expected to show bremsstrahlung x-radiation, UV and optical thermal radiation, and cyclotron emission in the optical and IR, with the latter component being strongly polarized. Whether we see mostly linear or circular polarization will depend on the orientation of the accretion column near the magnetic pole to our line of sight. We will observe linear polarization if the column is perpendicular to our line of sight, because the electrons will seem to be oscillating in a single plane. But if we look down the column, parallel to the magnetic field, then the spiralling motion of the electrons around the field lines will give rise to circular polarization, at least for non-relativistic speeds. For higher speeds, the situation becomes more complicated, requiring discussion of Doppler/relativistic beaming. Not all of the cyclotron radiation will be emitted at the fundamental frequency of $(2.8 \times 10^{14})B_8$ Hz, where B_8 is the magnetic-field strength in units of 10^8 G. For $B_8 = 1$, this frequency corresponds to about 1 μm in the near IR. Cyclotron radiation will be emitted also at higher harmonics in the visible part of the spectrum, and observations have confirmed the identification of the so-called *cyclotron humps* in the spectra of polars, which are asymmetric because of the Maxwellian distribution of the velocities of the electrons in the column. Broad-band photometry monitors radiation from more than one cyclotron harmonic, but they can be separated spectroscopically and used to estimate the magnetic-field strength.

Many theoretical models have been developed to explain the observations of polarized radiation from polars, and they have been referenced in the aforementioned texts. The first models considered only a single accreting pole, offset at an angle β from the rotation axis of the spherical white dwarf, with limited geometry and opacities for the accretion column. But they were able to demonstrate that the substantial variations in the degrees of linear and circular polarization and the variations of polarization angle with orbital phase observed in polars could be understood. Examples are shown in Figure 7.13, where the variations in the Stokes parameters Q and V and the polarization angle θ are plotted as functions of orbital phase. These curves were calculated from the equations given by Barrett and Chanmugam (1984), adopting a magnetic-field strength of $B = 10^8$ G, with some different values for the orbital inclination i and the angular displacement of the magnetic pole from the rotation axis, β; following Cropper (1985), the calculated values for Q and V have been reduced by half. It is assumed that the rotation axis of the white dwarf is perpendicular to the orbital plane, that the magnetic dipole is centred on the centre of the star, and that the star is rotating synchronously. With that geometry, the angle α between the magnetic-field direction and the line of sight is

$$\cos\alpha = \cos i \cos\beta - \sin i \sin\beta \cos\phi \tag{7.11}$$

where ϕ is the orbital phase. The direction of the linearly polarized component of the cyclotron radiation is perpendicular to the field, and the position angle θ of that direction projected onto the sky is given by

$$\tan\theta = \frac{\sin\beta \sin\phi}{\sin\beta \cos i \cos\phi + \sin i \cos\beta} \tag{7.12}$$

When $\alpha = 90°$, the accretion column is just appearing or disappearing over the limb of the star, and it follows that $\cos\phi = 1/(\tan i \tan\beta)$, and hence $d\theta/d\phi = \cos i$. It follows also that

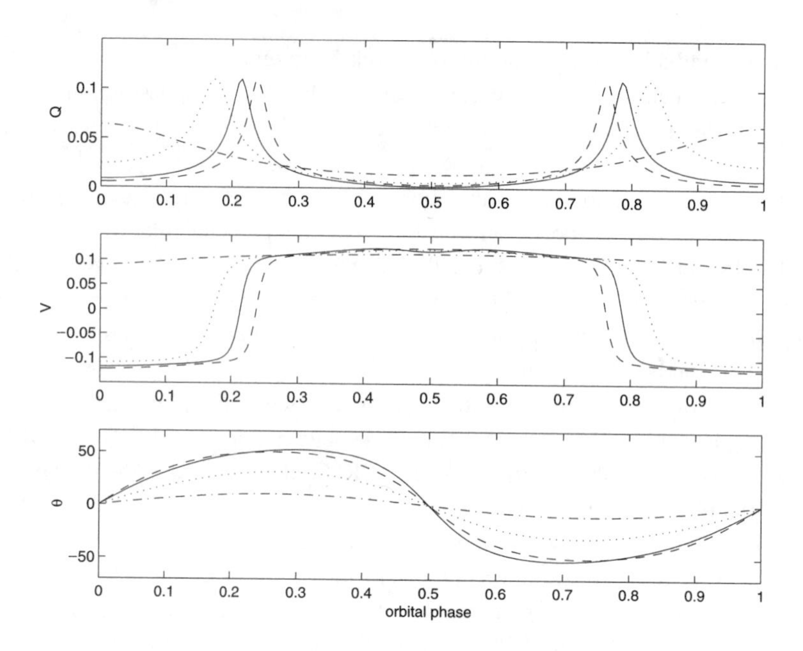

Fig. 7.13. Illustrative graphs of the variations with orbital phase of the Stokes parameters Q and V and the position angle of the linear polarization θ for a single-pole model of a polar. The model from Barrett and Chanmugam (1984) was used for these calculations. The different lines correspond to different values of (i, β), with $(75°, 50°)$, solid line; $(85°, 50°)$, dash line; $(75°, 30°)$, dotted line; $(75°, 10°)$, dash-dot line.

we expect to see peaks in the linear-polarization curve $Q(\phi)$ when $\alpha = 90°$, and a change of sign for the circular-polarization curve $V(\phi)$. These are illustrated in Figure 7.13 and confirm that when $(i + \beta) > 90°$, so that α passes through $90°$, the circular polarization changes sign, and the linear polarization exhibits sharp peaks. When $(i + \beta) < 90°$, there is no change of sign for $V(\phi)$, and the $Q(\phi)$ curve is much more shallow.

These factors, coupled with the $\theta(\phi)$ variation, permit determinations of the orbital inclination i and the angle β directly from polarization observations of sufficient accuracy and time resolution. The orbital periods for known polars are in the range from 80 minutes to a few hours, so that observations with time resolutions of a few seconds are necessary to follow the rapid changes in polarization exhibited by these sources. High-speed polarimeters are used for this work, with the result that thousands of observations are recorded on each system over several orbital periods, and usually they are presented in graphical form only in the published papers. Examples include Cropper and Warner (1986) on VV Pup, Cropper (1985) and Piirola, Reiz, and Coyne (1987) on EF Eri, and Wickramasinghe et al. (1991) on AM Her. Figure 7.14 shows the optical photometric and polarimetric data on the polar BL Hyi from Schwope et al. (1998). The circular- and linear-polarization curves are rather more complex than the idealized curves in Figure 7.13, particularly for circular polarization, but nevertheless show two significant maxima for the linear polarization. The accretion geometry for this system is not yet established.

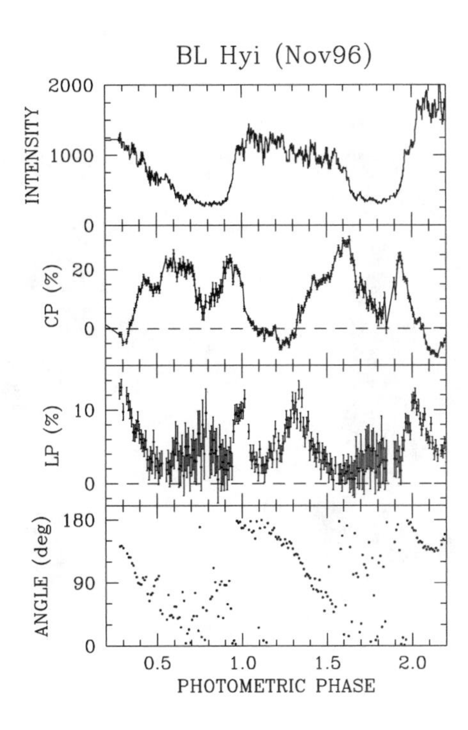

Fig. 7.14. Photometric and polarimetric light curves for the polar system BL Hyi, with two orbital cycles plotted to show the continuity. Whilst the major variation of the linear polarization (LP = Q) is sensibly double-peaked, there is some remarkable flickering in the orbital-phase range 0.6–0.9. The circular polarization (CP = V) is more complex than that in the simple models shown in Figure 7.13. (From Schwope et al., 1998, with permission.)

The early models for cyclotron emission from polars were soon found to be unable to make detailed agreements with high-quality observations. At first, the emission regions were extended into arc-shaped features around one magnetic pole, but later models had to introduce two-pole accretion, with elongated regions near the poles providing the observed emission. Those high-quality polarization curves showed much more detailed structure as a function of orbital phase, as illustrated by the previously referenced observations of AM Her, and the models had to become more general to explain that, as described in the same paper.

In addition to the polarization curves and the light curves for the eclipsing polars obtained from x-rays and UV and optical radiation, as summarized by Warner (1995), we are now seeing the results of Doppler tomography of polars from phase-resolved spectroscopy. The first application of this technique to polars was by Diaz and Steiner (1994), studying VV Pup, and they found substantial differences in the Doppler maps between polars and standard CVs. For polars, we would not expect to be able to convert a Doppler map in projected velocity coordinates straightforwardly into a spatial map, because the gas motions would not be expected to be taking place only in the orbital plane. The orientation of the magnetic field on the white-dwarf accretor can be at any arbitrary angle with respect to the orbital plane or the white-dwarf

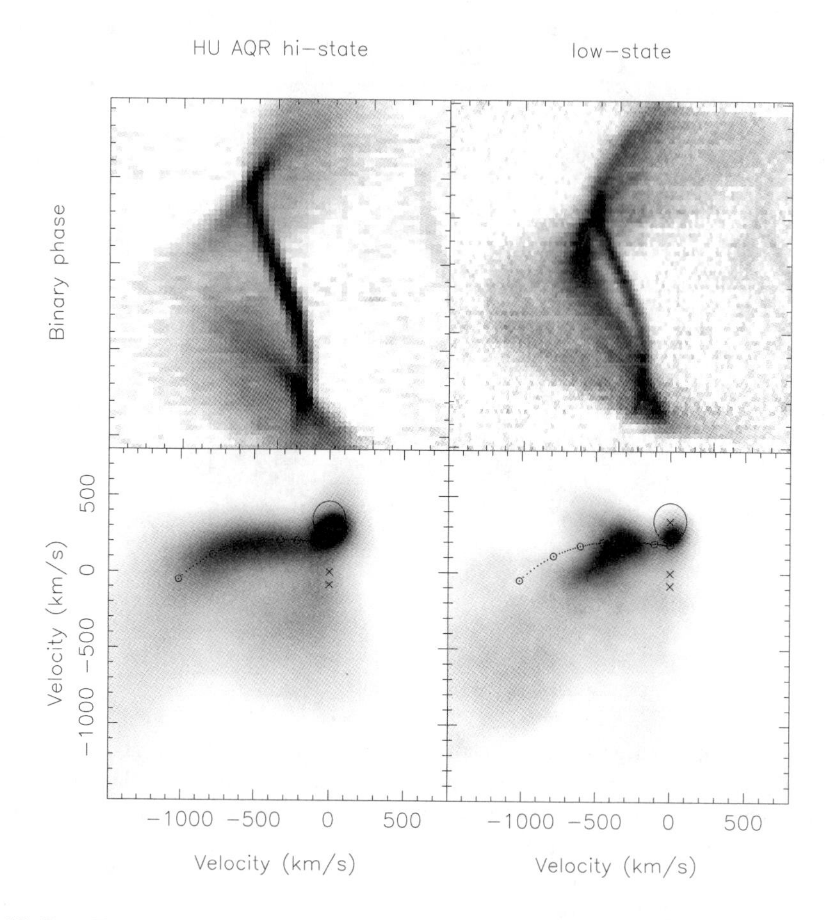

Fig. 7.15. Top: Phase-resolved trailed spectrograms for the region containing the He II λ4686 line of the polar HU Aqr seen in a high state and a low state of activity. The binary phase runs from 0.0 at the base of the figure to 1.0 at the top. Bottom: The corresponding Doppler maps have the same form as in Figure 7.11. The facing hemisphere of the secondary component is clearly visible, as is a ballistic accretion stream in the high-state image; there also are wide ranges of velocities at lower brightness levels that do not correspond to any normal accretion disc, but are due to an accretion curtain channelled towards a magnetic pole on the white dwarf. (From Schwope et al., 1998, with permission.)

rotation axis, and hence magnetically controlled flow will generally occur out of the orbital plane, with a significant z component of velocity. Schwope et al. (1999) have discussed these issues of interpreting Doppler maps in velocity coordinates for polars, showing how we might expect the various components of the accretion structure to be manifested in tomography.

One of the great occasions in this subject area was the discovery from ROSAT observations of a new polar, RXJ2107.9-0518, now known as HU Aqr, by Hakala et al. (1993). At $V \sim 15$ mag, that system was bright by polar standards, and phase-resolved spectra were soon acquired during that high state. Those observations were repeated 3 years later, when the system was in a low state and much fainter, and the trailed spectrograms are shown in Figure 7.15. Also shown

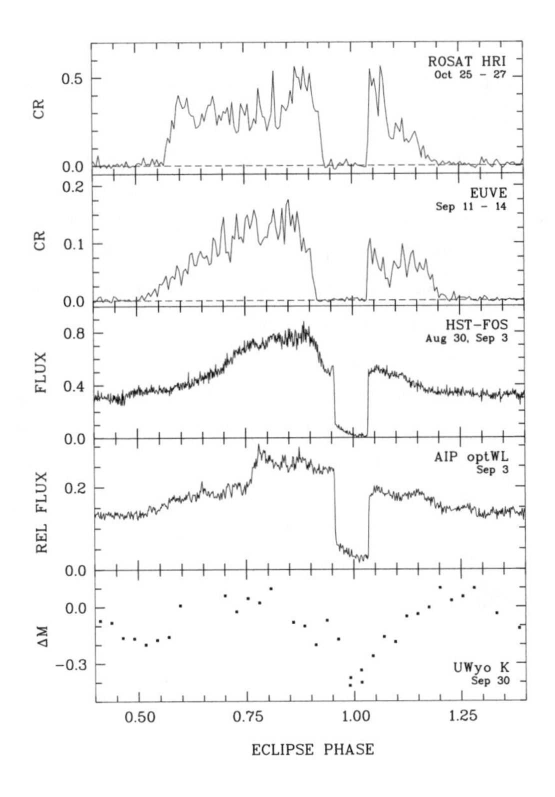

Fig. 7.16. From top to bottom: Light curves from x-rays, EUV, UV, and optical radiation for the polar HU Aqr obtained during a low state. Note the extremely rapid ingress/egress phases of the eclipses, lasting about 1 second in x-rays, indicating that the eclipsed source is very small. Note also that the pre-eclipse dip, seen in the high state in 1993 at phase 0.88 in the data in Figure 5.22, is missing. (From Schwope et al., 1998, with permission.)

in Figure 7.16 are the results from a photometric-monitoring campaign undertaken in autumn 1996 on HU Aqr with the ROSAT, EUVE, and HST satellites and ground-based telescopes; see Schwope, Thomas, and Beuermann (1993) and the references cited therein.

The trailed spectrograms show a clear S wave that is the orbital motion of the illuminated hemisphere of the secondary component. In the high-state spectrograms, the S wave is crossed at phases 0.12 and 0.70 by a bright narrow component due to the ballistic part of the stream and by a much broader, fainter component extending over a wide range of velocities. In the low-state spectrograms, the ballistic component is less extreme, but the broader component is still very evident. In the Doppler images, the secondary star is located at $(v_x, v_y) = (0,\ 290)$ km s^{-1}, providing a well-determined orbital semiamplitude after non-Keplerian corrections for the effect of its non-uniform illumination by the x-ray source (the reflection effect again) have been made. Further examinations of those Doppler images show that the radiation from the facing hemisphere of the secondary component is not symmetric with respect to the line of centres, but is displaced to the trailing side, indicating that the source of irradiation on the white dwarf is shielded from the leading side by the presence of the accretion column/curtain. So the calculations of non-Keplerian corrections will not be simple, though aided considerably by the information contained in the image.

In the Doppler images in velocity space, the ballistic component of the accretion structure is strongly evident in the high-state image, but much weaker and extending to lower velocities in the low-state image. In both images there is a very extended and lower luminosity distribution, which perhaps corresponds to the accretion curtains onto one or two poles. Theoretical models of such structures will have to be synthesized to predict the expected Doppler images in velocity space for direct comparison with observations. The locations of some of these structures are revealed by the eclipse timings and durations in the multi-wavelength light curves, which change with time as the mass-accretion rate varies. The sequences of x-ray light curves, in particular, on HU Aqr indicate that as the mass-accretion rate decreases, the location of the coupling region where the mass flow is threaded onto the magnetic-field lines moves towards the L_1 point. Further coordinated observing programmes are needed to provide photometry, polarimetry, and spectroscopy, particularly so when the new x-ray satellites, *Chandra* and XMM, are operational with their x-ray spectroscopic facilities.

A review of our current understanding of the *intermediate polars* is given by Hellier (1999a), based upon interpretations of observational data in the x-ray wavelengths through the optical wavelengths. The most obvious and distinctive difference between polars and intermediate polars is that the dominant periodicity in the x-ray photometry is at the white-dwarf spin period, rather than at the orbital period. The main sources of radiation are the accretion hot spots on the white dwarf that rotate in and out of view as the white dwarf spins on its axis and provide a modulation of the observed signal. There are additional frequencies in power spectra of x-ray and optical light curves, particularly at the beat frequency ($\omega - \Omega$) between the spin (ω) and orbital (Ω) frequencies, and rarely at others. There is an on-going debate about whether or not intermediate polars have disrupted accretion discs and whether or not the evidence from periodogram analyses of the x-ray and optical modulations supports the idea of the accretion stream overflowing the usual impact region on the disc and connecting directly onto the field lines. Hellier (1992) analysed the x-ray modulation data from several systems to show that various other beat frequencies predicted by discless accretion models were not present. Norton, Beardmore, and Taylor (1996) reported an extensive analysis of the various contributors to the emission and attenuation of x-rays within an intermediate polar and then constructed power spectra for the expected modulations of the x-ray emission as the white dwarf rotated and the binary revolved. Their conclusions, after testing their model against long strings of x-ray data from various intermediate polars, was that accretion discs and accretion streams onto two asymmetric regions near the magnetic poles were necessary, with the dominance of one component over another being dependent on the particular conditions in a specific binary.

Certainly there is plentiful evidence now that accretion occurs onto both polar regions in most intermediate polars. The recent applications of Doppler tomography to phase-resolved spectra for these sources support the evidence for discs and for two-pole accretion onto a rapidly rotating white dwarf (Hellier 1999a). A good example is the system RXJ0558+53, studied by Harlaftis and Horne (1999) by means of time-resolved spectroscopy. It was established by

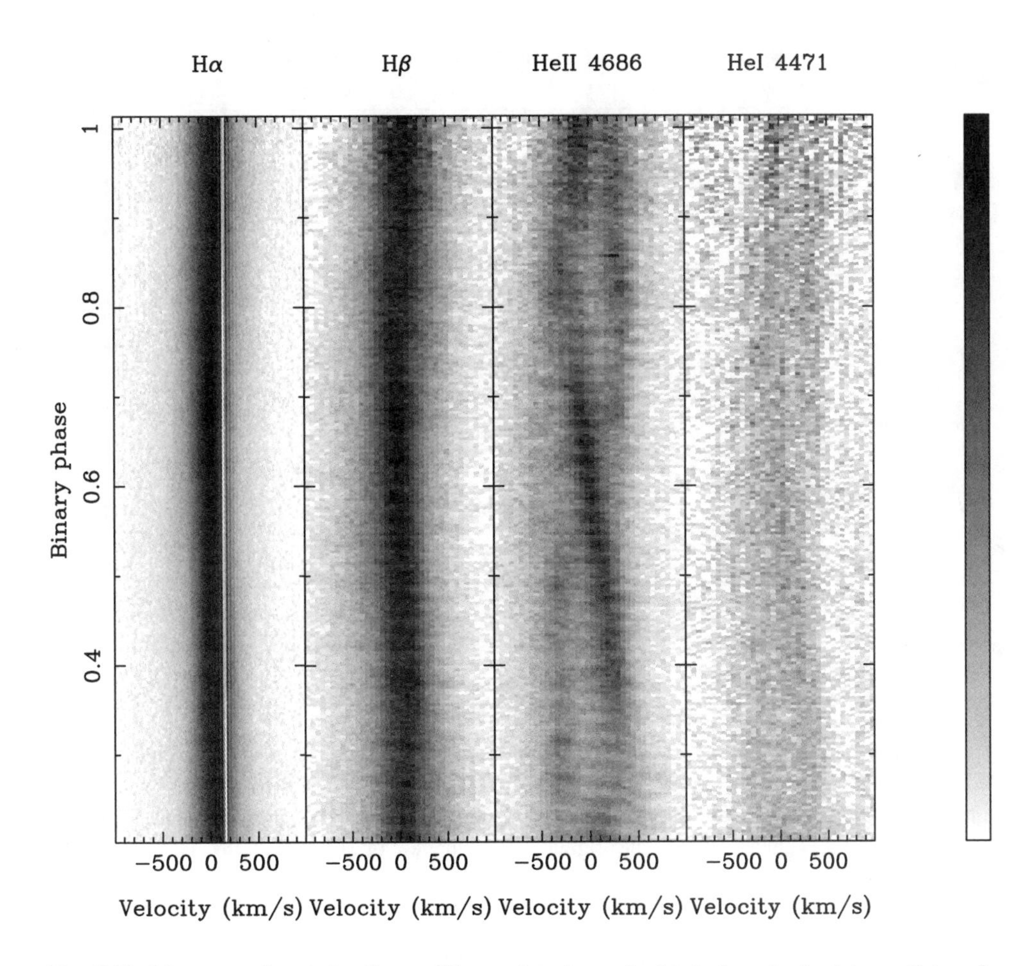

Fig. 7.17. Montages of emission-line profiles as functions of orbital phase in the intermediate polar RXJ0558+53. The darker regions indicate brighter emission. The trailed spectrogram in Hα is marred by a 3-pixel fault on the CCD. (From Harlaftis and Horne, 1999, with permission.)

Haberl et al. (1994), after the discovery x-ray observations, that the system had an orbital period of 4.15 hours, with an additional modulation at 272 seconds. Allan et al. (1996) showed that the spin period for the white dwarf was actually 545 seconds, and the first harmonic resulted from the fact that two poles were contributing to the signal. The amplitudes of the modulations were in the range 1–10% of the signal. Figure 7.17 shows the montage of spectral regions around Hα, Hβ, He II λ4686, and He I λ4471 ordered in orbital phase. The most striking feature of the trailed profiles is the obvious difference between the H lines and the He II line. In the latter, we see two components moving in antiphase with the orbital period, though displaced in phase from the adopted time of conjunction. The stronger of these two profiles is weakly evident in the He I line. Superposed on the He II profiles, and throughout the orbital cycle, we see an obvious modulation of the signal that is found to have the same period as the photometric or continuum spin period of the white dwarf. The resultant Doppler maps formed on the orbital period are

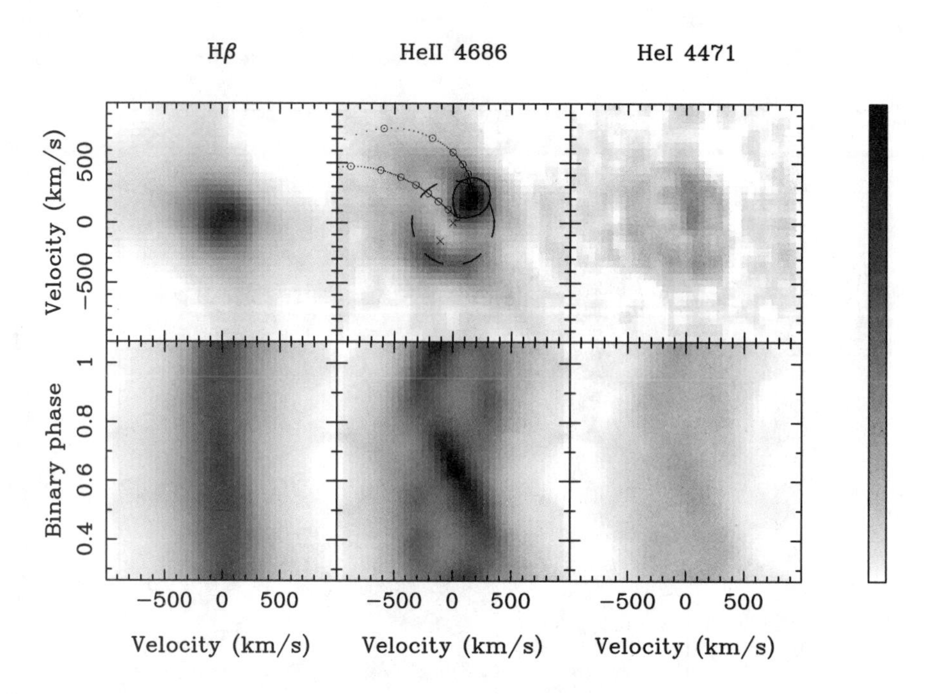

Fig. 7.18. Top: Doppler maps of RXJ0558+53 for the orbital period formed from the phase-resolved spectroscopy shown in Figure 7.17. In the He II image, the location of the secondary star is identified by the Roche-lobe shape, tilted because of the phase offset. Also shown is the ballistic-stream trajectory and an arc-shaped emission feature. Bottom: The corresponding trailed spectrograms computed from these images are shown for comparison with the original data in Figure 7.17. (From Harlaftis and Horne, 1999, with permission.)

shown in Figure 7.18. The He II image identifies the illuminated hemisphere of the secondary component, estimated to have an orbital velocity of 250 ± 30 km s^{-1}, and tilted because of the phase offset from the adopted ephemeris. The antiphase feature in the trailed spectrograms provides an emission arc in the He II image around 350 km s^{-1}, indicated by the dashed circle, and there is much ill-defined structure at other velocities. The antiphase arc is taken as evidence of the accretion disc around the white dwarf and on the far side from the companion star. The Hβ line provides an image that is nearly axisymmetric and centred on zero velocity.

When the spectral-line profiles are phased according to the spin period and placed into 12 phase bins, the result is as shown in Figure 7.19, with the pattern repeated over two spin cycles. Here the mean spectral profiles have been subtracted to leave only that component that varies on the spin cycle, and in the figure the white regions correspond to stronger emission, and the dark regions to less emission (opposite to that in Figure 7.17). The corkscrew effect is obvious, particularly in the He II line, showing that the white dwarf has two accretion hot spots, and the resultant tomogram from the data (not shown here) confirms that picture. The amplitude of this modulated line emission is 1–2%, compared with 5–10% for the continuum, and the line emission exhibits a phase lag of 0.12 relative to the continuum

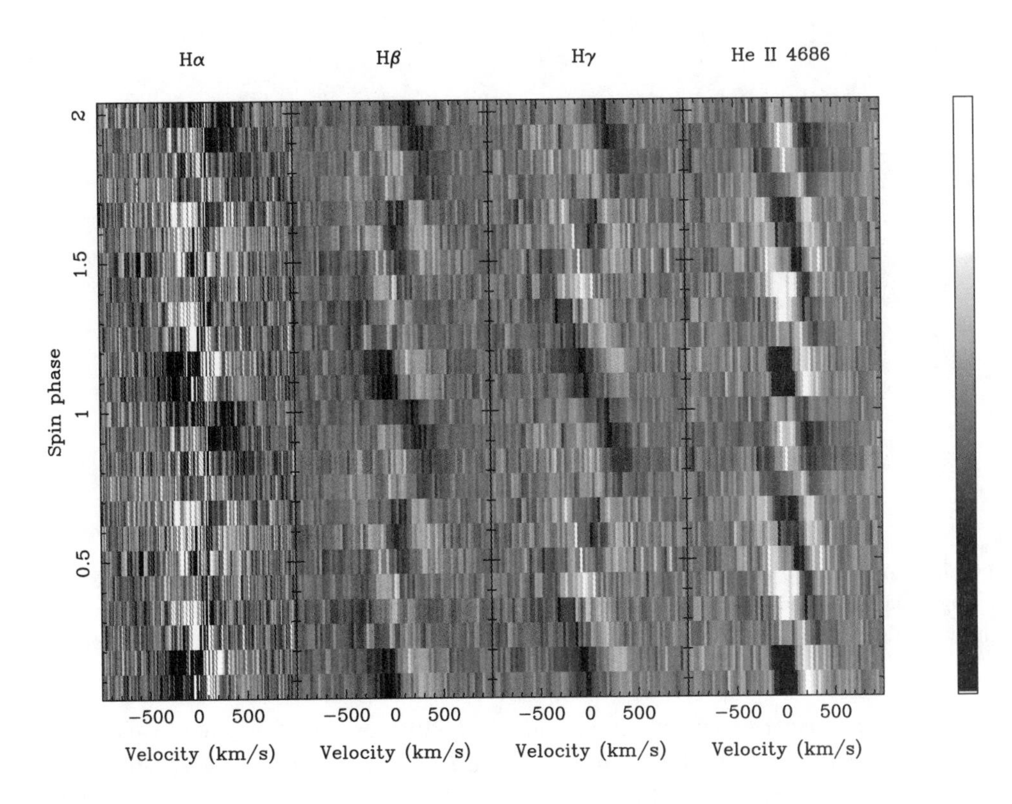

Fig. 7.19. The spectral-line profiles of RXJ0558+53, with the mean profile subtracted, summed into 12 phase bins around the spin period of the white dwarf; the plot is repeated over two cycles to emphasize the corkscrew pattern seen in the data (white is brighter, black is darker). One region is notably stronger in He II than the other, although the continuum pulses have similar brightnesses. (From Harlaftis and Horne, 1999, with permission.)

modulation. Hellier (1997, 1999b) has also derived Doppler tomograms from spin cycles in the systems AO Psc and PQ Gem, and both show clear two-pole accretion onto the white dwarf.

A schematic diagram of the accretion geometry in RXJ0558+53 is shown in Figure 7.20, with an accretion disc between the outer co-rotation radius $R_{\rm co}$ and the magnetospheric radius $R_{\rm mag}$ and the two accretion curtains directing the flows to the two active poles on the white dwarf. (The co-rotation radius in the disc is that distance at which the orbital period around the accretor is equal to the spin period of the white dwarf.) The phase lag between the line modulation and the continuum modulation is perhaps explained by the azimuthal spread in the accretion curtains.

A final example of multi-wavelength studies of intermediate polars is that for YY Dra, by Haswell et al. (1997), based upon their own HST UV spectroscopy, *UBVRI* photometry, and Hα spectroscopy, as well as earlier work referenced in that paper. The white-dwarf spin period is found to be $P_{\rm spin} = 529.31 \pm 0.02$ seconds, and the orbital period, at 3.9689755 hours, is

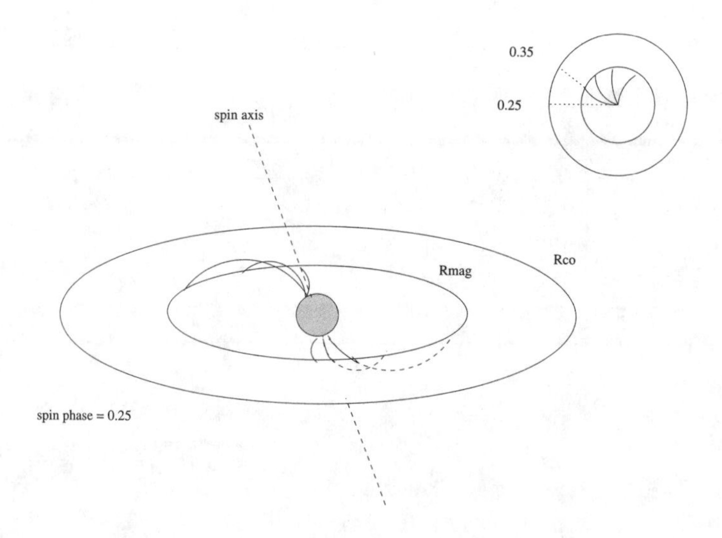

Fig. 7.20. A schematic diagram of the accretion geometry in the intermediate polar RXJ0558+53 indicating an outer disc and two accretion curtains leading from the inner disc to the two accreting poles on the white dwarf. (From Harlaftis and Horne, 1999, with permission.)

now specified to an uncertainty of 0.01 second. The *UBVRI* photometry was obtained with the Steining 5-channel high-speed photometer, and the *I*-band data clearly show standard ellipsoidal variations from the Roche-lobe-filling dM4 secondary star, which constrains the orbital inclination to the range $36° \leq i \leq 50°$. The shorter-wavelength data are progressively more dominated by the hotter components in the system, so that in the continuum UV data we see a dominant apparent flickering of ±50% about the mean light level. Within those data, the spin pulse of the white dwarf becomes obvious to the eye when plotted at sufficient time resolution, and the periodogram analysis shows a single dominant periodicity at 529.31 seconds, of amplitude about 15%, with the remainder of the flickering being random scatter. Published radial velocities for the secondary star give $K_2 = 202 \pm 3$ km s^{-1}, whilst HST spectroscopy shows orbital modulation from the C IV emission line associated with the white dwarf at $K_1 = 91 \pm 10$ km s^{-1}. With an assumed Roche-lobe-filling secondary, the masses of the two stars are determined to be $m_1 = 0.83 \pm 0.10\ M_\odot$ and $m_2 = 0.375 \pm 0.014\ M_\odot$. Detailed studies of the C IV emission line have shown that its line profile is modulated on the spin cycle of the white dwarf, with the broadest line wings appearing at the time of maximum brightness in the continuum spin pulses. Pulse maximum appears to occur when the two accretion columns are most closely aligned to our line of sight. The temperatures associated with the accretion columns are uncertain, because even these mid-UV data are sampling only the Rayleigh-Jeans tail of that energy distribution.

Even amongst the seemingly most intractable binary systems, with multiple components at widely differing temperatures, it is possible to determine the underlying stellar structure of the binary and then to use that information to probe further the accretion structures within the system.

7.3 Boundary layers in accreting CVs and x-ray binaries

Oscillations in the brightness of CVs, with periods in the range of tens of seconds and amplitudes of about 0.5% in the optical region, were first discovered in some CVs nearly 30 years ago by means of high-speed photometry. These oscillations are quite coherent, but their source remains uncertain. The time scales suggest an association with the innermost parts of the accretion disc, or the boundary layer between the disc and the white dwarf, because the Keplerian period at, say, $1.1 R_{wd}$ is about 25 seconds. Recent UV spectroscopy with time resolutions of 4–5 seconds from the HST for the old nova DQ Her, by Silber et al. (1996), and the nova-like CV UX UMa, by Knigge et al. (1998b), have provided further discriminating observations to aid in understanding the source(s) of these oscillations. In these systems, there is UV-continuum pulsation and some spectral-line pulsation with the same period as the optical light, but not necessarily coincident in phase, and having amplitudes of 5–20%. The hotter component, perhaps at 80,000–100,000 K, is eclipsed in UX UMa, but there remains always a non-eclipsed component of substantially lower temperature, about 25,000 K, that would seem to be a result of reprocessing of the radiation from the hotter, more compact source by the irradiated disc. A bright accretion spot on the white dwarf is obviously a candidate explanation, but the observed fast changes in oscillation period that occur in UX UMa create difficulties for that model.

In earlier sections we considered the observational evidence for accretion structures around the white dwarfs in CVs and the neutron stars and black holes in x-ray binaries. Neutron stars have radii that are a thousand times smaller than those for white dwarfs, that is, around 10 km, and black holes of stellar mass have event horizons of a few kilometres. Matter spiralling into such potential wells via a disc, or being captured from a wind, would be expected eventually to convert its potential energy into electromagnetic radiation, predominantly in the x-ray region, with frequencies in the range 1 keV $\leq \nu \leq$ 50 MeV and corresponding temperatures of about 10^7 K. The accretion luminosity is defined by $L_{\mathrm{acc}} = Gm\dot{m}/r$ for a mass-accretion rate of $\dot{m}$ onto a mass m of radius r. The *Eddington limit* L_{Edd} is the name given to the upper limiting luminosity from a mass-accreting source due to the effects of radiation pressure inhibiting the inward flow of matter, and $L_{\mathrm{Edd}} \sim (1.3 \times 10^{31})(m/\mathrm{M}_{\odot})$ W (Frank et al. 1992).

If the accretor is non-magnetic, then perhaps an accretion disc can extend very close to the accreting surface; but if there is a significant magnetic field, then the Alfvén or magnetospheric radius will intervene, and the magnetic field will control the flow, as we have seen. For neutron stars, the magnetic fields can range from small values up to $B \sim 10^8$ T, or 10^{12} G, about 10^4 times stronger than that for a magnetic white dwarf, but the stellar radius will be smaller by the same factor, so that the magnetic moment will be $\mu = BR^3 \sim 10^{20}$ T m^3 = 10^{30} G cm^3, rather less than for a magnetic white dwarf. In the LMXBs, most of the neutron stars do not display pulses, and it is concluded that their magnetic fields must be very weak. In the HMXBs, however, most of the neutron stars are x-ray pulsars, and the long-accepted explanation for their behaviour is that the accreting matter is channelled by the magnetic field towards the magnetic poles, where accretion hot spots are formed on the neutron star that give rise to the observed pulses as the star rotates. The RLOF systems, with the mass loser being an O–B supergiant filling its Roche lobe, create an accretion disc around the neutron star – the so-called

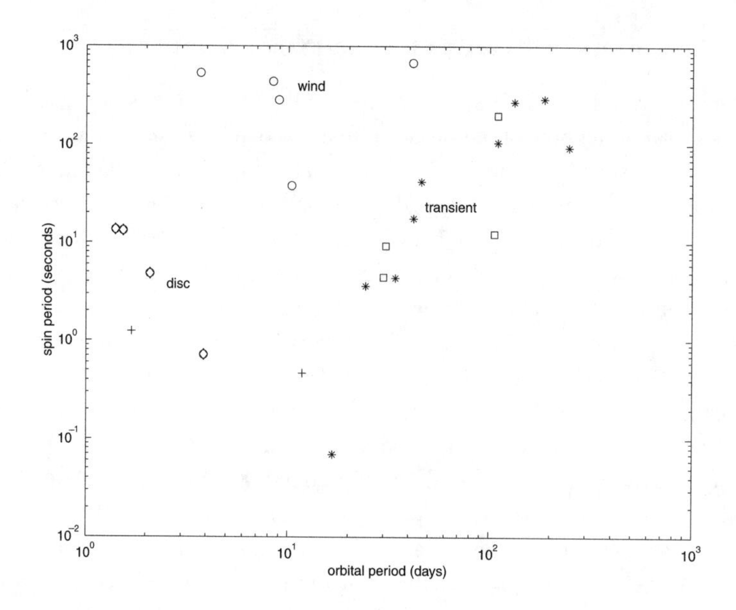

Fig. 7.21. The P_{orb}–P_{spin} diagram for x-ray binaries with accreting pulsars, from data collated by Bildsten et al. (1997). The HMXB systems with RLOF supergiants feeding accretion discs are shown as diamonds and are labelled *disc*. Wind-fed HMXBs are shown as open circles and are labelled *wind*. Transient Be systems are shown as asterisks. Transient systems with unknown companions are shown as squares and are labelled *transient*.

disc-fed HMXBs. Other HMXBs are *wind-fed* systems, with a radiatively driven wind from an O–B supergiant that does not fill its Roche lobe, or an episodic outflow from Be-star losers in eccentric-orbit binaries.

An established procedure for comparing the orbital and spin properties of the different x-ray binaries is to place them in a P_{orb}–P_{spin} diagram, as in Figure 7.21, using the data collated by Bildsten et al. (1997).

The different systems are certainly segregated, with the most efficient disc-fed accretors showing an unexplained anti-correlation between P_{spin} and P_{orb}. The transient systems show a quite well-defined positive correlation, with the neutron stars closest to their companions having the shortest spin periods. The wind-fed HMXB systems show no particular correlation, as would be expected from an inefficient mode of accretion.

The angular momentum brought by accreting matter onto a central star will exert a torque to spin the star up to faster rotation speeds. From the earlier discussion on the magnetospheric radius,

$$r_m^7 = \mu^4/(8Gm\dot{m}^2) \tag{7.13}$$

or

$$r_m = \mu^{4/7} L_{acc}^{-2/7} R_\star^{-2/7} (Gm)^{1/7}/8 \tag{7.14}$$

In a Keplerian disc, the specific angular momentum of matter at r_{m} will be $r_{\mathrm{m}} v_{\phi}(r_{\mathrm{m}})$, where $v_{\phi} = (Gm/r_{\mathrm{m}})^{1/2}$ is the Keplerian azimuthal velocity. Thus the torque exerted by the accretion of matter at a rate $\dot{m}$ will be $N = I\omega \sim \dot{m}(Gmr_m)^{1/2}$ for a star with moment of inertia I and $\omega = 2\pi/P_{\mathrm{spin}}$. The corresponding change in the rotation period of the accretor will be

$$-\dot{P}_{\mathrm{spin}} \propto P_{\mathrm{spin}}^2 L_{\mathrm{acc}}^{6/7} (Gm)^{-3/7} \mu^{4/7} R^{6/7} \tag{7.15}$$

For neutron stars with the same mass and radius, it follows that the observed changes in the spin periods of x-ray pulsars should follow a relationship of the form $-\dot{P}_{\mathrm{spin}} \propto P_{\mathrm{spin}}^2 L_{\mathrm{acc}}^{6/7}$, provided that the differences in magnetic moments are not too extreme. That was shown to be the case (Ghosh and Lamb 1979) for the set of x-ray binaries with sufficient data at that time to establish the spin-up rates. The time scale for spinning up a neutron star is about 10^5 years, which is notably shorter than its evolution age, so we might expect to see x-ray pulsars in an oscillating state about some sort of equilibrium.

As a neutron star spins up, the co-rotation radius decreases until it reaches the magnetospheric radius, $r_{\mathrm{co}} \sim r_{\mathrm{m}}$, at the inner edge of the accretion disc. The spin-up torque reduces to zero, and essentially a centrifugal barrier will be formed to prevent further accretion. But theoretical models of the magnetospheric boundary layer at the inner edge of the disc show that a negative torque is imposed upon the star to force it to spin down, whilst, at the same time, continuing to accrete matter (Ghosh and Lamb 1979). In the corresponding $\dot{P}_{\mathrm{spin}} \sim P_{\mathrm{spin}}^2 L_{\mathrm{acc}}^{6/7}$ relationship, this spin-down causes a change of slope at low x-ray luminosities that is in accordance with observations of systems like Her X-1. White et al. (1995) have discussed observed examples of x-ray binaries where the values for $\dot{P}_{\mathrm{spin}}$ have been monitored through transient outbursts and declines, showing confirmation of this simple relationship. Some systems show continous monotonic decreases in spin period, and others show episodic changes, with alternating spin-up and spin-down intervals.

A major survey of the properties of accreting x-ray pulsars has been completed by Bildsten et al. (1997), reporting 5 years of monitoring x-ray binaries with the Burst and Transient Source Experiment (BATSE) on the Compton Gamma-Ray Observatory. Their observations were secured at the hard-x-ray end of the spectrum, between 20 and 70 keV, and for the first time provided a comprehensive sampling of the time intervals in the range 10–100 days, as well as over longer periods. Their results demand a significant revision of the earlier picture derived from less extensive data. In the disc-fed system Cen X-3, the pulsar is found to have *strictly alternating* intervals of spin-up and spin-down, with average torques of $+7 \times 10^{-12}$ Hz s^{-1} and -3×10^{-12} Hz s^{-1}, respectively, in time scales ranging from 10–100 days. These rates are consistent with the simple accretion-powered model described earlier, even though the longer-term trend for spin-up in that system is slower. Some switches occur in less than 10 days, which is not consistent with the simple model. Other x-ray pulsars, both disc-fed and wind-fed, show similar strict alternating spin rates, with the torques more nearly equal and opposite. The consistency between these short-term observed and theoretical spin torques does not seem to extend to the predicted relationships for changes in x-ray luminosity on these 10–100-day intervals, but the available data from BATSE cover only the spectral-energy range 20–70 keV and are not bolometric. In the transient sources, there are strong correlations between the flux changes and the torques, but these are not evident in persistent sources. A recent theoretical model by Li and Wickramasinghe (1998) proposes an explanation for these BATSE results that removes the dependence upon mass-accretion rate that is central to the Ghosh and Lamb model and requires the pulsar-disc systems to switch between two magnetospheric states.

The links between this area of research and the existence of the *millisecond pulsars* is clear. Bhattacharya (1995) has extensively reviewed the properties of these rapidly rotating neutron stars in binaries. If the accreting neutron star does not possess a large magnetic field (and one may question how it could have lost it), then the foregoing accretion theory might suggest that the lower the magnetic field, the faster the pulsar would be made to spin by further accretion. These millisecond pulsars in the galactic-disc fields seem to be confined to the LMXBs, with about 90% of them in binaries; in globular-cluster fields, about 50% of millisecond pulsars are in binaries. In such dense stellar configurations, other influences on binary properties are operative, such as collisions between stars forming and disrupting binaries. The binary fraction of radio pulsars is only around 5%, and the accepted explanation for the occurrence of millisecond pulsars in binaries seems to be that they are *recycled pulsars*, old neutron stars that have been regenerated as active x-ray pulsars as a consequence of mass accretion from their low-mass companions, with the companion of each x-ray pulsar having reached an RLOF stage due to angular-momentum loss from the binary orbit caused by a magnetic stellar wind from that star, or caused by gravitational radiation.

With the masses for neutron stars having been established at 1.4 $M_\odot$, the gravitational-potential well is much deeper than that for a white dwarf, with the result that the orbital periods for particles in the innermost layers of an accretion disc are of the order of milliseconds or less. The dynamical time scale is given by $t_{dyn} = (r^3/Gm)^{1/2}$, so that $t_{dyn} \sim 0.1$ ms at the surface of a neutron star. For a black hole at 7 $M_\odot$, the potential well is more extreme again; the *last stable orbit* is taken to be at about six times the Schwarzschild radius, or $6(2Gm/c^2) \sim 18(m/M_\odot)$ km, and the corresponding dynamical time scale is $t_{dyn} \sim 0.6$ ms. These time scales are of interest because they coincide with the time scales for one other observed phenomenon, namely, the *quasi-periodic oscillations* (QPOs) in brightness seen in all types of x-ray binaries, as well as noise and other irregular flickerings and flares. In a standard periodogram analysis of brightness variations, any regular, periodic phenomenon will be manifested as well-defined sharp spikes in a power spectrum, whilst aperiodic noise will occupy many frequencies to some power level. The term *quasi-periodic oscillation* was coined to describe the features in the power spectra of x-ray binaries that are too broad to be regular periodic phenomena, but are nevertheless coherent and persistent features. Van der Klis (1995) has provided an excellent account of the state of the subject of rapid aperiodic variability in x-ray binaries, with many details being investigated. Suffice to say here that these QPOs are to be found at kilohertz frequencies, corresponding to periods of milliseconds, and are found in all the subtypes of x-ray binaries: the LMXBs in Z states and atoll states of accretion rate, the SXTs with their black-hole accretors, and the HMXBs with their neutron stars and black-hole accretors. With the high-quality data from the RXTE satellite providing x-ray fluxes at kilohertz sampling rates from these objects, the subject area has expanded dramatically within the past few years. A cursory scan of papers published during 1998 and 1999 reveals at least 50 papers within 12 months on the subject of QPOs in x-ray binaries. As just one example, the SXT, or *x-ray nova*, XTEJ1550-564 was observed to go through an outburst in September 1998 that was investigated by RXTE for spectral studies (Sobczak et al. 1999) and for QPO detections (Remillard et al. 1999) and by ground-based optical observations (Jain et al. 1999). The outburst started quite slowly over 10 days. The x-ray flare lasting about 2 days was very intense, at 6.8 crab, and was followed by a standstill lasting 30 days, before declining over a further 10 days. The spectra were dominated by a power-law component during the flare that weakened after the standstill and was then overtaken by a multiple-black-body disc component. Comparisons with spectral models suggested that the inner-disc radius had decreased during the flare from 33 km to just 2 km over an interval of 1 day! A QPO was detected at a fast rate of 185 kHz, with a width of 50 kHz, and RMS amplitudes of about 1% of the mean flux over the energy range of 2–30 keV. Such fast QPOs seem to be encountered only in outbursting systems, whereas most QPOs are seen at lower frequencies. The optical observations showed that the system had brightened by about 4 mag during the early part of the overall event, with a small response to the x-ray flare that was delayed by about 1 day.

Attempts are being made to characterize these phenomena in terms of correlations with x-ray fluxes, spectral distributions, and other evidence for accretion rates. The reason for the enormous interest is simply that these new x-ray observations provide some information about those parts of the accretor-disc/wind systems that are closest to the accreting surface or event horizon and therefore will at least place constraints on the properties of neutron stars and black holes and on the general theory of relativity.

7.4 Outflows (winds, jets, ejecta) from binary stars

The radiatively driven winds in O stars and WR stars were discussed in Chapter 5 in the context of their photometric-eclipse effects upon the observed optical light curves. It was assumed that the wind properties determined from studies of single stars of the same types could be applied to the binary problem, and we saw that significant success was achieved. There do appear to be some differences in the H-column densities found close to the surfaces of O stars in HMXBs when compared with standard wind models, but otherwise the wind properties in hot-star binaries seem to be unexceptional. There is the additional phenomenon of collisions between stellar winds, resulting in the formation of bow shocks between stars in O+O and WR+O binaries that may lie close to the star with the weaker wind. Among the WR binaries, there are a few systems that have been observed to undergo episodes of substantial dust formation that have been found to be periodic. Williams (1995) has reviewed the properties of the known dust-forming binaries, showing them to be eccentric-orbit systems of substantial periods, of which WR140 is the most well known. The dust-formation process takes place some time after periastron passage, when the companion has passed through the wind, causing ejection of matter that later cools to form dust grains.

The phenomenon of bow shocks in winds is evident in the HMXBs, where the line-of-sight H-column densities in the wind from the O supergiant increase by factors of 30 or more in the bow shock close to the neutron star that is travelling supersonically relative to the wind, as seen in systems like QV Nor and Cen X-3. Blondin et al. (1990) have presented some remarkable images of that kind from hydrodynamical simulations of the bow shocks and the trailing accretion wakes around compact stars moving through a stellar wind. But none of these observed phenomena in binaries is revealing anything particularly unusual about the outflows from hot stars; single stars also eject matter in the form of winds, seemingly in much the same way.

Other, more dramatic outflow phenomena do occur in binary stars, and these are observed in the form of collimated outflows or jets from many different types of binaries and in the form of ejecta from violent events occurring on the surfaces of white dwarfs (the nova phenomenon) or neutron stars (the x-ray bursters).

That most-studied of binary stars, β Lyr, was investigated by a large team of observers using a wide range of techniques from spectroscopy to interferometry (Harmanec et al. 1996). They found that this system exhibits emission lines of Hα and He I λ6678 that arise from matter being ejected perpendicular to the orbital plane, from the vicinity of the more massive star, which is surrounded by an accretion disc. More recently, Hoffman, Nordsieck, and Fox (1998) have presented ultraviolet and visual spectropolarimetry for β Lyr that also argues for the presence of a bipolar outflow. Amongst the planetary nebulae, many objects show bipolar shapes, with a central waist like an hour-glass, and weakly collimated outflows from the central source emerging perpendicular to the waist. Bell and Pollacco (1997) reported a survey of narrow-band, high-angular-resolution imaging for four planetary nebulae (Abell 41, 46, 63, 65), for which the central stars were well known to be binary systems in the post-common-envelope or pre-CV stage, all with orbital periods less than 1 day. They found that all four systems showed axisymmetric nebulae with a bipolar shape.

Livio (1997, 1998) has reviewed the evidence for the existence of collimated jets, as contrasted with the less dramatic bipolar outflows, in all types of objects, including binary stars. He has emphasized the fact that all of these types of objects showing jet ejections involve accretion onto a central source via an accretion disc. The ratio of the observed jet velocity to the escape velocity for all these objects is about unity, and it does seem to be necessary to have an accretion disc to produce a jet, though that seems not to be a sufficient condition. Amongst binary stars, jet-like features have been observed in planetary nebulae with binary central stars, and jets with velocities of about 1000 km s^{-1} have been discovered in some x-ray binaries, particularly the black-hole x-ray transients, and more recently the supersoft x-ray sources too. In the supersoft source RXJ0513.9-6951 in the LMC, Crampton et al. (1996) and Southwell et al. (1996) discovered weak satellite emission lines of He II and H in optical spectra that were just detectable above the noise level and were displaced on opposite sides of the main component at velocities of about $\pm$3800 km s^{-1}. Cowley et al. (1998) found that three out of six supersoft sources investigated by optical spectroscopy showed evidence for bipolar jets in the form of oppositely displaced satellite lines of H and He II, with velocities in the range 1000–4000 km s^{-1} and also indications of precession of those jets. Becker et al. (1998) found the spectroscopic signature of bipolar jets at $\pm$800 km s^{-1} in RXJ0019.8+2156, another supersoft source.

The most famous jet system is associated with the binary star SS433 in the supernova remnant W50, the properties of which have been reviewed by Margon (1984). Hjellming and Han (1995) reviewed the radio properties of x-ray binaries in general, detailing the more recent high-angular-resolution studies, from the Very Large Array (VLA), of the precessing jets in

SS433 in their 162.5-day precession period. The x-ray transient source GROJ1655-40 was observed to have had an x-ray outburst seen in 1994, and a weaker event seen in 1996. The system was observed from x-rays to radio to establish clear links between the x-ray transient event and the subsequent developments in the radio emissions. As reported by Hjellming (1997), there was ejection of relativistic jets at $0.92c$ in this system, and the VLA map shows radio emissions extending 0.3 arcsec in opposite directions from the central source. In the four systems discussed by Hjellming, there were bulk motions of gas at speeds in the range 0.26–$0.92c$ in outflows collimated by the accretion disc in the central binary system. The material in the jets was being replenished following each x-ray outburst, with the radio synchrotron emission from the new material not starting until the end of the hard-x-ray outburst, and then being observed to move outwards along the collimated flow. Mirabel and Rodriquez (1999) have reviewed the properties of sources of relativistic jets in binary systems, noting that the term *micro-quasar* is becoming adopted as a general term for such sources.

A final comment is in order regarding the violent phenomenon of a nova outburst on a white dwarf, or an x-ray burst on a neutron star. It is well established now that a nova outburst on a white dwarf in a CV occurs as a consequence of an accumulation of hydrogen-rich material on the surface of that star from its surrounding accretion disc. When the thickness of that additional layer has grown sufficiently to increase the temperature at the base of the layer to about 10^8 K, rapid thermonuclear fusion of H to He occurs on the white-dwarf surface. During this runaway process, matter from the core regions of the white dwarf is dredged up into these temporary fusion layers, and in the explosive outburst a rich variety of nuclei are expelled into the surrounding interstellar medium (ISM). Starrfield (1998) has reviewed the theoretical and observational studies of nova outbursts, showing that they occur on both C-O white dwarfs and on O-Ne-Mg white dwarfs, which are quite massive at about 1.25 $M_\odot$. The ejected matter is a mixture of the accreted material and that dredged up from the white-dwarf core, so that we can expect novae to be significant sources of increasing heavy-element abundances in the ISM. Gehrz (1998) has used IR spectrophotometry to investigate the ejecta from classical novae, with interesting results. Outbursts from C-O white dwarfs produce copious quantities of dust grains within 100–300 days of the event, causing the IR fluxes to decline rapidly after that time. By contrast, the O-Ne-Mg white dwarfs show an early decline, followed by a long tail in which there is little or no dust formation. These ejecta are rich in CNO, Ne, Mg, Al, and Si, and their estimated masses are substantially in excess of that expected from the standard thermonuclear-runaway calculations.

There are related phenomena in the supersoft-x-ray sources, where it is argued that continual fusion is occurring on the white-dwarf surface, and in the LMXBs, where bursts of x-ray emission are observed to occur in some systems. The most well known of these latter systems is the so-called *rapid burster*, which was discovered in 1975 and shows repeated x-ray bursts at intervals as short as 7 seconds. It has also been observed to act like a normal bursting LMXB. Lewin et al. (1995) have reviewed the properties of these systems, which are divided into two classes according to which physical mechanism is understood to power the outbursts: Type I have thermonuclear flashes on the surface of the neutron star in the LMXB, whilst Type II are thought to result from spasmodic accretion. The x-ray photometry for these sources shows remarkable changes in received fluxes, with count rates, for example, increasing from about

10 to about 600 in less than 10 seconds, and then declining over 100 seconds, or showing triple bursts, or repeated bursts at short time intervals.

There are also the type Ia supernova events (SNe Ia), which have assumed enormous cosmological importance in recent years as standard candles for determinations of cosmological distances. Branch (1998) has reviewed the evidence for the homogeneity of the absolute magnitudes of SNe Ia at maximum brightness, concluding that normal SNe Ia have M_V values of -19.4 to -19.5, with an uncertainty of about ± 0.2 mag. However, the type of object undergoing the outburst is still quite uncertain, either a CV with mass transfer pushing the mass of the accreting white dwarf over the Chandrasekhar limit of 1.44 $M_{\odot}$ and leading to the SNe and the formation of a neutron star, or the coalescence of white-dwarf + white-dwarf binaries to create the SNe. Our knowledge of the frequency of occurrence of such binaries in the galaxy is still uncertain, though work continues towards such data (e.g., Maxted and Marsh 1999).

These violent processes, together with the dwarf-nova outbursts and super-outbursts, are all intimately connected with the non-steady flow of matter through the entire accreting mechanism from the mass loser, through the stream and disc regions, and onto the surface of the compact star that is the mass gainer. There have been many substantial reviews of the observations and theoretical models used to investigate these curious phenomena, and appropriate references have already been cited. For some researchers, these events are the reasons that they find binary stars interesting to study, because they provide such excellent self-scanning sources in which the accretion structures are the dominant sources of radiation, particularly in the optical, UV, and x-ray regions. For others, these structures hide the interesting properties of the underlying stellar binary system that has reached that stage of evolution by a fairly complex and incompletely understood path. Whatever your preferences, I trust that you will agree with me that close binary stars are much more interesting than simply two mass points orbiting each other in a few minutes, or hours, days, weeks, years. And remember, you have been reading about points of light in the night-time sky.

Problems

Chapter 1

1.1 A binary system of total mass 2 $M_\odot$ and orbital period 3 days is at a distance of 20 pc from us. Calculate the maximum possible angular separation between the two stars as seen from the Earth. Compare that value with the angular separations currently attainable via speckle and direct interferometry, and comment on whether or not the binary is spatially resolvable.

1.2 The binary system Sirius has a main-sequence component of mass 2.20 $M_\odot$ (star A) and a white-dwarf component of mass 0.94 $M_\odot$ (star B). The orbital period is 50 years. Calculate the separation between the two stars in units of solar radii.

For round-figure estimates, assume that the Roche-lobe radius for each star does not exceed 50% of that separation. Hence confirm that each star has sufficient room to evolve (star A) or to have evolved (star B) independently and unaffected by the presence of its companion.

The progenitors of the current stars, Sirius A and B, probably were formed at the same time from the same star-formation region, with their contraction times to the main sequence not too different. In any event, these pre-main-sequence lifetimes (10^5–10^6 years) are very short compared with the main-sequence lifetimes of these stars of moderate mass. So let us assume, reasonably, that both stars arrived on the zero-age main sequence at the same time. The main-sequence lifetime is given approximately by $t_{\rm ms} \sim (7 \times 10^9)/m^3$ years for stars with initial masses less than $m \sim 10$ $M_\odot$, and the red-giant phase lasts for $t_{\rm rg} \sim 0.1 t_{\rm ms}$. All other time scales for transitions between these stages, and between the red-giant phase and the white-dwarf phase, are much shorter and can be ignored in this example. Thus, if Sirius A is currently half-way through its main-sequence lifetime, estimate the minimum initial mass for the current Sirius B to have had enough time to evolve to the observed white-dwarf stage.

1.3 Why would a W UMa-type binary, where the main-sequence primary component typically has a mass of about 1 $M_\odot$, not evolve into a cataclysmic variable of similar total mass and comparable orbital period?

1.4 A binary is composed of O-type and A-type main-sequence stars and has an orbital period of 10 days. Will it evolve into a contact binary, a cataclysmic variable, or a massive x-ray binary? Give all reasons for your conclusion.

1.5 Identify the categories of binary systems from the following very incomplete sets of observational data (i.e., a typical amount of initial-discovery data):
(1) emission-line spectra with linear and circular polarization; orbital period, hours
(2) continuous variations in brightness, with a period of hours; broad absorption-line spectra
(3) very small variations in radial velocities over many months; spectra show unusually strong lines of some elements
(4) spectra dominated by broad, double-peaked emission lines, variable on time scales of hours
(5) x-ray source with characteristic pulses of irregular amplitude, but regular periodicity; optical spectra have emission lines not all moving in phase

Chapter 2

2.1 The radio-pulsar binary PSRB1259-63 is found to have an orbital eccentricity of $e = 0.87$. Calculate the ratios of linear speeds (V) and angular speeds ($\dot{\theta}$) at the periastron and apastron positions in the orbit of the pulsar.

2.2 In an orbit of eccentricity $e = 0.63$ and period $P = 5.2$ days, calculate the value of the eccentric anomaly E 1.1 days after periastron passage. Hence determine the true anomaly θ.

2.3 Confirm the derived expressions for the magnitude of the total orbital angular momentum

$$J = m_1 m_2 \left[\frac{Ga(1-e^2)}{(m_1+m_2)} \right]^{1/2}$$

$$J = \frac{2\pi\, a^2(1-e^2)^{1/2} m_1 m_2}{P(m_1+m_2)}$$

2.4 The massive x-ray binary QV Nor has been observed in x-rays and via optical spectroscopy. The velocity semiamplitude of the x-ray source has been determined from x-ray pulse-timing observations to be $K_x = 309 \pm 11$ km s^{-1}, whilst that of the O-star companion has been determined from optical spectra to be $K_c = 19.8 \pm 1.1$ km s^{-1}. The orbital period is 3.72854 days, and the orbital eccentricity $e = 0.0$. If the orbital inclination is $i = 60°$, determined from the duration of the x-ray eclipse, calculate the masses of the two components.

2.5 Show that the mass function

$$f(m) \equiv m_2^3 \sin^3 i/(m_1+m_2)^2$$

can be rewritten in terms of the observable quantities K_1, P, and e in the form

$$f(m) = \frac{1}{2\pi\, G}(1-e^2)^{3/2} K_1^3 P$$

2.6 Write software code in your favourite language to calculate and plot radial-velocity curves as a function of orbital phase for orbits with different values of eccentricity e and orientation ω. Also plot out the shapes of the orbits projected onto the tangent plane of the sky (x, y).

Chapter 3

3.1 A telescope with an effective aperture of 4 m has a spectrograph at the Cassegrain focus. The total number of interfaces within the telescope and spectrograph is 8. It is planned to determine radial velocities with this instrument over a spectral range centred on 5000 Å and recorded on a CCD with a pixel size of 15 μm and a quantum efficiency of 75%. The target binary system has an apparent visual magnitude $V = 13.6$ and orbital period $P = 0.35$ day. Calculate a sensible upper limit to the integration time that will avoid velocity smearing due to the orbital motion in the binary. Hence estimate the expected S/N value that will be obtained in such an integration time at a spectral resolution of 10,000 and under conditions of normal seeing. Discuss whether or not this spectral resolution and S/N value will be sufficient to determine radial velocities for F stars and B stars accurate to $\pm 2\%$.

3.2 Repeat the calculations of Problem 3.1 for a cataclysmic variable with $V = 15.2$, orbital period $P = 3.5$ hours, and a white-dwarf spin period of 100 seconds. Allow also for a sky background equivalent to full Moon of $V = 18.0$ mag per square arcsecond, and a readout noise of 10 photons per pixel. If it is desired to secure spectroscopic observations that are time-resolved to 10% of the spin period, how will you maximize the S/N value per phase interval of the spin period?

3.3 If you have access to a set of spectra in standard format and a spectral-analysis package, such as the Starlink DIPSO, or perhaps IRAF, experiment with adding different relative proportions of spectra of different spectral classes, and at a range of velocity differences, to produce composite spectra for artificial binary stars. Then use a CCF routine with different template spectra to explore whether or not you can recover the input values of velocity differences for the various artificial binaries. This experiment is very valuable, particularly if you have access to spectra with a range of spectral resolutions as well.

3.4 Find a set of times of observation and radial velocities (t_i, V_i) for a binary system in the published literature, preferably where the radial-velocity observations have provided a determination of the orbital period – for example, one of the many such binaries in R. F. Griffin's papers in *The Observatory*. Use a period-analysis package, such as the Starlink PERIOD, to determine the orbital period via one or more of the techniques incorporated in such a package, and compare it with the final published values. Alternatively, try writing some code to evaluate the Lomb-Scargle normalized periodogram for a set of (t_i, V_i) values, and test it against some real data.

3.5 Confirm that the projection of a binary's x–y coordinates on the sky onto the north–south and east–west baselines of an interferometer are as given by equations (3.61) and (3.62). Hence calculate the variations of the squared visibility as functions of time for resolved binaries with different angular separations, at various declinations, and from various observing sites. Compare your results with those published in the references cited.

3.6 The binary systems HD197406 and CQ Cep show variations in linear polarization as measured by the Stokes parameters Q and U. The coefficients $q_{n=0...4}$ and $u_{n=0...4}$ of the Fourier sine/cosine series that fits the variations with orbital phase are

n	q(HD)	u(HD)	q(CQ)	u(CQ)
0	−0.809	−0.548	−3.111	4.060
1	0.002	0.039	−0.024	0.032
2	−0.001	−0.036	0.001	−0.059
3	0.132	0.098	−0.269	0.170
4	−0.163	0.167	0.241	0.057

Calculate the values of Q and U as functions of orbital phase for each binary, and plot the results in the (Q, U) plane to reveal the Lissajous-type figures that are traced out during each orbit. Show that the orbital inclinations determined from these data are $i = 71^\circ.4$ for HD197406 and $i = 102^\circ.1$ for CQ Cep. Experiment with other values of q_n and u_n.

Chapter 4

4.1 The gravitational potential of a star of mass m in its equatorial plane at a distance d is given by

$$\frac{Gm}{d}\left[1+\frac{b}{d^2}\right]$$

where b is a small constant. The companion star is small enough to have a point-mass potential. Show that in the absence of other perturbations, the apsidal rotation rate is given by

$$\dot{\omega} = \frac{3Gmb(1-e^2)^{-2}}{na^5}$$

where a and n are the semimajor axis of the relative orbit and the mean daily motion, respectively.

4.2 The mass of the Earth is 5.98×10^{24} kg, and its equatorial radius is 6378.2 km. The J_2 coefficient for the Earth is approximately 10^{-3}. The Moon is of mass 7.3×10^{22} kg, and its mean distance from the Earth is 3.84×10^5 km. Use these data together with the theory of gravitational potentials to compare the various accelerations acting on a particle on the surface of the Earth at the equator due to (1) the gravitational potential of the Earth treated as a point mass, (2) the disturbing potential due to the non-spherical shape of the Earth (the dominant term with J_2), (3) the rotational potential in a reference frame co-rotating with the Earth, and (4) the disturbing or tidal potential due to the Moon. What additional term would arise if the particle started to move?

4.3 Determine the apsidal period of the binary system AS Cam given that the masses of the two stars are 3.32 $M_\odot$ and 2.51 $M_\odot$, their radii are 2.60 $R_\odot$ and 1.98 $R_\odot$, the orbital period is 3.430971 days, the orbital eccentricity is 0.135, and the theoretical apsidal constant for both stars is $\log k \sim -2.3$. How much shorter would the apsidal period be if the radii were increased by a factor of 2? Determine also the relativistic apsidal motion for this system. What sets of observations would assist in understanding why the observed apsidal period for this system is less than one-third of the predicted rate?

4.4 Determine the total mass of the pulsar-binary system PSR1913+16 given that the rate of advance of periastron due to relativistic effects is $\dot{\omega} = 4^\circ.22 \text{ yr}^{-1}$, $e = 0.61717$, $a_1 \sin i = 7.0043 \times 10^5$ km, $P = 27906.98$ seconds. At what inclination would the masses of the two stars be equal?

4.5 Extend the discussion in the text regarding the observed rate of period decrease ($\dot{P}/P = -3.36 \times 10^{-6}$ per year) of the x-ray-binary system SMC X-1 by considering the use of equation (4.71) for the ejection of matter through L_2. The relevant masses are $m_1 = 17.2\ M_\odot$ and $m_2 = 1.6\ M_\odot$, and $d = 1.314a$ for $q = 0.076$. Examine also the option of wind-driven mass loss from the O star and the option of conservative mass transfer.

Chapter 5

5.1 The binary system GG Lup has been shown to be composed of two B stars and to have a total secondary eclipse of the smaller, cooler star (the secondary) by its larger, hotter companion (the primary). The *uvby* colour indices at the middle of secondary eclipse ($\phi = 0.5$) are $(b - y) = -0.047$, $m_1 = 0.097$, and $c_1 = 0.447$. Determine the interstellar reddening for the binary system and the intrinsic colours of the primary component.

Given that the observed colour indices of the binary at quadrature ($\phi = 0.25$) are $(b - y) = -0.042$, $m_1 = 0.107$, and $c_1 = 0.517$ and that the light ratio in y is $f_{\text{pri}}/f_{\text{sec}} = 3.125$, determined from a solution of the light curve, calculate the intrinsic colours of the secondary component. Use the $[u - b] \sim T_{\text{eff}}$ calibration of Napiwotski et al. (1993) to determine the effective temperatures of both components.

5.2 A set of photometric observations through an annular eclipse in the binary system AA Dor shows that the four times of contact, $t_{1,2,3,4}$, of the eclipse are respectively 0.3400, 0.3462, 0.3510, and 0.3576 in units of days, with an uncertainty of ± 0.0002. The orbital period of the binary is found to be 0.261540 day. Determine the radii of both stars in terms of the semimajor axis of the relative orbit, assuming that $i = 90°$. The depth of the annular eclipse is observed to be 0.38 mag in the V band. Determine the brightness of each component relative to the total brightness of the binary taken as unity, and hence estimate the depth of the alternating total eclipse, on the assumption that the larger star has a limb-darkening coefficient $u = 0.0$.

5.3 The optical companion in an x-ray binary has a velocity semiamplitude of $K_c = 20 \pm 2$ km s^{-1}, and the orbital period is $P = 3.5$ days. The pulsed x-ray flux reveals a light-time effect from which a velocity semiamplitude of $K_x = 400 \pm 10$ km s^{-1} is calculated. The x-ray-eclipse half-angle is measured to be $\phi_e = 28° \pm 3°$. Estimate the range of possible orbital inclinations, assuming that the optical companion fills its Roche lobe. Calculate also the projected rotational velocity for the optical companion that would be expected for synchronous rotation.

5.4 Follow the prescriptions given in the text (Section 5.5.5) for calculating the shapes of stars in the Roche model, and write a code to perform those calculations and plot out the three-dimensional results. Apply your code to several examples to generate detached, semidetached, and contact binaries.

5.5 If you have access to a light-curve synthesis code, such as the WD code, or *Binary Maker 2.0* by Bradstreet (1993), generate light curves with different input parameters of radii, inclinations, temperatures, mass ratios, and wavelengths of observations to explore the many different forms of light curves that are possible. Investigate how well you can generate light curves that resemble the many example light curves shown in Chapter 5.

Chapter 6

6.1 The double-lined spectroscopic eclipsing binary GG Lup has had the following parameters determined from analyses of multi-colour light curves and radial-velocity curves: $K_1 = 124.5 \pm 0.5$ km s^{-1}; $K_2 = 204.2 \pm 0.9$ km s^{-1}; $e = 0.150 \pm 0.005$; $\omega = 86°.2 \pm 0°.2$; $P = 1.849693 \pm 0.000003$ days. Mean relative radii: $R_1/a = 0.2000 \pm 0.0020$; $R_2/a = 0.1450 \pm 0.0015$. Inclination $i = 86°.75 \pm 0°.10$. Mean effective temperatures: $T_1 = 14{,}750 \pm 450$ K; $T_2 = 11{,}000 \pm 600$ K.

Calculate the values (with uncertainties) of the masses, radii, surface gravities, and luminosities for both stars. Calculate also the absolute bolometric magnitudes and absolute visual magnitudes, and hence the distance to the binary if the V magnitude at maximum is $V = 5.589 \pm 0.006$. Confirm that both stars are on the main sequence.

6.2 A current research programme seeks to establish the masses, radii, and luminosities of the stars in eclipsing-binary systems discovered from recent photometric surveys to be in the nearby galaxy M31. The distance modulus $(V - M_V)$ for M31, according to studies of other distance calibrators such as classical Cepheids, is 24.3 mag. These data on eclipsing binaries should provide independent measures of their distances. Use the equations of Chapter 6 to evaluate the levels of accuracy required in the determinations of the radii and temperatures of the stars in these binaries to ensure a distance to an individual binary that is accurate to $\pm 10\%$. Discuss the feasibility of such a programme with the currently available spectroscopic instrumentation on 4-m to 10-m optical telescopes, and photometric instruments on smaller telescopes.

Chapter 7

7.1 In order to make eclipse maps of stellar surfaces it is necessary to know the sizes, shapes, and temperatures of the two stars in an eclipsing binary and the orbital inclination. These quantities are derived from an initial analysis of the asymmetric light curves and therefore can be systematically in error. Discuss what artifacts might be seen in an eclipse map if some of the input parameters such as radii and inclination are incorrect.

7.2 Find information on the capabilities of different instruments on the Gemini 8-m telescopes and the Keck 10-m telescopes for conducting optical/near-infrared spectroscopic observations at time resolutions of about 1 second. Likewise, find information about the photometric and spectroscopic instruments on the recently launched *Chandra* and XMM x-ray satellites for similar time resolution. Hence evaluate and plan an observing programme to investigate the properties of accreting x-ray pulsars on time scales of their typical spin and orbit periods and over intervals of hundreds of days.

Outline answers

Chapter 1

1.1 Use Kepler's third law and the small-angle formula to obtain $a = 0.051$ AU and $\alpha = 2.5$ mas, still below the currently attainable spatial resolution.

1.2 Use Kepler's third law to obtain $a = 19.87$ AU, or $a = 4270.9\ R_\odot$. Thus both stars have space to evolve independently. The present age of Sirius A is $t = 3.28 \times 10^8$ years. Sirius B must be massive enough to evolve to a white dwarf in that time. Hence $m_B = 2.86\ M_\odot$ is the minimum mass for the progenitor of Sirius B, and it must have lost $1.92\ M_\odot$ through the red-giant wind-driven mass-loss phase and the planetary-nebula phase. Discuss whether or not these figures are in accordance with our knowledge of mass loss at different evolutionary stages.

1.3 A CV is composed of a white dwarf and a low-mass main-sequence star. The white dwarf is the degenerate core remnant of a star that was once a red giant. Hence the need for enough space in the binary to allow evolution through the red-giant phase undisturbed, requiring Roche lobes greater than about $100\ R_\odot$. Thus the initial orbital period must have been years, rather than a few as hours as for W UMa systems.

1.4 High-mass x-ray binary (HMXB): O star's main-sequence lifetime about 2×10^6 years; A star's, about 100 times longer. O star evolves to red-supergiant stage, losing mass to companion through RLOF and mass-ratio reversal. O-star remnant evolves faster to SN II and neutron star than does the now more massive companion on the main sequence – hence HMXB composed of neutron star and O-type supergiant.

1.5 (1) intermediate polar or polar; (2) W-UMa-type contact binary; (3) Ba- or CH-star binary; (4) cataclysmic variable; (5) HMXB.

Chapter 2

2.1 Use the linear-speed equation in Table 2.1 and Kepler's second law. $V_p/V_a = (1+e)/(1-e) = 14.4$; $\dot{\theta}_p/\dot{\theta}_a = [(1+e)/(1-e)]^2 = 206.9$ – large numbers!

2.2 Use Kepler's equation and solve it by Newton-Raphson iteration to obtain $E = 1.9209$ radians, $\theta = 2.498$ radians.

2.3 Use $J = m_1 L_1 + m_2 L_2$ and expressions for $L_{1,2}$. Replace G via Kepler's third law to obtain alternative expression.

2.4 Use equations for minimum masses to obtain $M_c = 19.9\ \mathrm{M}_\odot$ and $M_x = 1.28\ \mathrm{M}_\odot$.

2.5 Use Kepler's third law and the ratio of axes a_1/a expression to obtain equation for a_1^3. Then use equation for K_1, and rearrange to obtain result.

Chapter 3

3.1 An upper limit of 1% of the orbital period should be placed on the integration time, i.e., about 300 seconds. Resolution gives $\Delta\lambda = 0.5$ Å, equivalent to 2.5 pixels, or 0.2 Å per pixel. S/N calculation gives $S = 6.78 \times 10^4$ photons per second per angstrom. Hence S/N = 116 per pixel in 300 seconds, without allowing for slit losses, which depend crucially on the observing conditions. Good radial velocities attainable.

3.2 For the CV, S/N = 30 per pixel would be obtained in 100-second integrations, allowing for the background readout noise, but without allowing for slit losses. To resolve the white-dwarf spin period into 10 10-second bins would require 10-second integrations, with very accurate timing and an accurate ephemeris, repeated over many spin cycles in order to add up all the spectra in each phase bin to achieve a useful S/N value per bin.

Chapter 4

4.1 The disturbing potential is $S = Gmb/d^3$. Put this expression into Lagrange's planetary equation for $d\omega/dt$, noting that there is no dependence of S on i; use the expression for $\partial d/\partial e$ given in the text to obtain the final result.

4.2 (1) $Gm_E/a_{\mathrm{eq}}^2 = g = 9.84\ \mathrm{m\ s^{-1}}$; (2) the ratio of (2) to (1) is $-3J_2 \simeq 3 \times 10^{-3}$; (3) at Equator, $-\partial\Phi_{\mathrm{rot}}/\partial r = \omega^2 r \sin^2\phi'$, and transverse component is zero. Thus the ratio of (3) to (1) is $\omega^2 a_{\mathrm{eq}}/g = 0.0034$; (4) the ratio of (4) to (1) is $2m_{\mathrm{Moon}} a_{\mathrm{eq}}^3/(m_{\mathrm{Earth}} d_{\mathrm{Moon}}^3) = 1.1 \times 10^{-7}$. The additional term is the coriolis term $2\vec{\omega} \times \dot{\vec{r}}$.

4.3 Apsidal period for AS Cam is $U = 1116$ years. Increasing radii by a factor of 2 increases P/U by 2^5, so that $U = 35$ years. Relativistic apsidal period is 4253 years. Detailed spectroscopic observations – there is still only one radial-velocity study!

4.4 Total mass is $2.82\ \mathrm{M}_\odot$; $i = 46°$ for two equal masses.

4.5 For ejection of matter through L_2, the mass-loss rate would be about $1 \times 10^{-6}\ \mathrm{M}_\odot$ per year.

Chapter 5

5.1 Interstellar reddening $E(b-y) = 0.026$. Primary component: $c_0 = 0.442; m_0 = 0.105$; $(b-y)_0 = -0.073; (u-b)_0 = 0.586; [u-b] = 0.618$. Temperature of primary component: 14,980 K. Secondary component: $(b-y)_0 = -0.051; (u-b)_0 = 1.003; [u-b] = 1.081$. Temperature of secondary component: 11,200 K.

5.2 Radii are 0.135 and 0.075 of the separation. Larger star contributes 95% of the light of the system, smaller star 5%. Depth of the other eclipse about 0.05 mag.

5.3 Range of i is 60°–72°. $V_{\rm rot} \sin i = 237$ km s^{-1}.

Chapter 6

6.1 The values published by Andersen et al. (1993) are as follows:
$m_1 = 4.116 \pm 0.040\ \mathrm{M_\odot}; m_2 = 2.509 \pm 0.024\ \mathrm{M_\odot}$
$R_1 = 2.379 \pm 0.025\ \mathrm{R_\odot}; R_2 = 1.725 \pm 0.019\ \mathrm{R_\odot}$
$\log g_1 = 4.301 \pm 0.011$ (CGS units); $\log g_2 = 4.364 \pm 0.010$
$M_{\rm bol,1} = -1.26 \pm 0.13; M_{\rm bol,2} = 0.71 \pm 0.24$
$\log(L_1/\mathrm{L_\odot}) = 2.38 \pm 0.05; \log(L_2/\mathrm{L_\odot}) = 1.59 \pm 0.10$
$M_{V,1} = 0.08 \pm 0.13; M_{V,2} = 1.21 \pm 0.24$
Distance $= 1140 \pm 10$ pc

6.2 See Question 6.1 and its answer, as well as Hilditch (1996) and Guinan et al. (1996).

Chapter 7

7.1 See Collier Cameron (1997) and the several references cited therein.

7.2 Use the relevant Web sites for information on the various instruments available:
Keck: http://www2.keck.hawaii.edu
Chandra: http://chandra.harvard.edu
XMM: http://sci.esa.int/xmm
Gemini: http://www.gemini.edu

Bibliography

Abt, H.A., 1983, *A. Rev. A. & A.*, 21, 343.

Aitken, R.G., 1932, *New General Catalogue of Double Stars within 120° of the North Pole*, Carnegie Institution, Washington, DC.

Aitken, R.G., 1935, *The Binary Stars*, 2nd ed., McGraw-Hill, New York.

Alcock, C., and 19 other authors (MACHO), 1997, *Astron. J.*, 114, 326.

Allan, A., Horne, K., Hellier, C., Mukai, K., Barwig, H., Bennie, P.J., Hilditch, R.W., 1996, *MNRAS*, 279, 1345.

Aller, L.H., 1954, *Publ. Dom. Ap. Obs. Victoria*, 9, 321.

Andersen, J., 1983, *The Observatory*, 103, 165.

Andersen, J., 1991, *A. & A. Rev.*, 3, 91.

Andersen, J., 1997, In *Fundamental Stellar Properties: The Interaction between Observation and Theory*, eds. Bedding, T., Booth, A.J., Davis, J., IAU Symp. No. 189, Kluwer, Dordrecht, p. 99.

Andersen, J., Nordström, B., 1983, *A. & A.*, 122, 23.

Andersen, J., Clausen, J.V., 1989, *A. & A.*, 213, 183.

Andersen, J., Clausen, J.V., Nordström, B., 1980. In *Close Binary Stars: Observations and Interpretation*, eds. Plavec, M., Popper, D.M., Ulrich, R.K., IAU Symp. No. 88, Reidel, Dordrecht, p. 81.

Andersen, J., Clausen, J.V., Gustafsson, B., Nordström, B., VandenBerg, D.A., 1988, *A. & A.*, 196, 128.

Andersen, J., Clausen, J.V., Nordström, B., Tomkin, J., Mayor, M., 1991, *A. & A.*, 246, 99.

Andersen, J., Clausen, J.V., Giménez, A., 1993, *A. & A.*, 277, 439.

Anderson, L., Shu, F.H., 1979, *Ap. J. Suppl.*, 40, 667.

Anderson, L., Raff, M., Shu, F.H., 1980, In *Close Binary Stars: Observations and Interpretation*, eds. Plavec, M., Popper, D.M., Ulrich, R.K., IAU Symp. No. 88, Reidel, Dordrecht, p. 485.

Anderson, L., Stanford, D., Leininger, D., 1983, *Ap. J.*, 270, 200.

Andronov, I.L., Richter, G.A., 1987, *Astr. Nachr.*, 308, 235.

Applegate, J.H., 1992, *Ap. J.*, 385, 621.

Argelander, R., 1844, In *A Source Book in Astronomy*, eds. Shapley, H., Howarth, H.E., McGraw-Hill, New York (1929), p. 229.

Armstrong, J.T., Mozurkewich, D., Vivekanand, M., Simon, R.S., Denison, C.S., Johnston, K.J., 1992a, *Astron. J.*, 104, 241.

Armstrong, J.T., Hummel, C.A., Quirrenbach, A., Buscher, D.F., Mozurkewich, D., Vivekanand, M., Simon, R.S., Denison, C.S., Johnston, K.J., 1992b, *Astron. J.*, 104, 2217.

Armstrong, J.T., Hutter, D.J., Johnston, K.J., Mozurkewich, D., 1995, *Physics Today*, 48 (May), p. 42.

Armstrong, J.T., and 14 other authors, 1998, *Ap. J.*, 496, 550.

Baade, D., 1991, ed., *Rapid Variability of OB Stars: Nature and Diagnostic Value*, ESO Conference and Workshop Proc. No. 36, ESO, Garching.

Bagnuolo, W.G., Gies, D.R., 1991, *Ap. J.*, 376, 266.

Bagnuolo, W.G., Gies, D.R., Wiggs, M.S., 1992a, *Ap. J.*, 385, 708.

Bagnuolo, W.G., Mason, B.D., Barry, D.J., Hartkopf, W.I., McAlister, H.A., 1992b, *Astron. J.*, 103, 1399.

Bagnuolo, W.G., Gies, D.R., Hahula, M.E., Wiemker, R., Wiggs, M.S., 1994, *Ap. J.*, 423, 446.

Bailey, J., Hough, J.H., 1982, *PASP*, 94, 618.

Bailyn, C.D., 1996, In *The Origins, Evolution, and Destinies of Binary Stars in Clusters*, eds. Milone, E.F., Mermilliod, J.-C., ASP Conf. Ser. Vol. 90, p. 320.

Bailyn, C.D., Jain, R.K., Coppo, P., Orosz, J.A., 1998, *Ap. J.*, 499, 367.

Balona, L.A., Henrichs, H.F., Le Contel, J.M., 1994, eds., *Pulsation, Rotation and Mass Loss in Early-Type Stars*, IAU Symp. No. 162, Kluwer, Dordrecht.

Baptista, R., Horne, K., Hilditch, R.W., Mason, K.O., Drew, J.E., 1995, *Ap. J.*, 448, 395.

Baptista, R., Horne, K., Wade, R.A., Hubeny, I., Long, K., Rutten, R.G.M., 1998a, *MNRAS*, 298, 1079.

Baptista, R., Catalán, M.S., Horne, K., Zilli, D., 1998b, *MNRAS*, 300, 233.

Baranne, A., Mayor, M., Poncet, J.L., 1979, *Vistas in Astronomy*, 23, 279.

Barlow, D.J., Fekel, F.C., Scarfe, C.D., 1993, *PASP*, 105, 476.

Barnes, J.R., Collier Cameron, A., James, D.J., Donati, J.-F., 2000, *MNRAS*, 314, 162.

Barrett, H.H., Swindell, W., 1981, *Radiological Imaging*, vol. 2, Academic Press, New York, p. 375.

Barrett, P.E., Chanmugam, G., 1984, *Ap. J.*, 278, 298.

Bastien, P., Drissen, L., Menard, F., Moffat, A.F.J., Robert, C., St. Louis, N., 1988, *Astron. J.*, 95, 900.

Batten, A.H., 1970, *PASP*, 82, 574.

Batten, A.H., ed., 1973, *Extended Atmospheres and Circumstellar Matter in Spectroscopic Binary Systems*, IAU Symp. No. 51, Reidel, Dordrecht.

Batten, A.H., 1978, *Vistas in Astronomy*, 22, 265.

Batten, A.H., ed., 1989, *Algols*, IAU Coll. No. 107, Kluwer, Dordrecht.

Batten, A.H., Fletcher, J.M., 1971, *Ap. Sp. Sci.*, 11, 102.

Batten, A.H., Fletcher, J.M., MacCarthy, D.G., 1988, *Publ. Dom. Ap. Obs. Victoria*, vol. 17.

Becker, C.M., Remillard, R.A., Rappaport, S.A., McClintock, J.E., 1998, *Ap. J.*, 506, 880.

Bedding, T.R., Booth, A.J., Davis, J., eds., 1997, *Fundamental Stellar Properties: The Interaction between Observation and Theory*, IAU Symp. No. 189, Kluwer, Dordrecht.

Bell, S.A., Hilditch, R.W., 1984, *MNRAS*, 211, 229.

Bell, S.A., Malcolm, G., 1987, *MNRAS*, 226, 899.

Bell, S.A., Pollacco, D.L., 1997, In *Planetary Nebulae*, eds. Habing, H.J., Lamers, H.J.G.L.M., Kluwer, Dordrecht, p. 210.

Bell, S.A., Adamson, A.J., Hilditch, R.W., 1987, *MNRAS*, 224, 649.

Bell, S.A., Rainger, P.P., Hilditch, R.W., 1990, *MNRAS*, 247, 632.

Bell, S.A., Hill, G., Hilditch, R.W., Clausen, J.V., Reynolds, A.P., Giménez, A., 1991, *MNRAS*, 250, 119.

Bell, S.A., Hill, G., Hilditch, R.W., Clausen, J.V., Reynolds, A.P., 1993, *MNRAS*, 265, 1047.

Bell, S.A., Pollacco, D.L., Hilditch, R.W., 1994, *MNRAS*, 270, 449.

Bergeron, P., Liebert, J., Fulbright, M.S., 1995, *Ap. J.*, 444, 810.

Berriman, G., Reid, I.N., 1987, *MNRAS*, 227, 315.

Bessell, M.S., 1983, *PASP*, 95, 480.

Bessell, M.S., 1990, *PASP*, 102, 1181.

Bessell, M.S., 1995, *PASP*, 107, 672.

Bessell, M.S., Brett, J.M., 1988, *PASP*, 100, 1134.

Bessell, M.S., Castelli, F., Plez, B., 1998, *A. & A.*, 333, 231 (plus minor erratum in *A. & A.*, 337, 321).

Bevington, P.R., 1969, *Data Reduction and Error Analysis for the Physical Sciences*, McGraw-Hill, New York.

Bevington, P.R., Robinson, D.K., 1992, *Data Reduction and Error Analysis for the Physical Sciences*, 2nd ed., McGraw-Hill, New York.

Bhattacharya, D., 1995, In *X-ray Binaries*, eds. Lewin, W.H.G., van Paradijs, J., van den Heuvel, E.P.J., Cambridge University Press, p. 233.

Bildsten, L., and 12 other authors, 1997, *Ap. J. Suppl.*, 113, 367.

Binnendijk, L., 1970, *Vistas in Astronomy*, 12, 217.

Binney, J., Tremaine, S., 1987, *Galactic Dynamics*, ch. 2, Princeton University Press.

Blandford, R., Teukolsky, S.A., 1976, *Ap. J.*, 205, 580.

Blondin, J.M., Kallman, T.R., Fryxell, B.A., Taam, R.E., 1990, *Ap. J.*, 356, 591.

Bohlin, R.C., 1996, *Astron. J.*, 111, 1743.

Böhm-Vitense, E., 1981, *A. Rev. A. & A.*, 19, 295.

Bolton, C.T., 1972, *Nature*, 235, 271.

Bond, H.E., 1989, In *Planetary Nebulae*, ed. Torres-Peimbert, S., IAU Symp. No. 131, Kluwer, Dordrecht, p. 251.

Bopp, B.W., 1974, *Ap. J.*, 193, 389.

Bowyer, S., Lieu, R., Lampton, M., Lewis, J., Wu, X., Drake, J.J., Malina, R.F., 1995, *A. & A.*, 297, 764.

Bradstreet, D.H., 1993, *Binary Maker 2.0 User Manual – A Light Curve Synthesis Program*, Eastern College, St. David's, PA, USA.

Branch, D., 1998, *A. Rev. A. & A.*, 36, 17.

Brett, J.M., Smith R.C., 1993, *MNRAS*, 264, 641.

Brickhouse, N.S., Dupree, A.K., 1998, *Ap. J.*, 502, 918.

Brouwer, D., Clemence, G.M., 1961, *Methods of Celestial Mechanics*, Academic Press, New York.

Brown, E.A., 1936a, *MNRAS*, 97, 56.

Brown, E.A., 1936b, *MNRAS*, 97, 62.

Brown, E.A., 1936c, *MNRAS*, 97, 116.

Brown, E.A., 1937, *MNRAS*, 97, 388.

Brown, J.C., McLean, I.S., Emslie, A.G., 1978, *A. & A.*, 68, 415.

Brown, J.C., Aspin, C., Simmons, J.F.L., McLean, I.S., 1982, *MNRAS*, 198, 787.

Budding, E., 1977, *Ap. Sp. Sci.*, 48, 207.

Budding, E., 1984, *CDS Bull.*, 27, 91.

Budding, E., 1986, *Ap. Sp. Sci.*, 118, 241.

Burch, S.F., Gull, S.F., Skilling, J., 1983, *Comp. Vis. Graph. Im. Proc.*, 23, 113.

Burnham, S.W., 1906, *Catalogue of Double Stars within 121° of the North Pole*, Carnegie Institution, Washington, DC.

Byrne, P.B., and 16 other authors, 1998, *A. & A. Suppl.*, 127, 505.

Carnochan, D.J., 1982, *MNRAS*, 201, 1139.

Casares, J., Charles, P.A., Naylor, T., 1992, *Nature*, 355, 614.

Casey, B.W., Mathieu, R.D., Vaz, L.P.R., Andersen, J., Suntzeff, N.B., 1998, *Astron. J.*, 115, 1617.

Chabrier, G., Baraffe, I., 1997, *A. & A.*, 327, 1039.

Chandrasekhar, S., 1946, *Ap. J.*, 103, 365.

Charles, P.A., Seward, F.D., 1995, *Exploring the X-ray Universe*, Cambridge University Press.

Cherepaschuk, A.M., 1973, In *Eclipsing Variable Stars*, ed. Tsesevich, V.P., Wiley, New York, p. 225.

Cherepaschuk, A.M., Eaton, J.A., Khaliullin, K.F., 1984, *Ap. J.*, 281, 774.

Claret, A., 1995, *A. & A. Suppl.*, 109, 441.

Claret, A., 1998, *A. & A. Suppl.*, 131, 395.

Claret, A., Giménez, A., 1992, *A. & A. Suppl.*, 96, 255.

Claret, A., Giménez, A., 1993, *A. & A.*, 277, 487.

Claret, A., Giménez, A., Martin, E.L., 1995, *A. & A.*, 302, 741.

Claria, J.J., Piatti, A.E., Lapasset, E., 1994, *PASP*, 106, 436.

Clark, G.W., Woo, J.W., Nagase, F., 1994, *Ap. J.*, 422, 336.

Claudius, M., Florentin-Nielsen, R., 1981, *A. & A.*, 100, 186.

Clausen, J.V., 1991, *A. & A.*, 246, 397.

Clausen, J.V., 1996, *A. & A.*, 308, 151.

Clausen, J.V., Andersen, J., Giménez, A., Helt, B.E., Jensen, K.S., Lindgren, H., Nordström, B., Reipurth, B., Vaz, L.P.R., 1991, *A. & A. Suppl.*, 88, 535.

Clausen, J.V., Garcia, J.M., Giménez, A., Helt, B.E., Vaz, L.P.R., 1993, *A. & A. Suppl.*, 101, 563.

Clausen, J.V., Larsen, S.S., Garcia, J.M., Giménez, A., Storm, J., 1997, *A. & A. Suppl.*, 122, 559.

Clemence, G.M., Szebehely, V., 1967, *Astron. J.*, 72, 1324.

Code, A.D., Davis, J., Bless, R.C., Hanbury Brown, R., 1976, *Ap. J.*, 203, 417.

Collier Cameron A., 1997, *MNRAS*, 287, 556.

Collier Cameron A., Hilditch, R.W., 1997, *MNRAS*, 287, 567.

Collier Cameron, A., Jianke, L., Mestel, L., 1991, In *Angular Momentum Evolution of Young Stars*, eds. Catalano, S., Stauffer, J.R., Kluwer, Dordrecht, p. 297.

Company, R., Portilla, M., Giménez, A., 1988, *Ap. J.*, 335, 962.

Cousins, A.W.J., 1980, *SAAO Circular*, 1, 234.

Cousins, A.W.J., Caldwell, J.A.R., 1996, *MNRAS*, 281, 522.

Cowley, A.P., 1992, *A. Rev. A. & A.*, 30, 287.

Cowley, A.P., Schmidtke, P.C., Crampton, D., Hutchings, J.B., 1998, *Ap. J.*, 504, 854.

Crampton, D., Hutchings, J.B., Cowley, A.P., Schmidtke, P.C., McGrath, T.K., O'Donoghue, D., Harrop-Allin, M.K., 1996, *Ap. J.*, 456, 320.

Cranmer, S.R., 1993, *MNRAS*, 263, 989.

Crawford, D.L., 1958, *Ap. J.*, 128, 185.

Crawford, D.L., 1966, In *Spectral Classification and Multicolour Photometry*, eds. Lodén, K., Lodén, O., Sinnersted, U., IAU Symp. No. 24, Academic Press, New York, p. 170.

Crawford, D.L., 1975a, *PASP*, 87, 481.

Crawford, D.L., 1975b, *Astron. J.*, 80, 955.

Crawford, D.L., 1978, *Astron. J.*, 83, 48.

Crawford, D.L., 1979, *Astron. J.*, 84, 1858.

Crawford, D.L., Mander, J.V., 1966, *Astron. J.*, 71, 114.

Crawford, D.L., Barnes, J.V., 1970, *Astron. J.*, 75, 978.

Crawford, D.L., Mandwewala, N., 1976, *PASP*, 88, 917.

Crawford, J.A., 1955, *Ap. J.*, 121, 71.

Cropper, M., 1985, *MNRAS*, 212, 709.

Cropper, M., 1998, In *Accretion Processes in Astrophysical Systems: Some Like It Hot!*, eds. Holt, S.S., Kallman, T.R., AIP Conf. Proc., 431, 447.

Cropper, M., Warner, B., 1986, *MNRAS*, 220, 633.

Davis, J., 1992, In *Complementary Approaches to Double and Multiple Star Research*, eds. McAlister, H.A., Hartkopf, W.I., IAU Coll. No. 135, ASP Conf. Ser. Vol. 32, p. 521.

Davis, J., Tango, W.J., Booth, A.J., ten Brummelaar, T.A., Minard, R.A., Owens, S.M., 1999a, *MNRAS*, 303, 773.

Davis, J., Tango, W.J., Booth, A.J., Thorvaldsen, E.D., Giovannis, J., 1999b, *MNRAS*, 303, 783.

De Greve, J.-P., 1989, In *Algols*, IAU Coll. No. 107, ed. Batten, A.H., Kluwer, Dordrecht, p. 127.

De Greve, J.-P., 1993, *A. & A. Suppl.*, 97, 527.

De Greve, J.-P., de Loore, C., 1992, *A. & A. Suppl.*, 96, 653.

De Greve, J.-P., Linnell, A.P., 1994, *A. & A.*, 291, 786.

de Landtsheer, A.C., 1983, *A. & A. Suppl.*, 53, 161.

de Loore, C.W.H., Doom, C., 1992, *Structure and Evolution of Single and Binary Stars*, Kluwer, Dordrecht.

de Loore, C., De Greve, J.-P., 1992, *A. & A. Suppl.*, 94, 453.

de Loore, C., Vanbeveren, D., 1994, *A. & A.*, 292, 463.

Demircan, O., Derman, E., Akalin, A., 1991, *Astron. J.*, 101, 201.

Demircan, O., Selam, S.O., 1993, *A. & A.*, 267, 107.

Dhillon, V.S., Jones, D.H.P., Marsh, T.R., 1994, *MNRAS*, 266, 859.

Dhillon, V.S., Privett, G.J., 1995, PPARC Starlink User Note No. 167.4.

Diaz, M.P., Steiner, J.E., 1994, *A. & A.*, 283, 508.

Djurasević, G., Zakirov, M., Erkapić, S., 1999, *A. & A.*, 343, 894.

Donati, J.-F., 1999, *MNRAS*, 302, 457.

Donati, J.-F., Collier Cameron, A., 1997, *MNRAS*, 291, 1.

Donati, J.-F., Semel, M., Carter, B.D., Rees, D.E., Cameron, A.C., 1997, *MNRAS*, 291, 658.

Dorman, B., Nelson, L.A., Chau, W.Y., 1989, *Ap. J.*, 342, 1003.

Downes, R.A., Shara, M.M., 1993, *PASP*, 105, 127.

Downes, R.A., Webbink, R.F., Shara, M.M., 1997, *PASP*, 109, 345.

Drechsel, H., Rahe, J., Wargau, W., Wolf, B., 1982, *A. & A.*, 110, 246.

Drechsel, H., Haas, S., Lorenz, R., Mayer, P., 1994, *A. & A.*, 284, 853.

Drissen, L., Lamontagne, R., Moffat, A.F.J., Bastien, P., Seguin, M., 1986a, *Ap. J.*, 304, 188.

Drissen, L., Moffat, A.F.J., Bastien, P., Lamontagne, R., 1986b, *Ap. J.*, 306, 215.

Duquennoy, A., Mayor, M., 1991, *A. & A.*, 248, 485.

Duquennoy, A., Mayor, M., 1992, *Binaries as Tracers of Stellar Formation*, eds. Duquennoy, A., Mayor, M., Cambridge University Press.

Dworetsky, M.M., 1983, *MNRAS*, 203, 917.

Eaton, J.A., Hall, D.S., 1979, *Ap. J.*, 227, 907.

Eaton, J.A., Wu, C.C., 1983, *PASP*, 95, 319.

Eaton, J.A., Wu, C.C., Rucinski, S.M., 1980, *Ap. J.*, 239, 919.

Eaton, J.A., Henry, G.W., Fekel, F.C., 1996, *Ap. J.*, 462, 888.

Eggleton, P.P., 1983, *Ap. J.*, 268, 368.

Eggleton, P.P., 1996, In *The Origins, Evolutions, and Destinies of Binary Stars in Clusters*, eds. Milone, E.F., Mermilliod, J.-C., ASP Conf. Ser. Vol. 90, p. 257.

Eggleton, P.P., Mitton, S., Whelan, J., eds., 1976, *Structure and Evolution of Close Binary Systems*, IAU Symp. No. 73, Reidel, Dordrecht.

Eggleton, P.P., Pringle, J.E., eds., 1985, *Interacting Binaries*, Proc. NATO ASI Series, Reidel, Dordrecht.

Elias, J.H., Frogel, J.A., Matthews, K., Neugebauer, G., 1982, *Astron. J.*, 87, 1029.

Eracleous, M., Horne, K., 1996, *Ap. J.*, 471, 427.

Etzel, P.B., 1980, *EBOP User's Guide*, 3rd ed., Department of Astronomy, University of California, Los Angeles.

Etzel, P.B., 1993, In *Light Curve Modelling of Eclipsing Binary Stars*, ed. Milone, E.F., Springer-Verlag, New York, p. 113.

Evans, N.R., Böhm-Vitense, E., Carpenter, K, Beck-Winchantz, B., Robinson, R., 1998, *Ap. J.*, 494, 768.

Fekel, F., 1980, *Bull. AAS*, 12, 500.

Fekel, F.C., Scarfe, C.D., Barlow, D.J., Duquennoy, A., McAlister, H.A., Hartkopf, W.I., Mason, B.D., Tokovinin, A.A., 1997, *Astron. J.*, 113, 1095.

Fellgett, P.B., 1953, *Optica Acta*, 2, 9.

Ferrario, L., Wickramasinghe, D.T., Tuohy, I.R., 1989, *Ap. J.*, 341, 327.

Figueiredo, J., De Greve, J.-P., Hilditch, R.W., 1994, *A. & A.*, 283, 144.

Flannery, B.P., 1976, *Ap. J.*, 205, 217.

Fletcher, J.M., Harris, H.C., McClure, R.D., Scarfe, C.D., 1982, *PASP*, 94, 1017.

Frank, J., King, A.R., Raine, D.J., 1985, *Accretion Power in Astrophysics*, Cambridge University Press.

Frank, J., King, A.R., Raine, D.J., 1992, *Accretion Power in Astrophysics*, 2nd ed., Cambridge University Press.

Fraunhofer, J., 1815, In *A Source Book in Astronomy*, eds. Shapley, H., Howarth, H.E., McGraw-Hill, New York (1929), p. 196.

Frost, E.B., 1906, *Ap. J.*, 24, 259.

Gehrz, R.D., 1998, In *Wild Stars in the Old West: Proceedings of the 13th North American Workshop on Cataclysmic Variables and Related Objects*, eds. Howell, S., Kuulkers, E., Woodward, C., ASP Conf. Ser. Vol. 137, p. 146.

Gezari, D.Y., Labeyrie, A., Stachnik, R.V., 1972, *Ap. J.*, 173, L1.

Ghosh, P., Lamb, F.K., 1979, *Ap. J.*, 234, 296.

Giacconi, R., Gursky, H., Paolini, F.R., Rossi, B.B., 1962, *Phys. Rev. Lett.*, 9, 439.

Gies, D.R., Bolton, C.T., 1986, *Ap. J.*, 304, 371.

Gies, D.R., Bagnuolo, W.G., Penny, L.R., 1997, *Ap. J.*, 479, 408.

Giménez, A., Garcia-Pelayo, J.M., 1982, In *Binary and Multiple Stars as Tracers of Stellar Evolution*, eds. Kopal, Z., Rahe, J., IAU Coll. No. 69, Reidel, Dordrecht, p. 37.

Giménez, A., Garcia-Pelayo, J.M., 1983, *Ap. Sp. Sci.*, 92, 203.

Giménez, A., Claret, A., 1992, In *Evolutionary Processes in Interacting Binary Stars*, eds. Kondo, Y., Sistero, R.F., Polidan, R.S., IAU Symp. No. 151, p. 277.

Giménez, A., Clausen, J.V., 1994, *A. & A.*, 291, 795.

Golay, M., 1974, *Introduction to Astronomical Photometry*, Reidel, Dordrecht.

Goodricke, J., 1783, *Phil. Trans. R. Soc. London*, 73, 474.

Gray, D.F., 1992, *The Observation and Analysis of Stellar Photospheres*, 2nd ed., Cambridge University Press.

Greiner, J., Remillard, R.A., Motch, C., 1998, *A. & A.*, 336, 191.

Griffin, R.F., 1967, *Ap. J.*, 148, 465.

Griffin, R.F., 1998, *The Observatory*, 118, 209 (paper 141 and the preceding 140 papers in that journal).

Griffin, R.F., Duquennoy, A., 1993, *The Observatory*, 113, 52.

Griffin, R.F., Mayor, M., Pont, F., Udry, S., 1997, *The Observatory*, 117, 288.

Grison, P., and 31 other authors (EROS), 1995, *A. & A. Suppl.*, 109, 447.

Grøsbol, P.J., 1978, *A. & A. Suppl.*, 32, 409.

Guinan, E.F., Maloney, F.P., 1985, *Astron. J.*, 90, 1519.

Guinan, E.F., Bradstreet, D.H., 1988, In *Formation and Evolution of Low Mass Stars*, eds. Dupree, A.K., Lago, M.T., Kluwer, Dordrecht, p. 345.

Guinan, E.F., Giménez, A., 1993, In *The Realm of Interacting Binary Stars*, eds. Sahade, J., McCluskey, G.E., Kondo, Y., Kluwer, Dordrecht, p. 51.

Guinan, E.F., Bradstreet, D.H., DeWarf, L.E., 1996, In *The Origins, Evolutions, and Destinies of Binary Stars in Clusters*, eds. Milone, E.F., Mermilliod, J.-C., ASP Conf. Ser. Vol. 90, p. 196.

Gursky, H., Giacconi, R., Gorenstein, P., Waters, J.R., Oda, M., Bradt, H., Garmire, G., Sreekantan, B.V., 1966, *Ap. J.*, 144, 1249.

Haberl, F., Thorstensen, J.R., Motch, C., Schwarzenberg-Czerny, A., Pakull, M., Shambrook, A., Pietsch, W., 1994, *A. & A.*, 291, 171.

Habing, H.J., Lamers, H.J.G.L.M., eds., 1997, *Planetary Nebulae*, IAU Symp. No. 180, Kluwer, Dordrecht.

Hadrava, P., 1995, *A. & A. Suppl.*, 114, 393.

Hakala, P.J., Watson, M.G., Vilhu, O., Hassall, B.J.M., Kellett, B.J., Mason, K.O., Piirola, V., 1993, *MNRAS*, 263, 61.

Hall, D.S., 1972, *PASP*, 84, 323.

Hall, D.S., 1989, *Space Sci. Rev.*, 50, 219.

Hall, D.S., 1991, *Ap. J.*, 380, L85.

Hanbury-Brown, R., 1974, *The Intensity Interferometer: Its Application to Astronomy*, Taylor-Francis, London.

Hardie, R.H., 1962, In *Astronomical Techniques*, ed. Hiltner, W.A., University of Chicago Press, p. 178.

Harlaftis, E., Horne, K., 1999, *MNRAS*, 305, 437.

Harlaftis, E., Steeghs, D., Horne, K., Martin, E., Magazzu, A., 1999, *MNRAS*, 306, 348.

Harmanec, P., Scholz, G., 1993, *A. & A.*, 279, 131.

Harmanec, P., and 16 other authors, 1996, *A. & A.*, 312, 879.

Harries, T.J., Hilditch, R.W., 1997a, *MNRAS*, 291, 544.

Harries, T.J., Hilditch, R.W., 1997b, In *Boulder-Munich II: Properties of Hot, Luminous Stars*, ed. Howarth, I.D., ASP Conf. Ser. Vol. 131, p. 401.

Harries, T.J., Howarth, I.D., 1996a, *A. & A.*, 310, 235.

Harries, T.J., Howarth, I.D., 1996b, *A. & A. Suppl.*, 119, 61.

Harries, T.J., Howarth, I.D., 1997, *A. & A. Suppl.*, 121, 15.

Harries, T.J., Hilditch, R.W., Hill, G., 1998, *MNRAS*, 295, 386.

Harrington, R.S., 1968, *Astron. J.*, 73, 190.

Harrington, R.S., 1969, *Cel. Mech.*, 1, 200.

Hartkopf, W.I., 1992, In *Complementary Approaches to Double and Multiple Star Research*, eds. McAlister, H.A., Hartkopf, W.I., IAU Coll. No. 135, ASP Conf. Ser. Vol. 32, p. 459.

Hartkopf, W.I., McAlister, H.A., Franz, O.G., 1989, *Astron. J.*, 98, 1014.

Hasinger, G., van der Klis, M., 1989, *A. & A.*, 225, 79.

Haswell, C.A., Patterson, J., Thorstensen, J.R., Hellier, C., Skillman, D.R., 1997, *Ap. J.*, 476, 847.

Hatzes, A.P., Kürster, M., 1999, *A. & A.*, 346, 432.

Hecht, E., 1992, *Optics*, 2nd ed., Addison-Wesley, Reading, MA.

Heemskerk, M.H.M., van Paradijs, J., 1989, *A. & A.*, 223, 154.

Heintz, W.D., 1978, *Double Stars*, Reidel, Dordrecht.

Hellier, C., 1992, *MNRAS*, 258, 578.

Hellier, C., 1997, *MNRAS*, 288, 817.

Hellier, C., 1999a, In *Annapolis Workshop on Magnetic Cataclysmic Variables*, eds. Hellier, C., Mukai, K., ASP Conf. Ser. Vol. 157, p. 1.

Hellier, C., 1999b, *Ap. J.*, 519, 324.

Henden, A.A., Kaitchuk, R.H., 1982, *Astronomical Photometry*, Van Nostrand Reinhold, New York.

Henden, A.A., Honeycutt, R.K., 1997, *PASP*, 109, 441.

Hendry, P.D., Mochnacki, S.W., Collier Cameron, A., 1992, *Ap. J.*, 399, 246.

Henry, G.W., Eaton, J.A., Hamer, J., Hall, D.S., 1995, *Ap. J. Suppl.*, 97, 513.

Herschel, W., 1802, *Phil. Trans. R. Soc. London*, p. 477.

Herschel, W., 1803, *Phil. Trans. R. Soc. London*, p. 339.

Hilditch, R.W., 1973, *MNRAS*, 164, 101.

Hilditch, R.W., 1981, *MNRAS*, 196, 305.

Hilditch, R.W., 1989, In *Algols*, ed. Batten, A.H., Kluwer, Dordrecht, p. 289.

Hilditch, R.W., 1996, In *The Origins, Evolutions, and Destinies of Binary Stars in Clusters*, eds. Milone, E.F., Mermilliod, J.-C., ASP Conf. Ser. Vol. 90, p. 207.

Hilditch, R.W., Hill, G., 1975, *Memoirs RAS*, 79, 101.

Hilditch, R.W., King, D.J., 1986, *MNRAS*, 223, 581.

Hilditch, R.W., Bell, S.A., 1987, *MNRAS*, 229, 529.

Hilditch, R.W., King, D.J., 1988, *MNRAS*, 232, 147.

Hilditch, R.W., Collier Cameron, A., 1995, *MNRAS*, 277, 747.

Hilditch, R.W., Hill, G., Barnes, J.V., 1983, *MNRAS*, 204, 241.

Hilditch, R.W., Skillen, I., Carr, D.M., Aikman, G.C.L., 1986, *MNRAS*, 222, 167.

Hilditch, R.W., King, D.J., McFarlane, T.M., 1988, *MNRAS*, 231, 341.

Hilditch, R.W., Hill, G., Khalesseh, B., 1992, *MNRAS*, 254, 82.

Hilditch, R.W., Harries, T.J., Hill, G., 1996, *MNRAS*, 279, 1380.

Hilditch, R.W., Collier Cameron, A., Hill, G., Bell, S.A., Harries, T.J., 1997, *MNRAS*, 291, 749.

Hilditch, R.W., Bell, S.A., Hill, G., Harries, T.J., 1998, *MNRAS*, 296, 100.

Hill, G., 1979, *Publ. Dom. Ap. Obs. Victoria*, 15, 297.

Hill, G., 1982a, *Publ. Dom. Ap. Obs. Victoria*, 16, 43.

Hill, G., 1982b, *Publ. Dom. Ap. Obs. Victoria*, 16, 59.

Hill, G., 1989, *A. & A.*, 218, 141.

Hill, G., 1993, In *New Frontiers in Binary Star Research*, eds. Leung, K.C., Nha, I.-S, ASP Conf. Ser. Vol. 38, p. 127.

Hill, G., Hutchings, J.B., 1970, *Ap. J.*, 162, 265.

Hill, G., Khalesseh, B., 1993, *A. & A.*, 276, 57.

Hill, G., Holmgren, D.E., 1995, *A. & A.*, 297, 127.

Hill, G., Rucinski, S.M., 1993, In *Light Curve Modelling of Eclipsing Binary Stars*, ed. Milone, E.F., Springer-Verlag, New York, p. 135.

Hill, G., Barnes, J.V., Hutchings, J.B., Pearce, J.A., 1971, *Ap. J.*, 168, 443.

Hill, G., Hilditch, R.W., Younger, F., Fisher, W.A., 1975, *Memoirs RAS*, 79, 131.

Hill, G., Fisher, W.A., Holmgren, D.E., 1989a, *A. & A.*, 211, 81.

Hill, G., Fisher, W.A., Holmgren, D.E., 1989b, *A. & A.*, 218, 152.

Hill, G., Perry, C.L., Khalesseh, B., 1993, *A. & A. Suppl.*, 101, 579.

Hill, G., Hilditch, R.W., Aikman, G.C.L., Khalesseh, B., 1994, *A. & A.*, 282, 455.

Hjellming, R.M., 1997, In *Accretion Phenomena and Related Outflows*, eds. Wickramasinghe, D.T., Bicknell, G.V., Ferrario, L., IAU Coll. No. 163, ASP Conf. Ser. Vol. 121, p. 53.

Hjellming, R.M., Webbink, R., 1987, *Ap. J.*, 318, 794.

Hjellming, R.M., Han, X., 1995, In *X-ray Binaries*, eds. Lewin, W.H.G., van Paradijs, J., van den Heuvel, E.P.J., Cambridge University Press, p. 308.

Hoffman, J.L., Nordsieck, K.H., Fox, G.K., 1998, *Ap. J. Suppl.*, 115, 1576.

Holmgren, D.E., Hill, G., Fisher, W.A., 1990a, *A. & A.*, 236, 409.

Holmgren, D.E., Hill, G., Fisher, W.A., Scarfe, C.D., 1990b, *A. & A.*, 231, 89.

Holmgren, D.E., Hill, G., Fisher, W.A., 1991, *A. & A.*, 248, 129.

Holmgren, D.E., Hill, G., Scarfe, C.D., 1995, *The Observatory*, 115, 188.

Horne, J.H., Baliunas, S.L., 1986, *Ap. J.*, 302, 757.

Horne, K., 1983, In *Interacting Binaries*, eds. Eggleton, P.P., Pringle, J.E., NATO ASI Ser., Kluwer, Dordrecht, p. 327.

Horne, K., 1985, *MNRAS*, 213, 129.

Horne, K., 1986, *PASP*, 98, 609.

Horne, K., 1998, In *Accretion Processes in Astrophysical Systems: Some Like It Hot!*, eds. Holt, S.S., Kallman, T.R., AIP Conf. Proc., Vol. 431, p. 426.

Hoskin, M., 1982, *Stellar Astronomy – Historical Studies*, Science History Publications Ltd., Chalfont St. Giles, UK.

Howarth, I.D., 1997, *The Observatory*, 117, 335.

Howarth, I.D., Siebert, K.W., Hussain, G.A.J., Prinja, R.K., 1997, *MNRAS*, 284, 265.

Hsu, J.-C., Breger, M., 1982, *Ap. J.*, 262, 732.

Hubeny, I., 1988, *Comp. Phys. Comm.*, 52, 103.

Hubeny, I., Lanz, T., 1992, *A. & A.*, 262, 501.

Hubeny, I., Lanz, T., 1995, *Ap. J.*, 439, 875.

Hubeny, I., Lanz, T., Jeffery, C.S., 1994, In *Newsletter on Analysis of Astronomical Spectra*, No. 20, ed. Jeffery, C.S., University of St. Andrews, p. 30.

Huggins, Sir W., 1897, In *A Source Book in Astronomy*, eds. Shapley, H., Howarth, H.E., McGraw-Hill, New York (1929), p. 290.

Hulse, R.A., Taylor, J.H., 1975, *Ap. J.*, 195, L51.

Hummel, C.A., Armstrong, J.T., Quirrenbach, A., Buscher, D.F., Mozurkewich, D., Simon, R.S., Johnston, K.J., 1993, *Astron. J.*, 106, 2486.

Hummel, C.A., Armstrong, J.T., Quirrenbach, A., Buscher, D.F., Mozurkewich, D., Elias, N.M., 1994, *Astron. J.*, 107, 1859.

Hummel, C.A., Mozurkewich, D., Armstrong, J.T., Hajian, A.R., Elias, N.M., Hutter, D.J., 1998, *Astron. J.*, 116, 2536.

Hutchings, J.B., Hill, G., 1971a, *Ap. J.*, 166, 373.

Hutchings, J.B., Hill, G., 1971b, *Ap. J.*, 167, 137.

Hynes, R.I., Haswell, C.A., Shrader, C.R., Chen, W., Horne, K., Harlaftis, E., O'Brien, K., Hellier, C., Fender, R.P., 1998a, *MNRAS*, 300, 64.

Hynes, R.I., O'Brien, K., Horne, K., Chen, W., Haswell, C.A., 1998b, *MNRAS*, 299, L37.

Ibanoglu, C., ed., 1990, *Active Close Binaries*, NATO ASI Conf. Ser., Kluwer, Dordrecht.

Iben, I., Livio, M., 1993, *PASP*, 105, 1373.

Iben, I., Tutukov, A.V., 1993, *Ap. J.*, 418, 343.

Innes, R.T.A., 1926, In *Orbital Elements of Double Stars*, ed. van den Bos, W.H., Union Obs. Circ. 68, p. 354; (and Vol. 86, p. 261, 1932).

Irwin, J.B., 1947, *Ap. J.*, 106, 380.

Jain, R.K., Bailyn, C.D., Orosz, J.A., Remillard, R.A., McClintock, J.E., 1999, *Ap. J.*, 517, L131.

Jarad, M.M., Hilditch, R.W., Skillen, I., 1989, *MNRAS*, 238, 1085.

Jaschek, C., Jaschek, M., 1987, *The Classification of Stars*, Cambridge University Press.

Jeffers, H.M.,van den Bos, W.H., Greeby, F.M.,

1963, *Index Catalogue of Visual Double Stars*, Publ. Lick Obs., 21.

Jeffery, C.S., Simon, T., 1997, In *The Third Conference on Faint Blue Stars*, eds. Philip, A.D.G., Liebert, J.W., Saffer, R.A., L. Davis Press, Schenectady, p. 189.

Jeffries, R.D., 1998, *MNRAS*, 295, 825.

Johnson, H.L., 1966, *A. Rev. A. & A.*, 4, 193.

Johnson, H.L., Morgan, W.W., 1953, *Ap. J.*, 117, 353.

Johnston, S., Manchester, R.N., Lyne, A.G., Bailes, M., Kaspi, V.M., Goujun, Q. & D'Amico, N., 1992, *Ap. J.*, 387, L37.

Johnston, S., Manchester, R.N., Lyne, A.G., Nicastro, L., Spyromilio, J., 1994, *MNRAS*, 268, 430.

Joss, P.C., Rappaport, S.A., 1984, *A. Rev. A. & A.*, 22, 537.

Joy, A.H., 1942, *PASP*, 54, 35.

Joy, A.H., 1947, *PASP*, 59, 171.

Joy, A.H., 1954, *Ap. J.*, 120, 377.

Joy, A.H., van Biesbroeck, G., 1944, *PASP*, 56, 123.

Kahabka, P., van den Heuvel, E.P.J., 1997, *A. Rev. A. & A.*, 35, 69.

Kalimeris, A., Rovithis-Livaniou, H., Rovithis, P., 1994, *A. & A.*, 282, 775.

Kallrath, J., 1993, In *Light Curve Modelling of Eclipsing Binary Stars*, ed. Milone, E.F., Springer-Verlag, New York, p. 39.

Kallrath, J., Linnell, A.P., 1987, *Ap. J.*, 313, 346.

Kallrath, J., Milone, E.F., 1999, *Eclipsing Binary Stars: Modelling and Analysis*, Springer-Verlag, New York.

Kaluzny, J., 1985, *Acta Astr.*, 35, 313.

Kaluzny, J., 1990, *Astron. J.*, 99, 1207.

Kaluzny, J., 1991, *Acta Astr.*, 41, 17.

Kaluzny, J., Stanek, K.Z., Krockenberger, M., Sasselov, D.D., Tonry, J.L., Mateo, M., 1998, *Astron. J.*, 115, 1016.

Kaluzny, J., Mochejska, B.J., Stanek, K.Z., Krockenberger, M., Sasselov, D.D., Tonry, J.L., Mateo, M., 1999, *Astron. J.*, 118, 346.

Kaspi, V.M., Johnston, S., Bell, J.F., Manchester, R.N., Bailes, M., Bessell, M., Lyne, A.G., D'Amico, N., 1994, *Ap. J.*, 423, L43.

Keenan, P.C., 1963, In *Basic Astronomical Data*, ed. Strand, K.A., University of Chicago Press, p. 78.

Kelley, R.L., Rappaport, S., Clark, G.W., Petro, L.D., 1983, *Ap. J.*, 268, 790.

Kelz, A., 1996, In *The Origins, Evolutions, and Destinies of Binary Stars in Clusters*, eds. Milone, E.F., Mermilliod, J.-C., ASP Conf. Ser. Vol. 90, p. 51.

Kenyon, S.J., 1986, *The Symbiotic Stars*, Cambridge University Press.

Khaliullin, K.F., Khodykin, S.A., Zakharov, A.I., 1991, *Ap. J.*, 375, 314.

Khodykin, S.A., Vedeneyev, V.G., 1997, *Ap. J.*, 475, 798.

Kholopov, P.N., Samus, N.N., Frolov, M.S., Goranskij, V.P., Gorynya, N.A., Kireeva, N.N., Kukarkina, N.P., Kurochkin, N.E., Medvedeva, G.I., Perova, N.B., Shugarov, S.Y., 1985, *General Catalogue of Variable Stars*, 4th ed., Nauka, Moscow.

Kilkenny, D., Laing, J.D., 1992, *MNRAS*, 255, 308.

Kilkenny, D., Spencer Jones, J.H., Marang, F., 1988, *The Observatory*, 108, 88.

Kilkenny, D., O'Donohugue, D., Koen, C., Stobie, R.S., Chen, A., 1997, *MNRAS*, 287, 867.

Kilkenny, D., O'Donohugue, D., Lynas-Gray, A.E., van Wyk, F., 1998, *MNRAS*, 296, 329.

Kippenhahn, R., Weigert, A., 1967, *Zts. f. Ap.*, 65, 251.

Kippenhahn, R., Weigert, A., 1991, *Stellar Structure and Evolution*, Springer-Verlag, Berlin, p. 210.

Kitamura, M., 1965, In *Advances in Astronomy & Astrophysics*, vol. 3, ed. Kopal, Z., Academic Press, London, p. 27.

Kitamura, M., 1967, *Tables of the Characteristic Functions of the Eclipse and the Related Delta-Functions for Solutions of Light Curves of Eclipsing Binary Stars*, University of Tokyo Press.

Knigge, C., Long, K.S., Wade, R.A., Baptista, R., Horne, K., Hubeny, I., Rutten, R.G.M., 1998a, *Ap. J.*, 499, 414.

Knigge, C., Drake, N., Long, K.S., Wade, R.A., Horne, K., Baptista, R., 1998b, *Ap. J.*, 499, 429.

Kondo, Y., Sisteró, R.F., Polidan, R.S., eds., 1992, *Evolutionary Processes in Interacting Binary Stars*, IAU Symp. No. 151, Kluwer, Dordrecht.

Koornneef, J., 1983, *A. & A.*, 128, 84.

Kopal, Z., 1946, Harvard Obs. Monograph No. 6, Harvard College Observatory.

Kopal, Z., 1950, Harvard Obs. Monograph No. 8, Harvard College Observatory.

Kopal, Z., 1955, *Ann. d'Ap.*, 18, 379.

Kopal, Z., 1959, *Close Binary Systems*, Chapman & Hall, London.

Kopal, Z., 1978, *Dynamics of Close Binary Stars*, Reidel, Dordrecht.

Kopal, Z., 1979, *Language of the Stars*, Reidel, Dordrecht.

Kopal, Z., Kitamura, M., 1968, *Adv. A. & A.*, 6, 125.

Kraft, R.E., 1963, *Adv. A. & A.*, 2, 43.

Kruszewski, A., 1966, In *Advances in Astronomy & Astrophysics*, vol. 4, ed. Kopal, Z., Academic Press, New York.

Kuiper, G.P., 1941, *Ap. J.*, 93, 133.

Kurucz, R.L., 1979, *Ap. J. Suppl.*, 40, 1.

Kurucz, R.L., 1991, Harvard Preprint 3348.

Labeyrie, A., 1970, *A. & A.*, 6, 85.

Lacy, C.H., 1977, *Ap. J.*, 218, 444.

Lacy, C.H.S., 1992, *Astron. J.*, 104, 2213.

Lacy, C.H.S., 1993, *Astron. J.*, 105, 1096.

Lacy, C.H.S., 1997, *Astron. J.*, 114, 2140.

Lacy, C.H.S., Torres, G., Latham, D.W., Zakirov, M.M., Arzumanyants, G.C., 1997, *Astron. J.*, 114, 1206.

Lamers, H.J.G.L.M., Cassinelli, J.P., 1999, *An Introduction to Stellar Winds*, Cambridge University Press.

Lamontagne, R., Moffat, A.F.J., Drissen, L., Robert, C., Matthews, J.M., 1996, *Astron. J.*, 112, 2227.

Landau, L.D., Lifshitz, E.M., 1962, *The Classical Theory of Fields*, 2nd ed. Pergamon, Oxford.

Landolt, A.U., 1983, *Astron. J.*, 88, 439.

Landolt, A.U., 1992, *Astron. J.*, 104, 340.

Lanza, A.F., Rodonó, M., Rosner, R., 1998, *MNRAS*, 296, 893.

Lee, Y.-S., 1989, *Ap. J.*, 338, 1016.

Leggett, S.K., 1992, *Ap. J. Suppl.*, 82, 351.

Lehmann-Filhés, R., 1894, *Astron. Nachr.*, 136, 17.

Lehmann-Filhés, R., 1908, *Publ. Allegheny Obs.*, 1, 33.

Leinert, C., Haas, M., Allard, F., Wehrse, R., McCarthy, D.W., Jahreiss, H., Perrier, C., 1990, *A. & A.*, 236, 399.

Lester, J.B., Gray, R.O., Kurucz, R.L., 1986, *Ap. J. Suppl.*, 61, 509.

Leung, K.-C., Schneider, D.P., 1978, *Astron. J.*, 83, 618.

Levine, A., Rappaport, S., Deeter, J.E., Boynton, P.E., Nagase, F., 1993, *Ap. J.*, 410, 328.

Lewin, W.H.G., van den Heuvel, E.P.J., eds., 1983, *Accretion-Driven Stellar X-ray Sources*, Cambridge University Press.

Lewin, W.H.G., van Paradijs, J., van den Heuvel, E.P.J., eds., 1995, *X-ray Binaries*, Cambridge University Press.

Li, J., Wickramasinghe, D.T., 1998, *MNRAS*, 300, 1015.

Linnell, A.P., 1984, *Ap. J. Suppl.*, 54, 17.

Linnell, A.P., Kallrath, J., 1987, *Ap. J.*, 316, 754.

Lister, T.A., Collier Cameron, A., Hilditch, R.W., 2000a, *MNRAS*, submitted.

Lister, T.A., Hilditch, R.W., Collier Cameron, A., James, D.J., 2000b, *MNRAS*, submitted.

Liu, N., and 12 other authors, 1997, *Ap. J.*, 485, 350.

Livio, M., 1997, In *Planetary Nebulae*, eds. Habing, H.J., Lamers, H.J.G.L.M., IAU Symp. No. 180, Kluwer, Dordrecht, p. 74.

Livio, M., 1998, In *Wild Stars in the Old West: Proceedings of the 13th North American Workshop on Cataclysmic Variables and Related Objects*, eds. Howell, S., Kuulkers, E., Woodward, C., ASP Conf. Ser. Vol. 137, p. 264.

Livio, M., Govarie, A., Ritter, H., 1991, *A. & A.*, 246, 84.

Lohmann, A.W., Weigelt, G., Wirnitzer, B., 1983, *Applied Optics*, 22, 4028.

Lomb, N.R., 1976, *Ap. Sp. Sci.*, 39, 447.

Lorenz, R., Mayer, P., Drechsel, H., 1999, *A. & A.*, 345, 531.

Lu, W.-X., Rucinski, S.M., 1992, *Astron. J.*, 106, 361.

Lu, W., Rucinski, S.M., 1999, *Astron. J.*, 118, 515

Lub, J., Pel, J.W., 1977, *A. & A.*, 54, 137.

Lubow, S.H., Shu, F.H., 1975, *Ap. J.*, 198, 383.

Lucy, L.B., 1967, *Zts. f. Ap.*, 65, 89.

Lucy, L.B., 1968a, *Ap. J.*, 151, 1123.

Lucy, L.B., 1968b, *Ap. J.*, 153, 877.

Lucy, L.B., 1976, *Ap. J.*, 205, 208.

Lyne, A.G., 1996, In *Compact Stars in Binaries*, eds. van Paradijs, J., van den Heuvel, E.P.J., Kuulkers, E., IAU Symp. No. 165, Kluwer, Dordrecht, p. 225.

Lyne, A.G., Graham-Smith, F., 1990, *Pulsar Astronomy*, Cambridge University Press.

Maceroni, C., van't Veer, F., 1992, *A. & A.*, 277, 515.

Maceroni, C., van't Veer, F., 1996, *A. & A.*, 311, 523.

Makishima, K., Koyama, K., Hayakawa, S., Nagase, F., 1987, *Ap. J.*, 314, 619.

Malina, R.F., and 20 other authors, 1994, *Astron. J.*, 107, 751.

Maloney, F.P., Guinan, E.F., Boyd, P.T., 1989, *Astron. J.*, 98, 1800.

Manchester, R.N., Peters, W.L., 1972, *Ap. J.*, 173, 221.

Manchester, R.N., Taylor, J.H., Van, Y.Y., 1974, *Ap. J.*, 189, L119.

Manset, N., Bastien, P., 1995, *PASP*, 107, 483.

Margon, B., 1984, *A. Rev. A. & A.*, 22, 507.

Marquardt, D.W., 1963, *SIAM J. Appl. Math.*, 11, 431.

Marschall, L.A., Stefanik, R.P., Lacy, C.H., Torres, G., Williams, D.B., Agerer, F., 1997, *Astron. J.*, 114, 793.

Marsh, T.R., 1999, In *Astrophysical Discs*, eds. Sellwood, J.A., Goodman, J., ASP Conf. Ser. Vol. 160, p. 3.

Marsh, T.R., Horne, K., 1988, *MNRAS*, 235, 269.

Marsh, T.R., Horne, K., Schlegel, E.M., Honeycutt, R.K., Kaitchuck, R.H., 1990, *Ap. J.*, 364, 637.

Marsh, T.R., Robinson, E.L., Wood, J.H., 1994, *MNRAS*, 266, 137.

Mason, B.D., McAlister, H.A., Hartkopf, W.I., 1996, *Astron. J.*, 112, 276.

Mason, B.D., Gies, D.R., Hartkopf, W.I., Bagnuolo, W.G., ten Brummelaar, T., McAlister, H.A., 1998, *Astron. J.*, 115, 821.

Massey, P., Gronwall, C., 1990, *Ap. J.*, 358, 344.

Massey, P., Strobel, K., Barnes, J.V., Anderson, E., 1988, *Ap. J.*, 328, 315.

Mateo, M., 1996, In *The Origins, Evolutions, and Destinies of Binary Stars in Clusters*, eds. Milone, E.F., Mermilliod, J.-C., ASP Conf. Ser. Vol. 90, p. 21.

Mathieu, R.D., 1992, In *Binaries as Tracers of Stellar Formation*, eds. Duquennoy, A., Mayor, M., Cambridge University Press, p. 155.

Mathieu, R.D., 1994, *A. Rev. A. & A.*, 32, 465.

Mathieu, R.D., Stassun, K., Basri, G., Jensen, E.L.N., Johns-Krull, C.M., Valenti, J.A., Hartmann, L.W., 1997, *Astron. J.*, 113, 1841.

Maxted, P.F.L., 1994, Ph.D. thesis, University of St. Andrews.

Maxted, P.F.L., Hilditch, R.W., 1996, *A. & A.*, 311, 567.

Maxted, P.F.L., Marsh, T.R., 1999, *MNRAS*, 307, 122.

Maxted, P.F.L., Hill, G., Hilditch, R.W., 1994a, *A. & A.*, 282, 821.

Maxted, P.F.L., Hill, G., Hilditch, R.W., 1994b, *A. & A.*, 285, 535.

Mayer, P., 1987, *Bull. Ast. Inst. Czech.*, 38, 58.

Mayer, P., 1990, *Bull. Ast. Inst. Czech.*, 41, 231.

Mayer, P., Drechsel, H., 1987, *A. & A.*, 183, 61.

Mayor, M., Queloz, D., 1995, *Nature*, 378, 355.

McAlister, H.A., 1977, *Ap. J.*, 215, 159.

McAlister, H.A., 1992, In *Complementary Approaches to Double and Multiple Star Research*, eds. McAlister, H.A., Hartkopf, W.I., IAU Coll. No. 135, ASP Conf. Ser. No. 32, p. 527.

McAlister, H.A., Hartkopf, W.I., eds., 1992, *Complementary Approaches to Double and Multiple Star Research*, IAU Coll. No. 135, ASP Conf. Ser. Vol. 32.

McAlister, H.A., Bagnuolo, W.G., ten Brummelaar, T., Hartkopf, W.I., Turner, N.H., 1995, *Proc. SPIE*, 2524, 180.

McClure, R.D., 1973, In *Spectral Classification and Multicolour Photometry*, eds. Fehrenbach, C., Westerlund, B.E., IAU Symp. No. 50, Reidel, Dordrecht, p. 162.

McClure, R.D., 1976, *Astron. J.*, 81, 182.

McClure, R.D., 1997a, *PASP*, 109, 536.

McClure, R.D., 1997b, *PASP*, 109, 256.

McClure, R.D., van den Bergh, S., 1968, *Astron. J.*, 73, 313.

McClure, R.D., Woodsworth, A.W., 1990, *Ap. J.*, 352, 709.

McCluskey, G.E., 1993, In *The Realm of Interacting Binaries*, eds. Sahade, J., McCluskey, G.E., Kondo, Y., Kluwer, Dordrecht, p. 39.

McLean, B.J., 1981, *MNRAS*, 195, 931.

McLean, B.J., Hilditch, R.W., 1983, *MNRAS*, 203, 1.

McLean, I.S., 1997, *Electronic Imaging in Astronomy*, Wiley, Chichester.

Menzies, J.W., Banfield, R.M., Cousins, A.W.J., Laing, J.D., 1989, *SAAO Circular*, 13, 1.

Menzies, J.W., Marang, F., Laing, J.D., Coulson, I.M., Engelbrecht, C.A., 1991, *MNRAS*, 248, 642.

Merrill, J.E., 1950, *Contr. Princeton Univ. Obs.*, No. 23.

Merrill, J.E., 1953, *Contr. Princeton Univ. Obs.*, No. 24.

Metcalfe, T.S., 1999, *Astron. J.*, 117, 2503.

Metcalfe, T.S., Mathieu, R.D., Latham, D.W., Torres, G., 1996, *Ap. J.*, 456, 356.

Michell, J., 1767, *Phil. Trans. R. Soc. London*, 57, 234.

Miller, G.E., Scalo, J.M., 1979, *Ap. J. Suppl.*, 41, 513.

Milone, E.F., ed., 1993, *Light Curve Modelling of Eclipsing Binary Stars*, Springer-Verlag, Berlin.

Milone, E.F., Mermilliod, J.-C., eds., 1996, *The Origins, Evolution, and Destinies of Binary Stars in Clusters*, ASP Conf. Ser. Vol. 90.

Mirabel, I.F., Rodriquez, L.F., 1999, *A. Rev. A. & A.*, 37, 409.

Mochnacki, S.W., 1981, *Ap. J.*, 245, 650.

Mochnacki, S.W., 1984, *Ap. J. Suppl.*, 55, 551.

Mochnacki, S.W., Doughty, N.A., 1972a, *MNRAS*, 156, 51.

Mochnacki, S.W., Doughty, N.A., 1972b, *MNRAS*, 156, 243.

Moon, T.T., Dworetsky, M.M., 1985, *MNRAS*, 217, 305.

Morbey, C.L., 1975, *PASP*, 87, 689.

Morbey, C.L., 1992, In *Complementary Approaches to Double and Multiple Star Research*, eds. McAlister, H.A., Hartkopf, W.I., IAU Coll. No. 135, ASP Conf. Ser. Vol. 32, p. 127.

Morgan, W.W., Keenan, P.C., 1973, *A. Rev. A. & A.*, 11, 29.

Morgan, W.W., Keenan, P.C., Kellman, E., 1943, *An Atlas of Stellar Spectra*, University of Chicago Press.

Muckerjee, J., Peters, G.J., Wilson, R.E., 1996, *MNRAS*, 283, 613.

Nagase, F., 1989, *PASJ*, 41, 1.

Napier, W.M., 1981, *MNRAS*, 194, 149.

Napiwotski, R., Schönberner, D., Wenske, V., 1993, *A. & A.*, 268, 653.

Nather, R.E., Warner, B., 1971, *MNRAS*, 152, 209.

Nordlund, A., Vaz, L.P.R., 1990, *A. & A.*, 228, 231.

Nordström, B., 1989, *Ap. J.*, 341, 934.

Nordström, B., Johansen, K.T., 1994, *A. & A.*, 291, 777.

Norton, A.J., Beardmore, A.P., Taylor, P., 1996, *MNRAS*, 280, 937.

O'Connell, D.J.K., 1951, Riverview Publ. No. 10.

Oke, J.B., 1990, *Astron. J.*, 99, 1621.

Oláh, K., Kövári, Z., Bartus, J., Strassmeier, K.G., Hall, D.S., Henry, G.W., 1997, *A. & A.*, 321, 811.

Olsen, E.H., 1983, *A. & A. Suppl.*, 54, 55.

Olsen, E.H., 1993, *A. & A. Suppl.*, 102, 89.

Orosz, J.A., Bailyn, C.D., 1998, *Ap. J.*, 477, 876.

Orosz, J.A., Jain, R.K., Bailyn, C.D., McClintock, J.E., Remillard, R.A., 1998, *Ap. J.*, 499, 375.

Orosz, J.A., Wade, R.A., Harlow, J.J.B., Thorstensen, J.R., Taylor, C.J., Eracleous, M., 1999, *Astron. J.*, 117, 1598.

Pantazis, G., Niarchos, P.G., 1998, *A. & A.*, 335, 199.

Parmar, A.N., White, N.E., Giommi, P., Gottwald, M., 1986, *Ap. J.*, 308, 199.

Patterson, J., 1994, *PASP*, 106, 209.

Penny, L.R., Gies, D.R., Bagnuolo, W.G., 1997, *Ap. J.*, 483, 439.

Pereira, C.B., Ortega, V.G., Monte-Lima, I., 1999, *A. & A.*, 344, 607.

Perry, C.L., Olsen, E.H., Crawford, D.L., 1987, *PASP*, 99, 1184.

Perryman, M.A.C., and the ESA Hipparcos Team, 1997, European Space Agency Special Report SP-1200, Vols. 1–17.

Petrie, R.M., 1962, In *Astronomical Techniques*, ed. Hiltner, W.A., University of Chicago Press, p. 63.

Petrie, R.M., Andrews, D.H., Scarfe, C.D., 1967, IAU Symp. No. 30, p. 221.

Phinney, E.S., 1996, In *The Origins, Evolutions, and Destinies of Binary Stars in Clusters*, eds. Milone, E.F., Mermilliod, J.-C., ASP Conf. Ser. Vol. 90, p. 163.

Pigulski, A., Boratyn, D.A., 1992, *A. & A.*, 253, 178.

Piirola, V., Linnaluoto, S., 1989, In *Polarized Radiation of Circumstellar Origin*, eds. Coyne, G.V., et al., University of Arizona Press, p. 655.

Piirola, V., Reiz, A., Coyne, G.V., 1987, *A. & A.*, 186, 120.

Piskunov, N.E., 1991, In *The Sun and Cool Stars: Activity, Magnetism, Dynamos*, eds. Tuominen, I., Moss, D., Rüdiger, G., IAU Coll. No. 130, Springer-Verlag, Heidelberg, p. 309.

Piskunov, N.E., Rice, J.B., 1993, *PASP*, 105, 1415.

Pollacco, D.L., Ramsay, G., 1992, *MNRAS*, 254, 228.

Pollacco, D.L., Bell, S.A., 1993, *MNRAS*, 262, 377.

Pollacco, D.L., Bell, S.A., 1994, *MNRAS*, 267, 452.

Popper, D.M., 1973, *Ap. J.*, 185, 265.

Popper, D.M., 1980, *A. Rev. A. & A.*, 18, 115.

Popper, D.M., 1987, *Ap. J.*, 313, L81.

Portegies Zwart, S.F., Verbunt, F., 1996, *A. & A.*, 309, 179.

Pounds, K.A., and 55 other authors, 1993, *MNRAS*, 260, 77.

Press, W.H., Rybicki, G.B., 1989, *Ap. J.*, 338, 277.

Press, W.H., Teukolsky, S.A., Vetterling, W.T., Flannery, B.P., 1992, *Numerical Recipes*, 2nd ed., Cambridge University Press.

Primini, F., Rappaport, S.A., Joss, P.C., 1977, *Ap. J.*, 217, 543.

Pringle, J.E., Wade, R.A., eds., 1985, *Interacting Binary Stars*, Cambridge University Press.

Provoost, P., 1980, *A. & A. Suppl.*, 40, 129.

Quataert, E.J., Kumar, P., Ao, C.O., 1996, *Ap. J.*, 463, 284.

Rabe, W., 1958, *Astron. Nachr.*, 284, 101.

Rafert, J.B., Twigg, L.W., 1980, *MNRAS*, 193, 79.

Reddish, V.C., 1966, *Sky & Telescope*, 32, 124.

Refsdal, S., Roth, M.L., Weigert, A., 1974, *A. & A.*, 36, 113.

Remillard, R.A., McClintock, J.E., Sobczak, G.J., Bailyn, C.D., Orosz, J.A., Morgan, E.H., Levine, A.M., 1999, *Ap. J.*, 517, L127.

Reynolds, A.P., Bell, S.A., Hilditch, R.W., 1992, *MNRAS*, 256, 631.

Reynolds, A.P., Hilditch, R.W., Bell, S.A., Hill, G., 1993, *MNRAS*, 261, 337.

Reynolds, A.P., Quaintrell, H., Still, M.D., Roche, P., Chakrabarty, D., Levine, S.E., 1997, *MNRAS*, 288, 43.

Ribas, I., Giménez, A., Torra, J., Jordi, C., Oblak, E., 1998, *A. & A.*, 330, 600.

Richards, M.T., Albright, G.E., Bowles, L.M., 1995, *Ap. J.*, 438, L103.

Ritter, H., Kolb, U., 1998, *A. & A. Suppl.*, 129, 83.

Robert, C., Moffat, A.F.J., Bastien, P., St. Louis, N., Drissen, L., 1990, *Ap. J.*, 359, 211.

Roy, A.E., 1978, *Orbital Motion*, Adam Hilger Ltd., Bristol.

Rubenstein, E.P., Patterson, J., Africano, J.L., 1991, *PASP*, 103, 1258.

Rucinski, S.M., 1969, *Acta Astr.*, 19, 125.

Rucinski, S.M., 1973, *Acta Astr.*, 23, 79.

Rucinski, S.M., 1974, *Acta Astr.*, 24, 119.

Rucinski, S.M., 1976a, *PASP*, 88, 244.

Rucinski, S.M., 1976b, *PASP*, 88, 777.

Rucinski, S.M., 1989, *Comments on Astrophysics*, 14, 79.

Rucinski, S.M., 1992, *Astron. J.*, 104, 1968.

Rucinski, S.M., 1993, In *The Realm of Interacting Binaries*, eds. Sahade, J., McCluskey, G.E., Kondo, Y., Kluwer, Dordrecht, p. 111.

Rucinski, S.M., 1997a, *Astron. J.*, 113, 407.

Rucinski, S.M., 1997b, *Astron. J.*, 113, 1112.

Rucinski, S.M., 1998a, *Astron. J.*, 115, 303.

Rucinski, S.M., 1998b, *Astron. J.*, 116, 2998.

Rucinski, S.M., 1999, In *Magnetic Activity in Cool Stars*, ed. Demircan, O., *Turkish J. Phys.*, 23, 271.

Rucinski, S.M., Lu, W., 1999, *Astron. J.*, 118, 2451.

Rucinski, S.M., Baade, D., Lu, W.X., Udalski, A., 1992, *Astron. J.*, 103, 573.

Rucinski, S.M., Lu, W.-X., Shi, J., 1993, *Astron. J.*, 106, 1174.

Rudy, R.J., Kemp, J.C., 1978, *Ap. J.*, 221, 200.

Russell, H.N., 1912a, *Ap. J.*, 35, 315.

Russell, H.N., 1912b, *Ap. J.*, 36, 54.

Russell, H.N., Shapley, H., 1912a, *Ap. J.*, 36, 239.

Russell, H.N., Shapley, H., 1912b, *Ap. J.*, 36, 385.

Russell, H.N., Merrill, J.E., 1952, *Contr. Princeton Univ. Obs.*, No. 24.

Rutten, R.G.M., 1998, *A. & A. Suppl.*, 127, 581.

Rutten, R.G.M., Dhillon, V.S., Horne, K., Kuulkers, E., 1994, *A. & A.*, 283, 441.

Sahade, J., 1960, In *Stellar Atmospheres*, Stars and Stellar Systems Series, vol. VI, ed. Greenstein, J.L., University of Chicago Press, p. 466.

St. Louis, N., Moffat, A.F.J., Lapointe, L., Efimov, Y.S., Shakhovskoy, N.M., Fox, G.K., Piirola, V., 1993, *Ap. J.*, 410, 342.

Sandage, A.R., and 11 other authors, 1966, *Ap. J.*, 146, 310.

Sarna, M.J., 1989, *A. & A.*, 224, 98.

Sarna, M.J., 1992, *MNRAS*, 259, 17.

Sarna, M.J., 1993, *MNRAS*, 262, 534.

Sarna, M.J., Fedorova, A.V., 1989, *A. & A.*, 208, 111.

Sarna, M.J., De Greve, J.-P., 1996, *QJRAS*, 37, 11.

Scarfe, C.D., 1979, *JRAS Canada*, 73, 258.

Scargle, J.D., 1982, *Ap. J.*, 263, 835.

Schaller, G., Schaerer, D., Meynet, G., Maeder, A., 1992, *A. & A. Suppl.*, 96, 269.

Schechter, P.L., Mateo, M., Saha, A., 1993, *PASP*, 105, 1342.

Schild, H., Mürset, U., Schmutz, W., 1996, *A. & A.*, 306, 477.

Schmid, H.M., 1989, *A. & A.*, 211, L31.

Schmid, H.M., 1992, *A. & A.*, 254, 224.

Schmid, H.M., Schild, H., 1994, *A. & A.*, 281, 145.

Schmid, H.M., Dumm, T., Mürset, U., Nussbaumer, H., Schild, H., Schmutz, W., 1998, *A. & A.*, 329, 986.

Schmidt-Kaler, T., 1982, *Landolt-Börnstein*, vol. 2b, Springer-Verlag, Berlin, p. 1.

Schmutz, W., Schild, H., Mürset, U., Schmid, H.M., 1994, *A. & A.*, 288, 819.

Schönberner, D., Harmanec, P., 1995, *A. & A.*, 294, 509.

Schreier, E., Levinson, R., Gursky, H., Kellogg, E., Tananbaum, H., Giacconi, R., 1972, *Ap. J.*, 172, L79.

Schwarzschild, M., 1958, *Structure and Evolution of the Stars*, Princeton University Press.

Schwope, A.D., Thomas, H.-C., Beuermann, K., 1993, *A. & A.*, 271, L25.

Schwope, A.D., and 14 other authors, 1998, In *Wild Stars in the Old West: Proceedings of the*

13th North American Workshop on Cataclysmic Variables and Related Objects, eds. Howell, S., Kuulkers, E., Woodward, C., ASP Conf. Ser. Vol. 137, p. 44.

Schwope, A.D., Schwarz, R., Staude, A., Heerlein, C., Horne, K., Steeghs, D., 1999, In *Annapolis Workshop on Magnetic Cataclysmic Variables*, eds. Hellier, C., Mukai, K., ASP Conf. Ser. Vol. 157, p. 71.

Sellwood, J.A., Goodman, J., eds., 1999, *Astrophysical Discs*, ASP Conf. Ser. Vol. 160.

Shao, M., Colavita, M.M., Hines, B.E., Staelin, D.H., Hutter, D.J., Johnston, K.J., Mozurkewich, D., Simon, R.S., Hershey, J.L., Hughes, J.A., Kaplan, G.H., 1988, *A. & A.*, 193, 357.

Shara, M.M., Livio, M., Moffat, A.F.J., Orio, M., 1986, *Ap. J.*, 311, 163.

Shaw, J.S., Caillault, J.-P., Schmitt, J.H.M.M., 1996, *Ap. J.*, 461, 951.

Shore, S.N., Kenyon, S.J., Starrfield, S., Sonneborn, G., 1996, *Ap. J.*, 456, 717.

Shortridge, K., Meyerdierks, H., Currie, M., Clayton, M., 1997, PPARC Starlink User Note No. 86.13.

Siegel, N., Reinsch, K., Beuermann, K., van der Woerd, H., Wolff, E., 1989, *A. & A.*, 225, 97.

Silber, A.D., Anderson, S.F., Margon, B., Downes, R.A., 1996, *Ap. J.*, 462, 428.

Silk, J., 1995, *Ap. J.*, 438, L41.

Simkin, S.M., 1974, *A. & A.*, 31, 129.

Simon, K.P., Sturm, E., 1994, *A. & A.*, 281, 286.

Simon, K.P., Sturm, E., Fiedler, A., 1994, *A. & A.*, 292, 507.

Sion, E.M., 1999, *PASP*, 111, 532.

Skilling, J., 1981, In *Workshop on Maximum Entropy Estimation and Data Analysis*, Reidel, Dordrecht.

Skilling, J., Bryan, R.K., 1984, *MNRAS*, 211, 111.

Skilling, J., Gull, S.F., 1985, In *Maximum Entropy and Bayesian Methods in Inverse Problems*, eds. Smith, C.R., Grandy, W.T., Reidel, Dordrecht, p. 83.

Skumanich, A., 1972, *Ap. J.*, 171, 565.

Slettebak, A., Collins, G.W., II, Parkinson, T.D., Boyce, P.B., White, N.M., 1975, *Ap. J. Suppl.*, 29, 137.

Smak, J., 1981, *Acta Astr.*, 31, 395.

Smak, J., 1994, *Acta Astr.*, 44, 265.

Smart, W.M., 1953, *Celestial Mechanics*, Longmans, London.

Smeyers, P., Willems, B., Van Hoolst, T., 1998, *A. & A.*, 335, 622.

Smith, D.A., Dhillon, V.S., 1998, *MNRAS*, 301, 767.

Smith, L.F., Maeder, A., 1989, *A. & A.*, 211, 71.

Sobczak, G.J., McClintock, J.E., Remillard, R.A., Levine, A.M., Morgan, E.H., Bailyn, C.D., Orosz, J.A., 1999, *Ap. J.*, 517, L121.

Sobieski, S., 1965, *Ap. J. Suppl.*, 12, 263, 276.

Söderhjelm, S., 1975, *A. & A.*, 42, 229.

Söderhjelm, S., 1982, *A. & A.*, 107, 54.

Sohl, K.B., Watson, M.G., Rosen, S.R., 1995, ASP Conf. Ser. Vol. 85, p. 306.

Southwell, K.A., Livio, M., Charles, P.A., O'Donoghue, D., Sutherland, W.J., 1996, *Ap. J.*, 470, 1065.

Stagg, C.R., Milone, E.F., 1993, In *Light Curve Modelling of Eclipsing Binary Stars*, ed. Milone, E.F., Springer-Verlag, New York, p. 75.

Stanek, K.Z., Kaluzny, J., Krockenberger, M., Sasselov, D.D., Tonry, J.L., Mateo, M., 1998, *Astron. J.*, 115, 1894.

Stanek, K.Z., Kaluzny, J., Krockenberger, M., Sasselov, D.D., Tonry, J.L., Mateo, M., 1999, *Astron. J.*, 117, 2810.

Starrfield, S., 1998, In *Wild Stars in the Old West: Proceedings of the 13th North American Workshop on Cataclysmic Variables and Related Objects*, eds. Howell, S., Kuulkers, E., Woodward, C., ASP Conf. Ser. Vol. 137, p. 352.

Steeghs, D., Stehle, R., 1999, *MNRAS*, 307, 99.

Steeghs, D., Harlaftis, E., Horne, K., 1997, *MNRAS*, 290, L28.

Steeghs, D., Marsh, T.R., Horne, K., 1999, in preparation.

Stella, L., Priedhorsky, W., White, R.E., 1987, *Ap. J.*, 312, L17.

Stellingwerf, R.F., 1978, *Ap. J.*, 224, 953.

Sterken, C., Manfroid, J., 1992, *Astronomical Photometry: A Guide*, Kluwer, Dordrecht.

Sterne, T.E., 1939, *MNRAS*, 99, 451.

Sterne, T.E., 1941, *Proc. Natl. Acad. Sci. USA*, 27, 175.

Stetson, P.B., 1987, *PASP*, 99, 191.

Stickland, D.J., 1997, *The Observatory*, 117, 37.

Stickland, D.J., Lloyd, C., 1999, *The Observatory*, 119, 16.

Still, M.D., Quaintrell, H., Roche, P.D., Reynolds, A.P., 1997, *MNRAS*, 292, 52.

Stobie, R.S., and 27 other authors, 1997, *MNRAS*, 287, 848.

Stobie, R.S., Koen, C., Kilkenny, D., O'Donoghue, D., Chen, A., 1997, In *The Third Conference on Faint Blue Stars*, eds. Philip, A.D.G., Liebert, J.W., Saffer, R.A., L. Davis Press, Schenectady, p. 69.

Strassmeier, K.G., 1988, *Ap. Sp. Sci.*, 140, 223.

Strassmeier, K.G., 1994, *A. & A.*, 281, 395.

Strassmeier, K.G., Bopp, B.W., 1992, *A. & A.*, 259, 183.

Strassmeier, K.G., Hall, D.S., Fekel, F.C., Scheck, M., 1993, *A. & A. Suppl.*, 100, 173.

Strassmeier, K.G., Hall, D.S., Henry, G.W., 1994, *A. & A.*, 282, 535.

Strömgren, B., 1963, In *Basic Astronomical Data*, ed. Strand, K.A., University of Chicago Press, p. 123.

Strömgren, B., 1966, *A. Rev. A. & A.*, 4, 433.

Sturm, E., Simon, K.P., 1994, *A. & A.*, 282, 93.

Sybesma, C.H.B., 1986, *A. & A.*, 159, 108.

Szabados, L., 1992, In *Complementary Approaches to Double and Multiple Star Research*, eds. McAlister, H.A., Hartkopf, W.I., IAU Coll. No. 135, ASP Conf. Ser. Vol. 32, p. 358.

Szkody, P., and 12 other authors, 1999, *Ap. J.*, 521, 362.

Tananbaum, H., Gursky, H., Kellogg, E.M., Levinson, R., Schreier, E., Giacconi, R., 1972, *Ap. J.*, 174, L143.

Tapia, S., 1977, *Ap. J.*, 212, L125.

Tatum, J.B., 1968, *MNRAS*, 141, 43.

Taylor, B.J., 1986, *Ap. J. Suppl.*, 60, 577.

Taylor, J.H., Hulse, R.A., Fowler, L.A., Gullahorn, G.E., Rankin, J.M., 1976, *Ap. J.*, 206, L53.

Taylor, J.H., Manchester, R.N., Lyne, A.G., 1993, *Ap. J. Suppl.*, 88, 529.

Thaller, M.L., Bagnuolo, W.G., Gies, D.R., Penny, L.R., 1995, *Ap. J.*, 448, 878.

Thiele, T.N., 1883, *Astron. Nachr.*, 104, 245.

Thorsett, S.E., Chakrabarty, D., 1999, *Ap. J.*, 512, 288.

Thorstensen, J.R., Ringwald, F.A., Wade, R.A., Schmidt, G.D., Norsworthy, J.E., 1991, *Astron. J.*, 102, 272.

Tinbergen, J., 1979, *A. & A. Suppl.*, 35, 325.

Tinbergen, J., 1996, *An Introduction to Astronomical Polarimetry*, Cambridge University Press.

Tinbergen, J., Rutten, R., 1992, *A User Guide to WHT Spectropolarimetry*, Isaac Newton Group, La Palma, User Manual No. 21.

Todoran, I., 1972, *Ap. Sp. Sci.*, 15, 229.

Tomasella, L., Munari, U., 1998, *A. & A.*, 335, 561.

Tomkin, J., Lambert, D.L., 1978, *Ap. J.*, 222, L119.

Tonry, J., Davis, M., 1979, *Astron. J.*, 84, 1511.

Torres, G., Stefanik, R.P., Andersen, J., Nordström, B., Latham, D.W., Clausen, J.V., 1997, *Astron. J.*, 114, 2764.

Tout, C.A., Hall, D.S., 1991, *MNRAS*, 253, 9.

Tsesevich, V.P., 1939, *Bull. Astr. Inst. USSR Acad. Sci.*, No. 45.

Tsesevich, V.P., 1940, *Bull. Astr. Inst. USSR Acad. Sci.*, No. 50.

Turnshek, D.A., Bohlin, R.C., Williamson, R.L., Lupie, O.L., Koornneef, J., 1990, *Astron. J.*, 99, 1243.

Twarog, B.A., 1984, *Astron. J.*, 89, 523.

Unsöld, A., 1955, *Physik der Sternatmospheren*, 2nd ed., Springer-Verlag, Berlin.

Vanbeveren, D., 1995, *A. & A.*, 294, 107.

Vanbeveren, D., de Loore, C., van Rensbergen, W., 1998, *A. & A. Rev.*, 9, 63.

VandenBerg, D.A., 1985, *Ap. J. Suppl.*, 58, 711.

van der Hooft, F., Heemskerk, M.H.M., Alberts, F., van Paradijs, J., 1998, *A. & A.*, 329, 538.

van der Kamp, P., 1967, *Principles of Astrometry*, Freeman, San Francisco.

van der Klis, M., 1995, In *X-ray Binaries*, eds. Lewin, W.H.G., van Paradijs, J., van den Heuvel, E.P.J., Cambridge University Press, p. 252.

van Duinen, R.J., Aalders, J.W.G., Wesselius, P.R., Wildeman, K.J., Wu, C.C., Luinge, W., Snel, D., 1975, *A. & A.*, 39, 159.

van Gent, R.H., 1989, *A. & A. Suppl.*, 77, 471.

Van Hamme, W., 1993, *Astron. J.*, 106, 2096.

Van Hamme, W., Wilson, R.E., 1985, *A. & A.*, 152, 25.

van Kerkwijk, M.H., van Paradijs, J., Zuiderwijk, E.J., Hammerschlag-Hensberge, G., Kaper, L., Sterken, C., 1995a, *A. & A.*, 303, 483.

van Kerkwijk, M.H., van Paradijs, J., Zuiderwijk, E.J., 1995b, *A. & A.*, 303, 497.

van Paradijs, J., Kuiper, L., 1984, *A. & A.*, 138, 71.

van Paradijs, J., van den Heuvel, E.P.J., Kuulkers, E., eds., 1996, *Compact Stars in Binaries*, IAU Symposium No. 165, Kluwer, Dordrecht.

van't Veer, F., Maceroni, C., 1992, In *Binaries as Tracers of Stellar Formation*, eds. Duquennoy, A., Mayor, M., Cambridge University Press, p. 237.

Vaz, L.P.R., 1985, *Ap. Sp. Sci.*, 113, 349.

Vaz, L.P.R., Nordlund, A., 1985, *A. & A.*, 147, 281.

Vaz, L.P.R., Cunha, N.C.S., Vieira, E.F., Myrrha, M.L.M., 1997, *A. & A.*, 327, 1094.

Vennes, S., Szkody, P., Sion, E.M., Long, K.S., 1995, *Ap. J.*, 445, 921.

Verbunt, F.,1993, *A. Rev. A. & A.*, 31, 93.

Verbunt, F., 1999, In *Astrophysical Discs*, eds. Sellwood, J.A., Goodman, J., ASP Conf. Ser. Vol. 160, p. 21.

Vincent, A., Piskunov, N.E., Tuominen, I., 1993, *A. & A.*, 278, 523.

Vinko, J., 1993, *MNRAS*, 260, 273.

Vogel, H.C., 1890, *Astron. Nachr.*, 123, 290.

Vogel, M., 1991, *A. & A.*, 249, 173.

Vogt, S.S., 1981, *Ap. J.*, 250, 327.

Vogt, S.S., Penrod, G.D., 1983, *PASP*, 95, 565.

Vogt, S.S., Hatzes, A.P., 1991, In *The Sun and Cool Stars: Activity, Magnetism, Dynamos*, eds. Tuominen, I., Moss, D., Rüdiger, G., IAU Coll. No. 130, Springer-Verlag, Heidelberg, p. 297.

Vogt, S.S., Penrod, G.D., Hatzes, A.P., 1987, *Ap. J.*, 321, 496.

Vogt, S.S., Hatzes, A.P., Misch, A.A., Kürster, M., 1999, *Ap. J. Suppl.*, 121, 547.

von Zeipel, H., 1924, *MNRAS*, 84, 665.

Wade, R.A., Rucinski, S.M., 1985, *A. & A. Suppl.*, 60, 471.

Wade, R.A., Hubeny, I., 1998, *Ap. J.*, 509, 350.

Walker, M.F., 1954, *PASP*, 66, 230.

Walker, M.F., 1956, *Ap. J.*, 123, 68.

Warner, B., 1988, *High Speed Astronomical Photometry*, Cambridge University Press.

Warner, B., 1995, *Cataclysmic Variable Stars*, Cambridge University Press.

Warren, J.K., Sirk, M.M., Vallerga, J.V., 1995, *Ap. J.*, 445, 909.

Warwick, R.S., Barber, C.R., Hodgkin, S.T., Pye, J.P., 1993, *MNRAS*, 262, 289.

Weber, E.J., Davis, L., 1967, *Ap. J.*, 148, 217.

Webster, B.L., Murdin, P., 1972, *Nature*, 235, 37.

Weigelt, G., 1977, *Optics Communications*, 21, 55.

Weigelt, G., Wirnitzer, B., 1983, *Optics Letters*, 8, 389.

Weisberg, J.M., Taylor, J.H., 1984, *Phys. Rev. Lett.*, 52, 1348.

Welsh, W.F., Horne, K., Gomer, R., 1998, *MNRAS*, 298, 285.

White, N.E., Nagase, F., Parmar, A.N., 1995, In *X-ray Binaries*, eds. Lewin, W.H.G., van Paradijs, J., van den Heuvel, E.P.J, Cambridge University Press, p. 1.

Whittet, D.C.B., Martin, P.G., Hough, J.H., Rouse, M.F., Bailey, J.A., Axon, D.J., 1992, *Ap. J.*, 386, 562.

Wickramasinghe, D.T., Bailey, J., Meggitt, S.M.A., Ferrario, L., Hough, J., Tuohy, I.R., 1991, *MNRAS*, 251, 28.

Wickramasinghe, D.T., Bicknell, G.V., Ferrario, L., eds., 1996, *Accretion Phenomena and Related Outflows*, IAU Coll. No. 163, ASP Conf. Ser. Vol. 121.

Wijnands, R., Homan, J., van der Klis, M., Méndez, M., Kuulkers, E., van Paradijs, J., Lewin, W.H.G., Lamb, F.K., Psaltis, D., Vaughan, B., 1997, *Ap. J.*, 490, L157.

Williams, P.M., 1995, In *Wolf-Rayet Stars: Binaries, Colliding Winds, Evolution*, eds. van der Hucht, K.A., Williams, P.M., IAU Symp. No. 163, Kluwer, Dordrecht, p. 335.

Williams, P.M., van der Hucht, K.A., Spoelstra, T.A.T., Swaanenvelt, J.P., 1995, In *Wolf-Rayet Stars: Binaries, Colliding Winds, Evolution*, eds. van der Hucht, K.A., Williams, P.M., Kluwer, IAU Symp. No. 163, Dordrecht, p. 504.

Wilson, R.E., 1979, *Ap. J.*, 234, 1054.

Wilson, R.E., 1990, *Ap. J.*, 356, 613.

Wilson, R.E., Devinney, E.J., 1971, *Ap. J.*, 166, 605.

Wilson, R.E., Terrell, D., 1998, *MNRAS*, 296, 33.

Wolf, G.W., Kern, J.T., 1983, *Ap. J. Suppl.*, 52, 429.

Woo, J.W., Clark, G.W., Blondin, J.M., Kallman, T.R., Nagase, F., 1995, *Ap. J.*, 445, 896.

Wood, D.B., 1971, *Astron. J.*, 76, 701.

Wood, J.H., Saffer, R., 1999, *MNRAS*, 305, 820.

Wood, J.H., Horne, K., Berriman, G., Wade, R.A., 1989, *Ap. J.*, 341, 974.

Worley, C.E., Heintz, W.D., 1983, *Fourth Catalogue of Orbits of Visual Binary Stars*, Publ. U.S. Naval Obs. 24, Part VII.

Wu, C.C., Faber, S.M., Gallacher, J.S., Peck, M., Tinsley, B.M., 1980, *Ap. J.*, 237, 290.

Wynn, G.A., King, A.R., Horne, K., 1997, *MNRAS*, 286, 436.

Zahn, J.-P., 1975, *A. & A.*, 41, 329.

Zahn, J.-P., 1977, *A. & A.*, 57, 383.

Zahn, J.-P., 1978, *A. & A.*, 67, 162.

Zahn, J.-P., 1989, *A. & A.*, 220, 112.

Zahn, J.-P., 1992, In *Binaries as Tracers of Stellar Formation*, eds. Duquennoy, A., Mayor, M., Cambridge University Press, p. 253.

Zahn, J.-P., Bouchet, L., 1989, *A. & A.*, 223, 112.

Zhang, E.-H., Robinson, E.L., Nather, R.E., 1986, *Ap. J.*, 305, 740.

Zucker, S., Mazeh, T., 1994, *Ap. J.*, 420, 806.

Zycki, P.T., Done, C., Smith, D.A., 1999, *MNRAS*, 305, 231.

Index